Advances in Optics and Optoelectronics

Series Editor

Perry Ping Shum, Southern University of Science and Technology, Shenzhen, China

The Advances in Optics and Optoelectronics series focuses on the exciting new developments in the fast emerging fields of optics and optoelectronics. The volumes cover key theories, basic implementation methods, and practical applications in but not limited to the following subject areas:

AI Photonics
Laser Science and Technology
Quantum Optics and Information
Optoelectronic Devices and Applications
Fiber-Based Technologies and Applications
Near-infrared, Mid-infrared and Far-infrared Technologies and Applications
Biophotonics and Medical Optics
Optical Materials, Characterization Methods and Techniques
Spectroscopy Science and Applications
Microscopy and Adaptive Optics
Microwave Photonics and Wireless Convergence
Optical Communications and Networks

Within the scopes of the series are monographs, edited volumes and conference proceedings. We expect that scientists, engineers, and graduate students will find the books in this series useful in their research and development, teaching and studies.

Jiasheng Tao

Space Optical Remote Sensing

Fundamentals and System Design

国防工业出版社
National Defense Industry Press

Springer

Jiasheng Tao
China Academy of Space Technology
Beijing, China

ISSN 2731-6009 ISSN 2731-6017 (electronic)
Advances in Optics and Optoelectronics
ISBN 978-981-99-3317-4 ISBN 978-981-99-3318-1 (eBook)
https://doi.org/10.1007/978-981-99-3318-1

Jointly published with National Defense Industry Press
The print edition is not for sale in China (Mainland). Customers from China (Mainland) please order the print book from: National Defense Industry Press.

Translation from the Chinese Simplified language edition: "Hang Tian Guang Xue Yao Gan Xi Tong Zong Ti She Ji" by Jiasheng Tao, © National Defense Industry Press 2019. Published by National Defense Industry Press. All Rights Reserved.

This Springer imprint is published by the registered company Springer Nature Singapore Pte Ltd.
The registered company address is: 152 Beach Road, #21-01/04 Gateway East, Singapore 189721, Singapore

Preface

Optics is both an ancient tradition and a rapidly developing subject. Based on it, space optical remote sensing is constantly innovating with the rapid development of space technology. Since 1960, the United States launched the sun-synchronous meteorological satellite, beginning the true long-term optical remote sensing of the earth from spacecraft. Space optical remote sensing research has developed from imaging to hyperspectral imaging, stereo mapping and other optical remote sensing forms, from large integrated satellites weighing 20 tons to tiny satellites weighing less than 10 kg, from low orbit optical remote sensing to geosynchronous orbit optical remote sensing, from earth orbit to beyond the solar system; the space optical remote sensor has different functions and poses.

Space optical remote sensing enables people to stand higher and see farther, breaking down borders and boundaries, and can realize the global imaging. Its efficient imaging of large areas makes maps of the quick update that becomes a reality, and it provides the foundation for high-precision automobile navigation, in which great places facilitate the life of modern people. It has become an important information source for weather forecast, disaster prevention and mitigation, resource survey, geological survey, urban planning, port railway and other large engineering construction.

Space optical remote sensing system is a large system involving orbit, space platform and optical remote sensor; it is a general subject based on space system, time system, space optics, atmospheric optics, light source, radiation measurement, optical measurement, photo detector and other disciplines and concepts. More subjects are involved, the compilation of this book strives to have clear basic concepts, complete content and comprehensive theory and technology, through the study of this book that can master the overall design of space optical remote sensing system at the level of parameter design.

In Chap. 1 introduction, the function of space optical remote sensing, the driving force of space optical remote sensing, the advantage of space optical remote sensing and the precision complexity of technology from the perspective of the development of space optical remote sensing were introduced. In Chap. 2, the main concepts of space optical remote sensing system was introduced. Space activities are characterized by wide region and coordination; for this, the space system, time system and their relationship were introduced from the perspective of the motion of the celestial bodies in the solar system. The main basic concepts of platform, payload and optical remote sensing were introduced to prepare the knowledge for the study of the following chapters. In Chap. 3, the orbit knowledge was introduced, the focus was on sun-synchronous orbit and geosynchronous orbit, how to determine remote sensing points on the earth from spacecraft, how to establish constellation were expounded, and how to realize global detection by space optical remote sensing satellite was explained. In Chap. 4, the light source of space optical remote sensing, the transmission and measurement of light energy, the basic structure of the atmosphere and the atmosphere window were introduced, and the influence of atmosphere on energy and positioning accuracy of space optical remote sensing were illustrated. In Chap. 5, the various types of photodetectors and their performance parameters, temperature adaptability and noise were introduced. CCD, CMOS, infrared devices and photon detectors were mainly introduced. In Chap. 6, the basic concept of optical system, six kinds of aberrations, the characteristics of refraction system, reflection system, catadioptric system and how to select the optical system of space optical remote sensing were introduced. Chapter 7 introduced the core chapter of space optical remote sensing, the basic concept, theory, parameter design and calculation of imaging camera, mapping camera, infrared camera and imaging spectrometer. In Chap. 8, the performance parameters and imaging modes of low- and high-orbit platforms closely related to optical remote sensing were introduced, and the development trend of optical remote sensing satellites was reviewed. In Chap. 9, on the basis of the previous chapters, the application, spectral segment selection, energy calculation, coordination and optimization of orbit parameters, optical remote sensing parameters, platform parameters, system comprehensive analysis and performance evaluation were discussed, in order to form a systematic and complete understanding of the overall parameter design of the large system. In Chap. 10, based on the basic theory of information optics, the performance design of optical remote sensor was discussed from the practical point of view, combined with the sampling discretization of photodetector, the performance characteristics, image resolution and image motion compensation of photoelectric optical remote sensor were discussed. This chapter focused on deepening the theoretical analysis of optical remote sensing.

Chapters 2 and 3 were reviewed by Weilian Yang, a senior expert on orbit design at the China Academy of Space Technology. Zhiqing Zhang, senior expert of National Satellite Meteorological Center, reviewed Chap. 4. Xingang Wang, a researcher at the Institute of Automation, Chinese Academy of Sciences, reviewed Chap. 5. Lin Li, a senior professor of Beijing Institute of Technology, reviewed Chaps. 6 and 7. Qingjun Zhang, senior expert of resource satellite of China Academy of Space Technology, reviewed Chaps. 8 and 9. I deeply appreciate the review and advice of experts.

This book was based on the handout "Design of Space Optical Remote Sensing System" written for Shenzhou Institute of China Academy of Space Technology. The book is compiled by Prof. Jiasheng Tao.

Thanks to the leaders of Shenzhou Institute of China Academy of Space Technology, teachers and leaders of the Institute of Telecommunication Satellite for their enthusiastic support and help. Thanks to my wife Shixiang Lu and my daughter Zhuo Tao for their support.

This book can be used as a reference book for engineers and students in the field of space optical remote sensing.

This book involves many subjects, and the authors' knowledge is limited. Readers are welcome to criticize and correct mistakes.

陶家生

Beijing, China
March 2022

Jiasheng Tao

Contents

1	**Introduction**	1
	1.1 Brief History of Optical Remote Sensing	1
	1.2 Advantages of Space Optical Remote Sensing Technology	9
	1.3 Demand for Space Optical Remote Sensing in Civil Field	10
	1.4 Military Application Promote to Development of Space Optical Remote Sensing	10
	1.5 Improvement of Optical Remote Sensing Technology	11
	References	15
2	**Fundamentals of Space Optical Remote Sensing System**	17
	2.1 Overview	17
	2.2 Basic Concepts of Spacecraft Orbit	18
	2.2.1 Space Concepts of Spacecraft Orbit	18
	2.2.2 Time Systems	25
	2.3 Basic Concepts of Spacecraft Platforms	30
	2.4 Basic Concepts of Spacecraft Payloads	31
	2.5 Basic Concepts of Space Optical Remote Sensing	32
	References	35
3	**Spacecraft–Earth and Time–Space Parameter Design**	37
	3.1 Proprieties of Two-Body Orbit	37
	3.1.1 In-Plane Motion	38
	3.1.2 Motion Along a Conical Curve	41
	3.1.3 Orbital Elements	47
	3.1.4 Satellite Entry Parameters	50
	3.1.5 Orbital Coordinate System and Transformation	52
	3.2 Satellite-Earth Geometry	56
	3.2.1 Ground Track of Satellite	56
	3.2.2 Field of View of Ground Station	57
	3.2.3 Communication Coverage of GEO	59
	3.2.4 Geo-positioning of Remote Sensing Image	61

3.3 Launch Window ... 65
3.3.1 Orbital Plane and Coplanar Window ... 66
3.3.2 Sunlight Window ... 72
3.4 Sun-Synchronous Orbit ... 78
3.5 Critical Inclination and Frozen Orbit ... 81
3.6 Regressive Orbit ... 83
3.7 Geosynchronous Orbit ... 90
3.7.1 Geostationary Orbit ... 93
3.7.2 Several Typical Geosynchronous Orbits ... 94
3.8 Constellation Orbit ... 102
3.8.1 Global Coverage Constellation ... 102
3.8.2 Geostationary Satellite Group ... 106
References ... 110

4 Radiation Source and Optical Atmospheric Transmission ... 111
4.1 Radiation and Units ... 112
4.1.1 Radiometry and Units ... 112
4.1.2 Photometry and Units ... 119
4.1.3 Blackbody Radiation ... 123
4.1.4 Radiation Calculation ... 129
4.1.5 Radiation from an Actual Object ... 136
4.1.6 The Electromagnetic Spectrum of Radiation ... 141
4.2 Solar Radiation ... 143
4.2.1 Solar Profile ... 143
4.2.2 Solar Radiation Spectrum ... 146
4.3 Earth Radiation ... 150
4.3.1 Shortwave Radiation ... 155
4.3.2 Longwave Radiation [2] ... 158
4.4 The Effect of Atmosphere on Radiation ... 168
4.4.1 Atmospheric Composition ... 168
4.4.2 Atmospheric Absorption and Atmospheric Windows ... 173
4.4.3 Atmospheric Scattering and Radiation ... 177
4.4.4 Atmospheric Turbulence ... 182
4.4.5 Atmospheric Refraction and Its Effect on Signal ... 183
4.5 The Moon and Planetary Radiation ... 187
References ... 188

5 Photoelectronic Detectors ... 189
5.1 Overview ... 189
5.2 Characteristic Parameters of Photoelectronic Detectors ... 191
5.3 Solid Image Components (CCD & CMOS) ... 195
5.3.1 CCD Image Components ... 195
5.3.2 CMOS Image Components ... 204
5.4 Infrared Image Components ... 208
5.4.1 Principle of IR Components ... 208
5.4.2 Characteristic Parameters of IR Components ... 212

5.5 Low Light Level Image Intensifiers 215
5.5.1 Principle and Function of Image Intensifiers 217
5.5.2 Major Characteristic Parameters of LLL Tubes 219
5.6 Photon Counters for Low Light Level Imaging 221
5.6.1 Operation Principle of Photon Counters 222
5.6.2 Characteristic Parameters of Photon Counters for Low Light Level Imaging 224
5.7 Television Camera Tubes 225
5.7.1 Operation Principle of Television Camera Tubes 226
5.7.2 Characteristic Parameters of Television Camera Tubes 227
5.8 X-Ray Imaging Devices 228
5.8.1 Principles of X-Ray Imaging Devices 228
5.8.2 Characteristics of X-Ray Imaging Devices 232
5.9 Other Photoelectronic Detectors 233
Reference 235

6 Optical System Selection of Remote Sensor 237
6.1 Introduction 237
6.2 Basic Concepts of Optical Systems 238
6.2.1 Basic Laws 238
6.2.2 Concepts of Imaging Optical System 239
6.3 Aberrations of Optical Systems 246
6.4 Introduction to Several Main Optical Systems 250
6.4.1 Transmission Optical System 251
6.4.2 Catadioptric Optical Systems 252
6.4.3 Reflective Optical System 253
6.5 Analysis and Comparison of Several Typical Optical Systems 259
References 261

7 Main Types of Optical Remote Sensors 263
7.1 Introduction 263
7.2 Optical Imaging Cameras 264
7.2.1 Main Parameters of the Imaging Cameras 264
7.2.2 Space Application of Imaging Cameras 266
7.3 Multiband Optical Camera 268
7.3.1 Bands Allocation 268
7.3.2 Applications of Multiband Camera 273
7.4 Mapping Optical Cameras 274
7.4.1 Fundamental Theory of Mapping Cameras 274
7.4.2 Stereo Mapping by Satellite Swing 277
7.4.3 Single-Lens Multi-Line Array Stereo Mapping 278
7.4.4 Multiple Single-Line Array Lens Combination Mode 279
7.4.5 Testing of Mapping Camera 282
7.4.6 Space Applications of Mapping Cameras 283

7.5 Infrared Optical Camera 284
7.5.1 System Composition of Infrared Camera 285
7.5.2 Basic Parameters of Infrared Camera 286
7.5.3 Scanning Mode 288
7.5.4 Common Scanning Mechanism 290
7.5.5 Several Optical Machine Scanning Schemes 303
7.5.6 Camera Mode 306
7.5.7 Refrigerating Methods 313
7.5.8 Comprehensive Evaluation of Infrared Detector Performance 319
7.5.9 Interpretation of Infrared Images 332
7.5.10 Typical Application of Infrared Cameras 333
7.6 Imaging Spectrometers 334
7.6.1 Dispersing Prism Splitting 336
7.6.2 Diffraction Splitting 339
7.6.3 Binary Optical Splitting 340
7.6.4 Interference Splitting 341
References 349

8 Platforms of Optical Remote Sensing 351
8.1 Main Types of Remote Sensing Satellites 351
8.1.1 Imaging Remote Sensing Satellites 352
8.1.2 Optical Stereo Mapping Satellites 354
8.1.3 Wide Coverage Optical Remote Sensing Satellites 360
8.2 LEO Optical Remote Sensing Platform 364
8.2.1 Specifications of CBERS Platform 366
8.2.2 On-Orbit Performances of CBERS Platform 369
8.3 Performances of GEO Optical Remote Sensing Platform 370
8.3.1 Characteristics of Optical Remote Sensing on GEO/LEO 371
8.3.2 Specifications of Optical Remote Sensing Platform on GEO 375
8.4 Development of Optical Remote Sensing Platforms 381
References 389

9 System Overall Parameters Design 391
9.1 Overview 391
9.2 Remote Sensing Method 392
9.3 Energy of Optical Remote Sensing Systems 395
9.3.1 Target Radiance 397
9.3.2 Remote Sensing Spectrum Bands 399
9.3.3 Optical System Parameters 400
9.4 Selection of System Design Objectives 405
9.5 Main Parameters Comprehensive Analysis and Performance Evaluation of the System 419
References 433

10 Resolution of CCD Sampling Imaging 435
10.1 Basic Theory of Optical Modulation Transfer Function 435
10.2 Example of MTF Calculation Under Various Obscuration Ratios 439
10.3 The Relationship of the Pixel Size of the CCD Detector, Obscuration Ratios and the MTF 443
10.4 Computer Simulation Analysis of Detector Static Image Resolution 450
References 463

Translation Contributors

Zhuo Tao (Chapters 1 and 5), Beijing Institute of Spacecraft System Engineering, China Academy of Space Technology, Beijing, China

Xiaoxiong Lin (Chapters 2 and 3), Institute of Telecommunication and Navigation Satellites, China Academy of Space Technology, Beijing, China

Wei Lu (Chapters 4 and 8), Beijing Institute of Spacecraft System Engineering, China Academy of Space Technology, Beijing, China

Lianying Dou (Chapters 6 and 7), Beijing Institute of Space Mechanics and Electricity, China Academy of Space Technology, Beijing, China

Haopan Wang (Chapters 9 and 10), Institute of Telecommunication and Navigation Satellites, China Academy of Space Technology, Beijing, China

Symbol Table of Main Parameters

$r_a = 6371.23$ km, Mean radius of the earth
$r_e = 6378.145$ km, Equatorial radius of the earth
$r_p = 6356.76$ km, Polar radius of the earth
$f = 1/298.257$, Oblateness of the earth
$e = 0.08182$, Eccentricity of the earth
$J_2 = 1.08263 \times 10^{-3}$, Second order coefficient of zonal harmonics of the gravitational potential of the earth
$J_{22} = 1.81222 \times 10^{-6}$, Second order field tesseral harmonic coefficients of the earth's gravitational potential
$\mu = 398600.44$ km^3s^{-2}, Gravitational constant
$\omega_e = 7.2921158 \times 10^{-5}$ rad/s, Rotational angular velocity of the earth
$J_3 = -2.5356 \times 10^{-6}$; The third order coefficient of zonal harmonics of the earth's gravitational potential
$m_e = 5.977 \times 10^{24}$ kg, Mass of the earth
$r_s = 42164.17$ km, Geostationary orbit radius
$h = 6.6260693 \times 10^{-34}$ J s, Planck constant
$k = 1.380658 \times 10^{-23}$ J K^{-1}, Boltzmann constant
$\sigma = 5.6704 \times 10^{-8}$ W m^{-2} K^{-4}, Stefan-Boltzmann constant
$c_0 = 299792458$ m/s, The speed of light in vacuum
$AU = 1.49597870 \times 10^8$ km, One astronomical unit (mean distance from the sun to the earth)
$E_0 = 1367$ W/m^2, Solar constant
$R_0 = 6.9599 \times 10^5$ km, Radius of the sun
$e^- = 1.6 \times 10^{-19}$ C, Electron charge
T = −273.15 °C, Zero absolute temperature
$n_0 = 1$, Refractive index of light in vacuum
$n = 1.00028$, Refractive index of light in air at standard pressure 760 mm Hg at 20 °C
$E = 23° 26'$, Obliquity of the ecliptic
$G = 6.670 \times 10^{-11}$ $\text{m}^3\text{kg}^{-1}\text{s}^{-2}$, Gravitation constant

Chapter 1
Introduction

1.1 Brief History of Optical Remote Sensing

The development of optical remote sensing has developed from the enlightenment stage, aerial remote sensing stage to space remote sensing stage. After entering the space optical remote sensing stage, it has developed vigorously. At this stage, the optical remote sensing of low earth orbit, high earth orbit and deep space observation have been developed.

1. **The enlightenment stage of optical remote sensing (1839–1909)**

In 1939, French Daguerre published the first aerial photograph. In 1858, Gaspard Felix Tournachon took a picture of the city of Paris, France, with a tethered balloon. In 1860, James Wallace Black and Sam King took off in a balloon to 630 m and successfully took pictures of Boston in the United States. In 1903, Julius Nenbronner designed a miniature camera tied to a flying pigeon. In 1906, G. R. Laurence took a photo with a kite in the sky and successfully recorded the scene after the famous San Francisco earthquake. In 1909, Wibour Wright took aerial photography over Centocelli, Italy [1]. These exploratory aerial photographies formed the enlightenment stage of optical remote sensing and builded the foundation for later practical aerial photography and remote sensing.

2. **Remote sensing stage (1910–1956)**

In 1913, aerial photography was applied to the survey of Bengashi oil field in Libya. C. Tardivo presented a paper at the conference of the international photogrammetry society in Vienna, describing the problem of aircraft photogrammetric mapping.

In 1924, the emergence of color film made the ground target information recorded more abundant by aerial photography.

3. **Space optical remote sensing (1957-until now)**

On October 4, 1957, the Soviet Union launched the first man-made satellite in history, which shocked the whole world. Figure 1.1 shows the first artificial satellite of the

J. Tao, *Space Optical Remote Sensing*, Advances in Optics and Optoelectronics,
https://doi.org/10.1007/978-981-99-3318-1_1

Soviet Union and its internal structure, indicating that mankind has entered a new era in observing the earth from space.

In 1958, the "Pioneer 2" probe launched by the United States took a cloud picture of the earth. Figure 1.2 shows that the Pioneer 2 probe launched by the United States, which weighs 38 kg. In October of the same year, the Soviet Union's "Moon 3" spacecraft took pictures of the back of the moon. Since the TIROS-1 and NOAA-1 solar synchronous meteorological satellites launched by the United States in 1960, the real long-term observation of the earth from spacecraft has been carried out. Since then, great progress has been made in space remote sensing. On March 2, 1972, Pioneer 10 took off. On June 13, 1983, it crossed the orbit of Pluto and flew to the boundary of the solar system. The earth received its last signal on January 22, 2003, in which there was no telemetry data. The Panchromatic resolution of DeoEye-1 commercial satellite launched by the United States in 2008 has reached 0.41 m, which makes the image quality of space optical remote sensing and aerial optical remote sensing almost the same.

In space, there are geosynchronous satellites, solar synchronous satellites, and some low orbit and high orbit satellites. The types of satellites are increasing, including optical imaging, stereo mapping, microwave remote sensing, communication, navigation, etc.

4. **Application of optical remote sensing**

During the First World War, aerial photography became an important means of military reconnaissance and formed a certain scale. At the same time, the interpretation level of photographs has also been improved to a certain extent. During the Second World War, Germany, Britain and other countries fully recognized the important

Fig. 1.1 Schematic diagram of the first artificial satellite of the Soviet Union and its internal structure

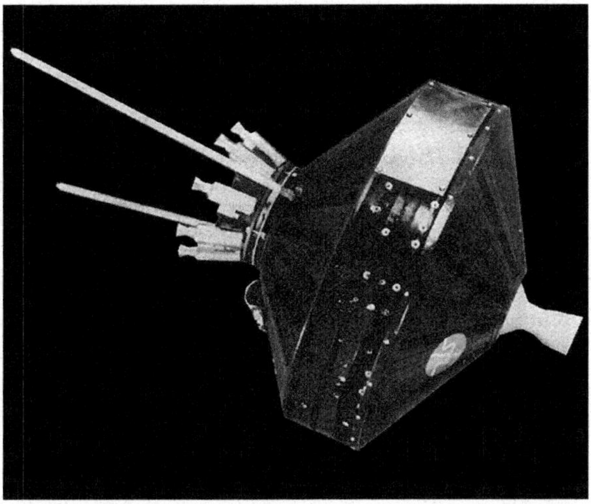

Fig. 1.2 Pioneer 2 probe launched by the United States

military value of aerial reconnaissance and aerial photography and achieved practical results in reconnaissance of the enemy's military situation and deployment of military operations. At the end of World War II, the aerial photography of the United States covered Eurasia and the Pacific coastal islands, including Japan, made maps and marked military targets, which became an important source of information for the United States in the Pacific War. Aerial photography played an important role in the decision making of military operations in major campaigns such as the defense of Stalingrad in the former Soviet Union.

Some works were published after World War II, such as "*aerial photographs: Their Use and Interpretation*" by A. J. Eardey in 1941, which discussed the geological application of aerial photographs and the characteristics of some ground features and vegetation. J. W. Bagley's "*Aerophotography and Aerosurveying*" focuses on the discussion of aerial survey methods. Since 1930, agriculture, forestry, animal husbandry and other government departments in the United States have applied aerial photography to planning.

Talent training and the publication of professional academic journals also were the characteristics of the *Remote Sensing* in this period. American universities offered courses in aerial photography and photo interpretation; the international Geographic Society established the Professional Committee on the application of aerial photography in 1949. In 1945, the United States founded the magazine "*Photogrammetric Engineering*" (changed to "*Photogrammetric Engineering and Remote Sensing*" in 1975, and has now become one of the international famous professional journals of remote sensing). All these have made full preparations and laid a foundation for the development of remote sensing into an independent discipline in theory and methods.

With the development of detection technology, the spectral range of detection continues to extend, from soft X-ray to thermal infrared; the segmentation of spectral segments is becoming more and more precise. It has developed from single spectral segment to multispectral segments, and the spectral resolution has reached the nanometer level. The emergence of imaging spectrum technology pushes the spectrum from hundreds to thousands, which more comprehensively reflects the nature of the target. It enables the undetectable substances in wide spectrum remote sensing to be detected. Various remote sensing technologies are becoming more and more mature. Also, laser radar, multispectral imaging, hyperspectral imaging and stereo mapping have entered the practical stage.

In terms of information processing, after more than half a century of development, optical remote sensing technology has penetrated into all fields of the national economy and played an important role in promoting economic construction, social progress, environmental improvement and national defense construction. The global change information obtained from earth observation by space remote sensing has proved irreplaceable. With the development of photoelectric remote sensing technology, space remote sensing adopts near real-time data transmission technology, and the remote sensing results are digital files, which makes the numerical image fusion more convenient and easy.

Space optical remote sensing makes rapid map updating a reality. Based on this, electronic map navigation and personal terminal navigation have become an indispensable part of modern social life. With the wide and deep development of optical remote sensing application, optical remote sensing detection has entered the practical, commercial and international and has become a new field of industrial development.

5. **Development of remote sensing in China**

About aerial remote sensing: Aerial photography was carried out in individual cities in 1930s, but systematic aerial photography began in the 1950s. It is mainly used in map mapping and updating and plays an important role in the investigation, survey and mapping in the fields of railway, geology and forestry.

In the application of remote sensing, great achievements have been made in the middle and late 1970s. Chinese government attaches great importance to the development of remote sensing technology and its application in national construction, and has completed a number of achievements with world advanced level and our own characteristics. The main achievements are as follows.

Extensive exploration and application experimental research have been carried out in the field of remote sensing applications, such as the comprehensive remote sensing experimental research of Tengchong Yunnan, the experimental research of Jingyuetan Changchun, the agricultural remote sensing experimental research of Taiyuan Basin Shanxi, the remote sensing experimental research of fishery in the East China Sea, the spectral experimental research of ground objects in the lower reaches of the Changjiang River, etc. These experimental studies are closely combined with the development and application of remote sensing technology, which has laid the foundation and played an exemplary role for large-scale multifield application research. Figure 1.3 are remote sensing aerial photos of Jingyuetan in Changchun during this period.

Fig. 1.3 Aerial mosaic of green spectrum section in Jingyuetan area, Changchun [2]

These remote sensing experiments are extensive and in-depth, and their application directions include geology, landform, water body, forestry, environment, atmosphere and other aspects.

As for remote sensing technology, such as multispectral color synthesis, colorimetric method of multispectral remote sensing image processing, application of color inversion aerial photos in Changchun experiment, improvement and enhancement of multispectral color synthetic image, ground multispectral photography, and identification method of effectiveness of multispectral remote sensing experiment.

As for geological aspects, such as the relationship between remote sensing information and geological structure, the application of aerial remote sensing image in geological structure research, the interference factors of lithologic geological interpretation of remote sensing image in covered and semi covered area, the diffuse reflectance spectral characteristics of rock powder and its significance in remote sensing interpretation.

As for the surface of ground aspects, such as the landform and image characteristics of the test area, soil types and interpretation of near-infrared aerial photos, interpretation of aerial photos of vegetation, interpretation of multispectral images of Xiaoying (Changchun) swamp, interpretation of vegetation spectra and images, research on swamp wasteland and land use changes in the middle and upper reaches of Belahong River in Sanjiang Plain, the interpretation and mapping experiment of land use types from multispectral remote sensing images, the representativeness of surface feature spectral determination and the interpretation of land types from multispectral images.

As for water conservancy aspects, such as, geological interpretation of water conservancy engineering and water body, groundwater is found with remote sensing technology, quantitative analysis of suspended sediment concentration of water body is made by remote sensing.

As for forestry aspects, the analysis of spectral characteristics of main tree species and the application of multispectral remote sensing in forestry investigation are made by remote sensing also.

As for environment aspects, the application of aerial remote sensing in environmental pollution monitoring, the preliminary interpretation of water pollution in Nanhu (Changchun) is made by using infrared images.

Atmospheric aspects, such as the study of the concentration and spectral distribution of atmospheric aerosols, are made by remote sensing also.

The results of remote sensing experimental research have been applied to various fields, including agricultural production conditions research, crop yield estimation, land and resources investigation, land use and vegetation cover, soil and water conservation, forest resources, mineral resources, grassland resources, fishery resources, environmental evaluation and monitoring, urban dynamic change monitoring, flood and fire monitoring, forest and crop pest monitoring, and meteorological monitoring, as well as the remote sensing research of port, railway, reservoir, power station and other engineering survey and construction, involving many business departments, which greatly expands the application field of remote sensing.

Remote sensing experiments have covered all parts of the country, from sparsely populated areas to densely populated and developed cities, such as Beijing, Shanghai, Tianjin, Guangzhou, Shenyang and other cities, and have carried out comprehensive remote sensing investigation and research as a regular monitoring means of urban construction and management, laying the foundation for "digital city". Different fields put forward different requirements for remote sensing applications, which promote the all-round development of remote sensing applications in China.

A number of large-scale application projects have been completed, such as national land area measurement and land resources survey, "Three North" shelter forest remote sensing comprehensive survey and research, Shanxi agricultural remote sensing, grassland resources remote sensing in Inner Mongolia Autonomous Region, soil and water loss and soil erosion remote sensing in the Loess Plateau in china, the Three Gorge Project remote sensing of the Changjiang River, Dongting Lake and Poyang Lake comprehensive remote sensing research. The scale, the comprehensive degree and professional application depth have reached the international advanced level, which provides a scientific basis for the decision of the state, relevant departments and local governments. On this basis, large-scale application systems such as national resource and environment dynamic service system, natural disaster monitoring and evaluation system and marine environment stereo monitoring system will be gradually formed. In these systems, remote sensing, as an important means of information collection, directly serves the national decision making.

Since 1970s, optical remote sensing has made great progress. Aerial photogrammetry has entered the operational stage, and aerial photogrammetry has been widely used in the updating of topographic maps across the country. On this basis, thematic remote sensing experiments and application research on different targets have been carried out, especially in the experiments of various new remote sensors and system integration experiments using aerial platforms.

China has successfully developed optical remote sensors such as airborne ground object spectrometer, multispectral scanner, infrared scanning camera, imaging spectrometer and laser altimeter, which has contributed to tracking the world's advanced level and promoting the localization of optical remote sensors. While developing new optical remote sensors, it is also paid attention to integrating several optical remote sensors into a comprehensive detection system.

On space remote sensing: China launched the "Dongfanghong 1" satellite on April 24, 1970, as shown in Fig. 1.4. Since then, a large number of artificial earth satellites have been launched one after another. The launch of the sun synchronous "Fengyun 1" (FY-1A, 1B), the earth synchronous "Fengyun 2" (FY-2A, 2B), the launch and recovery of the recoverable remote sensing satellite have provided, which made China had their own information source for space exploration, communication, scientific experiments and meteorological observation. On December 23, 2008, Fengyun-2 06 satellite was successfully launched at China Xichang Satellite Launch Center, as shown in Fig. 1.5.

Fig. 1.4 The "Dongfanghong 1" satellite

Fig. 1.5 Fengyun-2 06 satellite

On October 14, 1999, China Brazil Earth Resources Remote Sensing Satellite CBERS-1 was successfully launched. Since then, China has her own resource satellite. Chinese "Qinghua-1" and "Beidou-5" imaging satellites have been successfully launched on October 14, 2008, which shows that China has owed a wealth of satellite types. In 2008, imaging spectrometer has been launched successfully. With the further development of remote sensing in China, Chinese earth observation satellites and various satellites with different uses will also form a series of earth observation and enter the ranks of the world's advanced level.

In 2010, the second phase of China's lunar exploration took photos of the whole month.

On December 29, 2015, China's first geostationary orbit high-resolution optical imaging satellite was successfully launched. The satellite adopts frame imaging mode, and the resolution of subsatellite pixels is 50 m.

Since 1986, China has built a remote sensing satellite ground station, gradually forming a remote sensing center for receiving remote sensing data from Landsat of America, Spot of France, RADASAT of Canada and other resource satellites, and dozens of meteorological satellite receiving stations distributed all over the country, which can receive data from geosynchronous and solar synchronous meteorological satellites.

In terms of remote sensing image information processing, it has begun to move forward from the universal adoption of international advanced commercial software to the localization of software. It has launched relatively mature commercial software, such as photo, mapper, etc., and has gradually formed commercialization. New methods of image processing have also been widely explored, such as fractal geometry, artificial neural network, wavelet transform and so on, have entered the application phase.

In terms of remote sensing institutions, there is a National Remote Sensing Center to plan and manage the development of remote sensing science and the major projects; many institutes affiliated to the Chinese Academy of Sciences have remote sensing disciplines; military departments also have corresponding remote sensing research institutions; more than ten national societies of the China Association for Science and Technology have remote sensing branches or remote sensing professional committees. Tens of thousands of remote sensing scientific and technological personnel all over the country have formed a huge remote sensing scientific and technological team.

In terms of professional publications of remote sensing, there are works such as "Dictionary of Remote Sensing" and "Remote Sensing Geoscience Analysis". There are "Journal of Remote Sensing", "Remote Sensing Information" and other famous professional journals of remote sensing at home and abroad. "Opto-Electronic Engineering" and "Optics and Precision Engineering" are important journals in the field of optical remote sensing.

Chinese remote sensing education has made remarkable achievements, and the training of remote sensing talents has formed a series of undergraduate, master, doctor and postdoctoral.

In short, Chinese remote sensing industry has experienced the initial stage from 1970s to mid-1980s, the experimental application stage from late 1980s to early 1990s, and entered the practical and industrialized stage in the late 1990s. At present, it has made remarkable achievements in remote sensing theory, remote sensing platform, remote sensor development, system integration, application research, academic exchange, talent training and so on. It has made great contributions to the development of remote sensing discipline, national economic construction and national defense construction. Remote sensing has gone from remote sensing to the Earth, to the Moon and to Mars exploration.

1.2 Advantages of Space Optical Remote Sensing Technology

Advantage 1: better real-time performance of obtaining information. The advantage of space optical remote sensing lies in large-area synchronous observation and short repeated detection cycle of the same area. For example, in geostationary orbit, one satellite can cover more than 40% of the earth's surface; in terms of timeliness, when observing the earth in the geostationary orbit, the repeated observation cycle for the same area is the shortest, which is only limited by the integration time, data acquisition speed and data transmission capacity; in other orbits, the repeated observation period is also different according to the factors such as orbital height and camera field angle, which can range from a few days to dozens of days. For example, the revisit periods of the American Landsat satellite sand the French Spots satellites are 16 and 26 days, respectively. That is to say, if the method of space remote sensing imaging is used for mapping, it should be calculated in days, and if the traditional method of manual mapping is used, it should be calculated in years. The global information obtained by space remote sensing cannot be replaced by ground observation.

Advantage 2: More economical access to information. The benefits and cost investment ratio of information obtained by aerospace remote sensing imaging method is much better than that obtained by other traditional methods. Remote sensing has achieved good economic and social benefits in the application field: according to the data of regional land use remote sensing survey, compared with conventional ground survey, aerial remote sensing takes only 1/8 of the time, 1/3 of the capital investment and 1/16 of the manpower; if aerial remote sensing is applied to produce 1/100,000 land use map, the capital is 1/10 of the conventional method, the time is 1/24, and the manpower is only 1/36 [1].

Advantage 3: For the fusion of information, almost all the images obtained by modern space remote sensing are digital images, which is convenient for information fusion with information obtained by different means. For example, the images obtained from different spectral bands such as microwave remote sensing, infrared and visible light are fused to complement and confirm each other, so as to improve the reliability of information.

1.3 Demand for Space Optical Remote Sensing in Civil Field

It is just because of the real time and economy of space optical remote sensing imaging that it has been widely used in geological research, water resources utilization, vegetation protection, land monitoring and geographic information system.

It can realize the near real-time observation ability of small and medium-sized meteorological phenomena, realize the prediction and continuous observation of sudden changes and disastrous weather such as hurricanes and improve the rapid response ability to emergencies such as earthquakes and debris flows; it can realize long-term continuous monitoring of small time scale ocean phenomena with high time resolution and large time scale ocean phenomena; it can observe, monitor and track natural disasters, such as oil leakage, harmful algae flooding, fire and volcanic ash clouds; it can observe the growth state of crops, distinguish the types of rocks and soils and monitor plant diseases and pests.

Space optical remote sensing imaging technology has become an important means to obtain information in many civil fields, such as map mapping, resource survey, disaster reduction and prevention, pollution monitoring, urban planning and so on. Optical remote sensing imaging covers all aspects of scientific research, production and people's life. These applications also promote the rapid development of space remote sensing imaging technology.

1.4 Military Application Promote to Development of Space Optical Remote Sensing

Optical remote sensing has been accompanied by military applications since its birth. It has formed a certain scale in the First World War; it is used to draw maps and obtain information in the later stage of the Second World War. With the expansion of optical remote sensing imaging to space, space optical remote sensing has also become an important field of military struggle. In particular, high-resolution space remote sensing images have become an indispensable means in modern high-tech war in the field of space reconnaissance. Because the important feature of modern war is beyond visual range combat, obtaining high-resolution space remote sensing images has become an important condition for winning victory.

In peacetime, space optical remote sensing imaging is used to discover and track the armed forces and daily activities of the envisaged enemy and to supervise the implementation of the strategic weapons treaty.

In wartime, it is mainly used to determine the working state and working characteristics of the envisaged enemy. The first is the working state and working characteristics of missile nuclear weapons. Supplement reconnaissance of known enemy groups and targets and discover new enemy groups and targets. Provide target instructions for our own strike forces, evaluate the results of the strike against the enemy's targets, and find out the measures for the enemy to restore its combat capability. Determine the combat readiness of the enemy's aviation, the concentration of the enemy's campaign and strategic reserves and the enemy's camouflage measures to deal with our reconnaissance.

In the field of military struggle, space optical remote sensing imaging has two main applications: visible light photographic reconnaissance and infrared photographic reconnaissance.

In the aspect of visible light photographic reconnaissance, based on the statistical data, from late 1950s to the end of the last century, photographic reconnaissance satellites accounted for 30 and 40% of the military satellites launched by the United States and the former Soviet Union (Russia), respectively. From here, it can be seen that the importance of space optical remote sensing imaging in the field of military struggle.

In addition to the United States and the former Soviet Union, many countries in the world, such as Southeast Asia and other small- and medium-sized countries are also carrying out research in this filed.

China's research work in this field has also experienced a process similar to that of the United States and the former Soviet Union since 1960s. The ground resolution of the original film return remote sensing camera is about 10 m, and today's photoelectric transmission remote sensing camera is better than 1 m.

In the aspect of infrared photographic reconnaissance, reconnaissance satellites launched to monitor and discover enemy ballistic missiles often have the task of detecting nuclear explosions. Generally, it operates in geostationary orbit or large elliptical orbit with a period of 12 h. It is composed of several satellites, and the early warning network covers a wide range. Even for infrared cameras, their ground resolution has been continuously improved, from about 3 to 1 km.

Stereo mapping camera can quickly and accurately establish stereo topographic map through aerospace stereo mapping camera, which is not limited by territory, territorial sea and airspace, and can provide strong guarantee for terrain matching guidance. At present, the high-precision stereo mapping cameras launched by Japan, India and France are still for civilian purposes, but there is no qualitative difference as military applications. Nowadays, it is a fact that civilian optical remote sensors are used as military in wartime. Figure 1.6 shows the stereo mapping satellite of Japan, Fig. 1.7 shows the imagination of Indian stereo mapping satellite, and Fig. 1.8 shows Indian stereo mapping satellite.

1.5 Improvement of Optical Remote Sensing Technology

Since 1960s, with the development of computer technology, the numerical control grinding and polishing methods of optical aspheric surfaces have been studied and developed. Figure 1.9 is a schematic diagram of real-time interference measurement of the tested optical mirror by using Twyman laser interferometer, computer, etc. This method is used to process off-axis aspheric optical mirror with asphericity up

Fig. 1.6 Stereo mapping satellite of Japan

Fig. 1.7 Imagination of Indian stereo mapping satellite

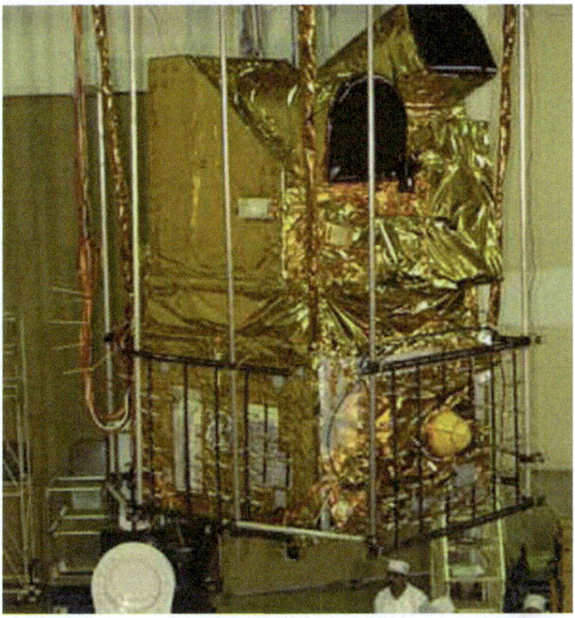

Fig. 1.8 Indian stereo mapping satellite

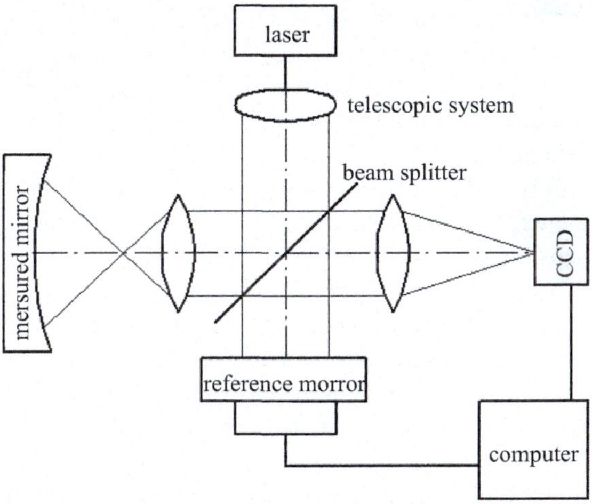

Fig. 1.9 Schematic diagram of interferometry

to 200 wavelengths, and the mirror surface shape error is up to l/40 wavelength root mean square (RMS).

In addition, diamond turning process is adopted, and the minimum cutting thickness is 1 nm. The surface shape accuracy of ion polishing is 0.1 μm, and the surface roughness is 0.01 μm.

In the ultra-smooth surface machining technology, the optical surface with surface roughness of 0.5 nm RMS can be obtained by ultra-precision machining with microelastic break method. The surface roughness of 0.27 nm can be obtained by water polishing technology. The optical surface with RSM value lower than 0.2 nm can be obtained by float polishing.

The progress of modern optical processing technology has effectively promoted the development of optical remote sensing imaging technology. QuickBird space remote sensing camera of the United States, main mirror aperture $\Phi = 600$ mm, when $\lambda = 632.8$ nm, the surface error after coating is 0.008 λ RMS value.

Due to the joint action of these three aspects, which are the demand of the civil field, the traction of military applications and the promotion of optical manufacturing technology, space optical remote sensing imaging technology had developed rapidly and the development trend also reflects diversification.

From the perspective of the performance of space optical remote sensing camera, it is developing toward long focal length and large field of view. This benefits not only from the traction of practical needs, but also from the development of optical manufacturing technology, so that the optical system of off-axis aspheric surface can be used in practice. At present, the focal length of space optical remote sensing camera is more than 10 m, which means that high ground resolution and clearer image can be obtained. The field of view of the space optical remote sensing camera can reach 20 degrees, which means that the wide ground coverage can greatly improve the observation efficiency. The field of view of the traditional space optical remote sensing camera is generally within 2–3°. Large relative aperture is mainly due to the improvement of receiver performance. Not only the quantum efficiency of CCD detector is improving, but also its imaging mode is changing, such as delay integral imaging.

From the perspective of volume and weight of space optical remote sensing camera, space optical cameras are developing in two directions, large-scale remote sensing camera and small-scale remote sensing camera. Large space optical remote sensing camera can obtain high-definition images, wide ground coverage and multispectral data. Small space optical remote sensing camera is suitable for small satellite transportation due to its small volume and light weight; therefore, it has the advantages of low manufacturing cost, short manufacturing cycle, low launch cost, relatively few mission items, convenient operation, simplicity, reliability and flexible launch.

From the perspective of the optical system form of space optical remote sensing camera, there are total reflection type, catadioptric hybrid type and refraction type. At present, the development trend is mainly total reflection type.

From the perspective of the receiver of the space optical remote sensing camera, the film remote sensing camera is gradually withdrawing from the market, mainly focusing on the photoelectric transmission remote sensing camera. This is mainly because with the reduction of the pixel size of the CCD detector, its adverse impact on the image resolution is decreasing, and the working life on orbit and real-time image advantage of the optical transmission remote sensing camera are unmatched by the film remote sensing camera.

References

1. Mei A. An introduction to remote sensing. Beijing: Higher Education Press; 1989.
2. Chinese Academy of Sciences, Changchun Branch *Symposium on Changchun Remote Sensing Experiment Editorial board*, *Symposium on Changchun Remote Sensing Experiment*. Changchun: Jilin People's Press; 1981.

Chapter 2
Fundamentals of Space Optical Remote Sensing System

2.1 Overview

The physical components of the space optical remote sensing system are the platform and payload of the satellite. In addition, it should also include the orbit of the spacecraft, which is the non-real objects component of the space optical remote sensing system, because this part is also necessary for the operation of the space optical remote sensing system which exists physically. For example, the orbit position of the geostationary orbit is the position of the geostationary orbit satellite when it runs. Due to the development of the geostationary orbit satellite, the reality is that there is one satellite about every one degree apart along this circle. Due to the overcrowding of satellites, the orbit position cannot meet the application requirements, and satellite orbit position lease business appears. From this point, the physical existence of orbit can be deeply understood.

In essence, the space optical remote sensing system is an information acquisition and transmission system. It uses electromagnetic wave as the medium to realize information acquisition and transmission. The payload of the system is some optical remote sensing equipment, such as imaging camera, spectrometer, etc. The payload achieves the acquisition of target information, which is configured on the platform. The platform is the carrier of the payload, which carries the payload and provides the working environment for the payload, such as thermal environment and dynamic environment. Generally, it also supports the payload to select the remote sensing target, information transmission and energy providing for the whole spacecraft. Target selection is to align the visual axis of the payload to aim the target through the attitude adjustment of the platform.

Sunlight emanates from the sun and hits the target, reaches to the optical remote sensing equipment through the target reflection. Generally, electrical signals are formed by photoelectric conversion of optical remote sensing equipment, which are converted into radio waves by the platform's data transmission system and transmitted to the ground receiving antenna. After being received by the receiving antenna, it is processed, stored and distributed to users by the ground station.

J. Tao, *Space Optical Remote Sensing*, Advances in Optics and Optoelectronics,
https://doi.org/10.1007/978-981-99-3318-1_2

2.2 Basic Concepts of Spacecraft Orbit

The spacecraft runs on the orbit, its operation involves space and time, so the spacecraft orbit concept involves the two concepts of orbital space and orbital time.

2.2.1 *Space Concepts of Spacecraft Orbit*

The reason why the spacecraft can stay in the space is that the centrifugal force produced by the orbital movement of the spacecraft and the celestial gravity of the two forces achieve the balance. For the spacecraft flying around the earth in low earth orbit and geostationary orbit, the main force they are subjected to is the earth's gravity. The sun, the moon on its force is far less than the earth's gravity. Figure 2.1 is diagram of orbit of the sun, the earth, the moon and spacecraft.

2.2.1.1 Celestial Bodies and Their Motion

(1) The sun is a hot ball of gas. The mass of the sun is 1.989×10^{30} kg, which is 332,770 times of the earth. The radius of the sun is 6.9599×10^{8} m, which is 109 times of the earth. The sun rotates every 27 days [1–6]. The sun is at the center of the solar system, which is in the Milky Way.
(2) The mass of the earth is 5.977×10^{24} kg, the average radius is 6371.23 km (1976 US Standard Atmosphere), which is a triaxial ellipsoid. As shown in

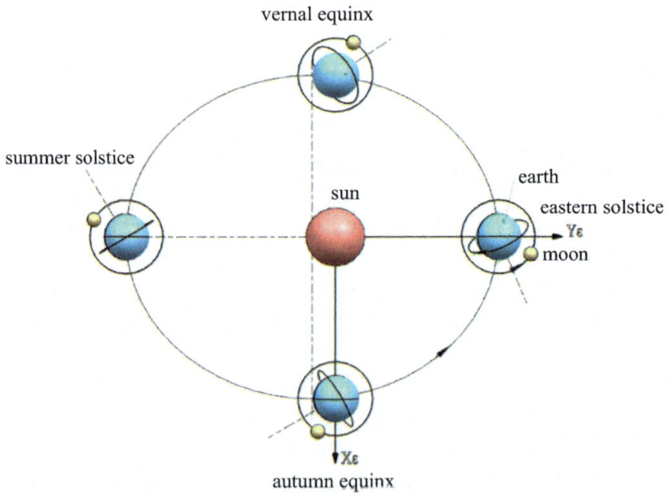

Fig. 2.1 Diagram of orbit of the sun, the earth, the moon and spacecraft

Fig. 2.2, the cross section of the earth's meridian is an ellipse, semi-major axis r_e is equatorial radius, semi-minor axis r_p is polar radius. This is called biaxiality, the second order perturbation coefficient ($J_2 = 1.08263 \times 10^{-3}$[7]) of the earth's gravitational field function with coefficient of zonal harmonics describes this property. As shown in Fig. 2.3, the cross section of the earth in the equatorial plane is similar to an ellipse, this may be called triaxiality. The second tesseral order harmonic coefficient ($J_{22} = 1.81222 \times 10^{-6}$) of the earth's gravitational field function describing this property. The semi-major axis is along 75.1°E–105.3°W, the semi-minor axis is along 11.5°W–161.9°E [7].

The oblateness and eccentricity of the ellipse in the meridional plane are:

$$f = \frac{r_e - r_p}{r_e}$$

$$e^2 = \frac{r_e^2 - r_p^2}{r_e^2}$$

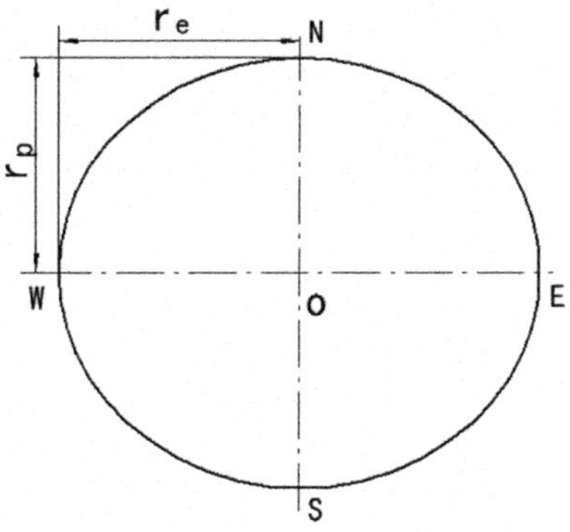

Fig. 2.2 Ellipse of the meridional section of the earth

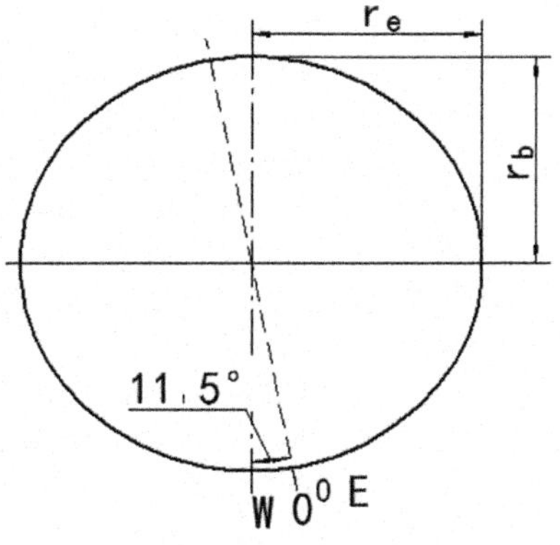

Fig. 2.3 Ellipse of the equatorial section of the earth

where the radius of semi-major axis $r_e = 6378.145$ km, the radius of semi-minor axis $r_p = 6356.76$ km. $f = 1/298.257$, e = 0.08182 can be obtained.

The earth's orbit around the sun is an ellipse called the ecliptic, and its plane is called the ecliptic plane. Perihelion distance of the earth is 1.4710×10^8 km, and aphelion distance of the earth is 1.5210×10^8 km. The average distance between sun and the earth is 1.4960×10^8 km, which is called an astronomical unit (AU).

(3) The mass of the moon is about 7.376×10^{22} kg, and the radius is 1738.2 km. The moon's orbit around the earth is an ellipse called the white path. The semi-major axis is about 384,748 km, perigee distance is about 364,397 km, apogee distance is about 406,731 km, average eccentricity is 0.0549006 (0.044–0.067). The mean orbital inclination of the white plane relative to the ecliptic is 5.145°, the angle between the moon's equatorial plane and the ecliptic plane is 1.543°. The angle between the white plane and the equatorial plane of the earth is 18.295–8.585°. The precession period of the nodal line between the white plane and the ecliptic plane around the yellow pole in space is 18.5996 years, which causes the ascending intersection of the white path to move westward along the ecliptic. The moon rotates every 27.32 days. This is also the time moon around the earth once, so the moon always faces the earth with the same side.

(4) The solar system is generally thought to have formed from a cluster of nebulae gradually condensed by its own gravity about 4.6 billion years ago. There are nine planets in the solar system, Mercury, Venus, Earth, Mars, Jupiter, Saturn, Uranus, Neptune and Pluto (Pluto is still listed as a planet) as shown in Table 2.1. They all orbit the sun from west to east; the angle between their respective orbital planes and the ecliptic plane is smaller than that of the earth.

(5) Precession and nutation. Precession and nutation are the result of the movement of celestial bodies in the solar system. The earth's spin axis precession around the vertical axis of the ecliptic plane is caused by the gradient moment caused by the earth's equatorial bulge due to the gravitational attraction of the sun and moon on the earth. This precession makes the line of the ecliptic plane intersects the equatorial plane (also known as the vernal equinox) move westward, about 0.65° every 50 years (one cycle every 26,000 years [6]). This movement is a long period movement called precession. The half cone angle of precession is the angle of 23° 26' between the ecliptic and the equatorial plane. Since the precession period of the nodal line of the lunar white plane and the ecliptic plane around the yellow pole in space is 18.5996 years, the period causing the change of the gravitational direction of the moon to the earth is also 18.5996 years. The change of gravitational direction is reflected in the constantly change of the direction of the earth's spin axis in space. That is, the spin axis of the earth orbits its average position with a period of 18.5996 years. This movement is called nutation, and the nutation angle is no more than 0.006° (Fig. 2.4).

(6) Pole shift and the inhomogeneity of the earth's rotation. The earth is not a rigid body, and the pole shift is the twist of the earth's spin axis during the rotation of the earth, which causes the movement of the earth's spin axis in the earth itself. The annual motion of the atmosphere causes the wobble of the annual period

Table 2.1 Comparison table of relative data of celestial bodies in the solar system [3]

Celestial bodies	Distances/AU	Revolution period	Obliquity/°	Eccentricity	Mass	Radius	Rotation period/d	Density g/cm^3	Number of satellites
Sun	0				332,770	109	25– 36*	1.410	9
Mercury	0.39	87.9 d	0.1	0.2056	0.05	0.38	58.8	5.46	0
Venus	0.72	224.7 d	177.4	0.0068	0.82	0.95	244	5.26	0
Earth	1.0	1 a	23.45	0.0167	1.00	1.00	1.00	5.52	1
Mars	1.5	1.9 a	25.19	0.0934	0.11	0.53	1.029	3.96	2
Jupiter	5.2	11.8 a	3.12	0.0483	317.94	11	0.411	1.33	16
Saturn	9.5	29.5 a	26.73	0.0560	95.18	9	0.428	0.70	18
Uranus	19.2	84.0 a	97.86	0.0461	14.63	4	0.748	1.24	15
Neptune	30.1	164.8 a	29.56	0.0097	17.22	4	0.802	1.66	8
Pluto	39.5	247.9 a	119.6	0.2482	0.0024	0.18	0.267	1.50	1

**Note* The rotation period at the sun's equator is about 25 days; the rotation period at the solar polar is about 36 days

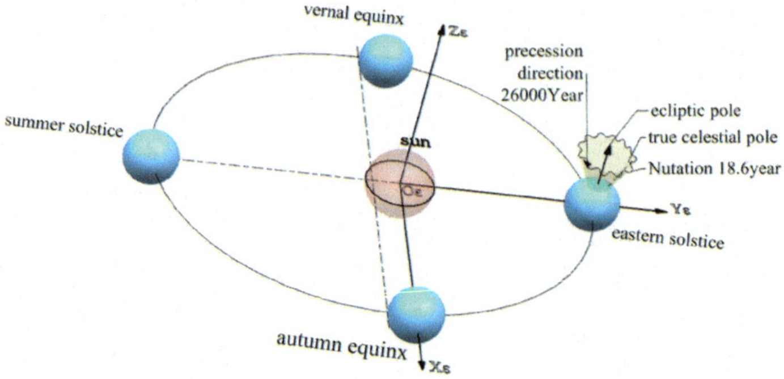

Fig. 2.4 Precession and nutation of the earth

of the earth and the free wobble of the non-rigid body about 14 months, and the pole shift is about 0.4 ". The non-uniformity of the earth's rotation, one side is the long-term slowing, 1–2 ms every 100 years. There are also massive air mass movements, tidal interactions of the sun and the moon, and the movement of material in the earth's interior that can cause variations in the length of the day.

2.2.1.2 Coordinate Systems

As the earth orbits the sun, in the meantime it spins from west to east on its spin axis, which is perpendicular to the equator. It is generally thought (ignoring small changes in the earth's axis of rotation) that the earth's spin axis points toward the north pole and is inertial. Therefore, the direction of the intersection between the equatorial plane and the ecliptic plane is also inertial, as shown in Fig. 2.5. The intersection line formed between the equatorial plane and the ecliptic plane at the vernal equinox is defined as x_i axis, which points from the center of the earth to the sun; z_i axis, namely the earth's spin axis, points to the north pole; the center of the earth as the origin of coordinates O; and y_i axis perpendicular to x_i axis and z_i axis to form the right hand system, which is called the geocentric equatorial inertial coordinate system $\mathbf{S_i}$.

As known from Fig. 2.5, the direction of the vernal equinox is the line between the earth and the sun at the vernal equinox (about March 21 or 22). When "pointing in the vernal equinox" is said, it is mean "pointing in this direction". The vernal equinox is understood as the point at which this line (x) intersects with the celestial surface centered on the earth (not just the position of the earth at the time of the vernal equinox).

In spacecraft orbit dynamics, the small oscillation of vernal equinox and the inertia force caused by the earth moving around the sun is thought to be in perfect equilibrium with the gravitational force between the sun and earth, so the inertial coordinate system of the local central equator can be used as the inertial system.

With the center of the sun as the origin of coordinates O_ε, the $x_\varepsilon y_\varepsilon$ coordinate plane coincides with the ecliptic plane (the plane of the earth around the sun), x_ε points to the vernal equinox, and the z_ε axis is along the direction of the angular velocity of the

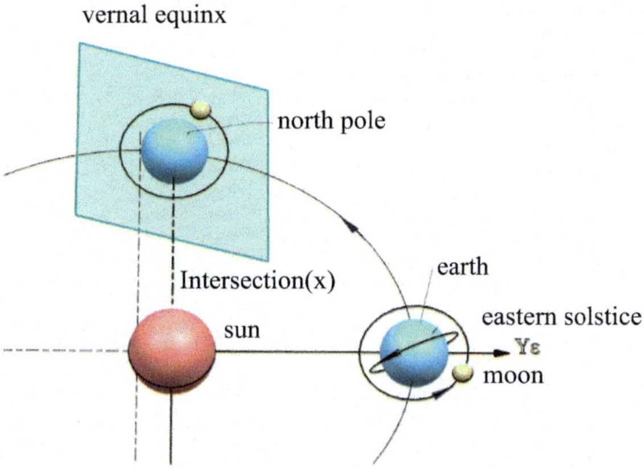

Fig. 2.5 Geocentric equatorial inertial coordinate system

earth moving around the sun. This coordinate system is called the heliocentric ecliptic coordinate system (ecliptic coordinate system), denoted as $O_\varepsilon x_\varepsilon y_\varepsilon z_\varepsilon$, as shown in Fig. 2.6. The heliocentric ecliptic coordinate system $O_\varepsilon x_\varepsilon y_\varepsilon z_\varepsilon$ can be used as an inertial coordinate system if slight variations in the vernal equinox are ignored, which are related to slight variations in the earth's rotation axis.

When the origin of heliocentric ecliptic coordinate system is shifted from the center of the sun to the center of the earth, a geocentric ecliptic coordinate system with the center of the earth as the origin $Ox_\varepsilon y_\varepsilon z_\varepsilon$ is formed, noted as S_ε.

The relationship between the geocentric ecliptic coordinate system S_ε and the geocentric equatorial inertial coordinate system S_i is shown in Fig. 2.7. The geocentric ecliptic coordinate system can be obtained by the rotation angle ε of the geocentric equatorial inertial coordinate system around the x_i axis, which is expressed as:

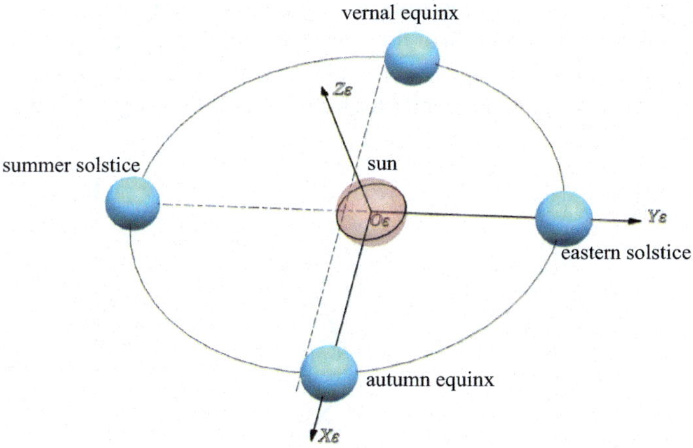

Fig. 2.6 Heliocentric ecliptic coordinate system

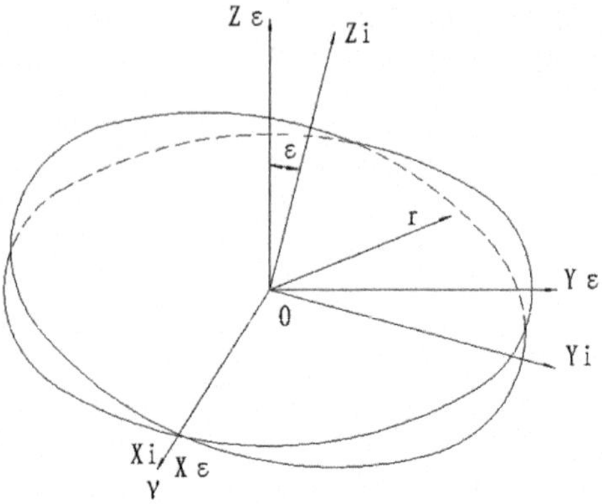

Fig. 2.7 Relationship between the geocentric ecliptic coordinate system S_ε and the geocentric equatorial inertial coordinate system S_i

$$S_{\mathrm{i}} \xrightarrow{R_x(\varepsilon)} S_\varepsilon$$

Obliquity of the ecliptic $\varepsilon = 23^\circ\ 26'$.

Due to the nutation of the earth's spin axis, the ε varies by ± 9" in a period of 18.6 years.

The center of the earth is the origin o of coordinates, the x_e axis is the intersection line between the equatorial plane and Greenwich meridian, the z_e axis is along the axis of the earth's spin, pointing to the north pole, the y axis is in the equatorial plane, this coordinate system is called the geocentric equatorial fixed coordinate system $\mathbf{S_e}$, with the earth's spin angular velocity ω_e.

With geocenter as the coordinate origin O, x_o points to the spacecraft, z_o points to the direction of the normal line of the orbital plane, which is consistent with moment of momentum **H**, y_o is in the orbital plane, as shown in Fig. 2.8. The coordinate system is called geocenter coordinate system, denoted as $\mathbf{S_o}$.

With geocenter as the coordinate origin o, x_p points to perigee p along the orbit arch line, and z_p points to the direction of the normal line of the orbit plane, consistent with moment of momentum **H**. y_p is in the orbit plane, as shown in Fig. 2.8. The established coordinate system is called geocenter arch line coordinate system, denoted as $\mathbf{S_p}$.

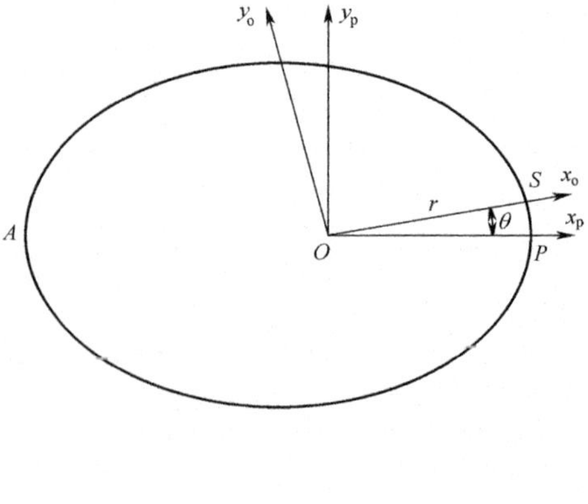

Fig. 2.8 Geocentric coordinate system

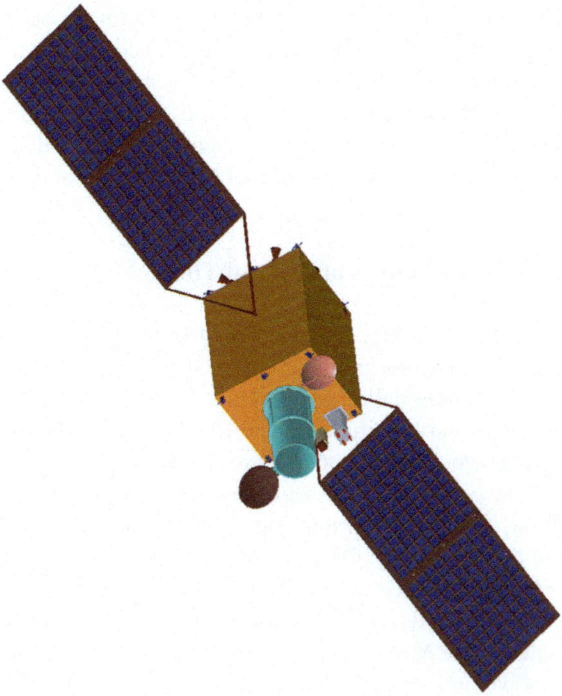

Fig. 2.9 Spacecraft body coordinate system

With the spacecraft centroid as the origin of coordinates o, x_b axis points to the flight direction of the spacecraft, z_b axis is one of the axis of symmetry of the spacecraft, pointing to the direction of the center of the earth, y_b axis and x_b axis, z_b axis constitute the right hand system. As shown in Fig. 2.8, the established coordinate system is called the spacecraft body coordinate system, denoted by $\mathbf{S_b}$. Generally speaking, for the satellite, the x_b axis points to the windward direction, namely the flight direction, the y_b axis is the pitching direction, and the z_b axis is the yaw direction. For the carrier rocket x_b axis pointing to the flight direction, y_b axis is yaw direction, and z_b axis is pitch direction (Fig. 2.9).

2.2.2 *Time Systems*

By watching the sun rise in the east and set in the west, people realized the passage of time. The difference between day and night gives people the understanding of Day, the change of the moon gives people the understanding of Month, and the change of the four seasons gives people the understanding of Year. These plain concepts of time are imprecise and limited. Such as Day defined by day-night observation and "date" defined on this basis, there will be different times and dates at different longitude locations around the globe. From the earth's elliptical motion around the sun, each position in the ecliptic is defined as a day in the interval between the sun's two consecutive transit times. Differences in the earth's linear velocity due to the earth's elliptical motion around the sun must result in different lengths of Day at different points of the ecliptic. Space activities, due to their high speed, global and deep space activities, must establish a precise and unified time system.

2.2.2.1 Definition of Units of Time

Local time is a time concept established by observing the sun rising east and setting west. The highest point of the local noon sun is defined as 12 o'clock, establishing a 24 h system called Local Time (LT). It can be known that places with different geographical longitude have different local time, but it is most consistent with the natural time concept of sunrise and sunset.

Universal time, for a wide range of human activities such as navigation and spaceflight, different geographical longitude has different local time, the application is very limited, obviously. On October 13, 1884, representatives of more than 20 countries held a conference in Washington, the United States, on the use of a unified international standard time and a unified meridian issue to make a resolution. The resolution states: "The Meeting recommends to the Participating Governments that the meridians passing through the Meridian Centre of the Royal Greenwich Observatory be defined as the prime meridian of longitude." As a result, the meridian through RGO was recognized as the prime meridian worldwide, as the starting point for calculating geographic longitude and the world's "time zones". Thus, Greenwich International Standard Time was born.

Mean solar day. Usually the so-called day refers to the time when the sun reaches its meridian twice in a row on a certain meridian, which can be called the natural day. Due to the existence of the obliquity of the ecliptic and the ellipticity of the ecliptic, the length of the natural day is not equal. As shown in Fig. 2.10, the earth's ecliptic is an ellipse, and the earth's rotation angle $\alpha 1$ in a day around the sun on the perihelion side is greater than the rotation angle $\alpha 2$ in a day around the sun on the aphelion side. Therefore, if the time of the sun's two consecutive transit times is recorded as one day, then a day on the perihelion side is longer than a day on the aphelion side dt of the earth's rotation of angle dα, that is, a day on the perihelion side is longer than a day on the aphelion side dt of a day. If the effects of precession and nutation are taken into account, the time of the natural day is not equal. This brings much inconvenience to the measurement of time. So imagine the earth moving around the sun and spinning at a uniform angular velocity. The time when the "mean sun" reaches the meridian twice in a row is defined as one mean solar day. The "mean solar day" is defined as 24 h a day equal to 86,400 s. One revolution of the earth around the sun is defined as 365 days a year. A day is thus defined as a mean solar day and a year as a mean year. The earth orbits the sun in a period of revolution exactly 365.2425 days (mean solar day). The extra 0.2425 days are balanced by a leap day every four years. The year with an extra day is called a leap year, and the year with an extra day is reduced by 3 leap days every 400 years. That is, every 400 years equals exactly 146,097 days [6]. It can be seen from Fig. 2.11 that one mean solar day is not the rotation period of the earth, but is longer than the rotation period of the earth. The extra time is caused by the earth's revolution around the sun. The cumulative length of time in a mean year is exactly one mean solar day, and the rotation period of the earth should be expressed by sidereal day.

Sidereal day. The "mean solar day" is defined as a mean solar day when the mean sun goes to the meridian twice in a row, the reference point is the sun, thus causes the difference between mean solar day and earth's rotation period. To calibrate the period of the earth's rotation stars in more distant are used than the sun as reference points, as shown in Fig. 2.11. A mean solar day is longer than a sidereal day. That is, the rotation period of the earth is the time of two consecutive transit times of distant stars, namely 1 sidereal day (Fig. 2.12).

Sidereal day is the movement of the earth within one revolution period, which causes the rotation of the earth 360° relative to the sun in 365.2425 mean solar days.

Fig. 2.10 Meridian of the Meridian center of RGO

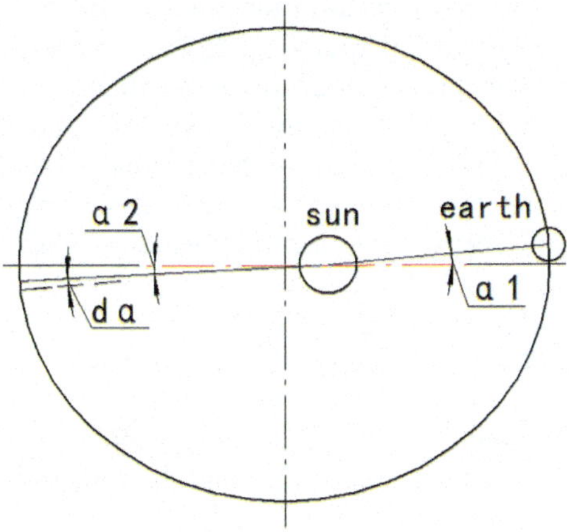

Fig. 2.11 Time difference of the natural day caused by the elliptic movement of the earth's ecliptic

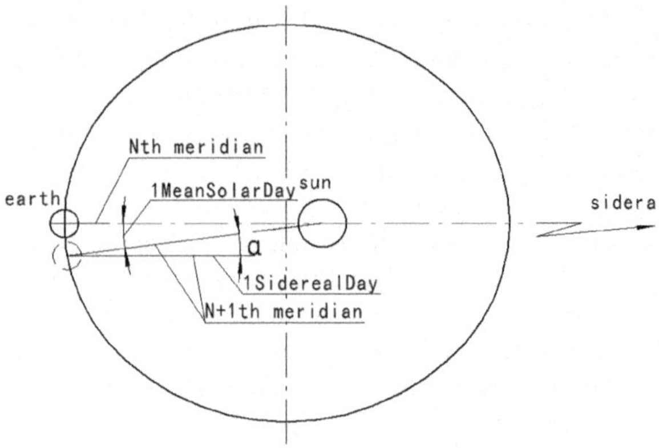

Fig. 2.12 Earth rotation period and sidereal day

Each mean solar day, the earth rotates more than the mean sidereal day by an angle of 360/365.2425 = 0.9856465°. The earth rotates in an average solar day is 360° + 0.9856465° = 360.9856465°, as 86,400 s. A more accurate unit of time is defined as a mean solar year of 31,556,925.9747 s. It takes 360 × 86,400/360.9856465 = 86,164.0907376 s while the earth rotates 360°. Thus, a mean sidereal day is 86164.0907 s [6].

Greenwich Mean Time is actually local time on the Greenwich meridian. In theory, noon at Greenwich Mean Time is the midday moment when the sun crosses the Greenwich meridian. The original use of Greenwich mean noon as the beginning of a mean solar day, but it brings some inconvenience in use. So in 1928, the International Astronomical Union decided to designate universal Time, commonly known as Greenwich Mean Time (GMT), as mean solar time, which starts at midnight GMT. Because the earth moves unevenly in its elliptical orbit, the winter solstice has the longest true solar day at perigee and the summer solstice has the shortest true solar day at apogee, with a difference of 16 s. Atomic time was introduced because the earth's daily rotation was irregular and was slowly slowing down.

Atomic time takes the time of 9,192,631,770 cycles of electromagnetic oscillation of the two hyperfined energy level transitions in the ground state of cesium atom Sc^{133} as the unit of second duration, which is atomic time (AT). Universal time (pole shift and seasonally modified universal time) 0 (difference of −0.0039 s) on January 1, 1958, was used as the starting point. In October 1967, the 13th International Conference on Metrology decided to take the length of atomic time at sea level as the length of international system of Units (SI) seconds.

2.2.2.2 Concepts of Time Systems

From the origin of time units mentioned above, it can be seen that with the development of social activities and the progress of science and technology, the definition of time is gradually clear and accurate, and the main time system comes from astronomical observation and artificial atomic time. The advantage of the time system of astronomical observation is that it is coordinated and unified with people's daily life. That is, different positions of the sun in the sky determine the time early, middle and late, which is advantageous. But it is not uniform, which also brings inconvenience to the continuous, accurate, large-scale, long-span space activities and other scientific research and production activities. The artificial atomic time system is a continuous, uniform and accurate time system. But it deviates from the time system of astronomical observation that people are used to. Hence, the coordinated universal time is generated, which is coordinated with atomic time.

The universal time system is a time system based on the earth's rotation and takes the earth's rotation period as the unit of measurement. Thus, different units of time are formed by observing the rotation period of the earth at different reference points.

Sidereal Time: take the vernal equinox as the reference point;

Apparent Time: take the real sun as the reference point;

Universal Time (UT, GMT) or Mean Time: Use the mean sun as a reference point.

In 1955, the International Astronomical Union decided to make two revisions to the directly observed universal time starting from 1965. Therefore, universal time (UT) has three forms:

UT0, the universal time obtained directly from astronomical observation, the universal time observed by different observatories varies slightly due to the existence of the pole shift;

UT1. UT0 polar shift corrected universal time. This is the unified universal time that can be observed by observatories around the world.

UT2. UT1 modified universal time through the ellipse of the ecliptic, is the internationally recognized time standard before 1972. It is still affected by the long period and irregular changes of the earth's rotation, and it is also an inhomogeneous time system.

Atomic time system, which is based on atomic time seconds (SI) as the basic time unit, is a continuous uniform time system.

Coordinated Universal Time (UTC) is a timing system that adopts atomic time seconds (SI) as the time unit. The atomic time system keeps the error with Universal Time UT1 within 0.9 s through "leap second". When the error between atomic time and universal time UT1 accumulates to 0.9 s, the time is automatically extended by one second. The normal second count is 57, 58, 59, and the jump second count is 57, 58, 59, 60 to 0. "Leap second" is announced six months in advance by the International Earth Rotation Service (IERS) system. Coordinated Universal time (UTC) was implemented from 0:00 UTC on January 1, 1972. It makes the ideal uniform time timing coordinate with the concept of natural time that people are used to and meets the unity of the uniform passage of time and people's natural feeling.

The GPS time system is an atomic time system established by the Global Positioning Satellite System (GPS) of the United States. The starting point of the GPS time system is 0 o'clock (UTC) on January 6, 1980. The length of the GPS time system is in atomic seconds (SI).

2.3 Basic Concepts of Spacecraft Platforms

The satellite is mainly composed of payloads and platforms. The payloads are mainly optical and microwave payloads, and the data transmission system is also included in the payload category.

The satellite platform is usually composed of structural subsystem, power subsystem, thermal control subsystem, control subsystem, telemetry and telecontrol subsystem, propulsion subsystem and digital management subsystem (or integrated electronic subsystem).

Structural subsystem, power supply and distribution subsystem and thermal control subsystem belong to service system. The structure subsystem provides positioning and support for the whole satellite including platform and payload, the power supply subsystem provides energy for the whole satellite including platform and payload, and the thermal control subsystem provides temperature environment for normal operation and on-orbit storage for the whole satellite including platform and payload.

The control subsystem provides attitude pointing control and orbit parameter control for the whole satellite. As shown in Fig. 2.13, in order to make the camera's visual axis point to target A, the control subsystem needs to adjust the satellite attitude to maneuver an angle around the pitch axis y_b and the rolling axis x_b of the platform respectively. During satellite imaging, the yaw axis z_b of the platform generally remains unchanged.

Telemetry and telecontrol subsystem is short for satellite telemetry and remote control subsystem. Remote control is the artificial control of the spacecraft from the ground, and the control information of the satellite is sent through the remote control system. Early simple satellites mainly send switching instructions such as power supply, power off, signal channel on, off and so on. With the development of satellite services and the adoption of on-board computers, more and more complex control information needs to be transmitted. Remote control adopts the form of data packets, such as the transmission of program-controlled data, the adjustment of satellite working parameters, software maintenance of on-board computers and

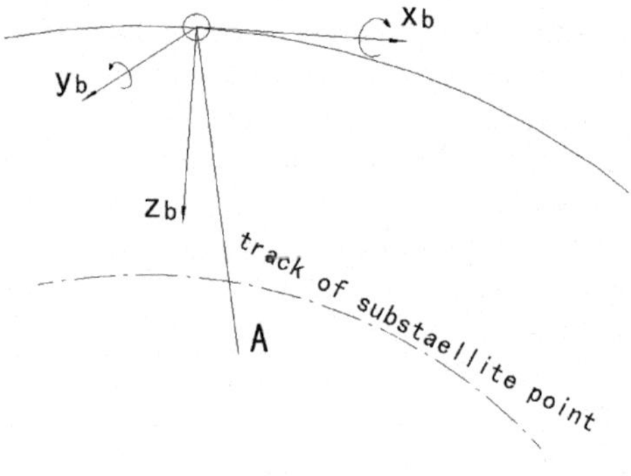

Fig. 2.13 Attitude pointing control diagram

system reconstruction. Telemetry is the measurement of the spacecraft on its own in-orbit state, such as the power supply voltage, working current, working temperature, switching on and off state of the satellite platform and payload equipment. Data form has simple frame format and packet format frame format, relatively simple and fixed, generally used for the important basic working state quantity measurement. Data packet format is mainly the transmission of data generated by the work of each device or highly random data such as fault data, while the payload such as the imaging data of the camera is transmitted through the data transmission system.

Propulsion subsystem is the power system of satellite in orbit. It mainly performs orbit keeping, orbit maneuvering, attitude adjustment with large angle and momentum wheel unloading. Attitude adjustment with small angle and attitude maintenance are generally completed by momentum wheel.

Data management subsystem is to compress and decompress data, encode and decode data, encrypt and decrypt data, distribute routes, isolate faults, restructure the system and degrade the operation. By integrating the functions of on-board equipment, software and hardware functions can be fully shared and developed into an on-board integrated electronic system.

2.4 Basic Concepts of Spacecraft Payloads

The payload of the spacecraft is on-board equipment that provides functional services directly to the user, such as cameras, microwave radiometers, and data transmission systems. The payload is relative to the platform equipment.

The payload of space optical remote sensing mainly includes optical camera and spectrometer. Figure 2.14 is an optical camera, which usually includes optical system, structure system, CCD imaging and calibration system, thermal control system, camera control system, focusing system, etc.

The optical system is the basic system of the camera, which is the basic channel for optical information transmission, imaging energy collection and transmission. It determines the field of view of the camera (unless there is a scanning system) and determines the volume and weight of the camera at a large extent.

Structural system is the basic supporting system of camera that is used to fix and support optical systems, CCD imaging and calibration systems. It is the basic system that determines the quality of the camera image.

CCD imaging system is a camera imaging photoelectric conversion system, which is the main system to determine the camera imaging quality.

Calibration system is a system to measure the on-orbit photoelectric conversion signal intensity relationship of camera, usually measuring the response uniformity of each pixel and the long-term on-orbit photoelectric conversion signal intensity slow variable value.

Thermal control system is an important system to maintain the in-orbit working temperature of camera and to determine the quality of in-orbit imaging.

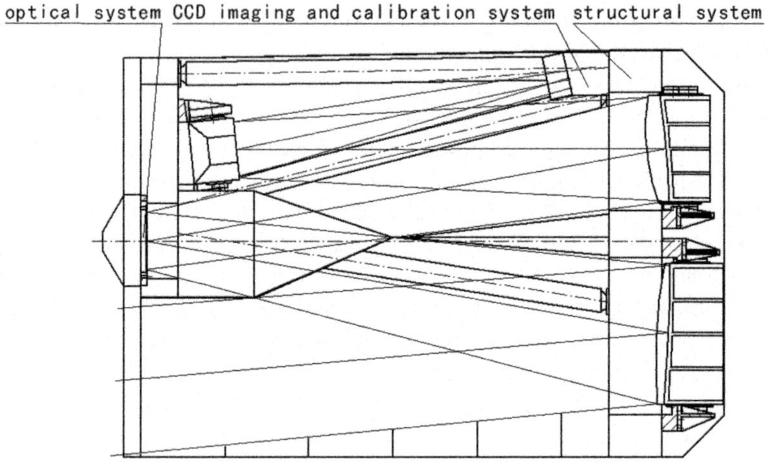

Fig. 2.14 Optical camera system composition

The camera control system is a system that manages the coordination of the camera internal imaging system, thermal control system, calibration system and focusing system, controls the camera imaging mode and coordinates the information exchange between the camera and the platform.

2.5 Basic Concepts of Space Optical Remote Sensing

Space optical remote sensing belongs to the category of space optics. Space optics is the theory of using light wave as detection medium to detect targets at a certain space distance in space. In order for optical instruments and equipment to work in space environment, it is necessary to study space environment conditions, adaptability of instruments and equipment to space environment, optical characteristics of targets in space environment conditions, storage, transmission and distribution of acquired information, and interpretation of information. The "space" in space optical remote sensing is relative to the "Earth", so the space optical remote sensing here mainly studies the content of using spacecraft to observe the earth, that is, using spacecraft as a carrier platform, using optical instruments to detect the ground target through the media of light wave information acquisition method, namely space optical remote sensing.

Due to the limitation of optical wavelength and aperture of optical instrument, the information acquisition ability of the detected target is also limited. There are many parameters to describe the information acquisition ability of space optical remote sensing, such as spatial resolution, time resolution, spectral resolution, and MTF.

Spatial resolution is the resolution ability of an optical instrument to sinusoidal contrast targets. It usually refers to the pairs of lines that can be resolved within a

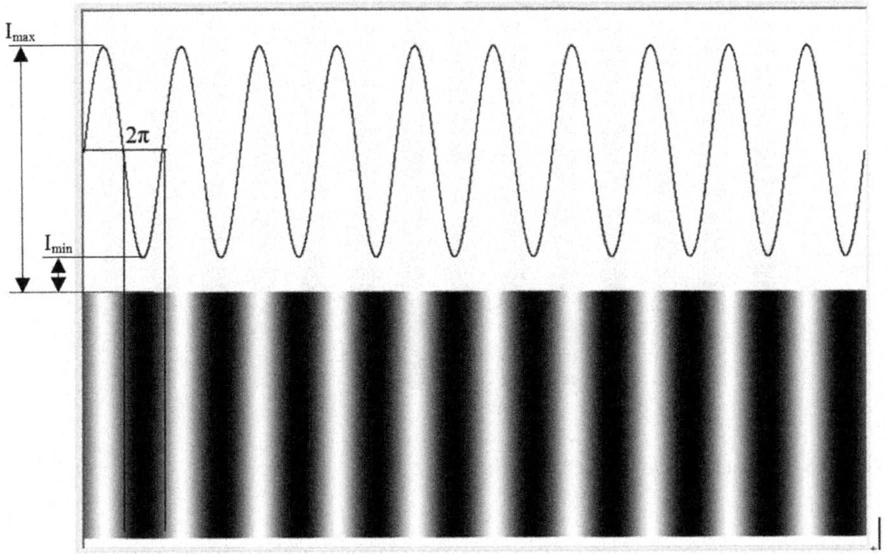

Fig. 2.15 Sinusoidal contrast target

distance of 1 mm on the image plane. Not simply called resolution, but called spatial resolution is relative to temporal resolution and spectral resolution.

The sinusoidal contrast target is the brightness distribution of the target as shown in Fig. 2.15. Within the 2π period of a sine wave, there are 10 pairs of lines (lp). If the graph is distributed within 1 mm, it is called 10 lp/mm.

What is often called photographic spatial resolution is imaging a sinusoidal luminance object like the one above to see the number of line pairs that can be resolved in 1 mm. The spatial scale of the object corresponding to the 1-line pair is the spatial resolution of imaging.

Set the brightness of the peaks as $I_{\max}$, the brightness of the troughs as $I_{\min}$ in Fig. 2.15, then contrast is:

$$c = \frac{I_{\max}}{I_{\min}}$$

Modulation is:

$$M = \frac{I_{\max} - I_{\min}}{I_{\max} + I_{\min}} = \frac{c - 1}{c + 1}$$

If the image in the lower half of Fig. 2.15 is taken as the target, there is a modulation M_o between the black and white stripes of this target. The image obtained by imaging this target is also a black and white strip graph with sinusoidal brightness distribution, the modulation of the strip graph after imaging is M_g, and the optical Modulation transfer function (MTF) is M_g/M_o.

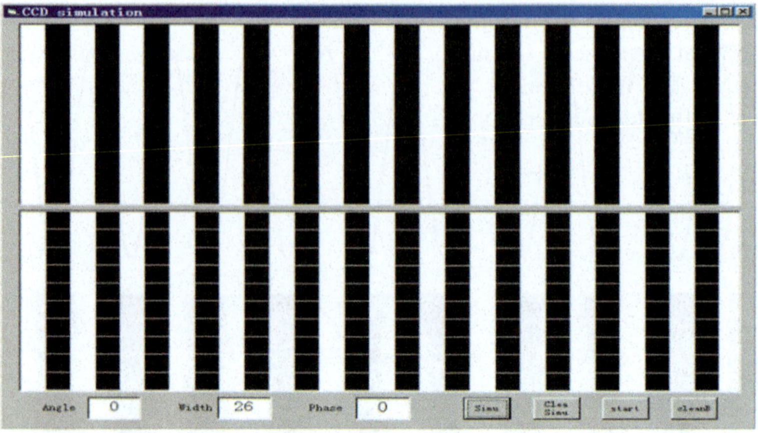

Fig. 2.16 Spatial resolution of CCD imaging

In the development of optical camera, the resolution identification rate plate used to test the performance of the camera adopts the black and white striped graph of the upper part of the graph in Fig. 2.16 instead of the sinusoidal brightness distribution graph introduced above. This is mainly for the convenience of graphic development. In the current commonly used CCD numerical imaging, it is the imaging form shown in Fig. 2.16. The lower half is the pixel of the CCD, and you can see that the pixel size is exactly equal to the width of the strip. CCD camera imaging resolution is usually said, when the optical axis perpendicular to of the object surface, take a image, a pixel width corresponding to the space scale of the object surface, this scale is pixel resolution.

The spatial resolution of sine wave imaging mentioned above is a line pair, namely the corresponding two stripes, one white and one black. In other words, in the spatial resolution unit of a sine wave, one white line and one black line are actually distinguished simultaneously. If the spatial resolution of sine wave imaging is 1 m, one white line and one black line can be distinguished in this 1 m, and the width of each line is 0.5 m. For CCD camera imaging, the resolution of 1 m pixels means that the width of one pixel corresponds to the scale of a 1 m target.

It can be seen that the spatial resolution of traditional film camera is two times of the spatial scale resolution of CCD camera pixel resolution. Therefore, in order to distinguish the imaging resolution of CCD camera from that of traditional film camera, the imaging of CCD camera is called resolving power or pixel resolution.

Time resolution for the camera is the time required for the camera to complete an image. For optical imaging satellites, it is the minimum time required from the completion of the previous image to the completion of the next image. During this period, in addition to the time required by the camera, there are also satellite attitude pointing and stabilization time, and data transmission time. In some applications, it can also refer to the time required from the completion of an exposure to the completion of the next exposure.

Spectral resolution is the minimum spectral band width that can be captured by an optical instrument. It can be expressed in two ways: wavenumber $\delta\nu$ and wavelength $\Delta\lambda$. Wave number, the number of wavelengths that can be arranged within a centimeter, is a concept of spatial frequency.

References

1. Mei A. An introduction to remote sensing. Beijing: Higher Education Press; 1989.
2. Chinese Academy of Sciences, Changchun Branch *Symposium on Changchun Remote Sensing Experiment* Editorial board, *Symposium on Changchun Remote Sensing Experiment*. Changchun: Jilin People's Press; 1981.
3. Jiancheng F, Xiaolin N. Principles and applications of astronomical navigation. Beijing: Beihang University Press; 2006.
4. Jiancheng F, Xiaolin N, Yucheng T. Principle and method of spacecraft autonomous astronomical navigation. Beijing: National Defense Industry Press; 2006.
5. Yan H. Spaceflight TT&C network. Beijing: National Defense Industry Press; 2004.
6. Li H. Geostationary satellite orbital analysis and collocation strategies. National Defense Industry Press Beijing; 2010.
7. Yinwei Z. Satellite orbital attitude dynamics and control. Beijing: National Defense Industry Press; 1998.

Chapter 3
Spacecraft–Earth and Time–Space Parameter Design

Global coverage forms of optical remote sensing systems, revisit time, imaging resolution, imaging width, local time of transit place, illuminance of sub-satellite point in a large extent depend on the design of the satellite orbit parameters. The orbits of optical remote sensing to earth usually fall into two categories, such as solar synchronous orbit and geosynchronous orbit. A single satellite in sun-synchronous orbit has global coverage except the poles, with high spatial resolution and low temporal resolution. A single satellite in geosynchronous orbit has the ability to cover a large range of a certain region of the world except the poles, with high temporal resolution and low spatial resolution, which are complementary to each other. Space–time parameters are closely related to optical remote sensing, including orbit height, orbit inclination, longitude of ascending point and local time of ascending point. There are a lot of references [1–8] to discuss orbit, and the first chapter of reference [5] is mainly referred here.

3.1 Proprieties of Two-Body Orbit

In the derivation of the basic equation of satellite orbit, only the interaction between satellite and earth is usually considered, and the effects of earth's non-sphericity and uneven density, solar and lunar gravity and atmospheric drag are ignored. Suppose, the satellite moves in the gravitational field at the center of the earth. This satellite orbit is called a two-body orbit. Analysis of its characteristics is called the two-body problem, and the two-body orbit represents the most important characteristics of satellite orbit motion.

J. Tao, *Space Optical Remote Sensing*, Advances in Optics and Optoelectronics,
https://doi.org/10.1007/978-981-99-3318-1_3

3.1.1 In-Plane Motion

According to the law of universal gravitation, the interaction between two objects of mass M and m in the inertial coordinate system is shown in Fig. 3.1.

$$\begin{aligned}
&\vec{F}_1 = \frac{GMm}{r^2}\frac{\vec{r}}{r} \\
&\vec{F}_2 = -\frac{GMm}{r^2}\frac{\vec{r}}{r} \\
&\vec{F}_1 = -\vec{F}_2 \\
&M\ddot{\vec{r}}_1 = \vec{F}_1 = \frac{GMm}{r^2}\frac{\vec{r}}{r} \\
&\ddot{\vec{r}}_1 = \frac{Gm}{r^2}\frac{\vec{r}}{r} \\
&m\ddot{\vec{r}}_2 = \vec{F}_2 = -\frac{GMm}{r^2}\frac{\vec{r}}{r} \\
&\ddot{\vec{r}}_2 = -\frac{GM}{r^2}\frac{\vec{r}}{r} \\
&\vec{r}_2 - \vec{r}_1 = \vec{r} \\
&\ddot{\vec{r}}_2 - \ddot{\vec{r}}_1 = \ddot{\vec{r}}
\end{aligned}$$

$$\begin{aligned}
\ddot{\vec{r}} &= -\frac{GM}{r^2}\frac{\vec{r}}{r} - \frac{Gm}{r^2}\frac{\vec{r}}{r} \\
&= -\frac{G(M+m)}{r^2}\frac{\vec{r}}{r} \\
&= -\frac{GM}{r^2}\frac{\vec{r}}{r} \\
&= -\mu\frac{1}{r^2}\frac{\vec{r}}{r}
\end{aligned} \tag{3.1}$$

In formula 3.1, set $M = m_e$, $M + m = M$ approximately.

$$\mu = Gm_e$$

where G is the universal gravitational constant, m_e is the mass of the earth, $\mu = 398{,}600.44\ \text{km}^3\ \text{s}^{-2}$ is gravitational constant.

In the geocentric equatorial inertial orthogonal coordinate system S_i, the x-axis points to the vernal equinox γ along the line of intersection between the ecliptic plane and the equatorial plane from the geocentric center, the z-axis refers to the north pole, and the y-axis is perpendicular to the x-axis in the equatorial plane, as shown in Fig. 3.2.

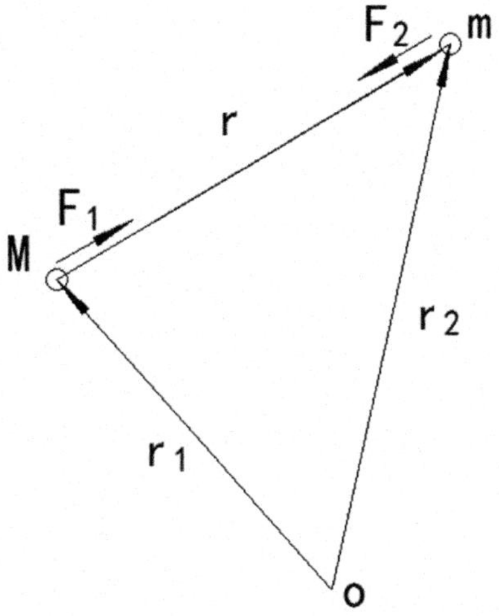

Fig. 3.1 Two-body gravitational interaction diagram

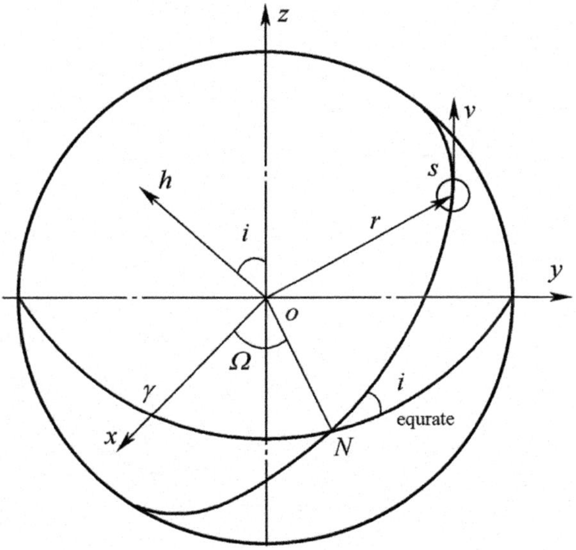

Fig. 3.2 Satellite orbit in geocentric equatorial inertial orthogonal coordinate system

Decompose the radial vector **r** of the satellite in formula 3.1 to x, y and z coordinate axes in the earth's equatorial inertial orthogonal coordinate system, and the components of the satellite motion equation are as follows:

$$\ddot{x} + \frac{\mu x}{r^3} = 0$$
$$\ddot{y} + \frac{\mu y}{r^3} = 0$$

$$\ddot{z} + \frac{\mu z}{r^3} = 0$$

where $r = (x^2 + y^2 + z^2)^{1/2}$.

Multiply above three equations by x, y and z and subtract the r^3 term to get:

$$x\ddot{y} - y\ddot{x} = 0$$
$$y\ddot{z} - z\ddot{y} = 0$$
$$z\ddot{x} - x\ddot{z} = 0$$

Rewrite $x\ddot{y} - y\ddot{x} = 0$ into

$$x\ddot{y} = y\ddot{x}$$

Integral over t,

$$\int x\ddot{y}\mathrm{d}t = \int y\ddot{x}\mathrm{d}t$$
$$\int x\mathrm{d}\dot{y} = \int y\mathrm{d}\dot{x}$$

Integration by parts,

$$x\dot{y} - \int \dot{y}\mathrm{d}x = y\dot{x} - \int \dot{x}\mathrm{d}y$$

$$\begin{aligned} x\dot{y} - y\dot{x} &= \int \dot{y}\dot{x}\mathrm{d}t - \int \dot{x}\dot{y}\mathrm{d}t \\ &= \int (\dot{y}\dot{x} - \dot{x}\dot{y})\mathrm{d}t \\ &= \int 0\mathrm{d}t \\ &= c \end{aligned}$$

Integrate the other two expressions in the same way, and so:

$$x\dot{y} - y\dot{x} = c_1$$
$$y\dot{z} - z\dot{y} = c_2$$
$$z\dot{x} - x\dot{z} = c_3$$

After multiplying above three equations by z, x and y, the sum can be obtained:

$$c_1 z + c_2 x + c_3 y = 0$$

By comparing $Ax + By + Cz + D = 0$ with the above equation, it can be known that $c_1z + c_2x + c_3y = 0$ is a plane equation that passes through the origin of the coordinate system. Thus, it is proved that the orbit of the satellite is not only in the same plane but also through the plane of the center of the earth.

3.1.2 *Motion Along a Conical Curve*

The satellite moves relative to the earth in the orbital plane, and its moment of momentum relative to the center of the earth is

$$
\begin{aligned}
&\vec{H} = \vec{r} \times m\vec{v} \\
&\vec{h} = \vec{r} \times \vec{v} \\
&\vec{h} = (h_x, h_y, h_z), \quad \vec{r} = (x, y, z) \quad \vec{v} = (\dot{x}, \dot{y}, \dot{z})
\end{aligned}
$$

$$
\begin{aligned}
\vec{h} &= \vec{r} \times \vec{v} \\
&= \begin{vmatrix} i & j & k \\ x & y & z \\ \dot{x} & \dot{y} & \dot{z} \end{vmatrix} \\
&= (y\dot{z} - \dot{y}z)i - (x\dot{z} - \dot{x}z)j + (x\dot{y} - \dot{x}y)k
\end{aligned}
$$

Let

$$
\begin{aligned}
h_x &= y\dot{z} - \dot{y}z = c_2 \\
h_y &= \dot{x}z - x\dot{z} = c_3 \\
h_z &= x\dot{y} - \dot{x}y = c_1
\end{aligned}
$$

According to the vector product formula, the left end of the first derivative equation set of x, y, z is the component of the moment vector **h** on the coordinate axis h_x, h_y, h_z and ${h_x}^2 + {h_y}^2 + {h_z}^2 = h^2$.

As shown in Fig. 3.2, the angle between the vector h and the z-axis is the orbital inclination i, the intersection line between the orbital plane and the equatorial plane is the nodal line ON, and N is the ascending intersection point. The angle between nodal line ON and x-axis is right ascension of ascending node Ω. i and Ω determine the spatial position of the orbital plane. The components of the vector h can be expressed as follows:

$$
\begin{aligned}
h_x &= h \sin i \sin \Omega \\
h_y &= -h \sin i \cos \Omega \\
h_z &= h \cos i
\end{aligned}
$$

Orbital inclination i and right ascension of ascending node Ω can be expressed as follows:

$$i = \arccos \frac{h_z}{h}$$

$$\Omega = \arctan \frac{-h_x}{h_y}$$

The position of the satellite (ξ, η) in the orbital plane is described as shown in Fig. 3.3. The equation of motion of the satellite is

$$\begin{cases} \ddot{\xi} + \dfrac{\mu}{r^3}\xi = 0 \\ \ddot{\eta} + \dfrac{\mu}{r^3}\eta = 0 \end{cases} \tag{3.2}$$

where

$$r = \left(\zeta^2 + \eta^2\right)^{1/2}$$

The polar form of this equation is

$$\xi = r\cos\theta$$

$$\eta = r\sin\theta$$

Derivation of t is

$$\dot{\xi} = \dot{r}\cos\theta - r\sin\theta \cdot \dot{\theta}$$

$$\dot{\eta} = \dot{r}\sin\theta + r\cos\theta \cdot \dot{\theta}$$

$$\ddot{\xi} = \ddot{r}\cos\theta - 2\dot{r}\dot{\theta}\sin\theta - \ddot{\theta}r\sin\theta - \dot{\theta}^2 r\cos\theta$$

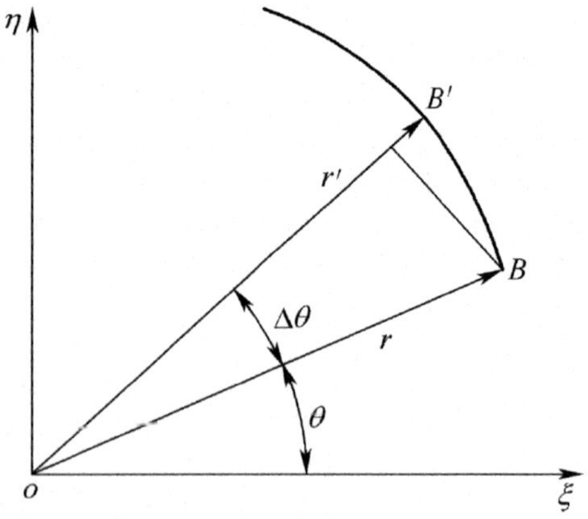

Fig. 3.3 Plane coordinates of a satellite's orbit

$$\ddot{\eta} = \ddot{r}\sin\theta + 2\dot{r}\dot{\theta}\cos\theta + \ddot{\theta}r\cos\theta - \dot{\theta}^2 r\sin\theta$$

Substitute into the formula 3.2, so:

$$\begin{cases} \ddot{r}\cos\theta - 2\dot{r}\dot{\theta}\sin\theta - \ddot{\theta}r\sin\theta - \dot{\theta}^2 r\cos\theta + \frac{\mu}{r^2}\cos\theta = 0 & (1) \\ \ddot{r}\sin\theta + 2\dot{r}\dot{\theta}\cos\theta + \ddot{\theta}r\cos\theta - \dot{\theta}^2 r\sin\theta + \frac{\mu}{r^2}\sin\theta = 0 & (2) \end{cases}$$

$(1)/\sin\theta + (2)/\cos\theta$

$$\ddot{r}(\cot\theta + \tan\theta) - \dot{\theta}^2 r(\cot\theta + \tan\theta) + \frac{\mu}{r^2}(\cot\theta + \tan\theta) = 0$$

$$\begin{aligned} \cot\theta + \tan\theta &= \frac{\cos\theta}{\sin\theta} + \frac{\sin\theta}{\cos\theta} \\ &= \frac{1}{\sin\theta\cos\theta} \\ &= \frac{2}{\sin 2\theta} \\ &\neq 0 \end{aligned}$$

Divide both sides of this equation by $(\cot\theta + \tan\theta)$ and so:

$$\ddot{r} - \dot{\theta}^2 r = -\frac{\mu}{r^2} \tag{3.3}$$

Similarly,

$$\begin{aligned} &(1)\sin\theta - (2)\cos\theta, \text{ then} \\ &r\ddot{\theta} + 2\dot{r}\dot{\theta} = 0 \end{aligned}$$

Namely,

$$\begin{cases} \ddot{r} - \dot{\theta}^2 r = -\frac{\mu}{r^2} \\ r\ddot{\theta} + 2\dot{r}\dot{\theta} = 0 \end{cases}$$

Integration of separated variables for $r\ddot{\theta} + 2\dot{r}\dot{\theta} = 0$

$$\begin{aligned} r\mathrm{d}\dot{\theta} &= -2\dot{\theta}\mathrm{d}r \\ \int \frac{\mathrm{d}\dot{\theta}}{\dot{\theta}} &= -2\int \frac{\mathrm{d}r}{r} \\ \log\dot{\theta} &= \log r^{-2} + c \\ \dot{\theta} &= 10^c r^{-2} \end{aligned}$$

The equation is rewritten as

$$r^2\dot{\theta} = h \tag{3.4}$$

The area that r sweeps over $\Delta\text{OBB}'$ in Δt is equal to

$$\Delta A = \frac{r\, r' \sin \Delta\theta}{2}$$

Time rate of ΔA is

$$\frac{\Delta A}{\Delta t} = \frac{r\, r' \Delta\theta}{2\, \Delta t} \frac{\sin \Delta\theta}{\Delta\theta}$$

Take the limit

$$\dot{A} = \frac{r^2\dot{\theta}}{2} = \frac{h}{2} \tag{3.5}$$

This is Kepler's second law, and the magnitude of the moment of momentum is equal to twice the velocity of this area.

Using $r^2\ \dot{\theta} = h$, the independent variable of formula 3.3 is converted from the orbital polar coordinate equation of time t to the polar angle θ

$$\begin{aligned}
\dot{r}_t &= \dot{r}_\theta \dot{\theta}_t \\
\ddot{r}_t &= \ddot{r}_\theta \dot{\theta}_t^2 + \dot{r}_\theta \ddot{\theta}_t \\
&= \ddot{r}_\theta \left(\frac{h}{r^2}\right)^2 + \dot{r}_\theta \left(-\frac{2\dot{r}_t}{r}\frac{h}{r^2}\right) \quad (\because r^2\dot{\theta} = h \;\; r\ddot{\theta} + 2\dot{r}\dot{\theta} = 0) \\
&= \ddot{r}_\theta \left(\frac{h}{r^2}\right)^2 + \dot{r}_\theta \left(-\frac{2\dot{r}_\theta \dot{\theta}_t}{r}\frac{h}{r^2}\right) \\
&= \frac{h^2}{r^4}\ddot{r}_\theta - \frac{2h}{r^3}\frac{h}{r^2}(\dot{r}_\theta)^2
\end{aligned}$$

Substitute into polar coordinate formula 3.3, so:

$$\begin{aligned}
\frac{h^2}{r^4}\ddot{r}_\theta - \frac{2h}{r^3}\frac{h}{r^2}(\dot{r}_\theta)^2 - \left(\frac{h}{r^2}\right)^2 r &= -\frac{\mu}{r^2} \\
-\frac{1}{r^2}\ddot{r}_\theta + \frac{2}{r^3}(\dot{r}_\theta)^2 + \frac{1}{r} &= \frac{\mu}{h^2}
\end{aligned}$$

And

$$\begin{aligned}
\frac{\mathrm{d}^2}{\mathrm{d}\theta^2}\left(\frac{1}{r}\right) &= \frac{\mathrm{d}}{\mathrm{d}\theta}\left(-r^{-2}\dot{r}_\theta\right) \\
&= \ddot{r}_\theta\left(-\frac{1}{r^2}\right) + 2\left(\frac{1}{r^3}\right)(\dot{r}_\theta)^2
\end{aligned}$$

There for

$$\frac{d^2}{d\theta^2}\left(\frac{1}{r}\right) + \frac{1}{r} = \frac{\mu}{h^2}$$

Then,

$$\frac{1}{r} = \frac{\mu}{h^2}[1 + e\cos(\theta - \omega)]$$

$$r = \frac{p}{1 + e\cos(\theta - \omega)} \tag{3.6}$$

where

$$p = \frac{h^2}{\mu}.$$

This is the conic curve of the polar coordinate orbital equation of the satellite. The satellite is at one of the focal points of the elliptic curve, as shown in Fig. 3.4, which is Kepler's first law.

$$c^2 = a^2 - b^2$$

$$e = \frac{c}{a}$$

e is eccentricity, $e < 1$, p is semi-latus rectum, it through the focus and perpendicular to the major axis of the elliptic.

$$p = a\left(1 - e^2\right) = b\sqrt{1 - e^2} \tag{3.7}$$

$$p = \frac{h^2}{\mu} \tag{3.8}$$

$$a = \frac{h^2}{\mu(1 - e^2)} \tag{3.9}$$

So a depends on h.

According to formula 3.6, when $\theta - \omega = 0$, geocentric distance r is the minimum, and this is perigee p; when $\theta - \omega = 180$, geocentric distance r is the maximum, and this is apogee A. The polar angle ω, called perigee angle, determines the direction of the major axis of the orbit.

The orbital equation is

$$r = \frac{a(1 - e^2)}{1 + e\cos f} \tag{3.10}$$

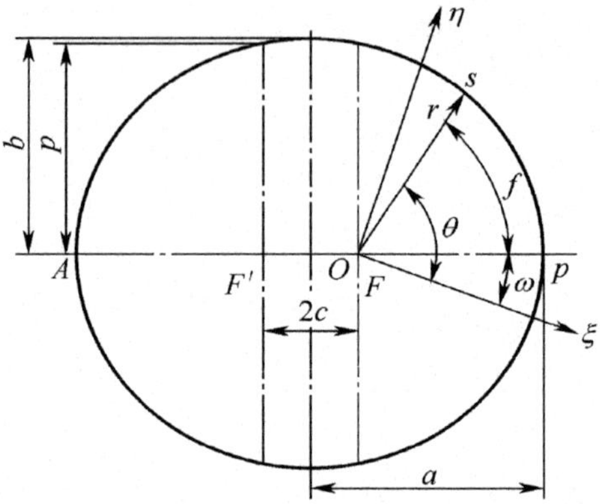

Fig. 3.4 Kepler orbit of the satellite

When $f = 0$, the perigee geocentric distance is r_{p}; when $f = 180$, the apogee geocentric distance is r_{a}.

$$r_{\mathrm{p}} = a(1 - e)$$
$$r_a = a(1 + e)$$
$$e = \frac{r_a - r_{\mathrm{p}}}{r_a + r_{\mathrm{p}}}$$

From formula 3.5, satellite orbit period T is gotten:

$$\frac{\pi ab}{T} = \frac{h}{2}$$

$$T = 2\pi\sqrt{\frac{a^3}{\mu}} \tag{3.11}$$

This is Kepler's third law.

Satellite orbit formula 3.6 reflects the properties of different types of satellite orbits:

When $e = 0$, it is a circular orbit around the center of the earth;

When $0 < e < 1$, it is an elliptical orbit around the center of the earth;

When $e = 1$, it is a parabolic escape orbit from the earth;

When $e > 1$, it is a hyperbolic escape orbit fast away from the earth.

3.1.3 Orbital Elements

The satellite moves with variable angular velocity in the elliptical orbit. In order to conveniently determine the in-orbit position of the satellite, it is assumed that the satellite moves in orbit with uniform angular velocity, which is called mean velocity n, then

$$n = \frac{2\pi}{T} = \sqrt{\frac{\mu}{a^3}} \tag{3.12}$$

As shown in Fig. 3.5, the position s of the satellite on the elliptical orbit, a vertical projection from s to the tangent outer circle as point Q, connecting point o and point Q to intersect the inner tangent circle with point R, connecting point S and point R. E is eccentric anomaly, and the satellite coordinate is $s(\xi, \eta)$.

To prove that R is the horizontal projection of s onto the inner tangent circle, prove that $sR//op$.

$$\xi = a\cos E = ae + r\cos f \tag{3.13}$$

$$\begin{aligned}\cos E &= e + \frac{r}{a}\cos f \\ &= e + \frac{1-e^2}{1+e\cos f}\cos f \\ &= \frac{e+\cos f}{1+e\cos f}\end{aligned} \tag{3.14}$$

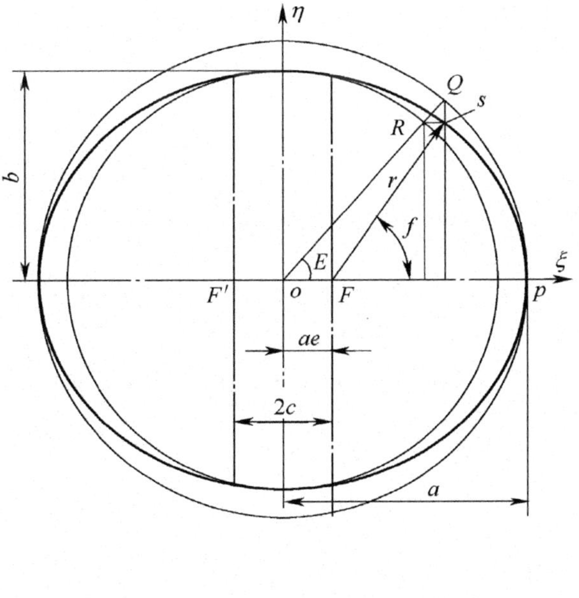

Fig. 3.5 Satellite's eccentric anomaly coordinate system

$$\sin E = \frac{\sqrt{1-e^2}\sin f}{1+e\cos f} \tag{3.15}$$

The y-coordinate of point R η_R.
Refer to formulas 3.7 and 3.15, so:

$$\begin{aligned}
\eta_R &= b\sin E \\
&= a\sqrt{1-e^2}\sin E \\
&= a\sqrt{1-e^2}\frac{\sqrt{1-e^2}\sin f}{1+e\cos f} \\
&= a\frac{1-e^2}{1+e\cos f}\sin f
\end{aligned}$$

The y-coordinate of point s η_S.

$$\begin{aligned}
\eta_S &= r\sin f \\
&= \frac{a(1-e^2)}{1+e\cos f}\sin f
\end{aligned}$$

$\eta_R = \eta_S$ is gotten, then

$$\begin{aligned}
&sR//op \\
&\eta = b\sin E = r\sin f
\end{aligned} \tag{3.16}$$

Refer to formulas 3.14 and 3.15, so:

$$\sin f = \frac{\sqrt{1-e^2}\sin E}{1-e\cos E} \tag{3.17}$$

$$\cos f = \frac{\cos E - e}{1-e\cos E} \tag{3.18}$$

Using double angle formula,

$$\tan\left(\frac{f}{2}\right) = \left(\frac{1+e}{1-e}\right)^{1/2}\tan\left(\frac{E}{2}\right)$$

Using formulas 3.13 and (3.18), a simple representation of the orbital equation r is derived.

$$r = a(1-e\cos E) \tag{3.19}$$

Derivative with respect to formulas 3.13 and (3.18), and using formulas 3.15 and (3.17), so,

$$\mathrm{d}f = \frac{\sqrt{1-e^2}}{1-e\cos E}\mathrm{d}E \tag{3.20}$$

$$\mathrm{d}E = \frac{\sqrt{1-e^2}}{1+e\cos f}\mathrm{d}f \tag{3.21}$$

Integral $r^2\theta' = h$, since ω is a constant in Fig. 3.4, $\theta' = f'$.

$$\int_0^f r^2\mathrm{d}f = h(t-t_p)$$

The moment a satellite passes perigee $t_\mathrm{p}, f = 0$.
Using formulas (3.9), (3.10), (3.19) and (3.20), so:

$$\begin{aligned} t - t_\mathrm{p} &= \frac{h^3}{\mu^2}\int_0^f \frac{\mathrm{d}f}{(1+e\cos f)^2} \\ &= \frac{h^3}{\mu^2}\int_0^E \frac{(1-e\cos E)}{(1-e^2)^{3/2}}\mathrm{d}E \\ &= \sqrt{\frac{a^3}{\mu}}(E - e\sin E) \end{aligned}$$

The Kepler equation describing the relationship between satellite position and time is obtained.

$$n\left(t-t_p\right) = E - e\sin E$$

$M = n\,(t - t_\mathrm{p})$ is defined as mean anomaly. n and e are given, and the solution E gives the satellite position. Iteratively solve the following equation,

$$\begin{aligned} E_1 &= M + e\sin E_0 \\ E_2 &= M + e\sin E_1 \end{aligned}$$

If $E_0 = M$, the series of E can be obtained by analyzing after each iteration:

$$E = M + \left(e - \frac{1}{8}e^3 + \frac{1}{192}e^5\right)\sin M + \left(\frac{1}{2}e^2 - \frac{1}{6}e^4\right)\sin 2M$$

$$+\left(\frac{3}{8}e^3-\frac{27}{128}e^5\right)\sin 3M+\frac{1}{3}e^4\sin 4M+\frac{125}{384}e^5\sin 5M+\cdots$$

Series solution of true anomaly f is

$$f=M+\left(2e-\frac{e^3}{4}\right)\sin M+\frac{5}{4}e^2\sin 2M+\frac{13}{12}e^3\sin 3M+\cdots$$

So far, six integral constants of the motion equation of the satellite are given: a, e, i, Ω, ω, t_p (or M), namely six orbital elements.

3.1.4 Satellite Entry Parameters

Given the position, velocity and direction r, v and γ of the satellite at a certain time, the orbital elements can be obtained. That is, the orbit can be determined. Or vice versa, the entry parameters required by the satellite can also be obtained from the orbital elements. As shown in Fig. 3.6, V is decomposed into radial velocity $\dot{r}$ and transverse velocity $r\dot{f}$.

Let

$$r=\frac{p}{1+e\cos f} \tag{3.22}$$

Take the derivative with respect to time t, and $r^2\theta'=h$, so:

$$\dot{r}=\frac{p\cdot p}{p(1+e\cos f)^2}e\dot{\theta}\sin f$$

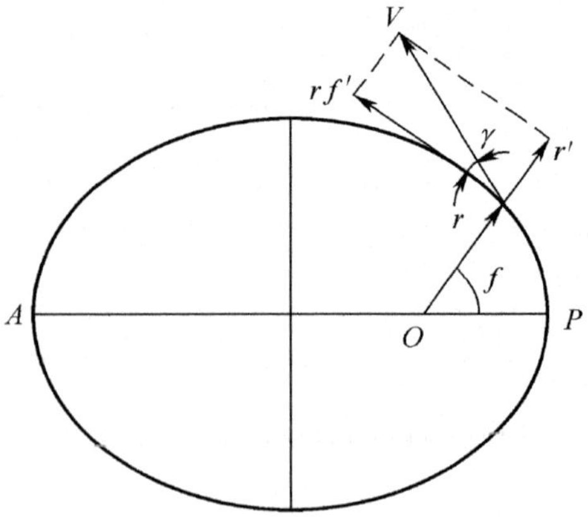

Fig. 3.6 Entry parameters of the satellite orbit

$$
\begin{aligned}
&= \frac{r^2 eh \sin f}{pr^2} \\
&= \frac{h}{p} e \sin f \\
r\dot{f} &= \frac{rh}{r^2} = \frac{h}{r} = \frac{h(1 + e\cos f)}{p}
\end{aligned}
$$

Sum the squares of the over two equations

$$
\begin{aligned}
V^2 &= \dot{r}^2 + r^2 \dot{f}^2 \\
&= \left(\frac{h}{p}\right)^2 (1 + 2e\cos f + e^2) \\
&= \frac{2h^2}{rp} - \left(\frac{h}{p}\right)^2 (1 - e^2) \\
&= \mu\left(\frac{2}{r} - \frac{1}{a}\right)
\end{aligned}
\tag{3.23}
$$

This is the relationship between satellite speed and position, which can be translated into

$$
\frac{V^2}{2} - \frac{\mu}{r} = -\frac{\mu}{2a} \tag{3.24}
$$

The first term is kinetic energy per unit mass, the second term is potential energy per unit mass, and the combined term is conservation of mechanical energy.

Substitute r_{p} and r_{a} into formula (3.24), the velocity of the satellite at perigee and apogee is gotten.

$$
V_{\mathrm{p}} = \sqrt{\frac{\mu(1+e)}{a(1-e)}} = \sqrt{\frac{\mu r_a}{a r_{\mathrm{p}}}}
$$

$$
V_a = \sqrt{\frac{\mu(1-e)}{a(1+e)}} = \sqrt{\frac{\mu r_{\mathrm{p}}}{a r_a}}
$$

Within the plane of the satellite's orbit, γ is called the flight angle, $\cos\gamma = r\dot{f}/V$, $\sin\gamma = \dot{r}/V$. With $r^2\dot{f} = h$, so:

$$
\begin{aligned}
\cos\gamma &= \frac{h}{r\sqrt{\mu\left(\frac{2}{r} - \frac{1}{a}\right)}} \\
&= \sqrt{\frac{a^2(1-e^2)}{r(2a-r)}}
\end{aligned}
$$

Replace r, so:

$$\cos\gamma = \frac{1 + e\cos f}{\sqrt{1 + 2e\cos f + e^2}} \tag{3.25}$$

Similarly,

$$\sin\gamma = \frac{e\sin f}{\sqrt{1 + 2e\cos f + e^2}}$$

From formulas (3.23) and (3.25), $V^2r^2\cos^2\gamma = \mu a(1 - e^2)$ is gotten.

From formulas (3.23), (3.22) and (3.25), the main parameters of satellite orbit are gotten.

$$\begin{aligned}
e &= \sqrt{\left(\frac{rV^2}{\mu} - 1\right)^2 \cos^2\gamma + \sin^2\gamma} \\
a &= \frac{r\mu}{2\mu - rV^2} \\
r_p &= \frac{r^2V^2\cos^2\gamma}{\mu(1+e)} \\
r_a &= \frac{r^2V^2\cos^2\gamma}{\mu(1-e)}
\end{aligned}$$

3.1.5 Orbital Coordinate System and Transformation

The space coordinates of satellite orbit are shown in Fig. 3.7.

As shown in Fig. 3.7, S refers to the satellite, P refers to perigee, f refers to true anomaly, **e** refers to eccentricity vector, **n** refers to unit normal vector of orbital plane, t_p refers to the moment a satellite passes perigee, N refers to ascending node, Ω refers to right ascension of ascending intersection point [0, 360°) [4], i refers to inclination [0, 180°), ω refers to argument of perigee [0, 360°), α refers to right ascension [0, 360°), δ refers to declination [±0, 90°), M refers to mean anomaly [0, 360°).

First, let's review the coordinate transformation.

As shown in Fig. 3.8, the original coordinates are $B(x, y, z)$, rotation angle is θ, and new coordinates are $B'(x', y', z')$.

$$\begin{aligned}
x' &= x \\
y' &= y\cos\theta + z\sin\theta \\
z' &= -y\sin\theta + z\cos\theta
\end{aligned}$$

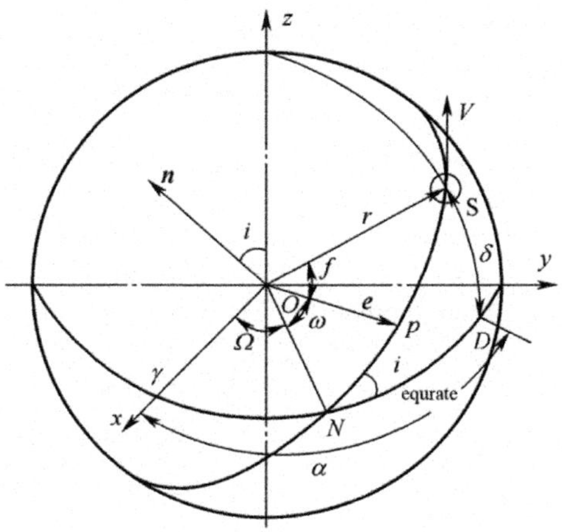

Fig. 3.7 Orbital coordinate parameters of a satellite

Rewrite the above formula in matrix form:

$$\begin{bmatrix} x' \\ y' \\ z' \end{bmatrix} = \begin{bmatrix} 1 & 0 & 0 \\ 0 & \cos\theta & \sin\theta \\ 0 & -\sin\theta & \cos\theta \end{bmatrix} \begin{bmatrix} x \\ y \\ z \end{bmatrix} = R_x(\theta) \begin{bmatrix} x \\ y \\ z \end{bmatrix}$$

$R_x(\theta)$ is called rotation matrix, which is similar to rotate angle φ around y and rotate angle ψ around z.

$$\begin{bmatrix} x'' \\ y'' \\ z'' \end{bmatrix} = \begin{bmatrix} \cos\varphi & 0 & -\sin\varphi \\ 0 & 1 & 0 \\ \sin\varphi & 0 & \cos\varphi \end{bmatrix} \begin{bmatrix} x \\ y \\ z \end{bmatrix} = R_y(\varphi) \begin{bmatrix} x \\ y \\ z \end{bmatrix}$$

$$\begin{bmatrix} x''' \\ y''' \\ z''' \end{bmatrix} = \begin{bmatrix} \cos\psi & \sin\psi & 0 \\ -\sin\psi & \cos\psi & 0 \\ 0 & 0 & 1 \end{bmatrix} \begin{bmatrix} x \\ y \\ z \end{bmatrix} = R_z(\psi) \begin{bmatrix} x \\ y \\ z \end{bmatrix}$$

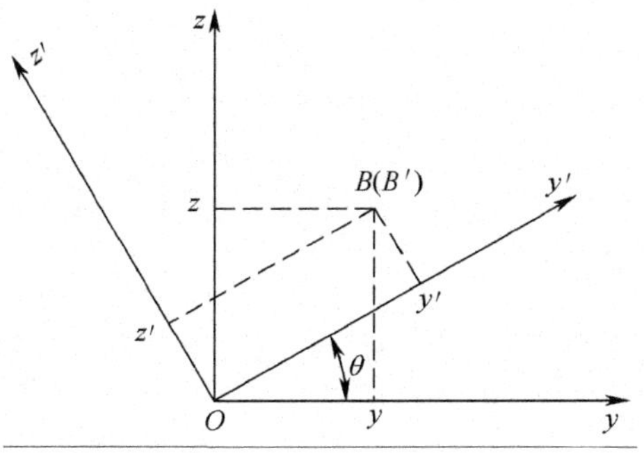

Fig. 3.8 Coordinate transformation diagram

If the rotation order is θ (along axis x), φ (along axis y'), ψ (along axis z''), then:

$$\begin{bmatrix} x''' \\ y''' \\ z''' \end{bmatrix} = R_z(\psi)R_y(\varphi)R_x(\theta)\begin{bmatrix} x \\ y \\ z \end{bmatrix}$$

Let x-, y-, z-axis unit vectors are $\mathbf{u_x}$, $\mathbf{u_y}$, $\mathbf{u_z}$, x', y', z' axis unit vectors are $\mathbf{u_x}'$, $\mathbf{u_y}'$, $\mathbf{u_z}'$, rotate angle θ around X, so:

$$\begin{aligned} \mathbf{u}_x' &= \mathbf{u}_x \\ \mathbf{u}_y' &= \mathbf{u}_y \cos\theta + \mathbf{u}_z \sin\theta \\ \mathbf{u}_z' &= -\mathbf{u}_y \sin\theta + \mathbf{u}_z \cos\theta \end{aligned}$$

After three rotations,

$$\begin{bmatrix} \vec{u}'''_x \\ \vec{u}'''_y \\ \vec{u}'''_z \end{bmatrix} = R_z(\psi)R_y(\varphi)R_x(\theta)\begin{bmatrix} \vec{u}_x \\ \vec{u}_y \\ \vec{u}_z \end{bmatrix}$$

Using the two unit vectors **e**/e and **n** representing orbital characteristics as the x_o axis and z_o axis of the orbital arch line coordinate system, the transformation from the orbital arch line coordinate system (x_o, y_o, z_o) to the equatorial inertial coordinate system (x, y, z) is as follows: rotate angle $-\omega$ along axis z_o, rotate angle $-i$ along axis x_o, rotate angle $-\Omega$ along axis z_o. The position coordinates of a satellite in an orbital coordinate system: $x_o = r$ cosf, $y_o = r$ sin$f$, $z_o = 0$.

The coordinates of the satellite in the equatorial inertial coordinate system are derived by using coordinate transformation,

$$\begin{bmatrix} x \\ y \\ z \end{bmatrix} = R_z(-\Omega)R_x(-i)R_z(-\omega)\begin{bmatrix} x_0 \\ y_0 \\ z_0 \end{bmatrix}$$

$$= \begin{bmatrix} \cos\omega\cos\Omega - \sin\omega\cos i\sin\Omega & -\sin\omega\cos\Omega - \cos\omega\cos i\sin\Omega & \sin i\sin\Omega \\ \cos\omega\sin\Omega + \sin\omega\cos i\cos\Omega & -\sin\omega\sin\Omega + \cos\omega\cos i\cos\Omega & -\sin i\cos\Omega \\ \sin\omega\sin i & \cos\omega\sin i & \cos i \end{bmatrix}$$

$$\begin{bmatrix} r\cos f \\ r\sin f \\ 0 \end{bmatrix} = \frac{a(1-e^2)}{1+e\cos f}\begin{bmatrix} \cos\Omega\cos(\omega+f) - \sin\Omega\sin(\omega+f)\cos i \\ \sin\Omega\cos(\omega+f) + \cos\Omega\sin(\omega+f)\cos i \\ \sin(\omega+f)\sin i \end{bmatrix}$$

where true anomaly f needs to be solved by Kepler equation.

Satellite position in the geocentric equatorial inertial coordinate system S_i using right ascension α and declination δ is as follows:

$$
\begin{aligned}
x &= r \cos\delta \cos\alpha \\
y &= r \cos\delta \sin\alpha \\
z &= r \sin\delta
\end{aligned}
$$

The relationship between right ascension α, declination δ and orbital elements can be obtained from spherical right triangle △NDS in Fig. 3.7.

$$
\begin{aligned}
&\sin\delta = \sin(\omega + f)\sin i \\
&\cos\delta\cos(\alpha - \Omega) = \cos(\omega + f) \\
&\cos\delta\sin(\alpha - \Omega) = \sin(\omega + f)\cos i
\end{aligned}
$$

Another way to describe the position of a satellite is as follows. Set the unit eccentricity vector **P**.

$$\boldsymbol{P} = e/e$$

Rotate **P** by 90° along the motion direction of the satellite, namely the semi-latus rectum direction, and write it as the unit vector **Q**, then the satellite position vector **r** is

$$
\begin{aligned}
\boldsymbol{r} &= r\cos f\,\boldsymbol{P} + r\sin f\,\boldsymbol{Q} \\
&= a(\cos E - e)\boldsymbol{P} + a\sqrt{(1 - e^2)}\sin E\,\boldsymbol{Q}.
\end{aligned}
$$

Similar to the coordinate conversion process from orbital coordinate system to equatorial inertial coordinate system, the components of **P** and **Q** in the equatorial inertial coordinate system can be obtained:

$$
\begin{aligned}
P_x &= \cos\omega\cos\Omega - \sin\omega\sin\Omega\cos i \\
P_y &= \cos\omega\sin\Omega + \sin\omega\cos\Omega\cos i \\
P_z &= \sin\omega\sin i
\end{aligned}
$$

Substituting $\omega + 90$ for ω in the above formula yields three components of **Q**.

$$
\begin{aligned}
Q_x &= -\sin\omega\cos\Omega - \cos\omega\sin\Omega\cos i \\
Q_y &= -\sin\omega\sin\Omega + \cos\omega\cos\Omega\cos i \\
Q_z &= \cos\omega\sin i
\end{aligned}
$$

These six elements are not independent; they should satisfy three constraints:

$$\mathbf{P}\cdot\mathbf{P} = 1,\quad \mathbf{Q}\cdot\mathbf{Q} = 1,\quad \mathbf{P}\cdot\mathbf{Q} = 0$$

So there are only three independent variables, and they are combined with *a*, *e* and *E* to form six new orbital elements.

3.2 Satellite-Earth Geometry

The system design and operation analysis of remote sensing satellite are related to the space–time relationship between the satellite and the earth. These spatial–temporal relations are represented by the interrelation of satellite orbital position, satellite attitude, ground observation position, earth rotation and revolution, remote sensor field of view and spectrum, etc.

3.2.1 Ground Track of Satellite

Sub-satellite track is the path of the sub-satellite point on the earth's surface, which is the combination of the satellite's orbital movement and the earth's rotation. A satellite sub-satellite point is a collection of longitude and latitude of the point where the satellite's radial path intersects the earth's surface. The orbit of the satellite is defined in the equatorial inertial coordinate system S_i as shown in Fig. 3.7. According to the position coordinates of the satellite (x, y, z), the right ascension α and declination δ can be obtained:

$$\alpha = \arctan\left(\frac{y}{x}\right)$$

$$\delta = \arcsin\left(\frac{z}{\sqrt{x^2 + y^2 + z^2}}\right)$$

Or right ascension α and declination δ from orbital elements can be gotten. The spherical right triangle $\triangle$NDS is

$$\alpha = \Omega + \arctan(\tan u \cdot \cos i)$$

$$\delta = \arcsin(\sin u \cdot \sin i)$$

where $u = \omega + f$ is the angular distance of the satellite from the ascending intersection point, the true anomaly f is obtained by solving the Kepler equation.

Geocentric equator fixed coordinate system $\mathbf{S_e}$.

The coordinate origin is the center of the earth, x_e axis is the intersection of equatorial plane and Greenwich meridian, z_e axis points to the north pole along the earth's spin axis, as shown in Fig. 3.9. This coordinate system has the earth's spin angular velocity ω_e.

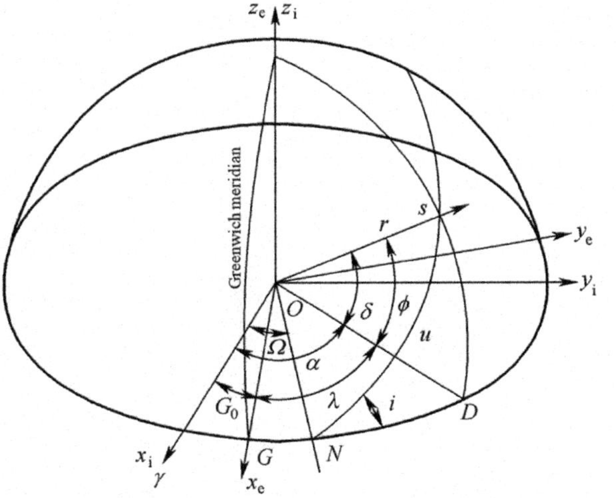

Fig. 3.9 Geocentric longitude of the satellite

The relationship between geocentric equatorial inertial coordinate system S_i and geocentric equatorial fixed coordinate system S_e:

$$\phi = \delta, \lambda = \alpha - G_0$$

When $t = 0$, the satellite is at point N, the Greenwich sidereal hour angle G_0, the range is 0–360° (to the west is positive, from the target to the vernal equinox, as shown in the figure is positive), right ascension of node N is ω, mean anomaly is M_0.

At time t, the satellite is at point s, and the satellite is rotated by an angle u (N to s). Greenwich sidereal hour angle $\alpha_G = G_0 + \omega_e\ (t - t_0)$, mean anomaly $M = M_0 + nt$, mean angular velocity is n.

The geocentric longitude λ of the satellite is equal to the difference between the right ascension of the satellite and the Greenwich sidereal hour angle, which is

$$\lambda = a - [G_0 + \omega_e(t - t_0)]$$
$$\omega_e = 7.2921158 \times 10^{-5}\ \text{rad/s}.$$

3.2.2 Field of View of Ground Station

Since the visible range of ground observation satellites is limited by the altitude angle (also known as elevation angle) of the satellite, the local elevation angle of the ground observer's line of sight to the satellite should be greater than 5°. The coverage area of ground stations is the visible area centered at the ground observation point P, and all satellites with sub-satellite points in this circle are observable.

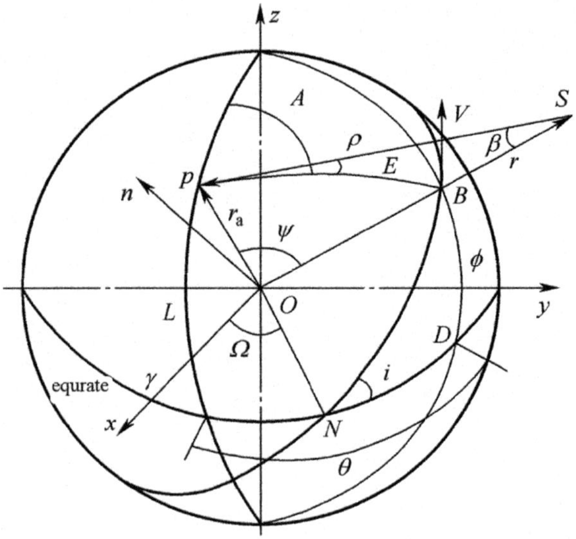

Fig. 3.10 Visible area of a satellite from a ground station

The altitude angle of the observation satellite in the ground station is the angle E between the direction of the satellite line of sight and the horizontal plane of the observation point P in the plane containing the observation point P, the geocentric o and the satellite S. Figure 3.10 shows the position diagram of sub-satellite point B relative to point P for a satellite whose elevation angle is greater than or equal to the given value E.

Angle ψ is the geocentric angular radius of the visible coverage of the satellite. 2 times of this is the maximum observable arc of the satellite, which is directly determined by the height of the satellite and the elevation E.

$$\psi = \frac{\pi}{2} - E - \arcsin\frac{r_a \cos E}{r} = \arccos\frac{r_a \cos E}{r} - E \tag{3.26}$$

According to formula (3.26), angle ψ of the visible coverage ring of the satellite is determined when the altitude and the minimum elevation angle E are given.

For a given elevation angle E, the longitude and latitude relationship between sub-satellite point B and point P in the covering circle can be obtained by the spherical triangle $\triangle$ZPB,

$$\cos\psi = \cos L\,\cos\varphi\,\cos\theta + \sin L\,\sin\varphi$$
$$\theta = \arccos\frac{\cos\psi - \sin\varphi\sin L}{\cos\varphi\cos L} \tag{3.27}$$

In formula (3.27), the geocentric latitude L of the observation point P is given, and the latitude φ of each point B on the covering ring and the corresponding longitude θ relative to the meridian of point P can be obtained. Given φ_i, the corresponding θ_i can be obtained. The set of (φ_i, θ_i) is the observable region of the earth station to the satellite, and all satellites with sub-satellite points in this area are visible.

From satellite observations to the earth, formation the satellite nadir angle β, defined as the angular distance between satellite relative to the observation point P on the ground and the sub-satellite point B. The relationship between satellite nadir angle β and the satellite elevation angle E can be obtained from triangle △OPS.

$$r \sin \beta = r_a \cos E$$

The azimuth angle formula is obtained from the spherical triangle △ZPB.

$$A = \arcsin \frac{\sin \theta \cos \varphi}{\sin \psi}$$

In the plane OPS, the oblique distance ρ and the elevation angle E are

$$\rho = \left(r_a^2 + r^2 - 2r\, r_a \cos \psi\right)^{1/2} \tag{3.28}$$

3.2.3 Communication Coverage of GEO

In the geostationary orbit, the service area where the communication satellite beam covers the ground is mainly determined by the beam pointing angle of the satellite antenna. Figure 3.11 is a schematic diagram of the relationship between the beam pointing of GEO communication satellite and the ground coverage service area.

In this figure, if the main axis of the communication antenna (beam center line) points to the center point P of the ground service area, the geocentric latitude of P is L, and the longitude difference between P and the sub-satellite point of the fixed-point satellite is θ, then the geocentric angle ψ of P relative to the sub-satellite point is

$$\psi = \arccos(\cos L \cos \theta)$$

The nadir angle β of point P relative to the satellite is

$$\beta = \arctan \frac{r_a \sin \psi}{r - r_a \cos \psi}$$

The direction of the antenna spindle relative to the ground target can be obtained from the above two formulas.

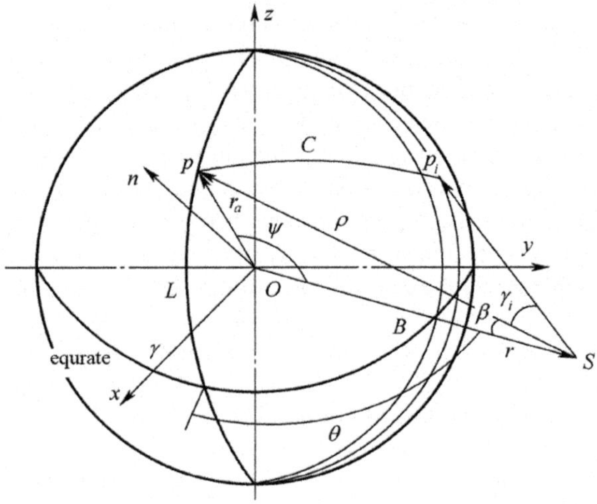

Fig. 3.11 GEO communications satellite beam ground coverage service area

Let point P_i be a point on the ground in the service area, and the angle between the beam pointing at the point and the main beam is the beam angle γ_i. The calculation formula of beam angle γ_i can be obtained from triangle ΔSPP$_i$ in Fig. 3.11.

$$\gamma_i = \arccos \frac{\rho^2 + \rho_i^2 - \overline{PP_i}^2}{2\rho\rho_i} \tag{3.29}$$

In the formula, ρ and ρ_i are the oblique distances of the satellite to point P and point P_i, and the calculation formulas of the geocentric angle ψ are, respectively.

$$\rho^2 = r^2 + r_a^2 + 2r\, r_a \cos L \cos \theta$$
$$\rho_i^2 = r^2 + r_a^2 + 2r\, r_a \cos L_i \cos \theta_i$$

where L_i and θ_i are the geocentric latitude of point P_i and the longitude difference with respect to the sub-satellite point B of the fixed satellite, respectively. In the formula, chord length PP_i between point P and point Pi can be obtained by converting arc length C.

$$\overline{PP_i} = 2r_a \sin \frac{C}{2}$$

The formula of arc length C in the above equation can be obtained from the spherical triangle ΔZPP$_i$ in the figure.

$$\cos C = \sin L \sin L_i + \cos L \cos L_i \cos(\theta - \theta_i)$$

There are two modes to describe the antenna beam service area, both of which take the antenna spindle point P as the center point.

Circle of equal gain. For conical beam, the equal gain circle corresponds to the equal beam angle. The beam angle $_{\gamma i}$ is obtained from the point P_i (L_i, θ_i) of formula (3.29), and then, γ_i is set to a fixed value. The combined solution (L_i, θ_i) is obtained from the ρ_i equation, and the equal gain circle is connected.

Circle of equal flux density. The flux density transmitted by the communication antenna at the ground service point P_i is related to the beam angle γ_i and oblique distance ρ_i satellite pointing to the point. The flux density reaching the ground is proportional to the transmitting power of the antenna and the gain in this direction and inversely proportional to the square of oblique distance satellite pointing to the point.

For the conical beam, a simple iterative method can be used to obtain the service circle of the ground flux density. Firstly, the intersection point of the longitude circle of the center point P of the beam and the circle of equal flux density is selected, and the flux density of this point is calculated. Then move the point according to the equal longitude step, and obtain the latitude of the point iteratively according to the requirements of equal flux density and so on.

3.2.4 Geo-positioning of Remote Sensing Image

On the remote sensing satellite, there are two ways for optical remote sensing instrument to obtain earth radiation image, namely scanning mode and area array imaging mode, while the scanning mode has two dimensional scanning mode of single pixel detector and push-sweep mode of line array detector. No matter which imaging method, the final result is a digital image. In order to obtain high-quality images, different imaging methods need different image data processing methods. The position of each pixel in the image corresponds to the earth coordinates of an observed point (or remote sensing point) on the earth surface, which is the intersection point between the line of sight of the pixel in the image and the earth surface. Due to the orbit motion of satellite, scanning motion of instrument and earth rotation, image positioning is a combination of spatial geometry and time sequence.

Suppose that the ground remote sensing point corresponding to a certain pixel S in the image is point P, SP is the line of sight vector $\boldsymbol{\rho}$ (simply called pixel line of sight vector) of P point on the image surface of the camera, and from the satellite pointing to the remote sensing point, as shown in Fig. 3.12. Assume that the geocentric vector of remote sensing point P in earth coordinates is $\mathbf{P}$, which has spatial geometric relation:

$$\boldsymbol{r} + \boldsymbol{\rho} = \boldsymbol{p} \tag{3.30}$$

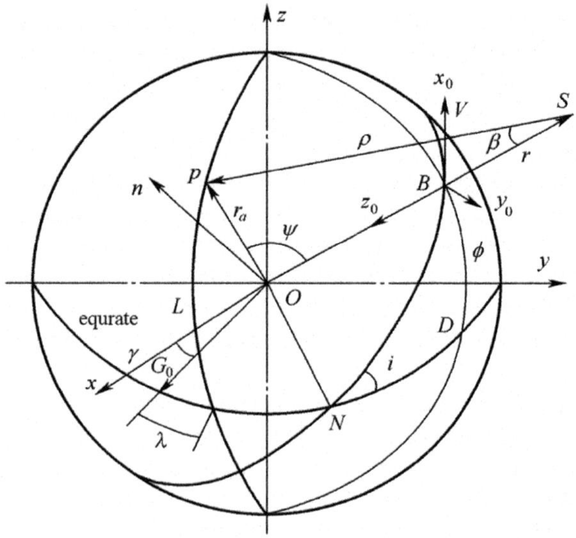

Fig. 3.12 Ground remote sensing point positioning principle diagram

The geocentric vector **r** of the satellite is defined in the equatorial inertial coordinate system. The above geometric relations must be solved in this coordinate system.

The pixel line of sight vector is expressed in the equatorial inertial coordinate system as

$$\boldsymbol{\rho}_i = \rho_i \cdot \boldsymbol{U}_i \tag{3.31}$$

where $\mathbf{U_i}$ is the unit vector of pixel line of sight. The coordinate position of the image plane is determined by the number of rows and columns of the pixel in the coordinate system of the instrument, and according to the focal length of the optical instrument, the unit vector $\mathbf{U_c}$ of the pixel line of sight in the instrument coordinate system (subscript *c* represents the instrument coordinate system) can be obtained. Based on the installation matrix *M* of the instrument on the satellite body, the unit vector $\mathbf{U}_\mathrm{b} = \mathbf{M}^{-1}\mathbf{U}_\mathrm{c}$ in the coordinate S_b of the satellite body is obtained. Based on the satellite attitude parameter matrix *A*, the unit vector $\mathbf{U}_\mathrm{o} = \mathbf{A}^{-1}\mathbf{U}_\mathrm{b}$ of pixel line of sight in satellite orbit coordinate S_o is obtained. Finally, the conversion matrix $\mathbf{R}_\mathrm{oi}$ between the orbits coordinates determined by the satellite orbit parameters, and the geocentric equatorial inertial coordinates are used to calculate the unit vector $\mathbf{U_i}$ of the pixel line of sight in the geocentric equatorial inertial coordinate system, i.e.,

$$\boldsymbol{U}_i = \boldsymbol{R}_{oi}^{-1}\boldsymbol{A}^{-1}\boldsymbol{M}^{-1}\boldsymbol{U}_\mathrm{c} \tag{3.32}$$

According to the definition of satellite orbit coordinate system: x_o and z_o axis are located in the orbit plane, z_o axis points from the satellite centroid to the center of the earth; x_o is in the velocity direction, perpendicular to the z_o axis; the y_o axis is right-handed orthogonal to the orbital plane. The direction of the orbital coordinate axis represented by the orbital motion parameters **r** and **v** is

$$\boldsymbol{x}_o = \boldsymbol{y}_o \times \boldsymbol{z}_o$$
$$\boldsymbol{y}_o = \frac{\boldsymbol{v} \times \boldsymbol{r}}{|\boldsymbol{v} \times \boldsymbol{r}|}$$
$$\boldsymbol{z}_o = -\frac{\boldsymbol{r}}{r}$$

The satellite's orbital coordinate axis is defined in the geocentric equatorial inertial coordinate system, and the transformation matrix $\mathbf{R}_{oi}$ can be directly expressed by the orbital coordinate axis vector as

$$\boldsymbol{R}_{oi} = \left[\boldsymbol{x}_o \boldsymbol{y}_o \boldsymbol{z}_o\right]^T$$

Thus, the unit vector $\mathbf{u}_i$ in the equatorial inertial coordinate system can be obtained from the unit vector $\mathbf{U}_c$ of the pixel line of sight in the instrument coordinate system, and then, the earth coordinates of the remote sensing point *P* on the ground corresponding to the pixel can be obtained.

The oblateness of the earth should also be considered when obtaining the earth coordinates of the remote sensing point *P* on the ground. The ellipsoid model in the meridional section of the earth is considered as shown in Fig. 3.13.

The geocentric latitude φ of the satellite is equal to the satellite declination δ, and the relationship between the geocentric latitude φ of the satellite and the satellite declination δ and the geographical latitude φ' is shown in Fig. 3.13. The ellipsoidal model of the earth is that the cross section of the earth's meridians is an ellipse, the major axis r_e is the equatorial radius, and the minor axis r_p is the polar radius of the earth.

The oblateness f and eccentricity e of an ellipse are defined as

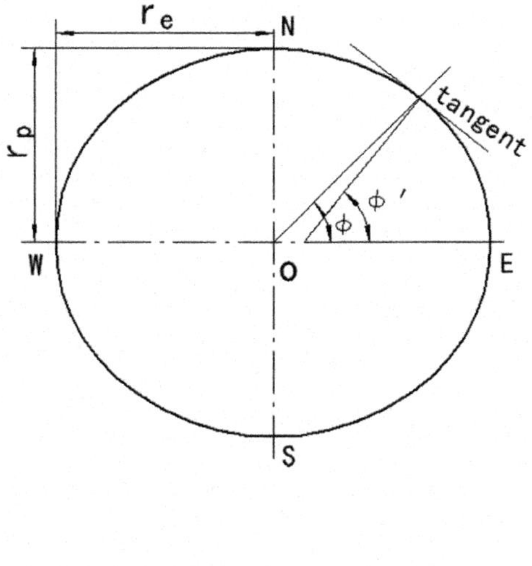

Fig. 3.13 Ellipsoid model of the earth

$$f = \frac{r_e - r_p}{r_e}$$

$$e^2 = \frac{r_e^2 - r_p^2}{r_e^2}$$

These are fundamental constants: $r_e = 6378.145$ km, $r_p = r_e(1 - f) = 6356.76$ km, $f = 1/298.257$, $e = 0.08182$.

The conversion relationship between geocentric latitude and geographic latitude is as follows:

$$\tan\varphi = (1 - f)^2 \tan\varphi'$$

Let the subscripts x, y and z represent the axis components of the vector **P** at the geocentric equator inertia coordinate system, and the coordinates P_x, P_y and P_z of P conform to the ellipsoid formula.

$$\frac{p_x^2 + p_y^2}{r_e^2} + \frac{p_z^2}{r_p^2} = 1 \tag{3.33}$$

where r_e and r_p are the semi-major axis and semi-minor axis of the earth, respectively.

From formulas (3.30), 3.31 and 3.32, so:

$$\frac{(\rho u_x + r_x)^2 + (\rho u_y + r_y)^2}{r_e^2} + \frac{(\rho u_z + r_z)^2}{r_p^2} = 1 \tag{3.34}$$

The distance ρ of pixel line of sight is obtained.

$$\rho = \frac{-B - \sqrt{B^2 - AC}}{A} \tag{3.35}$$

where

$$A = 1 + Du_z^2$$
$$B = \boldsymbol{r} \cdot \boldsymbol{u} + Dr_z u_z$$
$$C = r^2 + Dr_z^2 - r_e^2$$
$$D = \frac{r_e^2 - r_p^2}{r_p^2}$$

Let's substitute formulas 3.32 and 3.35 into 3.31, the earth coordinates of the ground remote sensing point P under the ellipsoid model can be obtained.

Then, it is obtained by the transformation matrix $\mathbf{R}_{ei}$ of the geocentric equatorial fixed coordinate system and the geocentric equatorial inertial coordinate system.

$$R_{ei} = \begin{bmatrix} \cos\alpha_G & \sin\alpha_G & 0 \\ -\sin\alpha_G & \cos\alpha_G & 0 \\ 0 & 0 & 1 \end{bmatrix}$$

Thus, the vector $\mathbf{P}_e$ of remote sensing point P in earth coordinates is obtained, and the geocentric longitude λ and latitude L of remote sensing point P are obtained.

$$\lambda = \arctan\left(\frac{P_y}{P_x}\right)_e$$

$$L = \arcsin\left(\frac{P_z}{|\boldsymbol{P}|}\right)_e$$

More accurate location determination of ground remote sensing points can also be further referenced in Chap. 2 of the earth's triaxiality.

3.3 Launch Window

The azimuth of the satellite orbital plane is determined by the position of the launching site, the launching direction and the launching time, which are called the three elements of the orbital plane azimuth. The determination of the three essential elements has two meanings. First, the nominal values of the three elements are determined, and then, the allowable tolerance of the three elements is determined according to the orbital plane precision required by engineering practice. The best choice for launching into a certain orbital plane should be the moment when the launch site enters that orbital plane with the rotation of the earth. Each specific technical requirement corresponds to a certain launch time, and the allowable range of the specific technical requirement corresponds to a certain range of launch time, which is commonly known as a "window". For the earth satellites, there are two types of factors that limit launch windows: those related to the sun are called sunlight windows, and those related to satellite networking are called plane windows.

The first task of satellite guidance is to determine the launch window after determining the specific requirements of the satellite flight mission and selecting the preset orbit of the satellite and the trajectory of the carrier rocket. The launch window of the spacecraft refers to the date, time and time interval that can launch the spacecraft. The establishment of the launch window should be based on the working conditions of the spacecraft and the ground, combined with the operation law of the spacecraft and the sun (or the moon and other celestial bodies), considering the location of the launch site and orbit parameters to determine the launch window.

The launch window of the spacecraft does not determine the shape of the spacecraft orbit but determines the angle between the sun and orbit, that is, the sunlight window, such as the inclination of the scheduled orbit is 90°, in the vernal or autumnal equinox, local time at 6:00 a.m. or 6:00 p.m. Launch spacecraft, the sunlight will be vertical irradiation of the orbital plane. In addition, the launch window, the position of launch site and launch azimuth jointly determine the right ascension (ascending or descent) intersection of the orbital plane.

3.3.1 Orbital Plane and Coplanar Window

As shown in Fig. 3.14, the spatial orientation of the orbital plane is determined by the inclination **i** and ascending intersection point Ω of the orbital plane. To launch into a predetermined orbit, it is necessary to obtain a definite inclination **i** and ascending point Ω.

3.3.1.1 Determination of Orbital Inclination

The conventional launch vehicle is used to send the satellite into the predetermined orbit. The trajectory of such launch vehicle does not make lateral maneuver during launch. The azimuth of the satellite's orbit plane in space is directly determined by the geocentric latitude φ of launch site L, launch azimuth A and launch time t_L.

In Fig. 3.14, from spherical triangle △NLD, so:

$$\cos i = \sin A \cos \varphi \tag{3.36}$$

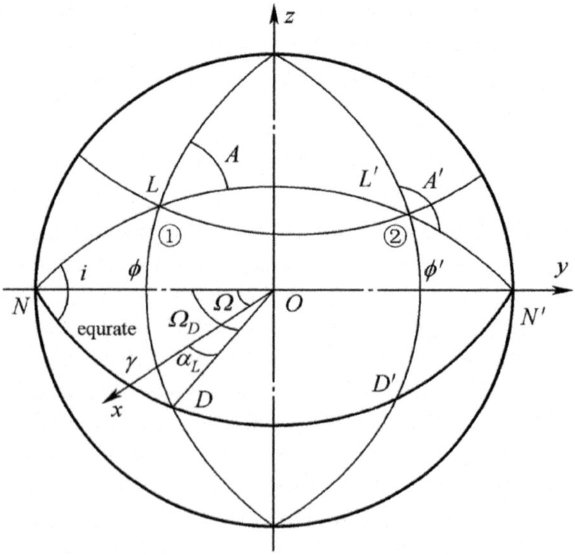

Fig. 3.14 Satellite launch elements diagram

That is, the orbital inclination i is determined by the geocentric latitude φ and launch azimuth angle A of the launch site L. If the spacecraft does not maneuver in orbit, the geographical latitude of the launching site directly determines the minimum inclination of the spacecraft orbit.

Launch azimuth A is defined as the clockwise angle between the north direction of the launch point and the horizontal direction of the launch vehicle velocity. There are two launching methods: One is to launch orbit from azimuth angle A, such as position ① on the sphere; another is to launch from orbit descent at A′ (= 180° – A) over A period of time as the launch site shifted to A spherical position ②. The same orbit can be gotten by the either way.

According to spherical trigonometry in formula 3.36, for launch site L, there are two solutions for launch azimuth in order to achieve a given orbital inclination, $A < 90°$ or $A > 90°$. That is, orbit ascending or descending launch, and the orbital inclination obtained is not less than the geocentric latitude of the launch site. If the orbital inclination angle is required to be greater than 90°, then $A < 0°$ (launched in the northwest direction); Or $A > 180°$, launch from orbit descent (southwest). If the orbital inclination is equal to 90°, then $A = 0°$ or 180°, descending orbit is generally adopted, that is, $A = 180°$.

3.3.1.2 Determination of Right Ascension of Ascending Node

From Fig. 3.14, it is known that right ascension of ascending node Ω is

$$\Omega = \alpha_{\mathrm{L}} - \Omega_D \tag{3.37}$$

where Ω_D is the angle between orbital node N (ascending intersection point) and node D on the meridian of the launch site on the spherical equator (positive in the east, negative in the west, positive in the figure). $\Omega_D \in [0, 360°]$, α_{L} is the sidereal hour angle of the launch site L, $\alpha_{\mathrm{L}} \in [0, 360°]$, and the westward direction is positive (pointing to the vernal equinox from the target).

$$\Omega_D = \arcsin\left(\frac{\tan\varphi}{\tan i}\right) \text{(The satellite is launched by ascending orbit)} \tag{3.38}$$

$$\Omega_D = 180 - \arcsin\left(\frac{\tan\varphi}{\tan i}\right) \text{(The satellite is launched by decending orbit)} \tag{3.39}$$

The above two expressions are used when $\varphi > 0$. When $\varphi < 0$, Ω_D has two values, ③ belongs to the third quadrant of southern latitude orbital launching, ④ belongs to the fourth quadrant of southern latitude orbital launching, as shown in Fig. 3.15.

According to formula 3.37, in order to obtain the predetermined ascending point right ascension Ω, the sidereal hour angle of the launching site (vernal equinox) must also be determined. From the perspective of timing sequence, the sidereal hour angle of the launch site at the launch time is equal to the sum of sidereal hour angle G_0

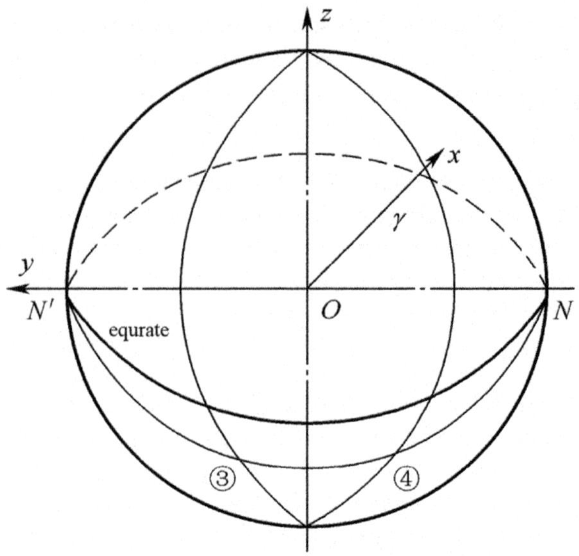

Fig. 3.15 Schematic diagram of satellite launch at southern latitude

at Greenwich midnight on the launch day, the longitude of the launch site λ and the universal time angle $\omega_e t_L$ at the launch time.

$$\alpha_L = G_0 + \lambda + \omega_e t_L$$

It takes time t_A from launch to enter orbit, so the actual launch time should be t_A earlier than the set time. For example, in order to achieve orbit with right ascension of ascending intersection point Ω, the universal time of launch time should be determined by the following formula (calculated in hours).

$$t_L = \frac{1}{15}\left[\Omega - G_0 - \lambda + \arcsin\left(\frac{\tan\varphi}{\tan i}\right)\right] - \frac{1}{60} t_A \tag{3.40}$$

The local time of the launch site is t_s.

$$t_s = \frac{\lambda}{15} + t_L$$

3.3.1.3 Coplanar Window

If two satellites share the same orbital plane, it is necessary that the inclination i of the two satellites and the ascending intersection point Ω are the same. After the launch site is determined, the geocenter latitude φ of the launch site L is also determined. The orbital inclination angle is determined by the launch azimuth angle A. The moment of coplanar launch is the moment when the launch site enters the target orbital plane with the rotation of the earth. Generally, there are two launch opportunities every day, one is for orbital launch *L*, and one is for orbital launch L', as shown in Fig. 3.14.

It is difficult to realize absolute coplanar launch in engineering practice. For example, at coplanar launch time, the launch site is located at point *L* when the target orbital plane passes through the launch site. Due to the launch delay time Δt, the actual launch site was located at L' point on the sphere, as shown in Fig. 3.16. If the launch azimuth was A', which is different from the predetermined azimuth point *A*, the acquired orbital plane angle i' and ascending intersection point right ascension Ω' would be different from the target orbital elements i and Ω of coplanar launch. Suppose, the two orbits are, respectively, Q and Q', the normal lines of the orbital plane are, respectively, k and k', and the intersection points of the longitude of *L* and L' and the equator are, respectively, m and m'. The angle between the two orbital planes is the angle γ of the normal lines k and k' of the two orbital planes, so that the two orbits are coplanar, that is, $\gamma = 0$.

Within the delay time Δt, the moving angle of the launch site is $w = \overset{\frown}{mm'}$, and the angle of the orbital plane formed is γ. In spherical triangle $\Delta kZk'$, as shown in Fig. 3.17, i refers to the target orbital inclination, i' refers to the actual orbital inclination when the launch azimuth angle is A', and γ refers to the allowable deviation angle between the actual orbital plane and the target orbital plane. The allowable delay time Δt is calculated below.

$$\tan\frac{z}{2} = \frac{m}{\sin(p-\gamma)}$$

where

Fig. 3.16 Coplanar windows

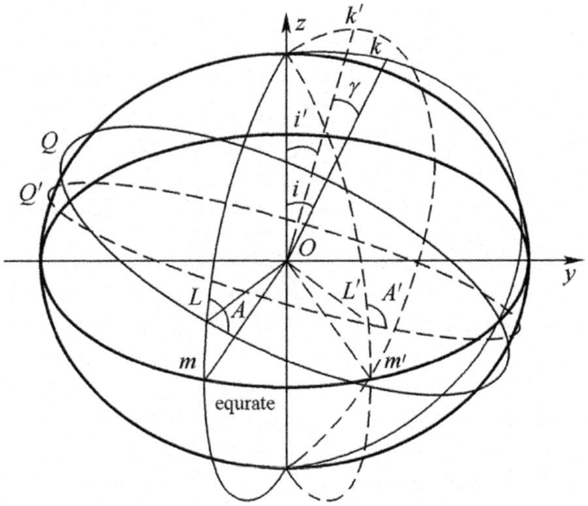

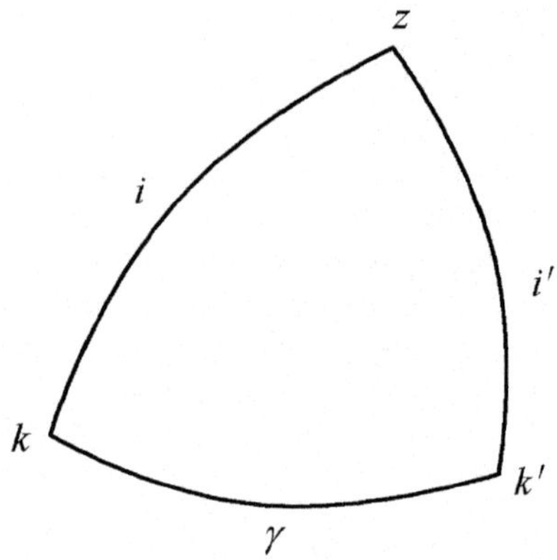

Fig. 3.17 Angle element between orbital planes

$$p = \frac{1}{2}(i + i' + \gamma), m = \sqrt{\frac{\sin(p-\gamma)\sin(p-i)\sin(p-i')}{\sin p}}$$

$$\tan\frac{z}{2} = \sqrt{\frac{\sin(p-i)\sin(p-i')}{\sin(p-\gamma)\sin p}}$$

$$z = 2\arctan\sqrt{\frac{\sin(p-i)\sin(p-i')}{\sin(p-\gamma)\sin p}} \tag{3.41}$$

In the spherical triangle ΔLZk in Fig. 3.16, $\overset{\frown}{LZ} = 90° - \varphi$, $\overset{\frown}{kZ} = \mathrm{i}$, $\overset{\frown}{Lk} = 90°$. Similarly,

$$\begin{aligned} p &= \frac{1}{2}(90° - \varphi + 90° + \mathrm{i}) \\ &= 90° - \frac{\varphi - i}{2} \end{aligned}$$

$$\begin{aligned} \tan\frac{\angle LZk}{2} &= \sqrt{\frac{\sin\left(90° - \frac{\varphi-i}{2} - 90° + \varphi\right)\sin\left(90° - \frac{\varphi-i}{2} - i\right)}{\sin\left(90° - \frac{\varphi-i}{2} - 90°\right)\sin\left(90° - \frac{\varphi-i}{2}\right)}} \\ &= \sqrt{\frac{\sin(i+\varphi)}{\sin(i-\varphi)}} \end{aligned}$$

$$\angle LZk = 2\arctan\sqrt{\frac{\sin(i+\varphi)}{\sin(i-\varphi)}} \tag{3.42}$$

In the same way,

$$p = 90^\circ - \frac{\varphi - i'}{2}$$

$$\angle L'Zk' = 2\arctan\sqrt{\frac{\sin(i'+\varphi)}{\sin(i'-\varphi)}} \tag{3.43}$$

And

$$\angle LZL' = \omega_e \Delta t \tag{3.44}$$

where ω_e is the angular velocity of the earth.

$$\begin{aligned} \angle kZk' &= \angle L'Zk' - \angle L'Zk \\ &= \angle L'Zk' - (\angle LZk - \angle LZL') \end{aligned} \tag{3.45}$$

If you substitute formulas 3.42, 3.43 and 3.44 into formula 3.45, so:

$$\angle kZk' = 2\arctan\sqrt{\frac{\sin(i'+\varphi)}{\sin(i'-\varphi)}} - 2\arctan\sqrt{\frac{\sin(i+\varphi)}{\sin(i-\varphi)}} + \omega_e \Delta t$$

Substitute 3.41 into the above equation,

$$\begin{aligned} 2\arctan\sqrt{\frac{\sin(p-i)\sin(p-i')}{\sin(p-\gamma)\sin p}} &= 2\arctan\sqrt{\frac{\sin(i'+\varphi)}{\sin(i'-\varphi)}} \\ &- 2\arctan\sqrt{\frac{\sin(i+\varphi)}{\sin(i-\varphi)}} + \omega_e \Delta t \\ \Delta t = \frac{2}{\omega_e}\Bigg(\arctan\sqrt{\frac{\sin(p-i)\sin(p-i')}{\sin(p-\gamma)\sin p}} &- \arctan\sqrt{\frac{\sin(i'+\varphi)}{\sin(i'-\varphi)}} \\ &+ \arctan\sqrt{\frac{\sin(i+\varphi)}{\sin(i-\varphi)}}\Bigg) \end{aligned} \tag{3.46}$$

The relationship between the error γ of the orbital plane inclination and delayed launch time Δt of launch site L is determined by formula 3.46.

The more practical case is to calculate maximum Δt when γ is fixed. When $i' = i$, then $A' = A$.

When $i' = i$ from formula 3.46 can be gotten:

$$\Delta t = \frac{2}{\omega_e}\arctan\sqrt{\frac{\sin(p-i)\sin(p-i')}{\sin(p-\gamma)\sin p}}$$

$$= \frac{2}{\omega_e} \arctan \frac{\sin \frac{\gamma}{2}}{\sqrt{\sin\left(i - \frac{\gamma}{2}\right) \sin\left(i + \frac{\gamma}{2}\right)}} \tag{3.47}$$

For example, if $\gamma = 1^\circ$ and $i = 50^\circ$, the launch can be advanced or delayed by 5.207 min, namely $\Delta t = 10.414$ min.

In general, when $i \neq i'$, when the target orbit and the actual launch orbit γ are fixed, the allowed launch time Δt can be obtained by solving the following equations.

$$\begin{cases} p = \frac{1}{2}(i + i' + \gamma) \\ \Delta t = \frac{2}{\omega_e}\left(\arctan \sqrt{\frac{\sin(p - i)\sin(p - i')}{\sin(p - \gamma)\sin p}} - \arctan \sqrt{\frac{\sin(i' + \varphi)}{\sin(i' - \varphi)}} \right. \\ \qquad \left. + \arctan \sqrt{\frac{\sin(i + \varphi)}{\sin(i - \varphi)}} \right) \end{cases} \tag{3.48}$$

Δt can be obtained by solving formula 3.48 numerically.

A' is obtained by using $\cos i' = \sin A' \cos\varphi$.

A is obtained by using $\cos i = \sin A \cos\varphi$.

3.3.2 *Sunlight Window*

In space optical remote sensing, in order to ensure the energy of satellite, it is necessary to have a certain angle between solar array and sunlight. In order to obtain good sub-satellite illuminance and ensure the required target illuminance, it is necessary to launch the satellite at a specific time, which is called the sunshine window.

3.3.2.1 Orbit Solar Angle

Sunlight is an important condition for satellites to survive and perform missions in space. The energy system and temperature control system of the satellite require sunlight to illuminate the satellite within a certain angle and time range. In order to avoid the interference of sunlight on the optical remote sensing instrument, sunlight is not required to enter a certain angle range of the satellite. In order to ensure the accuracy of attitude measurement for orbital maneuvering, good space geometry between the sun, the earth and the satellite is required.

Since the satellite attitude is stable in the orbital coordinate system, the angle between sunlight and orbital plane shows the radiation of sunlight on the satellite. The above technical requirements are reflected in the orbit solar angle β, which is defined as the angle between the orbit's normal direction and the direction of the

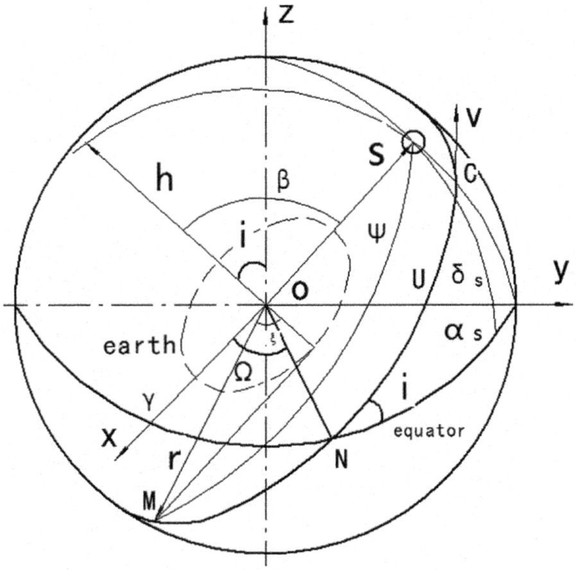

Fig. 3.18 Solar elevation angle

sun. In Fig. 3.18, S is the position of the sun on the geocentric celestial sphere. The elements of solar vector **S** in the geocentric equatorial inertial coordinate system are right ascension α_s, declination δ_s, orbital ascending point right ascension Ω, inclination i.

$$S = \begin{bmatrix} \cos\delta_s \cos\alpha_s \\ \cos\delta_s \sin\alpha_s \\ \sin\delta_s \end{bmatrix}$$

The normal vector of the orbital plane is **h**, and the orbital elements of this vector are determined by the orbital inclination and the right ascension of the ascending intersection point.

$$\vec{h} = \begin{bmatrix} \sin i \sin\Omega \\ -\sin i \cos\Omega \\ \cos i \end{bmatrix}$$

Thus, the cosine of the orbital solar angle β is

$$\cos\beta = (\boldsymbol{S}\cdot\boldsymbol{h}) = \cos\delta_s \sin(\Omega - \alpha_s)\sin\, i + \sin\delta_s \cos\, i \tag{3.49}$$

If the sun is required to be on the left of the satellite, make $\beta < 90°$ and conversely $\beta > 90°$. If $\beta = 90°$, the sun coincides with the orbital plane.

Sunlight shining the orbit also depends on the satellite's earth shadow. As shown in Fig. 3.19, **S** refers to solar vector direction, **r** refers to radial diameter of satellite,

Fig. 3.19 Earth shadow condition

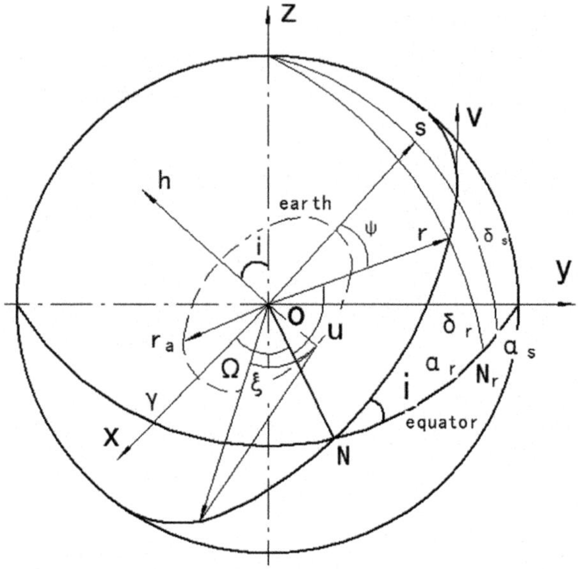

δ_r refers to latitude of satellite, α_r refers to longitude of satellite relative to point N. The included angle between the satellite position vector **R** and the solar vector **S** is defined as ψ, and the condition for the satellite to enter the earth's shadow is

From spherical right triangle ΔrNN_r

$$\begin{aligned}\tan\alpha_r &= \cos i \tan u\\ \sin\delta_r &= \tan\alpha_r \tan i \cos u\\ &= \cos i \tan u \tan i \cos u\\ &= \sin i \sin u\end{aligned}$$

$$\vec{r} = \begin{pmatrix}\cos\delta_r \cos(\Omega+\alpha_r)\\ \cos\delta_r \sin(\Omega+\alpha_r)\\ \sin\delta_r\end{pmatrix}$$

$$\begin{aligned}\cos\psi &= (\vec{s}\cdot\vec{r})\\ &= \begin{bmatrix}\cos\delta_s\cos\alpha_s\\ \cos\delta_s\sin\alpha_s\\ \sin\delta_s\end{bmatrix}\cdot\begin{pmatrix}\cos\delta_r\cos(\Omega+\alpha_r)\\ \cos\delta_r\sin(\Omega+\alpha_r)\\ \sin\delta_r\end{pmatrix}\\ &= \cos\delta_s\cos\delta_r[\cos\alpha_s\cos(\Omega+\alpha_r)+\sin\alpha_s\sin(\Omega+\alpha_r)]+\sin\delta_s\sin\delta_r\\ &= \cos\delta_s\cos\delta_r\cos(\Omega+\alpha_r-\alpha_s)+\sin\delta_s\sin\delta_r\end{aligned}$$

ξ refers to the earth shadow angle in Fig. 3.18, which is the angle between the satellitic radius vector M and the vertical line of the earth's shadow edge when the satellite exits the earth shadow.

$$\xi = \arcsin \frac{\sqrt{r^2 - r_a^2}}{r}$$

The condition for a satellite to enter the earth's shadow is

$$\psi \geq 90^\circ + \xi$$

The proportion of sunlight a satellite receives during one orbit is called the sun exposure factor. When the sunlight hits the orbital plane vertically, the sun exposure factor is the largest, and when the sunlight coincides with the orbital plane, the sun exposure factor is the smallest. The insolation factor is related to the orbital height and the orbital sun angle. In Fig. 3.18, C is the intersection point of the great circle of the vector **h**, **S** and the orbit. M is the position of the satellite leaving the shadowed area of the earth. The trigonometric relations are obtained from spherical triangle △SCM

$$\begin{aligned} \cos\psi &= \cos u \cos(90^\circ - \beta) \\ &= \cos u \sin\beta \end{aligned}$$

Thus, the relationship between sun exposure factor ε and orbital solar angle β can be expressed as

$$\varepsilon = \frac{u}{180} = \frac{\arccos \frac{\cos\psi}{\sin\beta}}{180^0}$$

In order to meet the technical requirements of orbital solar angle β, the longitude of the ascending intersection point of the launch orbit Ω, or the mean solar hour angle of the ascending intersection point Ω_s ($\Omega_s = \Omega - \alpha_s$), is shown in formula 3.49.

$$\Omega_s = \arcsin \frac{\cos\beta - \sin\delta_s \cos i}{\cos\delta_s \sin i}$$

From this, the mean solar time of the launch site is calculated.

In Fig. 3.18, point C is set as the launching site, the geocentric latitude is φ, and the longitude difference between the intersection of the longitude line C and the equator and the ascending intersection point N is Ω_C.

$$\Omega_C = \arcsin \frac{\tan\varphi}{\tan i}$$

Since Ω_C is positive to the east and Ω_S is positive to the west, the solar hour angle at point C is $\Omega_S + \Omega_C$, then the local mean solar time t_s at the launch time can be determined as

$$t_s = 12 + \frac{1}{15}(\Omega_S + \Omega_C) - \frac{t_A}{60} \tag{3.50}$$

The unit of time in the above formula is hours.

If it is a descending launch, the longitude difference between the launch site C and the ascending intersection point N is $180° - \Omega_C$, and the local mean solar time at the launch time is

$$t_s = 12 + \frac{1}{15}(\Omega_S + 180° - \Omega_C) - \frac{t_A}{60}$$

3.3.2.2 Solar Zenith Angle

Satellite observation of the earth by visible light requires that when the satellite passes through the observed area, the sub-satellite points should be under appropriate lighting conditions, so as to obtain high-quality remote sensing images. This technical requirement can be described by the solar zenith angle η. η is defined as the angle between the satellite's position vector **r** and the sun's vector **s**. As shown in Fig. 3.20, if the remote sensing observation area is on latitude circle φ, when the sub-satellite point C passes through this latitude circle, the longitude difference between the sub-satellite point's longitude circle and the node at the equator and the orbital ascending point N is Ω_C. From spherical triangle $\triangle CN\Omega_C$, Ω_C is gotten:

$$\Omega_C = \arcsin \frac{\tan \varphi}{\tan i} \tag{3.51}$$

In the geocentric equatorial inertial coordinate system, the unit vector **C** of the sub-satellite point C can be written as

$$\vec{C} = \begin{bmatrix} \cos \varphi \cos(\Omega + \Omega_C) \\ \cos \varphi \sin(\Omega + \Omega_C) \\ \sin \varphi \end{bmatrix}$$

The illuminance of sub-satellite point C, the solar zenith angle η, can be obtained by scalar product

$$\cos \eta = (\boldsymbol{S} \cdot \boldsymbol{C}) = \cos \varphi \cos \delta_s \cos(\Omega_s + \Omega_C) + \sin \varphi \sin \delta_s \tag{3.52}$$

where Ω_s is the mean solar hour angle of the ascending node N. Then:

$$\Omega_s = \Omega - \alpha_s$$

If $\Omega_s < 0°$, the ascending intersection point is in the morning; otherwise, afternoon.

Fig. 3.20 Solar zenith angle

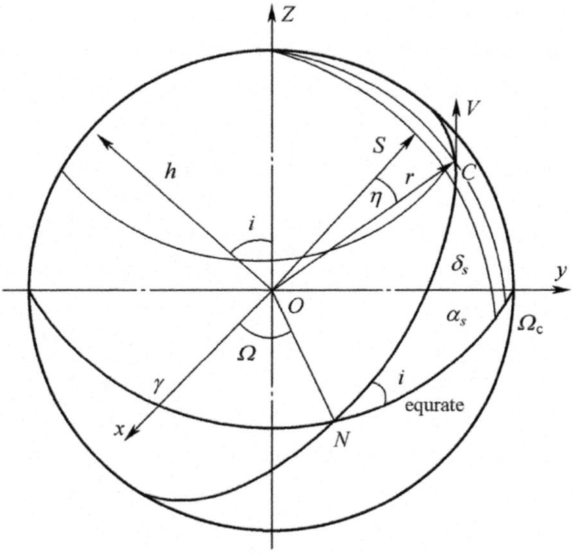

Set C as the launching site. In order to meet the technical requirements of solar illuminance at the specified latitude circle, the mean solar hour angle Ω_s of the ascending intersection point of the launch orbit is obtained by formulas 3.51 and 3.52.

$$\Omega_s = \arccos\frac{\cos\eta - \sin\varphi\sin\delta_s}{\cos\varphi\cos\delta_s} - \arcsin\frac{\tan\varphi}{\tan i}$$

The local mean solar time at the time of launch, t_s, can be determined as

$$t_s = 12 + \frac{1}{15}(\Omega_s + \Omega_C) - \frac{t_A}{60}$$

The unit of time in the above formula is hours.

If $i > 90°$, then $\Omega_C < 0°$. The node of launching site C is located on the left side of ascending intersection point N, as shown in Fig. 3.21. The longitude difference between the node from ascending intersection point N to C is $360° + \Omega_C$ (here $\Omega_C < 0$), solar right ascension $\alpha_s < 0$, and ascending intersection longitude $\Omega > 0$, then

$$\Omega_s + \Omega_C = \Omega - \alpha_s + 360° + \Omega_C = \Omega - \alpha_s + \Omega_C \quad \text{(Where } 2\pi \text{ should be deducted)}$$

$\Omega - \alpha_s$ is positive, Ω_C is negative, and $\Omega - \alpha_s + \Omega_C$ which refers to the angular distance from note C to the sun node is positive. The local time of launch in the figure should be afternoon, so that the local time of launch is still

$$t_s = 12 + \frac{1}{15}(\Omega_s + \Omega_C) - \frac{t_A}{60}.$$

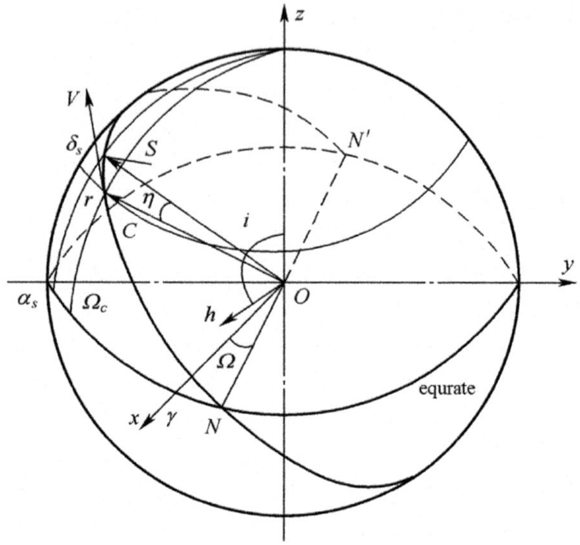

Fig. 3.21 Local time of launch site when $i > 90°$

3.4 Sun-Synchronous Orbit

In general, the earth is regarded as a sphere with uniform mass. In fact, the earth expands and rises near the equator, and the parts of these rises can be regarded as additional masses. For the satellites in the northern and southern hemispheres, additional gravity is generated that does not pass through the center of the earth, and additional torque **M** is formed for orbital motion, so that the satellite's moment of momentum **h** relative to the center of the earth precedes in space. The precession principle can be understood as the gyroscopic precession effect $\mathbf{M} = \dot{\mathbf{\Omega}} \times \mathbf{h}$, as shown in Fig. 3.22.

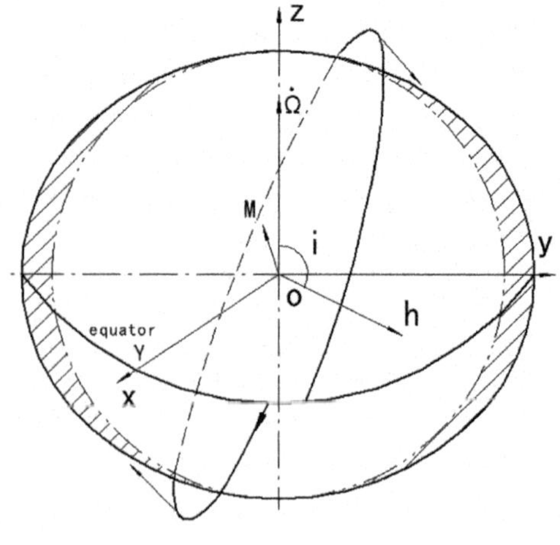

Fig. 3.22 Principle of sun-synchronous orbit

That is, the nodal line between the orbital plane and the equatorial plane of the satellite is no longer fixed in inertial space but rotates eastward or westward. The speed is not only related to the parameter J_2, which represents the flat distribution of the earth's mass, but also to the orbital height, inclination and eccentricity. For orbit perturbation, the mean value of right ascension change rate of ascending intersection point representing node-line precession within one orbit is

$$\dot{\Omega} = -\frac{3nJ_2 r_e^2}{2a^2(1-e^2)^2}\cos i \tag{3.53}$$

where n is average orbital angular velocity $n = \sqrt{\frac{\mu}{a^3}}$, μ is gravitational constant $\mu = 3.8960044 \times 10^5\ \text{km}^3\ \text{s}^{-2}$, $J_2 = 0.001082$, $r_e = 6378.145$ km, the unit of $\dot{\Omega}$ is rad/s.

For circular orbits, the above equation can be rewritten as the change in a day.

$$\Delta\Omega = -9.964\left(\frac{R_e}{a}\right)^{7/2}\cos i \tag{3.54}$$

where the unit $\Delta\Omega$ of is °/d. Obviously, if the orbital angle $i < 90°$, then $\dot{\Omega} < 0$, the orbit is the westward orbit. For $i > 90°$, then $\dot{\Omega} > 0$, the orbit is the eastward orbit, also known as retrograde orbit (contrary to the rotation of the earth).

If the combination of orbit semi-major axis a and inclination angle i is selected to make the satellite $\Delta\Omega = 0.985647°/\text{d}$, then the direction and rate of orbit precession are the same as the direction and rate of the earth's annual rotation around the sun (i.e., after 365.2425 mean solar day, the earth completes an annual movement of 360°). This specially designed orbit is called solar synchronous orbit.

Substitute formula 3.54, μ, J_2, R_e, into formula 3.53, and obtain:

$$\begin{aligned} \cos i &= -4.773724 \times 10^{-15} a^{7/2} \\ a_{\max} &= 12352.5\ \text{km} \end{aligned}$$

The maximum orbital height is about 5980 km, but this has no practical application. As can be seen from Fig. 3.22, when $\cos i = -1$, there is no precession moment **M**, so the maximum theoretical value of the height of the sun-synchronous orbit relative to the average radius of the earth does not exceed 5980 km.

The main characteristic of the sun-synchronous orbit is that the direction of the orbit plane from which the sun shines remain about a constant throughout the year. Precisely, the angle between the normal of the orbital plane and the projection of the sun's direction on the equatorial plane remains constant, that is, the local time at which the satellite passes the equatorial node does not change, as shown in Fig. 3.23.

In addition, low-earth sun-synchronous orbits are close to polar orbits, and the orbit and rotation of the earth allow satellites to fly over all parts of the globe except the very small polars regions. This type of orbit is particularly suitable for the earth

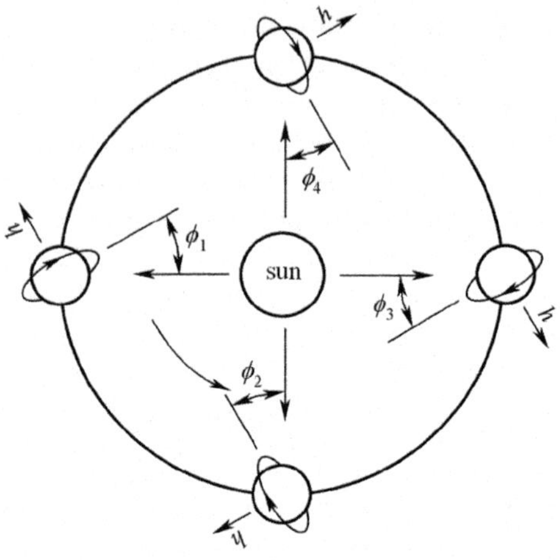

Fig. 3.23 Solar synchronous orbit annual illumination illustration

remote sensing satellites in low-earth orbit, and the main advantage is the minimum annual variation of the following important technical parameters. The parameters are

(1) Sun shining angle of satellite;
(2) Solar energy received;
(3) The local mean solar time of a substellar point at the same latitude;
(4) The illuminance of a sub-satellite point at the same latitude;
(5) Earth shadow time.

The first two items are directly determined by the orbital solar angle β (the angle between the orbital normal and the line of sight in the direction of the sun), as shown in formula 3.49.

$$\beta = \text{arcos}(\sin\,\Omega_s \cos\,\delta_s \sin\,i + \sin\,\delta_s \cos\,i) \tag{3.55}$$

where Ω_s refers to local mean solar hour angle of the ascending point ($\Omega_s = \Omega - \alpha_s$), Ω_s is constant for sun-synchronous orbit, and the value is determined by the launch window. In Eq. 3.55, only solar declination δ_s is a variable ($-23.5° \leq \delta_s \leq +23.5°$), so the orbital solar angle only varies with the season.

Items 3 and 4 are determined by the zenith angle of the substellar point η (the solar altitude angle of the substellar point $= 90° - \eta$)

$$\eta = arcos[\cos\varphi \cos\delta_s \cos(\Omega_s + \Omega_C) + \sin\varphi \sin\delta_s] \tag{3.56}$$

where φ is the latitude of the specified latitude circle, and Ω_C is the included angle between the node of the longitude circle where the satellite is located and the ascending intersection point, which is determined by formula 3.51.

Similarly, for sun-synchronous orbits, changes in local mean solar time and illuminance at sub-satellite points when the satellite crosses a given circle of latitude depend only on the season.

The circular orbit height and inclination angle of solar synchronous orbit remote sensing satellite are shown in the table below.

Altitude (km)	Inclination (°)
500	97.4
700	98.2
900	99.0
1000	99.5

Satellites with high inclination pass over the earth's polar regions, so sun-synchronous orbits are also called polar orbits.

Because the precession of the satellite orbit is synchronized with the sun's apparent motion, the variation of the earth shadow time is also minimal. If the orbital ascending intersection point is 6:00 am/pm, the earth shadow time is the shortest.

3.5 Critical Inclination and Frozen Orbit

The mass of the earth is distributed in the part of the bulge near the equator, so the effect of J_2 on satellite orbit has two effects. The first effect is to synchronize the precession of the satellite's orbit with the annual motion of the sun. The second effect, it causes the orbital arch to rotate in the orbital plane. It also causes changes in eccentricity. For orbit perturbation, if only the effect of J_2 is considered, the rate of change of perigee amplitude angle ω and eccentricity e is

$$\begin{aligned} \dot{\omega} &= -\frac{3nJ_2r_e^2}{2a^2(1-e^2)^2}\left(\frac{5}{2}\sin^2 i - 2\right) \\ \dot{e} &= 0 \end{aligned} \tag{3.57}$$

where r_e is the equatorial radius of the earth.

The rotation of the arch leads to the continuous change of the altitude of the satellite passing through the same latitude, which seriously affects the satellite application mission. For example, a large elliptical orbit is adopted for communication satellites serving high latitudes, with apogee altitude of 39,420 km, eccentricity of 0.72 and orbital period of 12 h. If the apogee is required to always be above the north pole, that is, the arch line must not rotate, then the orbital inclination of formula 3.57 should satisfy the formula:

$$\frac{5}{2}\sin^2 i - 2 = 0$$

Namely, $i = 63.43°$ ($i = 116.565°$). This inclination is called critical inclination, and such orbits are called critical orbits.

For the remote sensing satellite in low-earth orbit, the inclination of 63.4349° does not meet the inclination condition of sun-synchronous orbit. In theory, 116.565° meets the inclination condition of sun-synchronous orbit. If it is a circular orbit, the orbital height is 3445.7 km, and the spatial resolution of such high orbital images will be reduced. Only using this orbital height also has great restrictions on remote sensing satellites. If the effect of the higher-order perturbation term J_3 on the orbital elements ω and e is considered:

$$\dot{e} = \frac{3nJ_3r_e^3 \sin i}{2a^3(1-e^2)^2}\left(\frac{5}{4}\sin^2 i - 1\right)\cos\omega \tag{3.58}$$

$$\dot{\omega} = -\frac{3nJ_2r_e^2}{a^2(1-e^2)^2}\left(\frac{5}{4}\sin^2 i - 1\right)\left[1 + \frac{J_3r_e}{2J_2a(1-e^2)}\left(\frac{\sin^2 i - e^2\cos^2 i}{\sin i}\right)\frac{\sin\omega}{e}\right] \tag{3.59}$$

where $J_3 = -2.5356 \times 10^{-6}$.

To make the arch line not rotate, just make $\dot{e} = 0$ and $\dot{\omega} = 0$. According to formula 3.58, obviously $\dot{e} = 0$, $\omega = 90°$.

To make $\dot{\omega} = 0$, since $\omega = 90°$, then the square bracket items in equation are zero in formula 3.59. Then,

$$\frac{e^2\cos^2 i - \sin^2 i}{\sin i} = \frac{2J_2a(e - e^3)}{J_3r_e}$$

Then,

$$e^3 + \frac{J_3r_e}{2J_2a}\frac{\cos^2 i}{\sin i}e^2 - e - \frac{J_3r_e}{2J_2a}\sin i = 0 \tag{3.60}$$

If satellite remote sensing also requires that the eccentricity is small and constant, then the e^3 term is ignored.

$$pe^2 - e - q = 0$$

where

$$p = \frac{J_3r_e}{2J_2a}\frac{\cos^2 i}{\sin i}$$

$$q = \frac{J_3r_e}{2J_2a}\sin i$$

$$e = \frac{1 - \sqrt{1+4pq}}{2p}$$

Since $\dot{\omega} = \dot{e} = 0$, namely argument of perigee is held, or frozen at 90°, such orbit is called frozen orbit. But the inclination and altitude of the orbit can be chosen independently. For example, ocean satellite applications require $H = 800$ km, $i = 108°$, where the parameters related to frozen orbit are $\omega = 90°$, $e = 0.0009$.

Formula 3.59 is the expression for the slow change of orbit after short period perturbation, which has been ignored. For further elaboration, please refer to [9].

3.6 Regressive Orbit

The sun-synchronous orbit solves the problem of the stability of the sub-satellite illuminance of optical remote sensing satellites and basically achieves global coverage. The critical or frozen orbit solves the problem of the same spatial resolution of sub-satellite points at the same latitude, and the distance of adjacent tracks in the same latitude circle is the same. However, for remote sensing satellites, which are generally used for civil purposes, such as land and ocean satellites, circular orbits are adopted to ensure the same spatial resolution of sub-satellite images. In order to facilitate the application of satellite observation and the post-processing and application of remote sensing data, it is also expected that the sub-satellite point track (geographic coordinates) of the satellite repeats periodically. The orbit characteristic of this kind of satellite is that the sub-satellite point trajectory returns to the original path after a certain period of time, and this kind of orbit is called regression orbit or cyclic orbit.

An important analysis of the operation design of optical remote sensing satellite is the problem of orbit coverage. Generally, the following conditions should be met:

(1) Sun-synchronous orbit;
(2) Appropriate target illumination;
(3) Global coverage (except pole regions);
(4) Recursive orbit;
(5) In general, the spatial resolution is the same.

For a circular orbit satellite, the transversal movement of sub-satellite trajectories on the ground is a combination of earth rotation, orbital nodal line precession and satellite orbital motion. The traverse angle of the sub-satellite point trajectory across the equator within one orbit, that is, the interval of successive adjacent tracks on the equator, $\Delta\lambda$, is shown in Fig. 3.24.

$$\Delta\lambda = T_N\left(\omega_e - \dot{\Omega}\right)$$

where ω_e refers to rotational angular velocity of the earth, $\dot{\Omega}$ refers to the average precession rate of the orbital nodal line, T_N refers to node period of orbital motion.

T_N includes the action term of orbital average speed n and earth flat perturbation J_2, and its formula is (when $e = 0$)

$$T_N = 2\pi\sqrt{\frac{a^3}{\mu}}\left[1+\frac{3J_2r_e^2}{2a^2}\left(1 - 4\cos^2 i\right)\right]$$

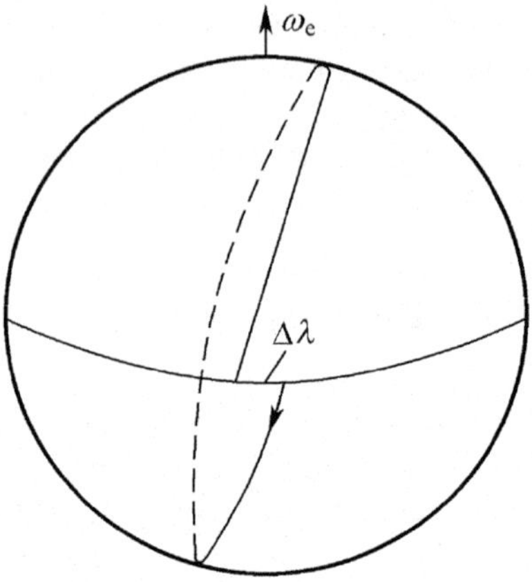

Fig. 3.24 Transverse angle of the sub-satellite trajectory at the equator

For example, the semi-major axis and inclination angle of the orbit are selected so that the orbital period satisfies the following formula:

$$R\ T_N\left(\omega_e - \dot{\Omega}\right) = R \cdot \Delta\lambda = 2\pi$$

where R is a positive integer.

Then, the regression period of this orbit is one day, and the number of orbit turns in a day is R.

Low-earth satellite orbit height is usually 600–1000 km. The loop number of a satellite in a day can be 13, 14 or 15 orbits. For a one-day regression orbit, the angular distance between adjacent orbits is 27.7°, 25.7° or 24°, and the distance between adjacent orbits is about 2850 km, which far beyond the wide coverage range of earth observation by remote sensing instruments on the satellite.

Therefore, in order to achieve global coverage and adapt to the field of view of remote sensing instruments, multiday regression orbit is usually used. That is, the semi-major axis and inclination angle of the designed orbit, so that the orbital period T_N satisfies the following formula:

$$R\ T_N(\omega_e - \Omega) = R\Delta\lambda = 2\pi\ N$$

or

$$R\ T_N = N\ D_N$$

$$D_N = \frac{2\pi}{\omega_e - \dot{\Omega}}$$

called node day.

In the formula, R and N are positive integers.

The formula above means that the orbit repeats once after N days and makes a total revolution of R in the regression cycle. The number of orbit turns per day is a non-integer Q, which is defined as

$$Q = \frac{R}{N} = \frac{2\pi}{\Delta\lambda} = I \pm \frac{C}{N} \tag{3.61}$$

Q is called regression coefficient, which is composed of integer and fraction. C is another positive integer. I, C and N constitute the three elements representing the regression orbit.

The phase shift angle of the I-th trajectory relative to the initial trajectory is denoted as α, obtained from formula 3.61.

$$2\pi = \left(I \pm \frac{C}{N}\right)\Delta\lambda = \Delta\lambda I \pm \frac{\Delta\lambda C}{N}$$

$$\alpha = \pm\frac{\Delta\lambda C}{N} = 2\pi - I\Delta\lambda$$

For example, if $\Delta\lambda C/N$ is "+", it indicates that the orbit moves eastward after a day; If $\Delta\lambda C/N$ is "−", it indicates that the orbit is moving westward after one day. In both cases, the phase shift trajectory of each passing day is inserted within the interval $\Delta\lambda$ of successive adjacent trajectory. N-day coverage also shows that the interval $\Delta\lambda$ of each successive adjacent trajectory is divided into N regions by the trajectory passing within N days. The angle of each interval is called the coverage angle, denoted by γ.

$$\gamma = \frac{\Delta\lambda}{N}$$

γ is the distance between any adjacent orbitals. According to the performance of the remote sensing instrument on board, the periodical coverage days N is selected so that the coverage angle γ of adjacent tracks is less than the observation width of the instrument.

The orbital period T_N is designed so that the orbital phase shift angle α is equal to the width angle γ, or an integer multiple of γ, so:

$$\alpha = C\gamma$$

From $RT_N = \mathrm{ND}_N$, $T_N = D_N N/R$ can be gotten, substitute into formula 3.61, further, the design formula of orbital period can be obtained.

$$T_N = \frac{D_N}{I \pm \frac{C}{N}}$$

The above equation shows that the period of the designed regression orbit is not only determined by the number of days requiring global coverage and the number of orbit circles per day, but also related to the selection of positive integer *C*. The integer value of *C* and its prefacing symbol "+" or "−" determines the coverage pattern within each successive adjacent orbital interval $\Delta\lambda$.

If $C = 1$, then the phase shift angle of one-day orbit is equal to the coverage angle, which is continuous coverage, that is, tracks passing $\Delta\lambda$ interval are continuously arranged according to the number of dates within *N* days, forming continuous coverage according to the date.

If $C > 1$, the phase shift angle of the daily trajectory is a multiple of the coverage angle. Within *N* days, the tracks through $\Delta\lambda$ interval are no longer arranged continuously by date number, forming intermittent coverage. In the denominator of the preceding equation, if "+" is taken, the trajectory moves east; If the "−" sign is taken, the trajectory shifts to the west.

For example, let regression orbit and sun-synchronous orbit, the coverage period is $N = 7$ days, and let regression coefficient $Q = 14 + 1/7$, $Q = 14 + 3/7$ and $Q = 14–3/7$ as three examples. The orbital height *H* and other orbital parameters are shown in Table 3.1.

Figure 3.25a–g shows the track of the start and end points of each day and *C*/*N* part of the distribution range, and Fig. 3.25h–j is a regression cycle track global coverage and the characteristics of the 1/*N* orbit by date arrangement, from the above chart can induce 1/*N* orbit by date as shown in Fig. 3.30.

Figure 3.26 illustrates the orbital alignment of a regression cycle when $Q = 14 + 3/7$. Figure 3.27 illustrates the track order by date in the +3/7 case. From the above two figures, the arrangement rule of $+C/N$ orbit is obtained by date in Fig. 3.31.

Figure 3.28 illustrates the order of tracks arranged within the first day with $Q = 14–3/7$.

Figure 3.29 illustrates the order of tracks arranged by date when $Q = 14–3/7$. The arrangement rule of $-C/N$ track by date can be summarized in Fig. 3.32.

What is important in the above orbital arrangement is the relative date arrangement of the orbits. It represents the situation of revisiting that optical remote sensing satellites can achieve, where revisiting means to see again, rather than recursive.

For the "No. orbit/No. day" in the preceding pictures, the date arrangement is obtained on the descending arrangement at equator under the condition of orbit entry at the ascending equator (declination equals zero). If the declination at the time of orbit entry changes, the date arrangement of the orbit will also change, and this requires specific analysis.

Table 3.1 Regression orbit parameter table

Parameter	*Q*	$\Delta\lambda$/°	γ/km	TN/min	H/km	*i*/°
1	14 + 1/7	25.45	404.80	101.7	839.2	98.7
2	14 + 3/7	24.95	396.78	99.7	743.4	98.3
3	14–3/7	26.53	421.84	106.0	1040.8	99.6

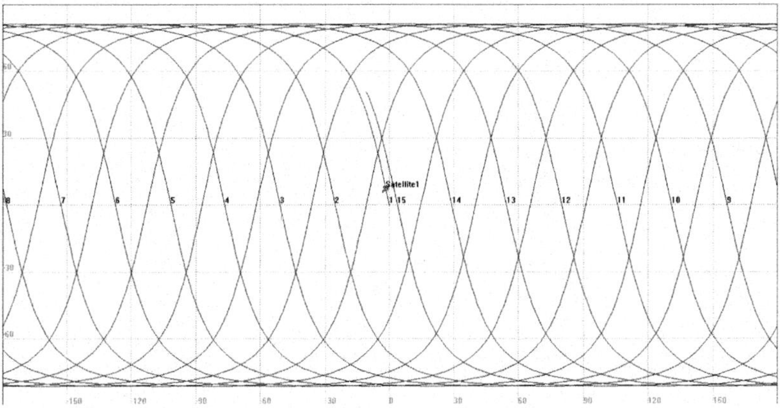

(a) Q=14+1/7 track alignment on day 1

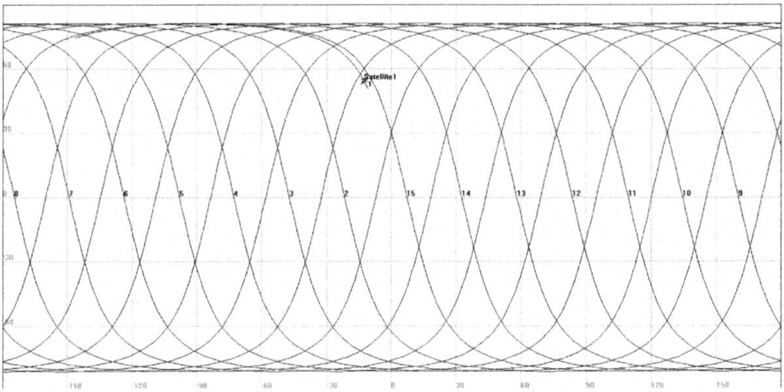

(b) Q = 14 + 1/7 track alignment on day 2

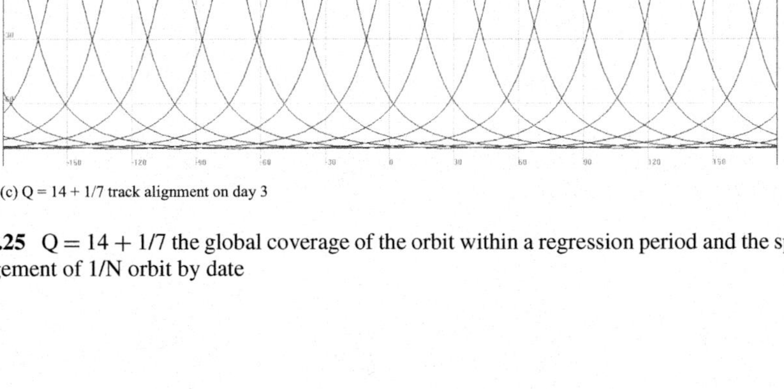

(c) Q = 14 + 1/7 track alignment on day 3

Fig. 3.25 Q = 14 + 1/7 the global coverage of the orbit within a regression period and the specific arrangement of 1/N orbit by date

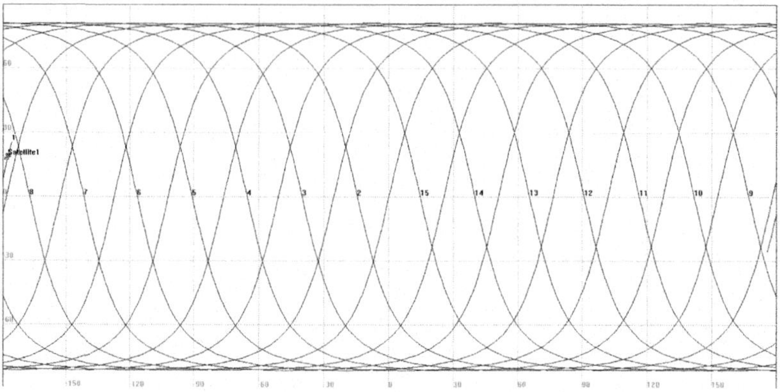

(d) Q = 14 + 1/7 track alignment on day 4

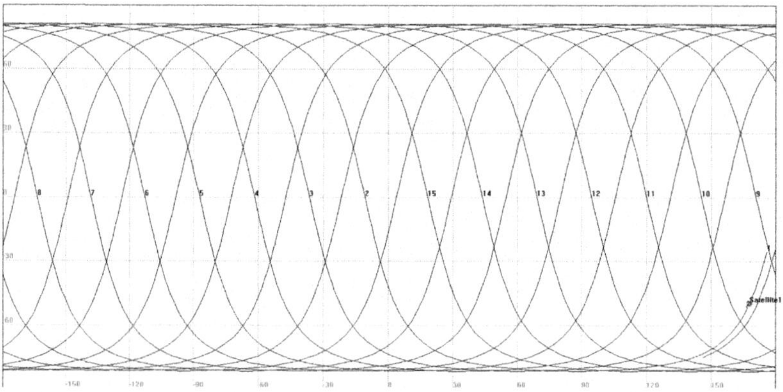

(e) Q = 14 + 1/7 track alignment on day 5

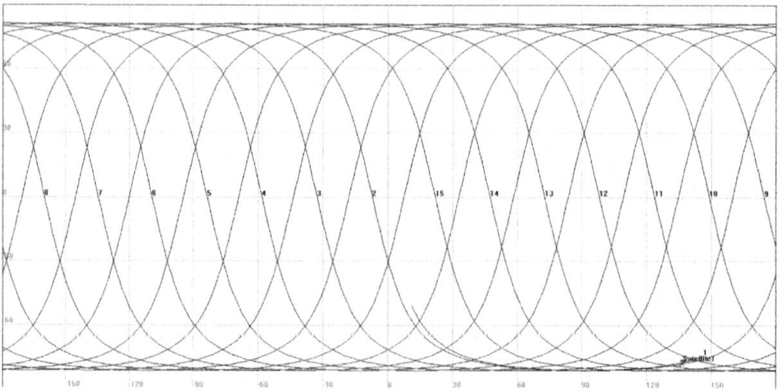

(f) Q = 14 + 1/7 track alignment on day 6

Fig. 3.25 (continued)

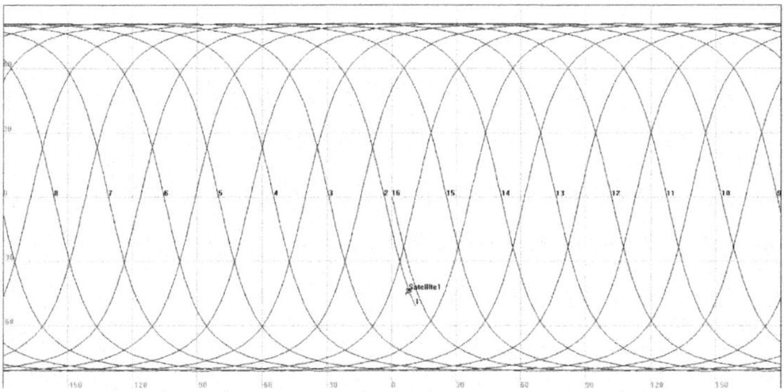

(g) Q = 14 + 1/7 track alignment on day 7

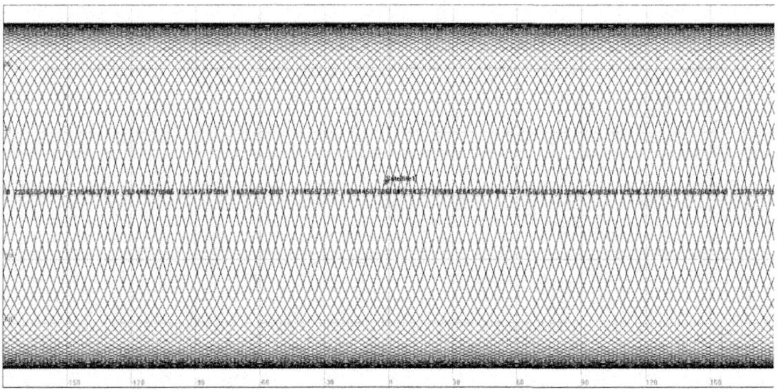

(h) Q = 14 + 1/7 A regression periodic orbital arrangement

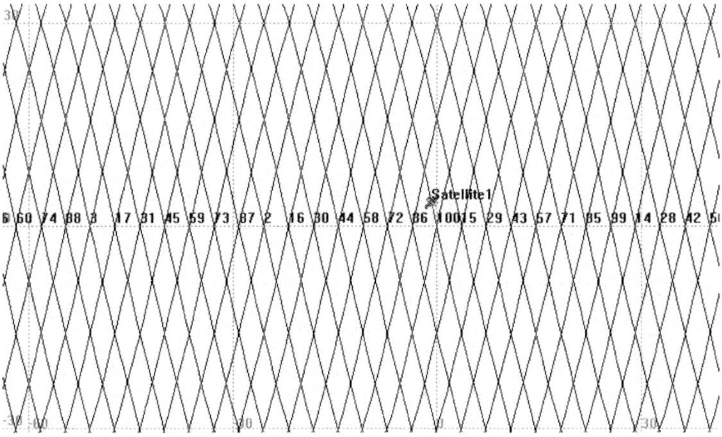

(i) Q = 14 + 1/7 A regression periodic orbital arrangement (Local amplification)

Fig. 3.25 (continued)

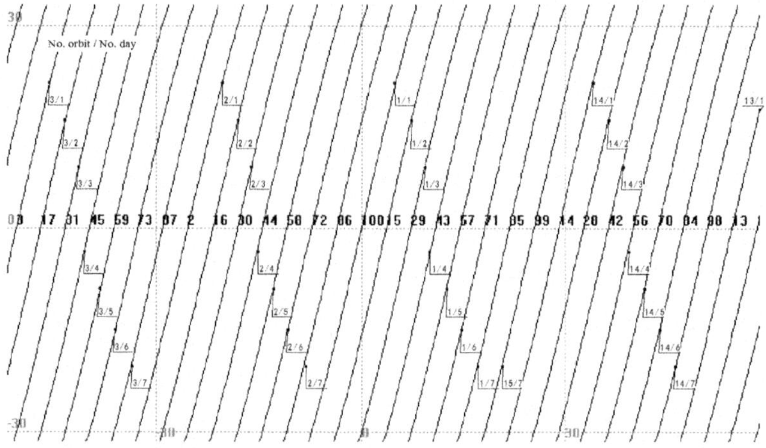

(j) Q = 14 + 1/7 A regression period to arrange track order by date (local amplification)

Fig. 3.25 (continued)

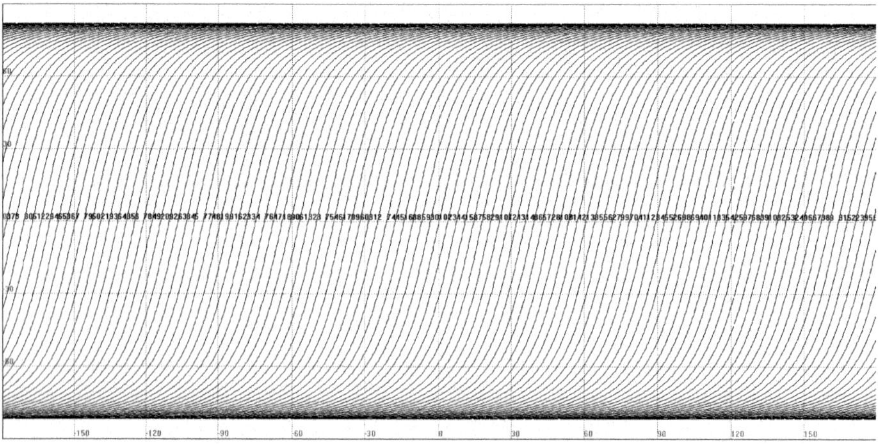

Fig. 3.26 $Q = 14 + 3/7$ descending orbit arrangement in a regression period

3.7 Geosynchronous Orbit

In the twentieth century, spaceborne optical remote sensing was limited by the manufacturing capacity of optical sensors, so solar synchronous orbit was mainly concerned. With the improvement of optical manufacturing ability and space resolution, geosynchronous orbit has also entered the application category of optical remote sensing to earth. It is feasible to achieve optical remote sensing in geostationary orbit with a spatial resolution better than 21.5 m at the current international optical remote sensing level [10]. The advantages of synchronous orbit for optical remote sensing

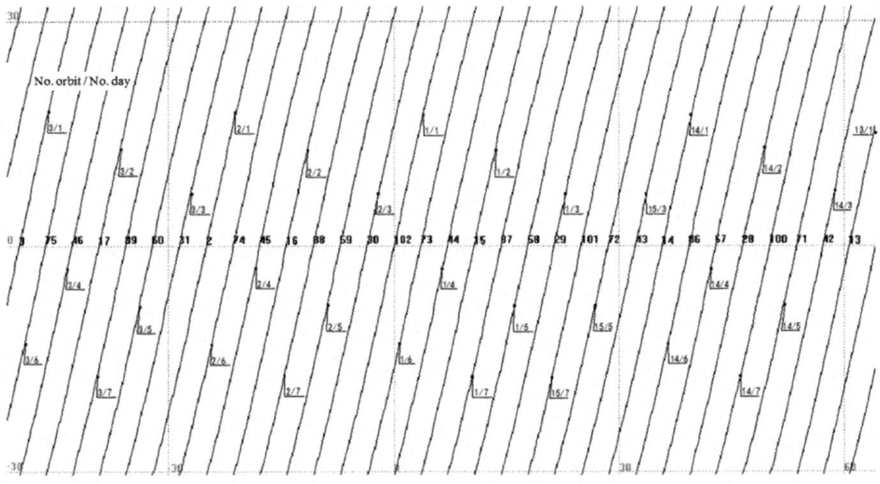

Fig. 3.27 $Q = 14 + 3/7$ a regression period to arrange track order by date (local amplification)

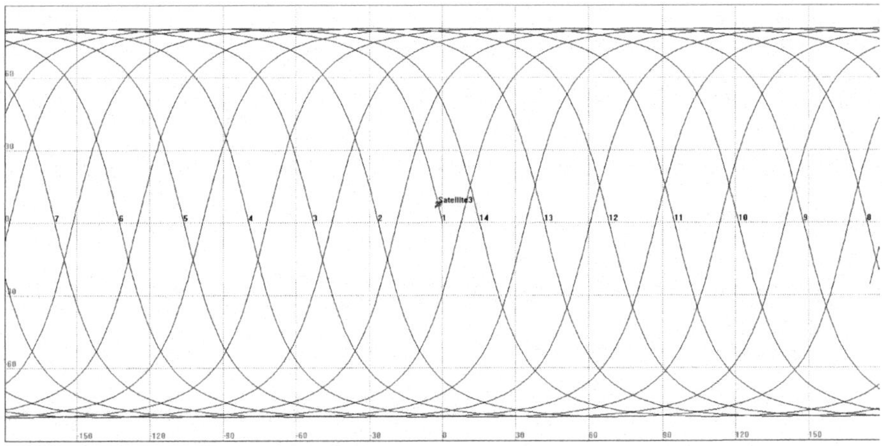

Fig. 3.28 $Q = 14 - 3/7$ track alignment on day 1

are high temporal resolution and low spatial resolution, which is complementary to the low temporal resolution and high spatial resolution of optical remote sensing in LEO.

The main characteristic of geosynchronous orbit is that the earth spins once in a sidereal day, and the spacecraft rotates once in a sidereal day around the earth in the same direction, that is, the two are synchronous, and if only the orbital period is equal to the earth's rotation period, this kind of orbit is called geosynchronous orbit. This is because they have the same period, and the synchronous orbit can have inclination $i \neq 0$ and eccentricity $e \neq 0$.

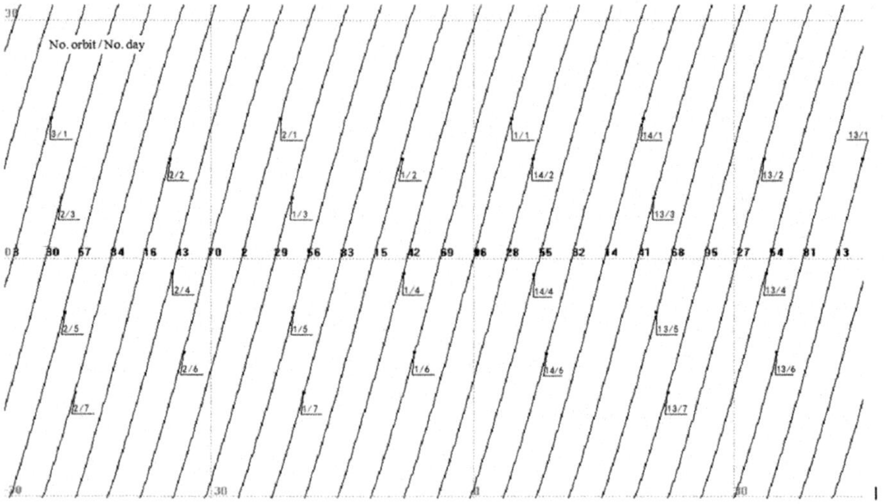

Fig. 3.29 $Q = 14 - 3/7$ a regression period to arrange track order by date (local amplification)

Fig. 3.30 + 1/7 the characteristic of tracks arranged by date

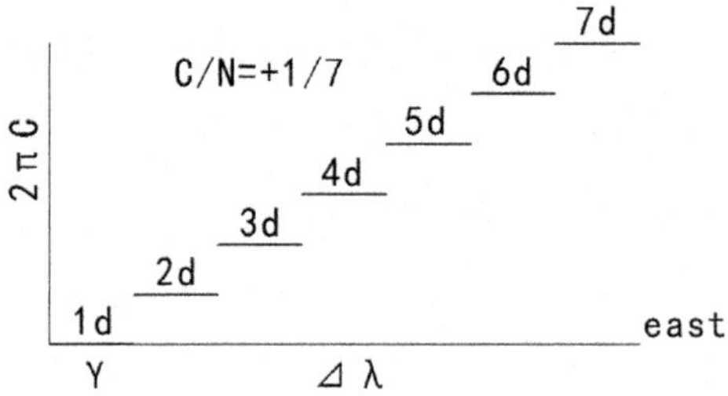

Fig. 3.31 + 3/7 the characteristic of tracks arranged by date

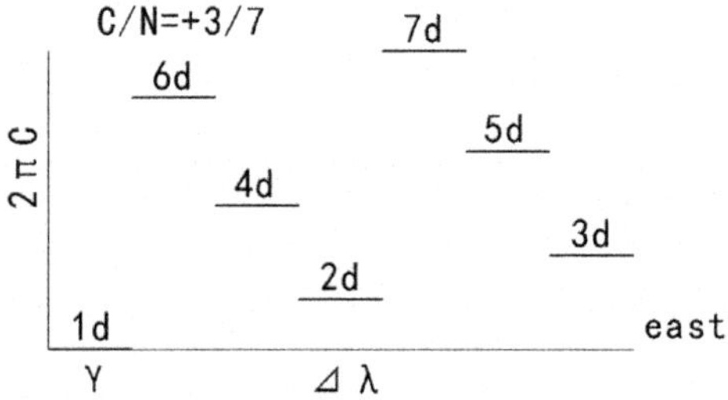

Fig. 3.32 − 3/7 the characteristic of tracks arranged by date

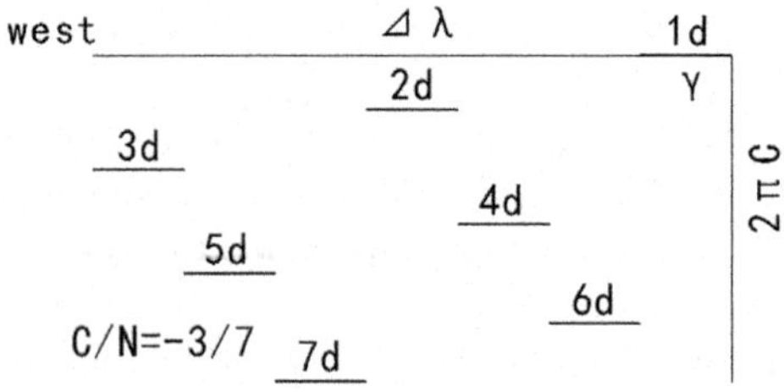

That is, geosynchronous orbit only limits the orbital period of the satellite to equal the rotation period of the earth, $T = 2\pi/\omega_e$.

If the geosynchronous orbit $i = 0$, $e = 0$, it is converted to a stationary orbit.

3.7.1 Geostationary Orbit

Geostationary orbit that is

(1) The period of a satellite's orbit is equal to the rotation period of the earth, $T = 2\pi/\omega_e$.
(2) The shape of the orbit is circular and the eccentricity $e = 0$.
(3) The orbit is in the equatorial plane of the earth with an inclination $i = 0$.

According to the law of universal gravitation, when the spacecraft and the earth are analyzed according to the two-body problem and $m_e \gg m$, the gravity of the earth on the spacecraft is

$$f = \frac{Gm_e m}{r^2} = \frac{\mu m}{r^2}$$

When the centrifugal force of the spacecraft flying around the earth at the speed of the earth's rotation ω_e is equal to the gravitational force f, can form a stationary orbit, so

$$f = \frac{\mu m}{r^2} = mr\omega_e^2$$

$$r = \left(\frac{\mu}{\omega_e^2}\right)^{1/3} = \left(\frac{398600.44}{7.2921158^2 \times 10^{-10}}\right)^{1/3} = 42164.17$$

That is, the geostationary orbit radius satellite 42,164.17 km.

The orbit with parameters $r_s = 42{,}164.17$ km, $i = 0$, $e = 0$ is geostationary orbit. A satellite in such an orbit is stationary at its sub-satellite position (latitude and longitude), and its azimuth and pitch angle relative to the observer on the ground are also constant. But a satellite in a geostationary orbit is not completely stationary with respect to the earth because of various disturbances.

The so-called earth's cycle of one rotation in a day is 24 h, which is measured by the local mean solar day, that is, the time interval between the mean sun continuously passing the same meridian twice. During this time, the earth moved an angle along the ecliptic (see Chap. 2 sidereal days), so the time interval is not equal to the time it takes the satellite to rotate once in the space. The time interval between the vernal equinox continuously passing through the same meridian twice is the earth's rotation period. This time interval is called sidereal day. The difference between the two is that the transit occurs when the sun is $4'$ behind the vernal equinox after one day, so the mean solar day is larger than the sidereal day, namely the rotation period of the earth is less than 24 h. According to astronomy, there are 365.2425 mean solar days and 366.2425 mean sidereal days in a regression year, namely

$$1 \text{ sidereal day} = (365.24/366.24) \cdot \text{ Solar day} = 23\text{h}56'4.09'' = 86164.09\,\text{s}$$

So the angular velocity of the earth's rotation is

$$\omega_e = 2\pi/\text{sidereal day} = 7.2921158 \times 10^{-5}\,\text{rad/s} = 360.9856\,°/\text{d}$$

The velocity of the geostationary orbit is

$$v_s = r_s \cdot \omega_e = 3.0747\,\text{km/s}$$

3.7.2 Several Typical Geosynchronous Orbits

In fact, there is no ideal stationary orbit. A satellite is subjected to various perturbations in orbit, such as the irregular shape, uneven distribution of density can make the satellite gravity changes, the attraction of the sun, the moon for the satellite and the solar radiation pressure on satellite, etc. All these make the inclination and eccentricity of the actual orbit change a little, and the orbital period is not completely synchronous with the earth. Seen from the ground, the satellite is not stationary but always drifting in the east–west longitude and north–south latitude directions.

Because the orbital period T, eccentricity e and inclination i are independent, there are four typical cases of whether they are equal to zero.

(1) The orbit is a synchronous circular orbit with a small inclination, namely $i \neq 0$, $e = 0$, $\Delta a = 0$.

As shown in Fig. 3.33, when the satellite passes through node N, it has $G_{0,}$ and its geocentric longitude is λ_N. This is the starting point t_0. After time t, when the satellite reaches point S, it turns the amplitude angle U and the Greenwich meridian turns $\omega_e t$, but the node N is fixed in inertial space. Using spherical triangle $\triangle$NDS, the geocentric longitude and latitude of the satellite can be obtained:

$$\begin{aligned} \lambda &= \lambda_N + \arctan(\cos\, i\ \tan\, u) - \omega_e t \\ \varphi &= \arcsin(\sin i \sin u) \end{aligned} \tag{3.62}$$

Let $\Delta\lambda = \lambda - \lambda_N$

$$\begin{aligned} \tan(\Delta\lambda + \omega_e t) &= \cos\, i\ \tan\, u \\ \cos\, i &= \tan(\Delta\lambda + \omega_e t)/\tan\, u \end{aligned}$$

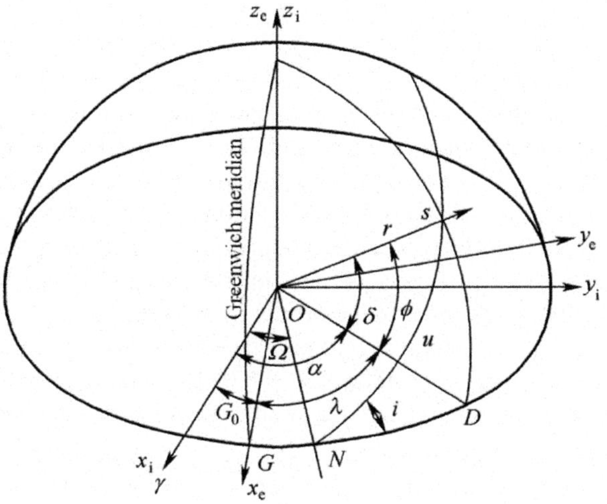

Fig. 3.33 Geosynchronous orbit parameter elements

Subtract 1 from both sides,

$$\begin{aligned}\cos i - 1 &= \frac{\sin(\Delta\lambda + \omega_e t)}{\cos(\Delta\lambda + \omega_e t)} \frac{\cos u}{\sin u} - 1 \\ &= \frac{\sin(\Delta\lambda + \omega_e t)\cos u - \cos(\Delta\lambda + \omega_e t)\sin u}{\cos(\Delta\lambda + \omega_e t)\sin u} \quad (i \text{ small enough } u \approx \omega_e t) \\ &= \frac{\sin\Delta\lambda}{\cos(\Delta\lambda + \omega_e t)\sin\omega_e t}\end{aligned}$$

When i is very small, $\Delta\lambda$ is also small.

$$\cos(\Delta\lambda + \omega_e t) \approx \cos\omega_e t$$
$$\sin\Delta\lambda \approx \Delta\lambda$$

The series expansion of $\cos i$

$$\cos i = 1 - \frac{i^2}{2} + \frac{i^4}{24} - \cdots$$

When i is small

$$\cos i \approx 1 - \frac{i^2}{2}$$

Then,

$$-\frac{i^2}{2} \approx \frac{\Delta\lambda}{\cos\omega_e t \sin\omega_e t} = \frac{2\Delta\lambda}{\sin 2\omega_e t}$$

$$\Delta\lambda \approx -\frac{i^2}{4}\sin 2\omega_e t$$

It can be seen from the above equation and $\varphi = \arcsin\,(\sin i\ \sin\omega_e t)$ that the longitude changes periodically with respect to the reference point $\Delta\lambda = \lambda - \lambda_N$ during one orbit of the satellite because of the orbital inclination, and the satellite drifts back and forth in the directions of east and west, north and south. The combined motion of the two causes the drift track to be a Fig. 3.8 in the local horizontal plane, as shown in Fig. 3.34. The maximum latitude of the Fig. 3.8 in the north–south direction is equal to the orbital inclination.

The maximum deviation of the east–west direction caused by the inclination angle is $i^2/4$. If $i = 1°$, then $\Delta\lambda = 0.0044°$. The longitude periodic drift caused by this small inclination is much smaller than inclination.

(2) The orbit is an unsynchronous equatorial circle, namely $i = 0$, $e = 0$, $a = r_s + \Delta a$.

Such orbit satellites will drift globally along the equator, but the drift period is long, and they have global observation capability (except for the high latitudes at the poles). However, compared with the global observation of solar synchronous orbit, they have

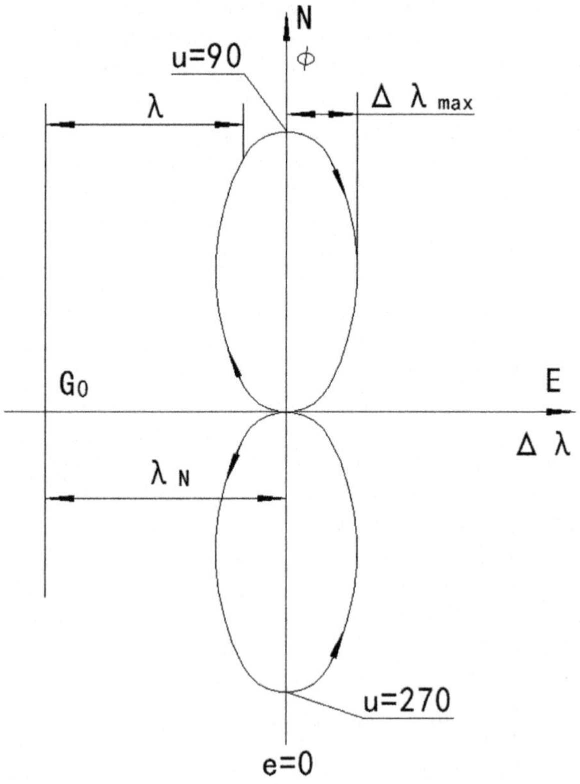

Fig. 3.34 Geosynchronous orbit "Fig. 3.8" relative track

no advantages in space resolution and time resolution, so they are listed only as a class of synchronous orbit reference.

The semi-major axis of the orbit a is not equal to the geosynchronous radius $r_s = 42{,}164.17$ km. If $a > r_s$, the orbital angular velocity n is less than synchronous speed ω_e, and the satellite drifts westward. If $a < r_s$, then the orbital angular velocity n is greater than the synchronous speed ω_e, and the satellite drifts eastward. The differential of the orbital angular velocity formula can be obtained from the satellite orbital speed law $n^2 = \mu/a^3$.

$$n = \omega_e + \Delta n = \omega_e(1 - 1.5\Delta a/r_s) \tag{3.63}$$

Approximate formula of mean anomaly,

$$M = n\left(t - t_{\mathrm{p}}\right) = \omega_e\left(t - t_{\mathrm{p}}\right)(1 - 1.5\,\Delta a/r_s)$$

where t_{p} is perigee time. According to the above equation, the non-synchronous drift amount of satellite caused by semi-major axis increment Δa in one day is obtained.

$$M = -0.013\,\Delta a\ (^\circ/\mathrm{km}).$$

(3) The orbit is an unsynchronous small eccentricity equatorial orbit, namely $i = 0$, $e \neq 0$, $\Delta a \neq 0$.

Such orbiting satellites will drift globally along the equator, but with long drift periods, similar (2). Except for special consideration, it does not have the advantage of space optical remote sensing and is only listed as a class of synchronous orbit.

According to the satellite radial distance formula:

$$r = \frac{a\left(1 - e^2\right)}{1 + e\cos f}$$

It can be linearized to the following formula:

$$r \approx a(1 - e\cos f) \approx r_s + \Delta a - r_s e\cos f$$

According to the satellite moment of momentum formula,

$$r^2 \cdot \dot{\theta} = h,\quad h^2 = \mu a$$

The approximate formula of the rate of change of true anomaly is obtained. ($\Delta a \ll r_{\mathrm{s}}$, $e \ll 1$).

$$\begin{aligned}\dot{f} &= \sqrt{\frac{\mu}{a^3}}(1 + e\cos f)^2 \\ &= \omega_e(r_s/a)^{3/2}(1 + e\cos f)^2\end{aligned}$$

$$\approx \omega_e\left(1 - \frac{3\Delta a}{2r_s}\right)(1 + 2e\cos f)$$

$$\approx \omega_e\left(1 - \frac{3\Delta a}{2r_s} + 2e\cos f\right)$$

When $\Delta a \ll r_s$, $e \ll 1$, the following approximation formula can be gotten:

$$\cos f \approx \cos M \approx \cos\left[\omega_e\left(t - t_\mathrm{p}\right)\right]$$

Therefore, the approximate integral formula of true anomaly f of the above equation (the integral is t_p initially) can be written as

$$f = \int_{t_\mathrm{p}}^{t} \omega_e\left(1 - \frac{3\Delta a}{2r_s} + 2e\cos f\right)\mathrm{d}t$$

$$= \omega_e\left(t - t_\mathrm{p}\right)\left(1 - \frac{3\Delta a}{2r_s}\right) + 2e\sin\left[\omega_e\left(t - t_\mathrm{p}\right)\right] \tag{3.64}$$

The satellite's geocentric longitude λ is equal to the satellite's sidereal hour angle (i.e., right ascension) minus the Greenwich sidereal hour angle. By referring to orbital elements, the following approximate formula can be obtained ($i \ll 1$).

$$\lambda = \Omega + \omega + f - [G_0 + \omega_e(t - t_0)]$$

where G_0 is the Greenwich sidereal hour angle at t_0.

Substitute formula 3.64 into the above formula,

$$\lambda = \lambda_0 - \frac{3\Delta a}{2r_s}\omega_e(t - t_0) + 2e\sin\left[\omega_e\left(t - t_p\right)\right]$$

where $\lambda_0 = \Omega + \omega + \omega_e\,(t_0 - t_\mathrm{p})\,(1 - 1.5\,\Delta a/r_s) - G_0$. It's called the mean longitude of the satellite at time t_0.

The deviation motion equation of satellite relative to mean longitude caused by eccentricity and semi-major axis deviation is

$$\Delta r = \Delta a - r_s\, e\, \cos\left[\omega_e\left(t - t_p\right)\right] \tag{3.65}$$

$$\Delta x = -1.5\Delta a\omega_e(t - t_0) + 2r_s\, e\, \sin\left[\omega_e\left(t - t_\mathrm{p}\right)\right] \tag{3.66}$$

where Δx represents tangential deviation relative to mean longitude.

The above equation shows that in the case of synchronous orbital period ($\Delta a = 0$), the eccentricity causes the satellite to move away from the fixed position and enter an elliptical track around the mean longitude (fixed position). It has a period of one day, and its long axis is in the east–west direction, with a length of $4r_s e$, and its short axis is in the radial direction, with a length of $2r_s e$ as shown in Fig. 3.35 (E for east).

The east–west drift of satellite longitude caused by eccentricity is $2e$, if $e = 10^{-3}$, then $\Delta\lambda = 0.11°$. The semi-major axis deviation causes the center of the ellipse to drift in an east (or west) direction, as shown in Fig. 3.36.

(4) Satellite orbit is a geosynchronous orbit with small inclination and ellipticity, namely $i \neq 0$, $e \neq 0$, $\Delta a = 0$.

In formulas (3.62), (3.68) and (3.66), $\Delta a = 0$, $\sin i \approx i$, $M = \omega_e\ (t - t_{\rm p})$ at a small inclination angle, and the motion equation of the satellite deviating from the fixed-point position is

$$\begin{aligned} \Delta r &= -r_s\, e \cos M \\ \Delta x &= 2r_s\, e \sin M \\ \Delta y &= r_s\, i \sin(M + \omega) \end{aligned}$$

where Δx and Δy represent tangential and lateral (normal) distances relative to the fixed position. In the orbital plane (Δr, Δx), the relative track is elliptic, as shown in Fig. 3.35. The relative orbit shape in the vertical plane of the orbit is obviously related to the perigee amplitude angle ω.

The relative track shape of the tangential vertical plane of the orbit (Δy, Δx) is shown in Fig. 3.37a. The relative track shape in the radial vertical plane of the orbit (Δy, Δr) is shown in Fig. 3.37b. The effect of orbital inclination is to twist the relative

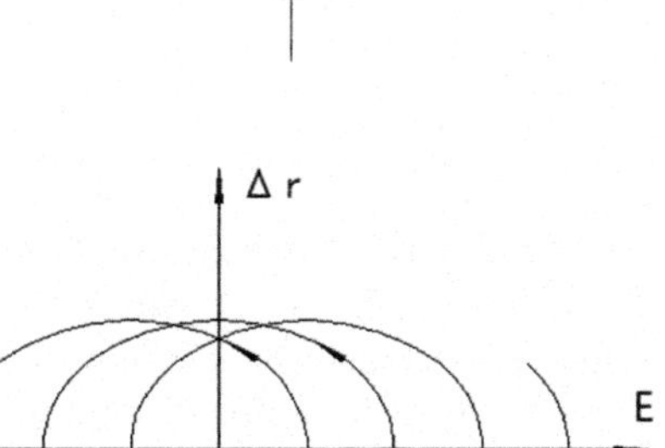

Fig. 3.35 Relative track diagram of geosynchronous orbit

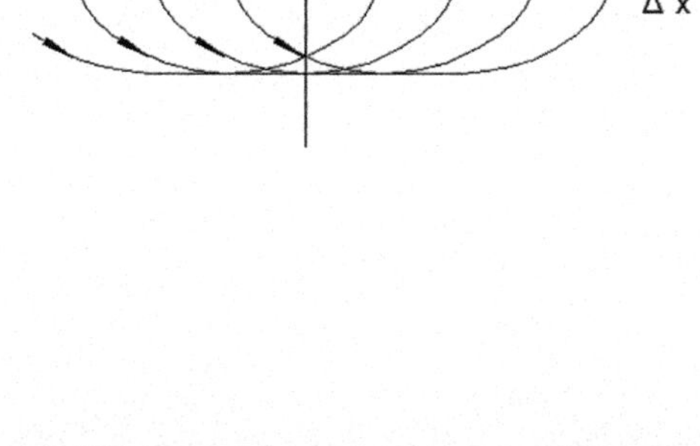

Fig. 3.36 $I = 0$ $e \neq 0$ $\Delta a \neq 0$ relative track of geosynchronous orbit

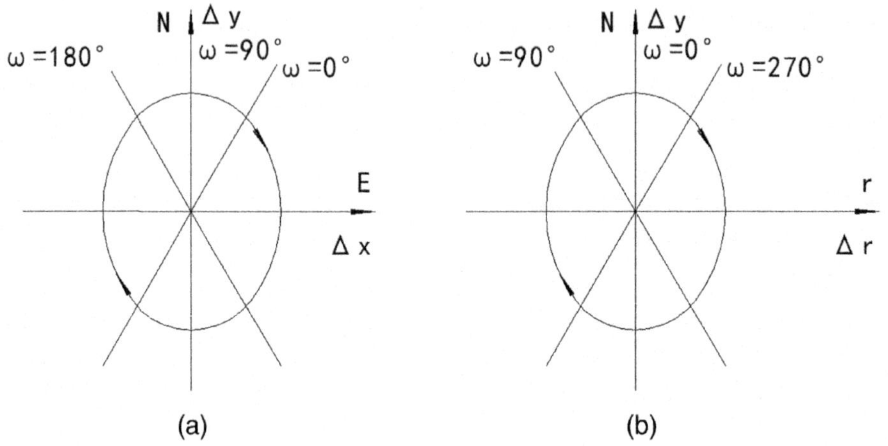

Fig. 3.37 $i \neq 0\ e \neq 0\ \Delta a = 0$ relative track of geosynchronous orbit

track of the ellipse out of the orbital plane, and the direction of the twist depends on the perigee angle.

If $\omega = 0°$ or 180°, the effect of the orbital inclination is to turn the original relative track plane out of the orbital plane around the radius vector of the fixed position. That is, the intersecting nodal line of the plane and the equatorial plane is consistent with the radial direction of the orbit, the short half axis of the relative track ellipse is still in the radial direction, its long axis is perpendicular to the radial direction of the orbit, and the inclination angle of the relative equatorial plane γ is

$$\gamma = \arctan \frac{i}{2e}$$

When the lateral deviation is maximum, the radial deviation is zero.

If $\omega = 90°$ or 270°, then the relative track plane turns tangentially out of the orbit plane around the orbit, the major axis of the relative track ellipse tangentially along the orbit, and the inclination angle of the minor axis of the ellipse γ is

$$\gamma = \arctan \frac{i}{e}$$

As inclination and eccentricity approach zero, the direction of ascending intersection and perigee becomes uncertain, and singularities appear in the orbital equation. In order to design orbit keeping strategy, another orbit element should be selected to describe the drift motion of satellite more conveniently. Define an inclination vector **i** whose length is equal to the inclination value and whose direction is consistent with the orbital normal. Define the eccentricity vector **e** with length equal to eccentricity and direction toward perigee. In Fig. 3.38, i_p and e_p are the projections of inclination and eccentricity on the equatorial plane of the equatorial inertial coordinate system.

Since $\cos(i) \approx 1$, the components of the vectors **e** and **i** can be listed and defined as

$$e_x = e\cos(\Omega + \omega) \tag{3.67}$$

$$e_y = e\sin(\Omega + \omega) \tag{3.68}$$

$$i_x = \sin i \sin \Omega \tag{3.69}$$

$$i_y = \sin c \cos \Omega \tag{3.70}$$

In the above formula, i_y is defined as the component of the inclination vector along the $(-y)$ axis.

For a stationary orbit, r_s is a constant value, see formula 3.63.

$$n = \omega_e + \Delta n = \omega_e(1 - 1.5\Delta a/r_s)$$

The mean longitude drift rate D is defined as an orbital element equivalent to semi-major axis a.

$$D = -1.5\Delta a/r_s \tag{3.71}$$

D is a measure of dimension 1 that causes the orbit's speed to deviate from geosynchrony, multiplied by the earth's speed of 361°/d, to obtain the drift angle in a day. If $\Delta a = 1$ km, then $D = 0.36 \times 10^{-4} = 0.0128$°/d. If $D > 0$, the satellite drifts eastward. If $D < 0$, the satellite drifts westward.

The mean right ascension of the satellite l is defined as follows:

$$l = \Omega + \omega + M \tag{3.72}$$

Fig. 3.38 Inclination vector and eccentricity vector of synchronous orbit

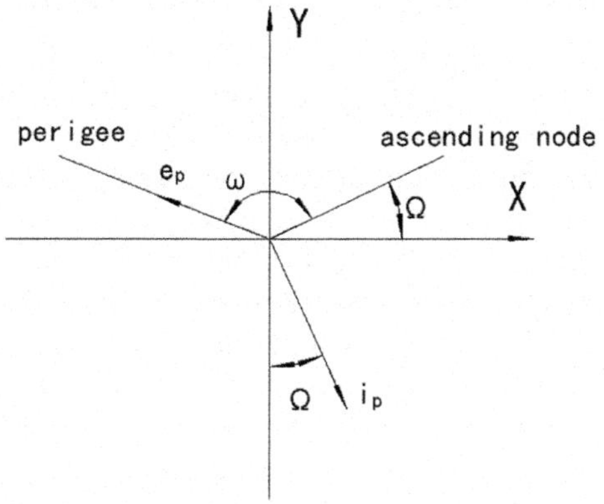

The mean right ascension $l_0 = \Omega + \omega + M_0$ at time t_0 can be used as the sixth orbital element to replace M_0. Quote drift rate D and formula 3.63, so:

$$M = n(t - t_p) = (1 + D)\omega_e(t - t_p)$$

And

$$D(l - l_0) = D(M - M_0) = \left(D + D^2\right)\omega_e(t - t_0) \approx D\omega_e(t - t_0)$$

Omit higher-order small quantities from the above equation D^2.

Using **e** and **i** vector projection of formulas (3.67), (3.68), (3.69), (3.70) and formulas (3.71), (3.72), from formulas (3.68) and (3.66), so:

$$r = r_s - r_s\left(1.5D + e_x \cos l + e_y \sin l\right) \tag{3.73}$$

$$\lambda = \lambda_0 + D(l - l_0) + 2e_x \sin l - 2e_y \cos l \tag{3.74}$$

And from $\Delta y = r_s\, i \sin(M + \omega)$, the equation of motion of a satellite deviating from the equatorial plane can be gotten.

$$\varphi = -\mathrm{i_x} \cos l + \mathrm{i_y} \sin l \tag{3.75}$$

Therefore, D, e_x, e_y, i_x, i_y and l_0 constitute the six elements of the stationary orbit. In the equation of motion, the time variable t is replaced by the satellite's mean sidereal hour angle (mean right ascension) l.

3.8 Constellation Orbit

3.8.1 Global Coverage Constellation

Global navigation, communications, environmental monitoring and other requirements of multisatellite network operation, the purpose is to any area of the earth at any time by a satellite in the system coverage, or by a number of satellite coverage. The study shows that the circular orbit network with equal height and inclination angle is the best configuration plan, and the solar synchronous orbit adopted by optical remote sensing is also circular orbit, each orbital plane is evenly distributed relative to the equatorial plane, and satellites are evenly distributed in each orbital plane. The most important factors in determining the constellation design are the inclination (a polar orbit or inclined orbit), the altitude of the orbit and the minimum elevation.

Here, only the geometric distribution of satellite groups covered by a single kind satellite is discussed. The coverage area of a single satellite depends on the altitude and field angle of the satellite. As shown in Fig. 3.39, the relation between coverage angle ψ, height H and elevation angle E is

$$\psi = \arccos \frac{r_a \cos E}{r_a + H} - E$$

Distribute the satellites uniform in each orbital plane. The angular distance between the sub-satellite points of the adjacent satellites is $2b$, and the width of the coverage band formed by the satellite group is $2c$. As shown in Fig. 3.40, there are spherical trigonometric relations:

$$\sin b = \frac{\tan c}{\tan \theta} \tag{3.76}$$

$$\sin c = \sin \psi \sin \theta \tag{3.77}$$

For a polar-orbiting satellite constellation, the orbital plane of each orbit coincides with some meridian plane of the earth. If the coverage belt of each track on the equator connects to cover the whole equator, then any latitude area is overlapped by adjacent orbital belt, and the number of tracks required for global coverage P is

$$P = \pi/2c$$

The number of satellites in each orbital plane q is

$$q = \pi/b$$

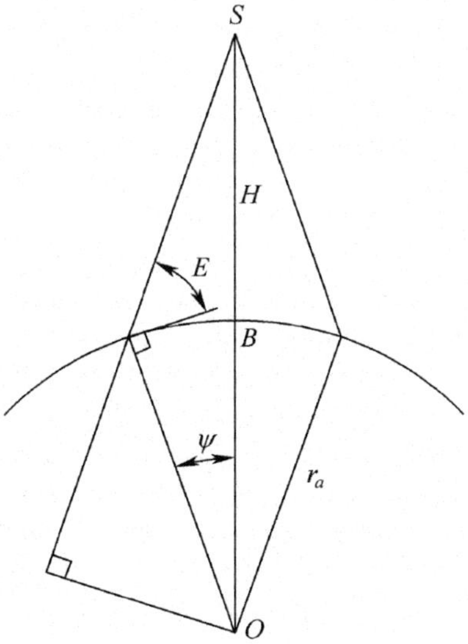

Fig. 3.39 Satellite–ground coverage

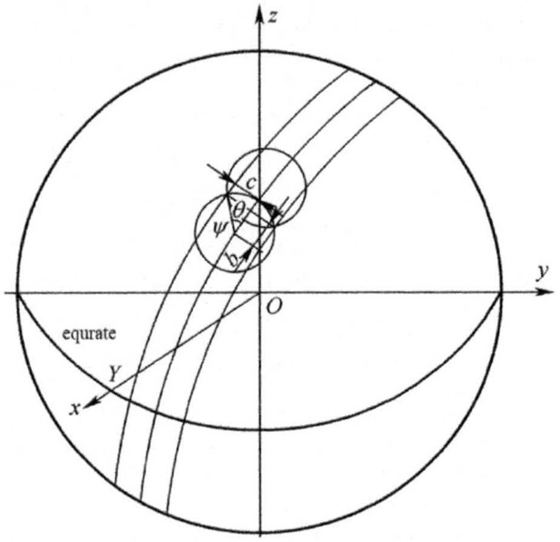

Fig. 3.40 Coverage belt of the satellite group

By referring to formulas (3.76) and (3.77), the total number of satellites in the polar-orbiting satellite group N is

$$N = P \times q = \frac{\pi^2}{2\arcsin(\sin\psi \sin\theta) \times \arcsin\left\{\frac{\tan[\arcsin(\sin\psi \sin\theta)]}{\tan\theta}\right\}}$$

According to the above formula and see Fig. 3.40, the optimal θ angle can be selected to minimize the total number of satellites. Since P, N and q are all positive integers, their differential is meaningless, but the approximate optimal angle θ can be obtained. It can be seen from the figure that increasing θ angle can enlarge the orbital coverage belt and decrease the number of orbits but increase the number of satellites in the orbit. If θ angle is decreased, the angular distance between satellites in the orbital plane is increased, and the number of satellites in the orbital plane is decreased, but the number of orbital planes should be increased. Therefore, for a given coverage ψ, the best angle θ can be selected, that is, the number P and q can be matched, and the coverage angle is directly determined by the minimum elevation angle E of the satellite.

Take the global communication network formed by small satellite groups as an example. Given the elevation angle $E = 5°$, the basic parameters of several typical polar-orbiting satellite groups are shown in Table 3.2.

For the non-polar inclined satellite constellation, adjacent orbits that join and cover at the equator and at high latitude regions, there may be gaps between adjacent track coverage bands. For an even orbital plane, let $P = 6$, as shown in Fig. 3.41. The crossing points of the cover bands of tracks ① and ②, ③ and ⑥ are A, C, D and F, and the holes are between C and D. The intersection point of tracks ① and ② is B, and the intersection point of tracks ③ and ⑥ is E. Because the even numbers are symmetric, point E is on the equator.

Table 3.2 Table of basic parameters of typical polar-orbiting satellite groups

N	p	q	$\theta/°$	$\psi/°$	H/km
12	3	4	39.23	52.24	5358
32	4	8	47.26	31.4	1514
48	8	6	21.69	31.86	1561
66	6	11	43.57	22.0	752

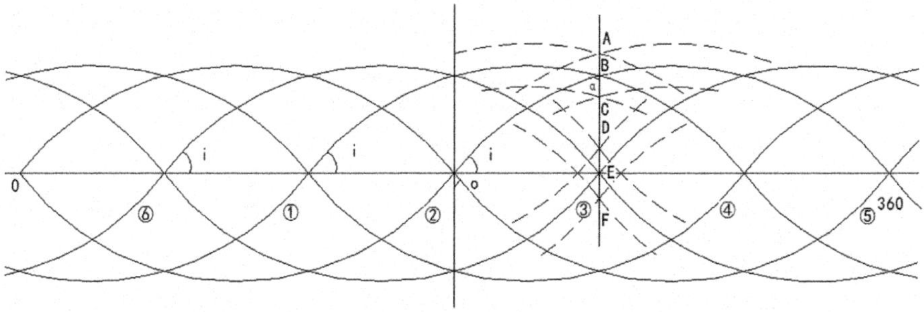

Fig. 3.41 Gap formation diagram

The latitude of each intersection is denoted by φ, and the trigonometric relation is obtained from spherical triangle $\triangle$OBE:

$$\tan \varphi_{\rm B} = \tan i \sin(2\pi/p) \tag{3.78}$$

$$\sin \alpha \sin \varphi_{\rm B} = \sin {\rm i} \sin(2\pi/p) \tag{3.79}$$

In the formula, angle α is the angle between track ② and arc BE.

It can be seen from Fig. 3.41 that the width of each track covering band is $2c$, and there are trigonometric relations at the intersection points B and C of track ① covering band:

$$\sin \alpha \sin(\varphi_B - \varphi_C) = \sin c \tag{3.80}$$

By eliminating the α and $\varphi_{\rm B}$ of formulas 3.79 and 3.80, the latitude formula of point C of gap is

$$\sin {\rm i} \cos \varphi_C \sin(2\pi/{\rm P}) - \cos {\rm i} \sin \varphi_C = \sin {\rm c} \tag{3.81}$$

Similarly, at the intersection points D and E of the coverage belt of track ③ and ⑥, there are trigonometric relations:

$$\sin\left(90^{\circ} - {\rm i}\right) \sin \varphi {\rm D} = \sin {\rm c} \tag{3.82}$$

Table 3.3 $E = 5°$ $i = 55°$ table of basic orbit parameters of satellite group

N	p	q	$\theta/°$	$\psi/°$	H/km
18	6	3	20	61.3	9419
36	6	6	32.8	34.4	1842

If the latitude of intersection points D and C is equal to $\varphi_D = \varphi_C$, gap coverage can be achieved. By eliminating φ_C and φ_D of formulas 3.81 and 3.82, the relationship between the half-width c and inclination angle i of the track coverage zone is obtained.

$$\sin c = \frac{\sin i \cos i \sin \frac{2\pi}{P}}{\sqrt{4\cos^2 i + \sin^2 i \sin^2\left(\frac{2\pi}{P}\right)}} \tag{3.83}$$

Let the satellite elevation $E = 5°$, orbital inclination $i = 55°$. The basic orbit parameters of satellite group are shown in Table 3.3.

When optical remote sensing satellite is in network operation, the number of orbit planes needed should also be considered illumination conditions, instead of blindly pursuing global simultaneous coverage. This is different from communication networking. Therefore, optical remote sensing satellite network does not necessarily cover the whole world simultaneously without omission.

3.8.2 *Geostationary Satellite Group*

The networking method and form of satellite communication can be used for networking of geostationary optical remote sensing satellites. One direction of development of satellite communication networking is to replace the functions of large satellite platforms with a number of sub-satellites, which are distributed around the parent satellite (or hypothetical parent satellite—fixed position), forming some form of constellation, some tens of kilometers apart. The main consideration of constellation design is to avoid the collision of sub-satellites and the mutual occlusion of communication to earth that affects radio transmission, and the limited range of the line of sight direction for interstellar communication between sub-satellites. Using small deviation motion equations of stationary satellites:

$$\Delta r = -r_s\, e \cos M$$
$$\Delta x = 2r_s\, e \sin M$$
$$\Delta y = r_s\, i \sin(M + \omega)$$

Substitute geostationary radius value into the above formulas, a quantitative equation for each satellite's deviation from its predetermined position can be obtained.

$$\Delta r = -42164.17\, e \cos M$$
$$\Delta x = 84328.5\, e \sin M$$
$$\Delta y = 42164.17\, i \sin(M + \omega)$$

In the formula, Δx is tangential deviation, Δy is lateral (deviation from the orbital plane) distance, λ_0 is the fixed position of the center of the satellite group (constellation), and λ_j is the predetermined position of a single satellite (longitude). It can be seen that the basic requirements of orbital elements of sub-satellites are $e < 10^{-3}$ and $i < 1°$. In addition, in order to maintain the synchronization of the constellation, the drift rate of each satellite is required to be basically the same. The geometric configuration of a geostationary satellite constellation is determined by the geostationary orbit elements of each satellite: λ_j, e_j, i_j. According to the principle of small deviation linearization, the relative motion relation between two satellites is similar to the small deviation motion formulas (3.73), (3.74) and (3.75) of a single satellite. Let the geostationary orbit elements of the two satellites be ($\lambda_{01}, D_1, e_1, i_1$) and ($\lambda_{02}$, D_2, e_2, i_2), respectively, the difference between them (marked with "δ") is

$$\delta\lambda_0 = \lambda_{01} - \lambda_{02}, \delta D = D_1 - D_2, \delta_e = e_1 - e_2, \delta i = i_1 - i_2$$

If the drift rate of the two satellites is the same and the relative distance between them is very small, then the mean right ascension of the two satellites is approximately the same. The relative distance equation of the two satellites in the radial, tangential and lateral orbits can be written as follows:

$$\Delta r = r_1 - r_2 = -r_s\left(\delta e_x \cos l + \delta e_y \sin l\right) \tag{3.84}$$

$$\Delta x = r_s(\lambda_1 - \lambda_2) - = r_s\left(\delta\lambda + 2\delta e_x \sin l - 2\delta e_y \cos l\right) \tag{3.85}$$

$$\Delta \mathrm{y} = \mathrm{r}_s(\varphi_1 - \varphi_2) = \mathrm{r}_s\left(-\delta i_x \cos l + \delta i_y \sin l\right) \tag{3.86}$$

The left end of the equation δx and δy represents the relative tangential and lateral distances along the orbital coordinates, and the right end subscripts x and y represent the components of the eccentricity and inclination vectors along the X and Y axes of the geocentric equatorial inertial coordinates. The above equation is also the relative distance equation between the child satellite and the parent satellite (or imaginary parent satellite).

In the configuration design of satellite group, the eccentricity and inclination of parent satellite can be zero. Therefore, the basic method of establishing satellite constellation is to set up the orbital elements of sub-satellites, respectively, to meet the requirements of different geometric configurations.

(1) Longitude split mode: This is the simplest split mode. Each sub-satellite is distributed along the orbital longitude circle, located on both sides of the central fixed position of the constellation, with different longitude. This simple

split requires a wide track window. Taking two satellites as an example, this separation is characterized by

$$\Delta\lambda_0 > 2(e_1 + e_2),\ \delta D = 0$$

(2) In the same plane eccentricity separation mode, each satellite enjoys the same fixed longitude, but the eccentricity e_j varies from satellite to satellite. A constellation formed by the phase difference of each satellite in the east—west direction. The characteristics of this mode are

$$\Delta\lambda_0 = 0, \delta D = 0, \delta e \neq 0$$

The relative distance equation between satellites is

$$\begin{aligned}\Delta r &= -r_s\left(\delta e_x \cos l + \delta e_y \sin l\right)\\ \Delta x &= 2r_s\left(\delta e_x \sin l - \delta e_y \cos l\right)\end{aligned}$$

The relative motion of one satellite around another forms an ellipse with the short axis along the radial direction and length $r_s{\cdot}\delta e$, the long axis along the tangential direction and twice as long as the short axis.

If the amplitude of the eccentricity of each satellite is the same, and the direction of the eccentricity is different, and if each satellite is on the same ellipse, separated by different phases, it rotates periodically around the common point of mean longitude. For example, the eccentricity of the four child satellites is 90° apart. Figure 3.42a shows the distribution of orbits on the equatorial inertial coordinate plane. The relative trajectories of each satellite are the same ellipse, and each sub-satellite has different phase on the ellipse.

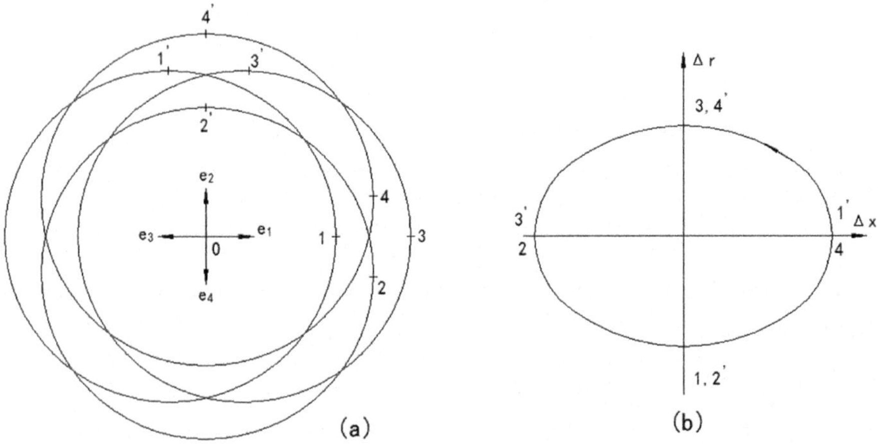

Fig. 3.42 Homoplanar eccentricity distribution configuration

For example, child satellites 1 and 3 are relative to each other on the minor axis of the ellipse. After 6 h, they rotate to the major axis of the ellipse, and after another 6 h, they return to the minor axis. The basic principle of eccentricity separation is to maximize the difference δe of eccentricity between any pair of satellites. The distribution of eccentricity vectors is shown.

As the satellite group is located on the same equatorial plane, the line of sight of sub-satellites 1 and 3 as well as 2 and 4 overlap twice a day, blocking each other and affecting radio communication. Therefore, it is necessary to set the orbital inclination angle to move the relative track out of the equatorial plane and form lateral separation.

(3) The separation mode of inclination and eccentricity synthesis: all sub-satellites share the same fixed-point longitude, and the inclination setting makes the relative track ellipse twist out of the equatorial plane. The equation of projected motion on the meridional plane of the earth can be written as follows:

$$\Delta r = -r_s\left(\delta e_x \cos l + \delta e_y \sin l\right)$$
$$\Delta y = -r_s\left(\delta i_x \cos l - \delta i_y \sin l\right)$$

In the meridional plane, the relative motion between satellites is similar to that in the equatorial plane. Thus, the characteristics of this mode are as follows:

$$\delta\lambda_0 = 0, \quad \delta D = 0, \quad \delta i_x = k\delta e_x, \quad -\delta i_y = k\delta e_y \tag{3.87}$$

where k is the constant coefficient. The relative trajectories of each satellite relative to the parent satellite are in the same inclined plane, and the line of intersection with the meridional plane of the earth is a straight line through the longitude of the fixed point.

According to the definitions of i_x and i_y (formula 3.69 and 3.70), i_y is defined on the $(-y)$ axis of the equatorial coordinate, so the eccentricity vector of each satellite represented by formula 3.87 is parallel to the projection of its own inclination vector on the equatorial plane (see formulas 3.84, 3.85 and 3.86). The projection of this model's "relative track plane" onto the equatorial plane's perpendicular tangential plane is also elliptical. The major axis is tangential to the orbit, and the minor axis is lateral.

The projection of the relative trajectories of the four sub-satellites configured according to formula 3.88 on the 3D plane is shown in Fig. 3.43, where E represents the eastern tangential direction, N represents the northern direction, and r represents the radial direction. As the "relative track plane" is relatively radial inclined, mutual occlusion of the sub-satellites' line of sight to earth is avoided in this direction.

Since the projection of the inclination vector on the equatorial plane lags behind the nodal line of the orbital plane by 90°, formula 3.87 of the split mode can be rewritten as follows:

$$\Delta\lambda = 0, \delta D = 0, \omega = 270^\circ, \Omega + \omega = \theta$$

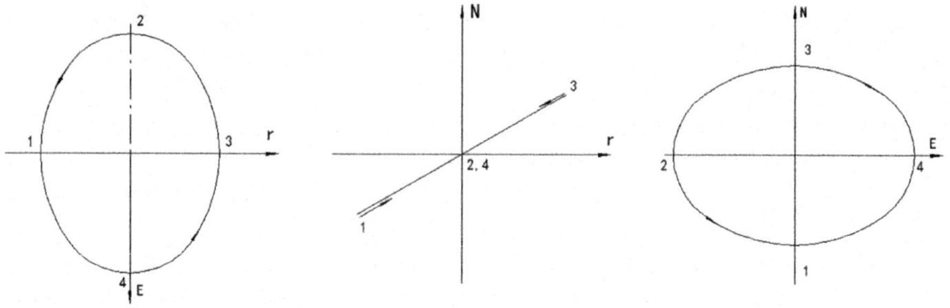

Fig. 3.43 Inclination and eccentricity are synthetically separated

That is, the perigee angle of each sub-satellite orbit should be 270°. The formula θ is the direction angle of the orbital eccentricity vector of each sub-satellite. Taking Fig. 3.42 as an example, if the mean right ascension of the fixed position of the constellation is 0°, the separated orbit elements of the four sub-satellites can be set as follows:

$$\begin{array}{lll} \Omega_1 = 90^\circ & \omega_1 = 270^\circ & M_{01} = 0^\circ \\ \Omega_2 = 180^\circ & \omega_2 = 270^\circ & M_{02} = 270^\circ \\ \Omega_3 = 270^\circ & \omega_3 = 270^\circ & M_{03} = 180^\circ \\ \Omega_4 = 0^\circ & \omega_4 = 270^\circ & M_{04} = 90^\circ \end{array} \tag{3.88}$$

References

1. Fang J, Ning X. Principles and applications of astronomical navigation. Beijing: Beihang University Press; 2006.
2. Fang J, Ning X, Tian Y. Principle and method of spacecraft autonomous astronomical navigation. Beijing: National Defense Industry Press; 2006.
3. Yan H. Spaceflight TT&C Network. Beijing: National Defense Industry Press; 2004.
4. Li H. Geostationary satellite orbital analysisand collocation strategies. Beijing: National Defense Industry Press; 2010.
5. Yinwei Z. Satellite orbital attitude dynamics and control. Beijing: National Defense Industry Press; 1998.
6. Jun Z. Spacecraft control principle. Xian: Northwestern Polytechnical University Press; 2001.
7. Yelun X. Aircraft flight dynamics principle. Beijing: Astronautic Publishing House; 1995.
8. Xie J. Remote sensing satellite orbit design. PLA Information Engineering University; 2005.
9. Yang W. Frozen ORBIT Basedon Brouwer's mean orbie elements. Chinese Space Scienceand Technology; 1998.
10. Tao J, Sun Z, Sun Y, Tao Z, Hu C. Exploration of high resolution optical remote sensing of the geostationary orbit. Opto-Electronic Engineering; 2012.

Chapter 4 Radiation Source and Optical Atmospheric Transmission

Radiation source is the energy source of light as the working medium in space optical remote sensing, in which the sun is the most important light source, followed by earth infrared radiation, atmospheric radiation and atmospheric glow which can also be included in this list, and they are unavoidable factors of space optical remote sensing. The others are the moon and stars. Moonlight is an important light source for low-light imaging in a clear night sky, and starlight can be used as the standard light source for on-orbit calibration of camera optical response under certain conditions.

Atmospheric transmission of light wave in space optical remote sensing is mainly concerned with atmospheric transmittance, backscattering and atmospheric window, because they are the important factors to determine the illumination of target and image plane. The factors involved in atmospheric transmission and scattering are complex and are usually calculated by LOWTRAN, MODTRAN and other software. This chapter focuses on the introduction of basic concept and theory. References to atmospheric optics include [1 , 2] and mainly Chap. 3 of [2].

As long as the temperature is not the absolute temperature 0 °K, any object is a source of radiation, and the object is irradiated by electromagnetic waves emitted by other objects at the same time. Optical remote sensing collects information from these light waves. The sun is the most important natural light source, and its spectral energy distribution depends on its absolute temperature, which is close to 6000 K. The earth is another important source of light, and its light wave radiation comes from two parts: the earth thermal radiation and the radiation reflected from the sun. Short spectrum, from about 0.3–2.5 μm, mainly to reflect solar radiation, and thermal radiation can be excluded; in the long spectrum, when the wavelength is larger than 6 μm, thermal radiation is dominant, while reflected radiation is negligible. In the 2.5–5 μm spectrum, both thermal and solar radiation reflections must be considered.

The light emission of an object can be monochromatic, such as the laser radiation spectrum; they can also be line spectrum, such as atomic radiation or continuous, such as blackbody radiation. The radiation sources of great significance to passive remote sensing are the sun and the earth, and their radiation spectrum is continuous

J. Tao, *Space Optical Remote Sensing*, Advances in Optics and Optoelectronics,
https://doi.org/10.1007/978-981-99-3318-1_4

spectrum. The most important continuous spectrum is the blackbody spectrum, and other continuous spectra are related to the blackbody spectrum by specific emissivity.

Both the earth thermal radiation and the solar radiation must pass through the earth's atmosphere to be picked up by remote sensors on space platforms. Due to the influence of atmospheric absorption, only some spectral segments can pass through the atmosphere, and these transparent spectral segments are called atmospheric windows. Atmospheric absorption, scattering, radiation and refraction affect space optical remote sensing, and the atmospheric effects must be corrected when analyzing optical remote sensing results.

4.1 Radiation and Units

4.1.1 Radiometry and Units

The detection results obtained by optical remote sensing need to be quantitatively analyzed, so the light wave must be strictly measured, and the necessary standards should be developed. The amount of radiation is in the international system of units, which applies to the entire electromagnetic spectrum. In radiometric and optical measurements, a basic unit is solid angle. Solid angle is usually represented by the Greek letter Ω, and its unit is steradian with symbol Sr. 1 Sr is one steradian, and the solid angle is shown that the vertex is located at the center of the sphere, and the area cutoff on the sphere is equal to the area of the square with the radius of the sphere as side length, as shown in Fig. 4.1.

If the radius of the sphere is 1, the solid angle Ω of the center O is equal to the area A_0 of the sphere intercepted by this solid angle. If the radius of the sphere is r and the area of the sphere intercepted by the solid angle is A, the following relation can be obtained:

$$\Omega = A_0 = \frac{A}{r^2} \tag{4.1}$$

Fig. 4.1 Solid angle

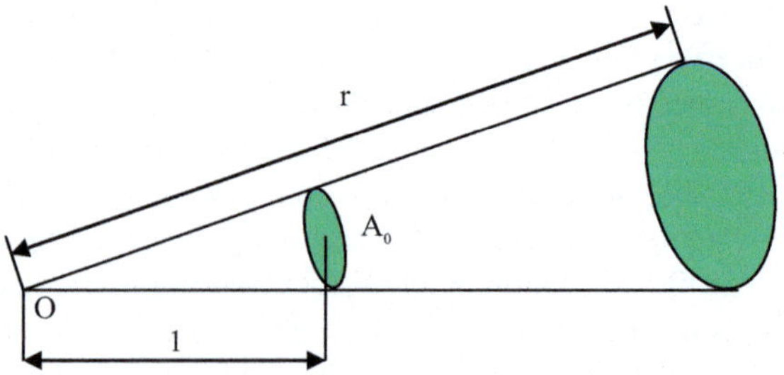

If the area opposite to the solid angle is not on the sphere, the area is projected onto the sphere. Solid angle is a dimensionless quantity. The solid angle corresponding to the entire sphere is 4π.

The solid angle stretched by the cone with half vertex angle α is

$$\Omega = 2\pi(1 - \cos\alpha) = 4\pi \sin^2\frac{\alpha}{2} \tag{4.2}$$

The solid angle enclosed by the conical surfaces of the co-vertices of half cone angle α_1 and α_2 is

$$\begin{aligned}\Omega &= 2\pi(\cos\alpha_1 - \cos\alpha_2)\\ &= 4\pi \sin\left(\frac{\alpha_1}{2} + \frac{\alpha_2}{2}\right)\sin\left(\frac{\alpha_2}{2} - \frac{\alpha_1}{2}\right)\end{aligned} \tag{4.3}$$

(1) Radiant Energy Q

Electromagnetic radiation is a form of energy transfer, which is manifested in various aspects such as raising the temperature of the irradiated object, changing the internal state of the object and making the charged object move under force. The unit is joule (J).

(2) Radiant Flux Φ

The radiant energy passing through an area in unit time is called radiant flux, which is a unit of radiant energy flow and has the same dimension as radiant power. The unit is watts = joules/second (W = J/s).

$$\Phi = \frac{\partial Q}{\partial t} \tag{4.4}$$

(3) Flux Density

The radiant flux per unit area is called the radiant flux density.

The radiant flux density on the surface of the irradiated object is called irradiance E.

$$E = \frac{\partial \Phi}{\partial A} \tag{4.5}$$

The radiant flux density on the surface of the radiant body is called the radiant emittance M.

$$M = \frac{\partial \Phi}{\partial A} \tag{4.6}$$

(4) Radiant Intensity I

Radiant intensity is the amount of radiation that describes the radiation characteristics of a point source, indicating the radiant flux within a unit solid angle in a certain direction. Radiant intensity is directional, so $I(\theta)$ is a function of direction angle θ, as shown in Fig. 4.2. Isotropic radiation source $I = \varphi/4\pi$.

$$I = \frac{\partial \Phi}{\partial \omega} \tag{4.7}$$

where ω is solid angle and the unit of I is W/Sr.

(5) Radiance L

The radiance L describes the radiant intensity of the surface source. L has directionality, as shown in Fig. 4.3. It refers to the radiant flux of the unit projection surface of the radiation source in a certain direction within a unit solid angle.

$$\begin{aligned} L(\theta) &= \frac{\partial^2 \Phi}{\partial \omega (\partial A \cos \theta)} \\ &= \frac{\partial I}{\partial A \cdot \cos \theta} \end{aligned} \tag{4.8}$$

where θ is the angle between the normal of the plane element and the radiation direction.

In general, the surface element $L(\theta)$ varies with the observed angle θ, but for some radiation sources, $L(\theta)$ is no correlation with θ, which means

$$I(\theta) = I_0 \cos(\theta) \tag{4.9}$$

This radiation source is called a Lambert source. Strictly speaking, only an absolute blackbody is a Lambert source. But some rough surfaces, such as those coated

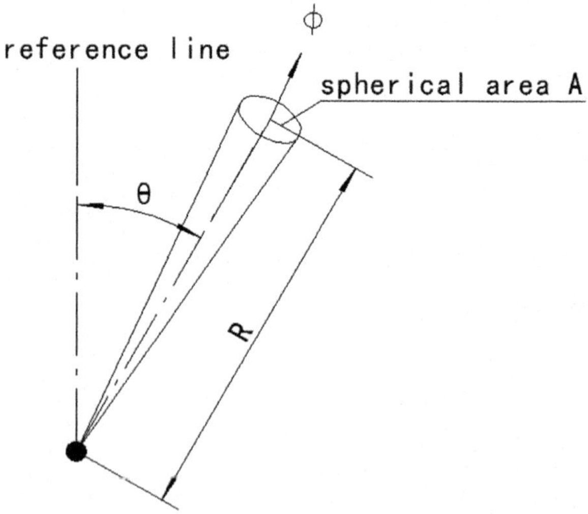

Fig. 4.2 Concept of radiant intensity

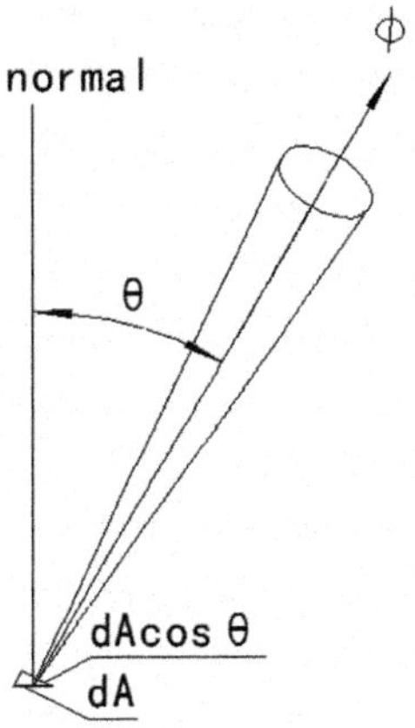

Fig. 4.3 Concept of radiance

with magnesium oxide, are a good Lambert source when irradiated. The sun also approximates a Lambert source.

According to the following formula:

$$L(\theta) = \frac{\partial}{\partial \omega}\left(\frac{\partial \Phi}{\partial A}\right)\frac{1}{\cos\theta}$$
$$M = \frac{\partial \Phi}{\partial A}$$

It is given

$$L(\theta) = \frac{\partial M}{\partial \omega}\frac{1}{\cos\theta}$$
$$\partial M = L(\theta)\cdot\cos\theta\cdot\partial\omega \tag{4.10}$$

$d\omega$ in Eq. 4.10 is shown in Fig. 4.4.

If the emitter is a Lambert body, that is, $L(\theta)$ is independent with θ, then the total power (W/(m^2Sr)) radiated by the emitter per unit area into the 2π space is

$$d\omega = \frac{r\sin\theta d\phi \cdot r d\theta}{r^2} = \sin\theta d\phi d\theta \tag{4.11}$$

$$M = \int\limits_{2\pi} dM$$
$$= L\int\limits_{2\pi}\cos\theta d\omega$$

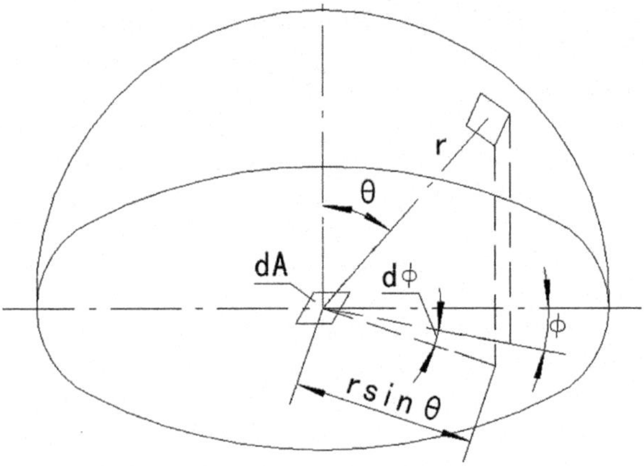

Fig. 4.4 Solid angle element calculation

$$= L \int_0^{2\pi} \int_0^{\frac{\pi}{2}} \cos\theta \cdot \sin\theta \quad d\theta d\varphi$$

$$= \pi L \tag{4.12}$$

(6) Photon number

The number of photons emitted by a radiation source, propagated in an optical medium or received by a detector, symbol Np, in units 1, the energy of a single photon is

$$Q = h\nu \tag{4.13}$$

where h is Planck's constant $h = 6.6260693 \times 10^{-34}$ J s [3] and ν is optical radiation frequency.

(7) Spectral radiant flux

Spectral radiant flux is a function of wavelength λ. Add subscript λ to ϕ to represent the spectral radiant flux ϕ_λ, which represents the radiant flux within the unit wavelength interval.

$$\phi_\lambda = \frac{\partial \phi}{\partial \lambda} \tag{4.14}$$

The unit of ϕ_λ is W/m or W/μm. Units of other spectral radiation are the same as the like E_λ, M_λ, I_λ, L_λ.

Sometimes it is necessary to calculate the radiant flux within a certain spectral interval $\phi(\lambda_1 \rightarrow \lambda_2)$. Its relationship with spectral radiant flux is

$$\phi(\lambda_1 \to \lambda_2) = \int_{\lambda_1}^{\lambda_2} \varphi_\lambda(\lambda)\mathrm{d}\lambda$$

And the total radiant flux ϕ is

$$\phi = \int_0^\infty \phi_\lambda(\lambda)\mathrm{d}\lambda$$

If the spectral radiant flux is represented by the radiant flux within a unit frequency interval, add subscript ν to ϕ to represent the spectral radiant flux ϕ_λ

$$\phi_\nu = \frac{\partial \phi}{\partial \nu}$$

The unit of ϕ_ν is (W/Hz).
And the total radiant flux ϕ is

$$\varphi = \int_\infty^\infty \varphi_\nu(\nu)\mathrm{d}\nu$$

For electromagnetic waves, the relation between ν and λ is

$$\nu \cdot \lambda = \mathrm{C}$$

Differentiate the above formula:

$$\mathrm{d}\nu = -\frac{c}{\lambda^2}\mathrm{d}\lambda$$

And

$$\varphi = \int_0^\infty \varphi_\nu(\nu) \cdot \frac{c}{\lambda^2}\mathrm{d}\lambda = \int_0^\infty \varphi_\lambda(\lambda)\mathrm{d}\lambda$$

$$\varphi_\nu = \frac{\lambda^2}{c}\varphi_\lambda \tag{4.15}$$

It represents the relationship between the radiant flux per unit frequency interval and the radiant flux per unit wavelength interval, as well as other radiation such as E.

(8) Hemispheric reflection, absorption and transmission

The effects of radiation on the interface can be divided into three parts: reflection, absorption and transmission. The intensity of these effects can be described by using reflectivity ρ, absorption α and transmittance τ. They represent the ratio of the total incident flux to the total reflected into the 2π space, or to the total absorbed flux, or to the total transmitted flux.

$\rho(\lambda)$, $\alpha(\lambda)$ and $\tau(\lambda)$ are all functions of wavelength λ. They are all dimensionless with value between 0 and 1.

$$\rho(\lambda) = \frac{\text{reflected flux of wavelength } \lambda}{\text{incident flux of wavelength } \lambda}$$

$$\alpha(\lambda) = \frac{\text{absorbed flux of wavelength } \lambda}{\text{incident flux of wavelength } \lambda}$$

$$\tau(\lambda) = \frac{\text{transmitted flux of wavelength } \lambda}{\text{incident flux of wavelength } \lambda}$$

$\rho(\lambda)$ can also be expressed as:

$$\rho(\lambda) = \frac{M(\lambda)}{E(\lambda)}$$

where $E(\lambda)$ is the irradiance of the radiation with wavelength λ on the target and $M(\lambda)$ is the radiation emission due to reflection.

The relationship between $\rho(\lambda)$, $\alpha(\lambda)$, $\tau(\lambda)$ is

$$\rho(\lambda) + \alpha(\lambda) + \tau(\lambda) = 1$$

For non-transparent body $\tau(\lambda) \equiv 0$, the equation above is

$$\rho(\lambda) + \alpha(\lambda) = 1$$

For a non-transparent body, if $\alpha(\lambda) \equiv 1$, then $\rho(\lambda) \equiv 0$, the object is called an absolute blackbody; while $\alpha(\lambda) \equiv 0$, then $\rho(\lambda) \equiv 1$, the object is called white body. In the visible and near infrared region, MgO is close to white body, and fresh MgO is often used as the reflectivity standard in the visible and near infrared region. If $\rho(\lambda)$ is independent of wavelength and it is a constant less than 1 and greater than 0, the object is called gray body.

For polychromatic radiation, the average reflectance $\overline{\rho}$, absorption $\overline{\alpha}$ and transmittance $\overline{\tau}$ are not only dependent on the spectral reflectance $\rho(\lambda)$, absorption $\alpha(\lambda)$ and transmittance $\tau(\lambda)$, but also related to the spectral distribution of the incident source $\phi(\lambda)$.

According to the definitions of $\overline{\rho}$, $\overline{\alpha}$, $\overline{\tau}$:

$$\overline{\alpha}(\lambda_1 \to \lambda_2) = \frac{\int_{\lambda_1}^{\lambda_2} \phi(\lambda)\alpha(\lambda)\mathrm{d}\lambda}{\int_{\lambda_1}^{\lambda_2} \phi(\lambda)\mathrm{d}\lambda}$$

$$\overline{\rho}(\lambda_1 \to \lambda_2) = \frac{\int_{\lambda_1}^{\lambda_2} \phi(\lambda)\rho(\lambda)\mathrm{d}\lambda}{\int_{\lambda_1}^{\lambda_2} \phi(\lambda)\mathrm{d}\lambda}$$

$$\overline{\tau}(\lambda_1 \to \lambda_2) = \frac{\int_{\lambda_1}^{\lambda_2} \phi(\lambda)\tau(\lambda)\mathrm{d}\lambda}{\int_{\lambda_1}^{\lambda_2} \phi(\lambda)\mathrm{d}\lambda}$$

In the formula, $\phi(\lambda)$ is the spectral radiant flux of incident radiation, and λ_1 to λ_2 is the wavelength range of incident radiation. The spectral distribution of incident sources $\phi(\lambda)$ is different, and the obtained are $\overline{\rho}$, $\overline{\alpha}$, $\overline{\tau}$ different.

4.1.2 Photometry and Units

Human eyes only sense the visible light in the electromagnetic wave from 0.385 μm to 0.789 μm, and it is accustomed to use photometric units in the visible spectrum.

In photometry, the definitions and symbols of various measures are the same as those of emissivity, only add subscript υ to the corresponding symbol to distinguish them. For example, the luminous flux ϕ_υ, the illuminance E_υ, the luminous emittance M_υ, the luminous intensity I_υ, the luminance L_υ and so on.

The basic unit of photometry is the Kendra cd. In 1979, it is defined as the luminous intensity of the light source in a given direction in the 16th International Metrology Congress, the monochromatic radiation at frequency 540×10^{12} Hz (wavelength 555.016 nm at standard atmospheric pressure, approximately 555 nm), and the radiation emitted at intensity 1/683 W/Sr is 1 cd.

The unit of luminous flux ϕ_υ is lumen, and the physically expressed is the power of energy flow. It is the luminous flux emitted by the light source with the luminous intensity of 1 cd within 1 sphere, that is, 1 cd = 1 lm/steradian.

The unit of illuminance E_υ is lux, $1\ \mathrm{lx} = 1\ \mathrm{lm/m^2}$

The unit of luminance L_υ is nit, $1\ \mathrm{nit} = 1\ \mathrm{lm/(m^2 \bullet\ steradian)}$.

The human eye has different responsivity to visible light of different wavelength; that is, the vision of light produced by the same radiant flux of different wavelength is different and represented by a visual function $K(\lambda)$

$$K(\lambda) = \frac{\Phi_\upsilon(\lambda)}{\Phi_\mathrm{e}(\lambda)}$$

where $\Phi_\upsilon(\lambda)$ is the unit of luminous flux and $\Phi_\mathrm{e}(\lambda)$ is the unit of radiant flux.

Relative visual function $V(\lambda)$ of standard human eye is

$$V(\lambda) = \frac{K(\lambda)}{K_{\mathrm{m}}}$$

When the luminance of daytime is greater than several nits ($L_\nu \geq 3$ nt), the human eye is in bright vision state, and the maximum value $K(\lambda) = K_{\mathrm{m}} = 683$ lm/W, $V(\lambda) = 1$ is obtained when the relative optical vision function $V(\lambda)$ curve is $\lambda = 0.555$ μm. For the luminance of nighttime less than a few parts per million nits ($L_\nu \leq 3 \times 10^{-5}$ nt) [2], the human eye is in dark vision state, when $\lambda = 0.507$ μm, $\mathrm{Km}' = 1700$ lm/W, the sensitivity of human eyes is the highest $V'(\lambda) = 1$, as shown in Fig. 4.5, and specific values are shown in Table 4.1. The difference between light vision of daytime and nighttime is that daytime vision relies on the action of conical cells in the retina, while nighttime vision relies on the action of rods in the retina. When the brightness is between light vision and dark vision, both types of visual cells work.

In the visible region, the influence of the optical visual function $V(\lambda)$ must be taken into account when the radiometric unit of measurement of radiation is converted to the photometric unit. For monochromatic radiation $\lambda = 0.555$ μm, the radiant flux of 1 W is equivalent to the luminous flux of 683 lumens. The conversion relationship between the luminous flux ϕ_υ and the radiant flux ϕ_{e} at any wavelength λ is

$$\phi_\upsilon(\lambda) = 683\phi_e(\lambda)\ V(\lambda) \tag{4.16}$$

where the unit of $\phi_\upsilon(\lambda)$ is lumen and the unit of $\phi_{\mathrm{e}}(\lambda)$ is watt.

And the conversion relation between λ_1 and λ_2 is

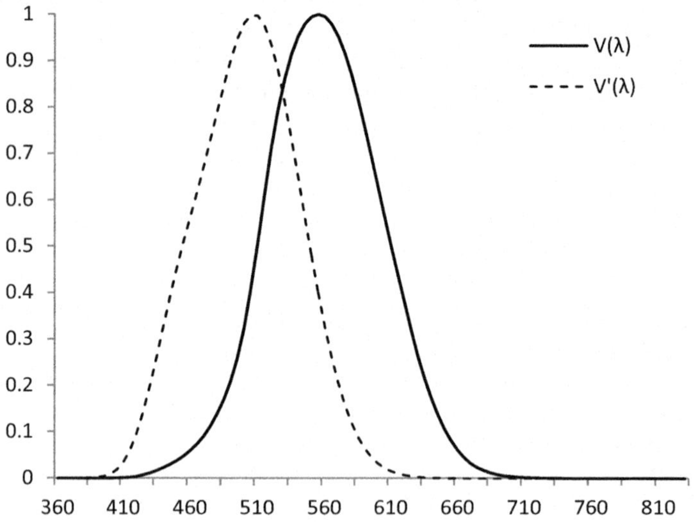

Fig. 4.5 Relative visual function curve

Table 4.1 Parameters of relative visual function [3]

λ/nm	$V(\lambda)$	$V'(\lambda)$	λ/nm	$V(\lambda)$	$V'(\lambda)$
360	0.000004		600	0.63100	0.03315
365	0.000007		605	0.56680	0.02312
370	0.000012		610	0.50300	0.01593
375	0.000022		615	0.44120	0.01088
380	0.000039	0.000598	620	0.38100	0.00737
385	0.000064	0.00111	625	0.32100	0.00497
390	0.000120	0.00221	630	0.26500	0.00334
395	0.00022	0.00453	635	0.21700	0.00224
400	0.00040	0.00929	640	0.17500	0.00150
405	0.00064	0.01852	645	0.13820	0.00100
410	0.00121	0.03484	650	0.10700	0.000677
415	0.00218	0.0604	655	0.08160	0.000459
420	0.00400	0.0966	660	0.06100	0.000313
425	0.00730	0.1436	665	0.04458	0.000215
430	0.01160	0.1998	670	0.03200	0.000148
435	0.01684	0.2625	675	0.02320	0.000103
440	0.02300	0.3281	680	0.01700	0.000072
445	0.02980	0.3931	685	0.01192	0.000050
450	0.03800	0.455	690	0.00821	0.000035
455	0.04800	0.513	695	0.00572	0.0000250
460	0.06000	0.567	700	0.00410	0.0000178
465	0.07390	0.620	705	0.00293	0.0000127
470	0.09098	0.676	710	0.00209	0.0000091
475	0.11260	0.734	715	0.00148	0.0000066
480	0.13902	0.793	720	0.00105	0.0000048
485	0.16930	0.851	725	0.00074	0.0000035
490	0.20802	0.904	730	0.00052	0.0000025
495	0.25860	0.949	735	0.000361	0.0000019
500	0.32300	0.982	740	0.000249	0.0000014
505	0.40730	0.998	745	0.000172	0.0000010
510	0.50300	0.997	750	0.000120	0.0000008
515	0.60820	0.975	755	0.000085	0.0000006
520	0.71000	0.935	760	0.000060	0.0000004
525	0.79320	0.880	765	0.000042	0.0000003
530	0.86200	0.811	770	0.000030	0.0000002
535	0.91485	0.733	775	0.0000212	0.0000002

(continued)

Table 4.1 (continued)

λ/nm	$V(\lambda)$	$V'(\lambda)$	λ/nm	$V(\lambda)$	$V'(\lambda)$
540	0.95400	0.650	780	0.0000150	0.0000001
545	0.98030	0.564	785	0.0000106	
550	0.99495	0.481	790	0.0000075	
555	1.00000	0.402	795	0.0000053	
560	0.99500	0.3288	800	0.0000037	
565	0.97860	0.2639	805	0.0000026	
570	0.95200	0.2076	810	0.0000018	
575	0.91540	0.1602	815	0.0000013	
580	0.87000	0.1212	820	0.0000009	
585	0.81630	0.0899	825	0.0000006	
590	0.75700	0.0655	830	0.0000005	
595	0.69490	0.0469			

Table 4.2 Comparison of radiometric units and photometric units

Radiometric	Definition	Unit		Photometric	Unit	
Radiant energy/Q_e		Joule	J	Luminous energy/Q_υ	Lumen·second	lm S
Radiant flux/ϕ_e	$\phi_e = \frac{\partial Q_e}{\partial t}$	Watt	W	Luminous flux/ϕ	Lumen	lm
Irradiance/ E_e	$E_e = \frac{\partial \phi_e}{\partial A}$	Watt/meter^2	W/m^2	illuminance/ E_υ	Lux	lx
Emittance/ M_e	$M_e = \frac{\partial \phi_e}{\partial A}$	Watt/meter^2	W/m^2	Luminous emittance/ M_υ	Lumen/meter^2	lm/ m^2
Radiant intensity/I_e	$I_e = \frac{\partial \phi_e}{\partial \omega}$	Watt/steradian	W/Sr	Luminous intensity/I_υ	Kendra	Cd
radiance/L_e	$L_e=\frac{\partial^2 \phi_e}{\cos\theta \partial A \partial \omega}$	Watt/(meter^2 steradian)	Wm^{-2} Sr^{-1}	Luminance/ L_υ	Nit	nt
Photon number/N_p	$h\upsilon$	Number	1			

$$\phi_\upsilon(\lambda_1 \to \lambda_2) = 683 \int_{\lambda_1}^{\lambda_2} \phi_e(\lambda) V(\lambda) \mathrm{d}\lambda \tag{4.17}$$

This relationship is also consistent with other optical measurements. For example, the conversion relation between the photometric intensity I_υ and the radiometric intensity I_e is $I_\upsilon = 683\, I_e\, V(\lambda)$, in which the unit of I_υ is lumen/steradian and the unit of I_e is watt/steradian.

4.1.3 Blackbody Radiation

At a certain temperature, any object radiates energy outward. The radiant emittance M_e of the object is a function of wavelength λ and temperature T, that is, $M_e(\lambda T)$. As early as 1860, Kirchhoff found that there was an internal relationship between the radiant emittance of an object (M_e) and the absorptionα of the object: an object with a higher absorption α also has a higher radiant emittance, which can be expressed by a quantitative relationship

$$\frac{M_e(\lambda T)}{\alpha(\lambda T)} = M_b(\lambda T)$$

In the formula, $M_b(\lambda T)$ is a universal function independent of the properties of the object, which only depends on the temperature T and wavelength λ of the object.

If the absorption rate of the object, α, is independent of temperature T and wavelength λ, is equal to 1, that is, the absolute blackbody, has

$$M_e(\lambda T) = M_b(\lambda T)$$

This formula shows that the emittance of the absolute blackbody is the universal function $M_b(\lambda T)$.

Absolute blackbody is an ideal absorber, an object that can completely absorb irradiation in any direction and wavelength at any temperature. The blackbody is a Lambert body as well as a diffuse reflector [3]. This is not easily found in nature. But a cavity can be seen as an ideal blackbody, as in Fig. 4.6. This is because the light enters the holes, need through the inner surface of the multiple reflections, possible to shoot from the small hole, as long as the inner surface is not absolutely white body ($\alpha \neq 0$), every reflection will weaken some light, multiple reflections to reduce the light greatly, when light exit from the holes, its intensity is close to zero, the absorption of the cavity is $\alpha \approx 1$.

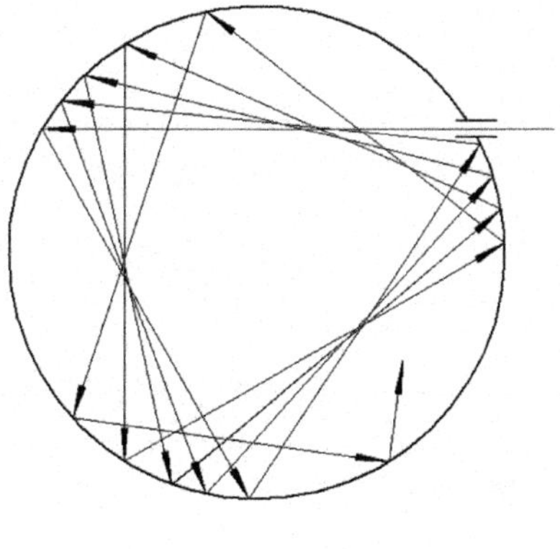

Fig. 4.6 Sketch of blackbody cavity

Therefore, a cavity with a small hole is an ideal absolute blackbody, and the inner surface of the cavity is not necessarily black. The radiant emittance $M_b(\lambda T)$ of a blackbody can be expressed by Planck's formula

$$M_b(\lambda T) = \frac{2\pi h c_0^2}{\lambda^5} \cdot \frac{1}{e^{hc_0/\lambda kT} - 1} \tag{4.18}$$

where

$h = 6.6260693 \times 10^{-34}$ Js is Planck's constant;

$k = 1.380658 \times 10^{-23}$ J K^{-1} is Boltzmann constant [3];

$c_0 = 299{,}792{,}458$ m/s is the light speed in vacuum [3];

λ wavelength with unit m;

T absolute temperature with unit K.

The unit of $M_b(\lambda T)$ is W/m^3.

And

$C_1 = 2\pi h c_0^2 = 3.741771 \times 10^{-16}$ J W m^2[3];

$C_2 = hc_0/k = 1.438775 \times 10^{-2}$ m K [3];

$M_b(\lambda T)$ can also be simplified to

$$M_b(\lambda T) = \frac{C_1}{\lambda^5 (e^{C_2/\lambda T} - 1)} \tag{4.19}$$

$M_b(\lambda T)$ represents the radiated power over a unit area (m^2) within a unit wavelength interval (m).

Since the absolute blackbody is a Lambert body, it can be known from $M = \pi L$ that the spectral radiance L_b of the blackbody is equal to M_b/π

$$L_b(\lambda T) = \frac{2hc_0^2}{\lambda^5 (e^{hc_0/\lambda kT} - 1)}$$

The unit of $L_b(\lambda T)$ is W/(m^3·Sr).

Substitute the constant of h, c_0, k into the formula

$$L_b(\lambda T) = \frac{1.1910428 \times 10^{-22}}{\lambda^5 (e^{1.438775 \times 10^{-2}/\lambda T} - 1)}$$

The unit of $L_b(\lambda T)$ is W m^{-2}·Sr^{-1}·μm^{-1}.

If the spectral radiance is expressed by the number of photons radiated per second, since $n_b(\lambda T) = L_b(\lambda T)/h\nu$, then

$$n_b(\lambda T) = \frac{2c_0}{\lambda^4 (e^{hc_0/\lambda kT} - 1)} \quad (4.20)$$

The unit of $n_b(\lambda T)$ is N_p $m^{-3}Sr^{-1}$ s^{-1}.

When the temperature T is different, the relation between $L_b(\lambda T)$ and λ is shown in Fig. 4.7. The vertical coordinate unit is $W \cdot m^{-2} \cdot Sr^{-1} \cdot \mu m^{-1}$, and the horizontal coordinate unit is μm. It is obvious that the blackbody radiation has a maximum value in the figure, which corresponds to the wavelength λ_{max}. Take the partial derivative of $L_b(\lambda T)$ and set it to zero, then

$$\frac{\partial M_b}{\partial \lambda} = \frac{-C_1\left[5\lambda^4(e^{C_2/\lambda T} - 1) + \lambda^5 e^{C_2/\lambda T}\left(-\frac{C_2}{\lambda^2 T}\right)\right]}{\lambda^{10}(e^{C_2/\lambda T} - 1)^2} = 0$$

This formula is zero only if its numerator is zero, and λ_{Max} can be found. If $x = C_2/(\lambda T)$

$$5\left(e^x - 1\right) - xe^x = 0$$

Simplified to

$$\left(1 - \frac{x}{5}\right)e^x = 1$$

$$x = \frac{C_2}{\lambda_{max} T} = 4.96511$$

Hence

$$\lambda_{max} T = b = 2.89777 \times 10^{-3} m\, k \quad (4.21)$$

This formula, also known as Venn's displacement law, states that the peak radiation moves in the shortwave direction as temperature increases. The table below gives the values of λ_{max} at different temperatures.

Substitute Eq. 4.21 into Eq. 4.19 [4].

$$\begin{aligned} M_b(\lambda_{max} T) &= \frac{C_1}{\lambda_{max}^5 (e^{C_2/\lambda_{max} T} - 1)} \\ &= \left(\frac{T}{b}\right)^5 \frac{C_1}{(e^{C_2/b} - 1)} \\ &= \frac{C_1}{b^5} \frac{1}{(e^{C_2/b} - 1)} T^5 \\ &= \frac{3.741771 \times 10^{-16}}{(2.89777 \times 10^{-3})^5} \frac{1}{(e^{1.438775 \times 10^{-2}/2.89777 \times 10^{-3}} - 1)} T^5 \\ &= 1.286702 \times 10^{-11} T^5 (Wm^{-2}\, \mu m^{-1}) \end{aligned} \quad (4.22)$$

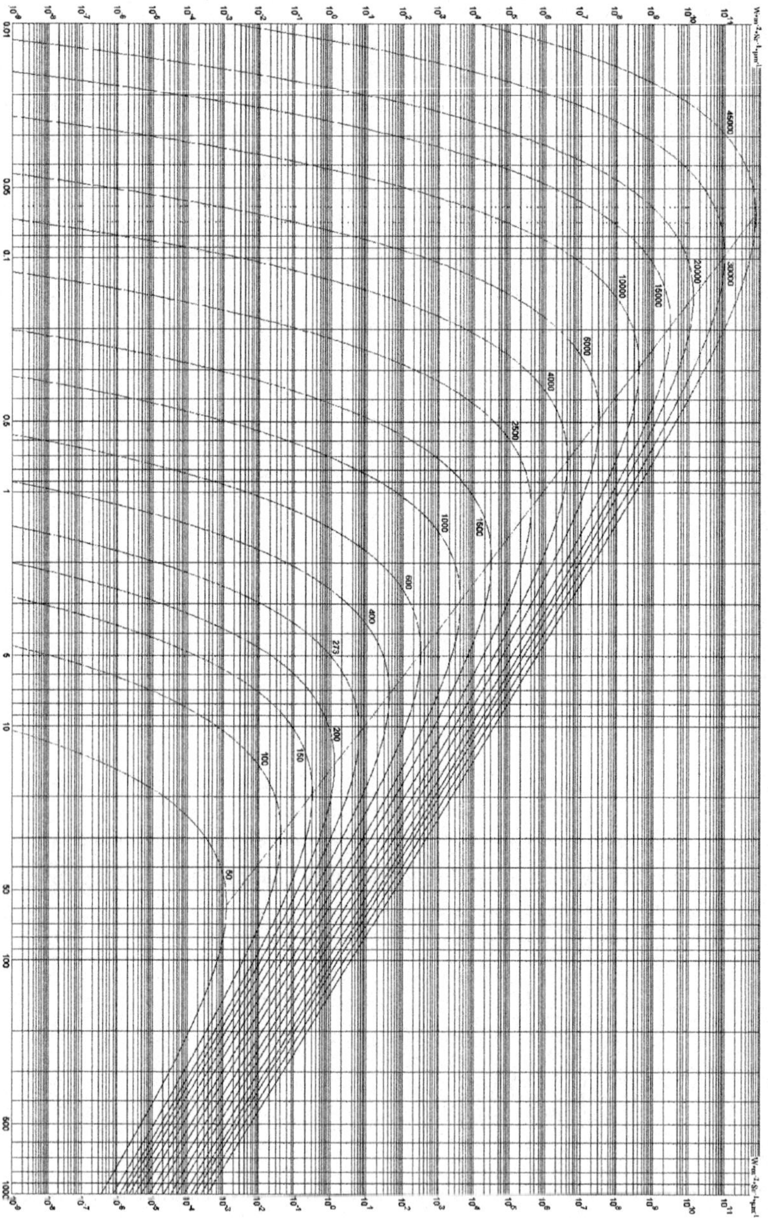

Fig. 4.7 Blackbody radiation curve

At peak wavelengths, the spectral radiance of the blackbody

$$L_{max} = 4.0957 \times 10^{-12} T^5 \tag{4.23}$$

The unit of L_{max} is W/(m^2·Sr·μm)

It can be seen from Table 4.3 that the maximum value of blackbody radiation at room temperature is about 10 microns. When the temperature reaches 1000 K, although λ_{max} is still in the infrared at 2.90 μm, it can be seen from the figure that some radiation has entered the visible area, and the composition of red light (0.62–0.76 μm) is overwhelmingly dominant, so the object looks dark red. With the increase of temperature, λ_{max} moves to the direction of shortwave, the proportion of shortwave in visible light composition increases, and the color changes from red to orange. When the temperature is 5000–6000 K, λ_{max} is located in the middle of the visible spectrum, the spectral brightness of the whole visible region is similar, and the object looks white. The maximum value of solar spectrum is 0.47 μm, and according to Wien's displacement law, the temperature of the sun is about 6150 K.

It can also be seen from the figure that the total power of blackbody radiation increases with the increase of blackbody temperature. It can be calculated from Planck's formula that the total power M_{bb} is proportional to the fourth power of temperature. The formula for calculating M_{bb} is

$$M_{bb} = \int_0^\infty M_b(\lambda) d\lambda$$

In order to calculate conveniently, the integral is converted from the wavelength space to the frequency space, hence

$$M_{bb}(T) = \int_0^\infty M_b(\lambda) \frac{\lambda^2}{c_0} d\nu$$

If λ in $M_b(\lambda)$ is also transformed to ν, then

$$M_{bb} = \int_0^\infty \frac{2\pi h\nu^3}{c_0^2} \cdot \frac{d\nu}{e^{h\nu/kT} - 1}$$

Table 4.3 Values of λ_{max} at different temperatures

$T(K)$	300	500	1000	2000	3000	4000	5000	6000	7000
λ(μm)	9.659	5.796	2.898	1.449	0.966	0.724	0.580	0.483	0.414

$$
\begin{aligned}
&= \frac{2\pi h}{c_0^2} \int_0^\infty \frac{\nu^3 e^{-h\nu/kT} \mathrm{d}\nu}{1 - e^{-h\nu/kT}} \\
&= \frac{2\pi h}{c_0^2} \int_0^\infty (e^{-h\nu/kT} + e^{-2h\nu/kT} + e^{-3h\nu/kT} + \cdots)\nu^3 \mathrm{d}\nu \\
&= \frac{2\pi^5 h}{15 c_0^2} \left(\frac{kT}{h}\right)^4 \\
&= \sigma T^4
\end{aligned} \tag{4.24}
$$

The above formula is called Stefan-Boltzmann law, where σ is called as Stefan-Boltzmann constant [3].

$$
\begin{aligned}
\sigma &= \frac{2\pi^5 k^4}{15 c_0^2 h^3} \\
&= 5.6704 \times 10^{-8} \ \mathrm{W/(m^2\,K^4)}
\end{aligned}
$$

$M_{bb}(T)$, in unit $\mathrm{W/m^2}$, is the total power radiated by the blackbody per unit area into 2π space and is proportional to T^4.

For the longer wavelength radiation region, that is, for the lower frequency spectrum, the Planck formula can be reduced to a simpler form if $h\nu \ll kT$

$$
e^{h\nu/kT} \approx 1 + h\nu/kT
$$

So

$$
M_b(\lambda T) = \frac{2\pi h c_0^2}{\lambda^5} \cdot \frac{1}{\left(1 + \frac{hc_0}{\lambda kT}\right) - 1} = \frac{2\pi c_0}{\lambda^4} kT \tag{4.25}
$$

The above formula is called the Sharp-Kings law. For $\lambda \gg \lambda_{max}$, the results of Sharp-Kings agree well with Planck's formula. For example, for an object at room temperature with $\lambda_{max} \approx 10\ \mu\mathrm{m}$, the distribution of its microwave radiation (λ ranging from 1 mm to 1 m) applies the Sharp-Kings formula. In the microwave region, frequency ν is commonly used instead of wavelength λ as an independent variable, then

$$
M_b(\nu, T) = M_b(\lambda T) \cdot \frac{c_0}{\nu^2} = \frac{2\pi}{\lambda^2} kT \tag{4.26}
$$

The above formula represents the power radiated by the blackbody to 2π space per square meter in a unit frequency interval (Hz). The unit of $M_b(\nu T)$ is $\mathrm{W/(m^2\,Hz)}$.

For the radiation source of blackbody, the International Commission on Illumination (CIE) recommends that the spectral data range be 380–780 nm, the wavelength interval is less than 10 nm, and the relative spectral temperature distribution difference will be less than 10% compared to blackbody when calculating spectral radiation.

4.1.4 Radiation Calculation

(1) Incidence and reflection

Object's reflectance $\rho(\lambda)$ is the function of wavelength, the same object has very big difference to the reflectance of different wavelength, and also it is the basis of vision. The reflectivity defined above is the ratio of incident flux to the flux reflected into the entire space and is independent of the spatial distribution of reflected flux. More accurate calculations also take into account such factors as the direction of incident and reflection and the state of the surface on the object. A ray of light has an incidence angle θ_i and an incidence azimuth angle φ_i. Angle of reflection θ_r, azimuth of reflection φ_r, incident irradiance E and reflection brightness L are as shown in Fig. 4.8.

The directional reflection factor $\rho'(\theta_i\varphi_i,\ \theta_r\varphi_r)$ is introduced to describe the directional reflection characteristics of objects.

$$\rho'(\theta_i\varphi_i,\ \theta_r\varphi_r) = \frac{L(\theta_r\varphi_r)}{E_i(\theta_i\varphi_i)}$$

In the formula, E_i is the irradiance caused by the incident beam in the direction of altitude angle θ_i and azimuth angle φ_i; L is the reflection brightness observed in the direction of altitude angle θ_r and azimuth angle φ_r, and the unit of ρ' is Sr^{-1}.

Fig. 4.8 Sketch of reflection

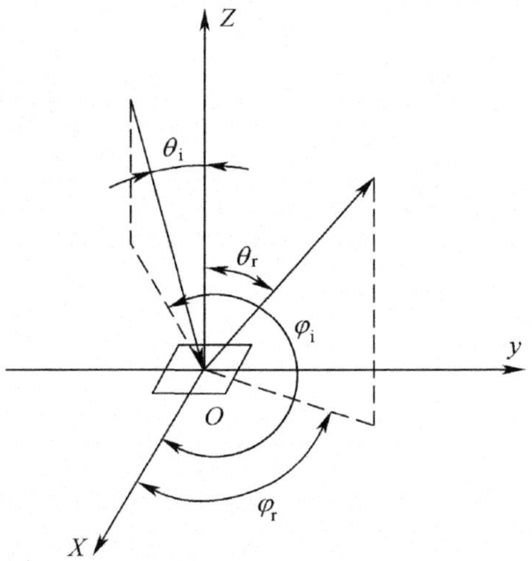

Generally, ρ' is a function of θ_i, φ_i, θ_r and φ_r, and any of the parameters changing may change ρ'. For diffuse incident beam (such as sky light), E_i is independent of $(\theta_i\ \varphi_i)$, so

$$\rho''(\theta_r\varphi_r) = \frac{L\ (\theta_r\varphi_r)}{E_i}$$

It can be seen from the equation that the directional reflection factor ρ'' is independent of the incident situation of the radiation source.

The brightness L of diffuse reflector or Lambert body is independent of observation angle θ_r and φ_r, as well as incidence direction θ_i and φ_i, so ρ' for Lambert body is constant.

$$\rho' = \frac{L}{E_i}$$

That is, the directional reflection factor ρ' of the Lambert body is independent of the direction of incident and observation.

Due to the definition of hemispheric reflectance $\rho(\lambda)$

$$\rho = \frac{M}{E_i}$$

It is given by Eq. 4.12 of Lambert

$$M = \int_{2\pi} L \cos\theta \mathrm{d}\omega \tag{4.27}$$

Deducing

$$\rho E_i = \int_{2\pi} E_i \rho' \cos\theta \mathrm{d}\omega \tag{4.28}$$

The relation between the directional reflection factor $\rho'(\theta_i\varphi_i,\ \theta_r\varphi_r)$ and the reflectance ρ can be obtained by comparing Eqs. 4.27 and 4.28

$$\begin{aligned}\rho &= \int \rho' \cos\theta \mathrm{d}\omega \\ &= \int_0^{2\pi} \mathrm{d}\phi \int_0^{\pi/2} \rho' \cos\theta \sin\theta \cdot \mathrm{d}\theta\end{aligned}$$

The constant ρ' of Lambert body is independent of θ and φ, so it can be obtained from the integral of the above equation

$$\rho = \pi \rho'$$

The brightness of an object when viewed from a certain direction

$$L(\theta_r \phi_r) = \rho'(\theta_i \phi_i,\ \theta_r \phi_r)\ E_i(\theta_i \phi_i)$$

Substitute $E_S = \cos\theta \cdot E_0 / D^2$, $E = E_S + E_D$, E_S direct solar light and E_D diffuse sky light into the above equation

$L(\theta_r \varphi_r) = \rho'(\theta_i \varphi_i, \theta_r \varphi_r) \cdot \cos\theta_i \cdot E_0 / D^2 + \rho''\ (\theta_r \varphi_r)\ E_D$, E_0 is the solar constant, and D is AU.

For Lambert body, ρ' is constant, so

$$L = \rho'\ E_i = \rho E_i / \pi \quad (4.29)$$

This formula is often used, but it must be used with the premise of Lambert.

It is known from life experience that when there are two reflective surfaces, one is smooth and the other is rough, the reflected light can only be observed in the mirror reflection direction of $\theta_i = \theta_r$ on the smooth surface, and the reflected light cannot be observed in other directions. The roughness h of this smooth surface should satisfy:

$$H \leq \lambda / (8 \cos\theta)$$

The smooth surface is a mirror, where λ is the wavelength of the incident beam and θ is the angle between the normal line of the plane and the incident beam.

The other rough surface is the diffuse reflection surface, and its reflection brightness L is a constant; that is, when this surface is observed from any angle, its brightness is independent of the observation angle as long as the incident irradiance is unchanged. If an observer looks at the two planes vertically, the first mirror appears "black"; when the observer looks from the specular reflection direction, the brightness of the specular surface is much greater than that of the diffuse surface, but the reflectivity of the two is the same. The actual surface is somewhere in between, neither specular nor diffuse and has a complex directional distribution.

(2) Irradiance of point source to microplane element [5]

As shown in Fig. 4.9, O is the point source, the distance of the surface element dA under irradiance is l, the included angle α between the normal line N of the surface element and the radiation direction, and the solid angle $d\omega$ between the surface element and the point source is

$$d\omega = \frac{dA \cos\alpha}{l^2}$$

Suppose the radiation intensity of the point source in the direction of the surface element is I, then the flux of the radiation from the point source to solid angle $d\omega$ is

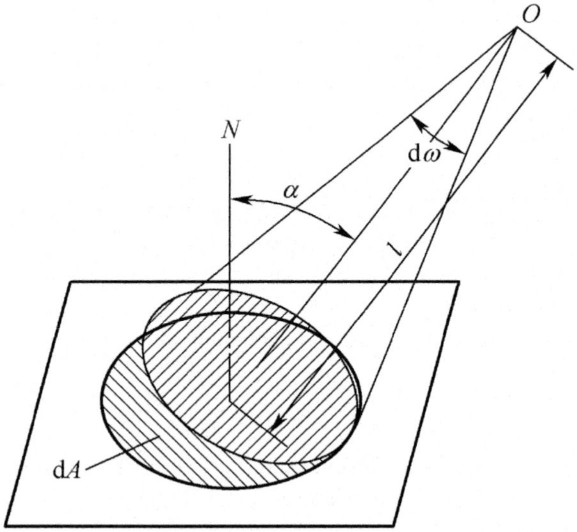

Fig. 4.9 Irradiance of a point source on a plane element

$$d\phi = I d\omega = \frac{I dA \cos\alpha}{l^2}$$

Ignoring the energy loss in light transmission, the irradiance on the microelement is

$$E = \frac{d\phi}{dA} = \frac{I \cos\alpha}{l^2} \tag{4.30}$$

(3) Radiant flux of point source to disk

As shown in Fig. 4.10, O is the point source, the distance between the irradiated disk and the point element is l_0, the radiation direction is perpendicular to the disk, and the radius of the disk is R. Irradiance varies at different radius of the disk.

Take the plane element dA on the disk, and the radiant flux on it is, see Eq. 4.30

$$d\phi = E dA = \frac{I dA \cos\alpha}{l^2}$$

$$dA = \rho d\theta d\rho$$

$$\cos\alpha = \frac{l_0}{\sqrt{\rho^2 + l_0^2}}$$

$$d\phi = I\rho d\theta d\rho \frac{l_0}{\left(\rho^2 + l_0^2\right)^{3/2}}$$

Fig. 4.10 Radiant flux from a point source to a disk

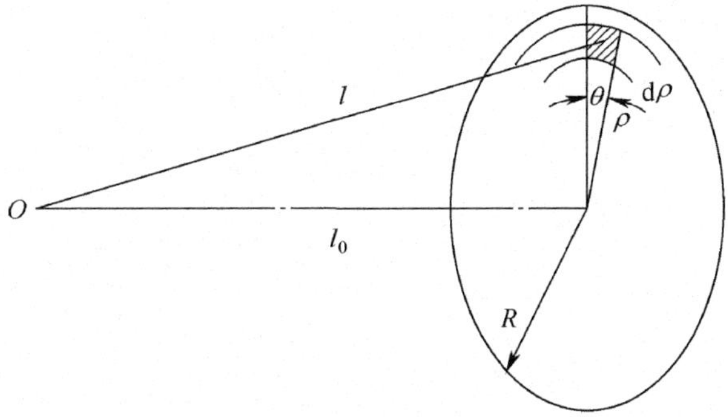

The total radiant flux on the disk is obtained by integrating the above equation with respect to ρ and θ

$$\begin{aligned}\phi &= I l_0 \int_0^{2\pi} \mathrm{d}\theta \int_0^R \frac{\rho}{\left(\rho^2 + l_0^2\right)^{3/2}} \mathrm{d}\rho \\ &= 2\pi I \left(1 - \frac{l_0}{\sqrt{l_0^2 + R^2}}\right)\end{aligned} \tag{4.31}$$

When $l_0 \gg R$ 时, $l \approx l_0$, cos$\alpha \approx 1$, as a plane element, the irradiance of the disk is equal everywhere, and its radiation flux is

$$\varphi = \frac{\pi I R^2}{l_0^2}$$

(4) Irradiance of surface source on microplane element

In Fig. 4.11, O is the irradiated plane element. A is the surface radiation source, and the distance between the plane element dA and the irradiated point O is l. The angle β is between the radiation direction and the normal line of the surface radiation source dA, the angle α is between the radiation direction and the normal line of the irradiated surface element O, the solid angle of dA to O is dω, and the irradiance of the point source on the surface element is shown in Formula 4.30.

$$\mathrm{d}E = \frac{I_\beta \cos\alpha}{l^2}$$

where I_β is the radiation intensity of dA in the direction of β.

According to the definition of radiation intensity and radiation radiance, suppose the radiation radiance of dA in the direction of β is L_β, and it is given

$$I_\beta = L_\beta \mathrm{d}A \cos\beta$$

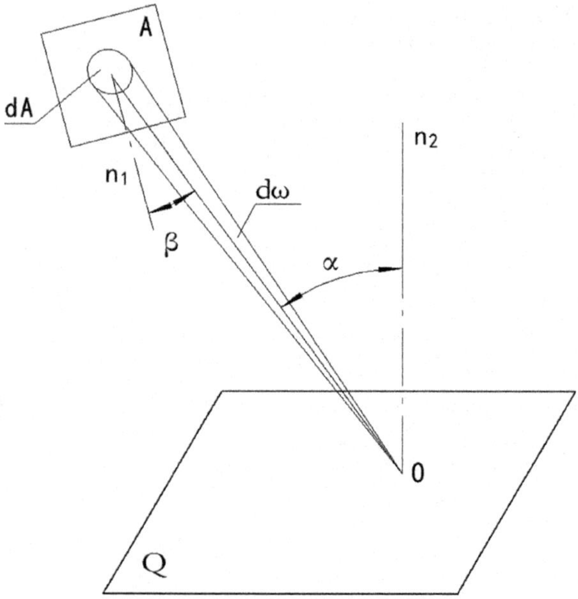

Fig. 4.11 Irradiance of surface source on microplane element

$$dE = \frac{L_\beta dA \cos\beta \cos\alpha}{l^2}$$

The solid angle $d\omega$ stretched by dA relative to O is as follows:

$$d\omega = \frac{dA \cos\beta}{l^2}$$
$$dE = L_\beta \cos\alpha d\omega$$

Irradiance of the whole surface source A to the surface element O

$$E = \int_A dE = \int_A L_\beta \cos\alpha d\omega$$

Generally, the brightness of plane sources is not equal in all directions, and it is difficult to integrate. If it is equal, L_β is constant

$$E = L_\beta \int_A \cos\alpha d\omega$$

(5) Irradiancc of Lambert radiation source

As shown in Fig. 4.12, the radius b of expanding lambert source and the brightness is L, and take its plane element $dA = x d\varphi dx$. Find the irradiance of a point h away

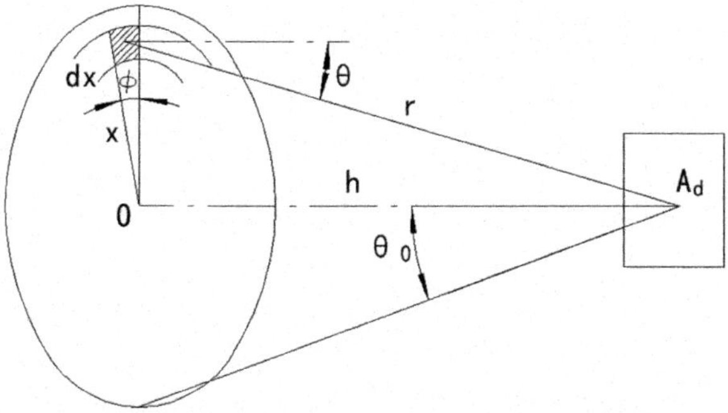

Fig. 4.12 Irradiance of Lambert radiation source

from Lambert source.

$$\mathrm{d}E = L\frac{\cos^2\theta}{r^2}x\mathrm{d}\phi\mathrm{d}x$$

As shown in the figure,

$$r = h/\cos\theta,\ x = h\tan\theta,\ dx = hd\theta/\cos^2\theta$$

The irradiance of a point on the source axis of the disk expansion can be obtained by integrating the above formula.

$$\begin{aligned} E &= L\int_0^{2\pi}\mathrm{d}\phi\int_0^{\theta_0}\sin\theta\cos\theta\mathrm{d}\theta \\ &= \pi L\sin^2\theta_0 \\ &= M\sin^2\theta_0 \end{aligned}$$

The condition that the extended source is approximate to a point source is shown in Fig. 4.12

$$\sin^2\theta_0 = \frac{b^2}{h^2+b^2}$$

The area of the disk $A = \pi b^2$

$$\begin{aligned} E &= \pi L\sin^2\theta_0 \\ &= \pi L\frac{b^2}{h^2+b^2} \\ &= \frac{LA}{h^2}\frac{1}{1+(b/h)^2} \end{aligned}$$

If the disk is a point source, the irradiance of that point on the axis is

$$E_0 = \frac{LA}{h^2}$$

If the Lambert source is regarded as a point source, then $E \approx E_0$

$$\frac{b}{h} \approx 0$$

It can be known that $h \geq 10\,b$ and the error <1%.

The error of irradiance is used to judge whether point source or surface source. This can make a big difference when the focus of concern is different. For example, when the brightness of satellite in GEO orbit on the ground is observed, it is intuitively thought it is a point source. When measuring its brightness with a long focal length CCD camera, it needs to be used as a surface source; otherwise, it will cause a big error.

4.1.5 Radiation from an Actual Object

Actual objects are neither blackbody nor white body, and the absorptivity$\alpha$of them is less than 1. And α is a function of wavelength and temperature. The radiant of the actual object, $M_e(\lambda T)$, is less than that of the blackbody $M_b(\lambda T)$. The ratio of $M_e(\lambda T)$ to the blackbody $M_b(\lambda T)$ at the same temperature is usually called specific emissivity $\varepsilon(\lambda T)$ to describe the radiation of actual objects.

$$\varepsilon(\lambda T) = \frac{M_e(\lambda T)}{M_b(\lambda T)}$$

By comparing the above formula with the definition of universal function, it can be known that

$$\varepsilon(\lambda T) = \alpha(\lambda T)$$

This formula shows that the specific emissivity $\varepsilon$of the object is equal to the absorption rate α at the same temperature and wavelength. The specific emissivity $\varepsilon(\lambda T)$ is a dimensionless quantity between 0 and 1, $\varepsilon \in (0\text{–}1)$.

By means of the specific emissivity $\varepsilon(\lambda T)$, the spectral power $W(\lambda T)$ of an object with area A can be obtained as

$$\begin{aligned} W(\lambda T) &= A\varepsilon(\lambda T)M_b(\lambda T) \\ &= A\varepsilon(\lambda T)\frac{C_1}{\lambda^5(e^{C_2/\lambda T}-1)} \end{aligned}$$

The average specific emissivity $\overline{\varepsilon}(\lambda_1 \sim \lambda_2)$ for a wavelength range (λ_1 to λ_2) is

$$\overline{\varepsilon}(\lambda_1 \rightarrow \lambda_2) = \frac{\int_{\lambda_1}^{\lambda_2} \varepsilon(\lambda T) \frac{C_1}{\lambda^5 (e^{C_2/\lambda T} - 1)} \mathrm{d}\lambda}{\int_{\lambda_1}^{\lambda_2} \frac{C_1}{\lambda^5 (e^{C_2/\lambda T} - 1)} \mathrm{d}\lambda}$$

Generally, $\overline{\varepsilon}(\lambda_1 \sim \lambda_2)$ is associated with the wavelength range, $\lambda_1 \sim \lambda_2$. Objects that $\overline{\varepsilon}$ are independent of the wavelength range λ_1 to λ_2 are called gray bodies, which means an object that $\varepsilon(\lambda T)$ does not change with λ.

The surface radiance L of the actual object is also related to the direction of the emittance θ. The specific emissivity ε is not only a function of λ and T, but also a function of the emittance angle θ. Figure 4.13 shows the angular distribution of the specific emissivity ε of a typical smooth dielectric[2].

If the dielectric is a non-transparent body, then

$$\varepsilon(\theta) = \alpha(\theta) = 1 - \rho(\theta)$$

For natural light, mirror surface reflectance $\rho(\theta)$ is a function of θ [2]

$$\rho(\theta) = \frac{1}{2}\left[\frac{\sin^2(\theta_i - \theta_t)}{\sin^2(\theta_i + \theta_t)} + \frac{\tan^2(\theta_i - \theta_t)}{\tan^2(\theta_i + \theta_t)}\right]$$

In the formula, θ_i is the angle of incidence, θ_t is the angle of refraction.
Hence,

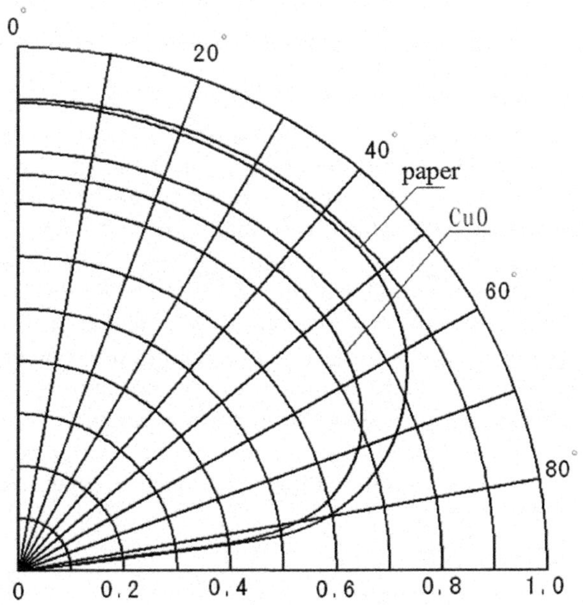

Fig. 4.13 Angular distribution of specific emissivity ε of the dielectric

$$\begin{aligned}\varepsilon(\theta) &= 1 - \rho(\theta) \\ &= 1 - \frac{1}{2}\left[\frac{\sin^2(\theta_i - \theta_t)}{\sin^2(\theta_i + \theta_t)} + \frac{\tan^2(\theta_i - \theta_t)}{\tan^2(\theta_i + \theta_t)}\right]\end{aligned}$$

When $\theta = 0$

$$\varepsilon(0) = \frac{4n}{(1+n)^2}$$

In the formula, n is the refractivity of the dielectric, which indicates that the normal specific emissivity of the dielectric is $4n/(1+n)^2$ times of blackbody.

In practical application, the angular distribution of the specific emissivity $\varepsilon(\theta)$ of smooth media is sufficient in the following form [2]:

$$\varepsilon(\theta) = \cos^\alpha \theta \frac{4n}{(1+n)^2}$$

where α is a number less than 1, and it can be obtained from experiments.

The angular distribution of conductor $\varepsilon(\theta)$ can be approximated by the following Formula [2]:

$$\varepsilon(\theta) = \frac{1}{n}\left(\cos\theta_i + \frac{1}{\cos\theta_i}\right)$$

When $\theta \to \pi/2$, the above formula is incorrect.

It is known from Planck's formula that the radiant emittance $M_b(\lambda T)$ of a blackbody is a function of temperature T and wavelength λ, and the Stefan-Boltzmann law of the total radiated power $M_{bb} = \sigma T^4$ becomes a function of temperature T only after integrating the wavelength. Therefore, considering the influence of $\varepsilon(\lambda, T, \theta)$, there are three ways to obtain the temperature of object from the perspective of remote sensing. The temperature of the object can be determined by measuring the total radiation power. Or with the radiant emittance $M_b(\lambda T)$ to measure the temperature of the object, which has two ways, when the temperature T of the object can be set, by $M_b(\lambda T)$ to obtain the spectrum $M_{b\lambda}(\lambda T)$ of each wavelength, is to determine the object's temperature with the spectral radiance distribution of the object, and graphically called "color temperature"; after setting the temperature and wavelength of the object, the spectrum $M_{b\lambda i}$ (λT) of the corresponding wavelength can be obtained by $M_b(\lambda T)$, and the temperature of the object can be determined by measuring $M_{b\lambda i}$ (λT) and vividly called "brightness temperature".

The conclusions of "color temperature", "brightness temperature" and total radiation power temperature (total radiation temperature) are strictly valid for blackbody, and there is a reciprocity relationship, while for non-blackbody, there is no reciprocity relationship due to the influence of specific emissivity.

(1) Color temperature

The color temperature is defined as: if the relative spectral distribution $L(\lambda T)$ of an object with temperature T is consistent with that of a blackbody with temperature T_c, then the temperature T_c of the blackbody is called the color temperature of the object, that is,

$$\frac{L_b(\lambda_1 T_c)}{L_b(\lambda_2 T_c)} = \frac{L(\lambda_1 T)}{L(\lambda_2 T)} = \frac{\varepsilon_1 L_b(\lambda_1 T)}{\varepsilon_2 L_b(\lambda_2 T)}$$

where $L_b(\lambda_1 T_c)$ and $L_b(\lambda_2 T_c)$ are the radiance of the blackbody with temperature of T_c at the wavelength λ_1 and λ_2. $L(\lambda_1 T)$ and $L(\lambda_2 T)$ are the radiance of the body with temperature of T at the wavelength λ_1 and λ_2. $L_b(\lambda_1 T)$ and $L_b(\lambda_2 T)$ are the radiance of the blackbody with temperature of T at the wavelength λ_1 and λ_2, as well as ε_1 and ε_2 are the corresponding specific emissivity. If the above relation is satisfied, the color temperature of the object between wavelength λ_1 and λ_2 is called T_c.

In different spectral segments, the relative spectral distribution of the same radiation source can be consistent with that of the blackbody at different temperatures, so there can be different color temperatures in different spectral segments. For the gray body $\varepsilon_1 = \varepsilon_2$, then $T_c = T$, which means the color temperature of the gray body is its actual temperature, but for non-gray body $T_c \neq T$.

The color temperature considers the relative distribution of radiated power, not the absolute radiated power. The radiated power of an object with a higher color temperature may be less than another object with a lower color temperature. Because the intensity of Rayleigh scattering is proportional to λ^{-4}, the proportion of blue light in sky light is relatively large, while the proportion of red light is small, and its relative spectrum is similar to that of 25,000 K blackbody. The color temperature of sky light is 25000 K, while the color temperature of the sun is 6000 K. Obviously, the radiation intensity of sky light is much weaker than that of the sun.

(2) Brightness temperature

If there is a blackbody whose radiance $L_b(\lambda T_b)$ at temperature T_b is same as that of $L(\lambda T)$ of an object, the brightness temperature of the object is called to be T_b, which is

$$L(\lambda T) = L_b(\lambda T_b)$$

The relationship between the radiance $L(\lambda T)$ of the actual radiation source and the radiance $L_b(\lambda T)$ of the blackbody at the same temperature satisfies the following equation:

$$L(\lambda T) = \varepsilon(\lambda)\, L_b(\lambda T)$$

Since ε is less than 1, the brightness temperature T_b of the object must be lower than the actual temperature T. As well as $\varepsilon(\lambda)$ is a function of wavelength λ, the brightness temperature T_b is also a function of wavelength λ.

Generally, the brightness temperature T_b has a complex relation with ε and T, but when $\lambda \gg \lambda_{max}$, under the Sharp-Kings approximation, the expression of L_b is simple: $L_b = kT{\cdot}2c_0/\lambda^4$

The formula can be obtained by $L(\lambda T) = \varepsilon(\lambda)\, L_b(\lambda T)$ and $L(\lambda T) = L_b(\lambda T_b)$

$$L(\lambda T) = \varepsilon\, L_b(\lambda T) = L_b(\lambda T_b)$$

Substitute $L_b = kT{\cdot}2c_0/\lambda^4$ into the above equation

$$T_b = \varepsilon T$$

This formula shows that when $\lambda \gg \lambda_{max}$, the brightness temperature T_b is equal to the product of the actual temperature T and the specific emissivity ε.

(3) Total radiation temperature

To measure the total radiation temperature is to measure the total radiation M_{bb} of the object, let the temperature of the object be T, specific emissivity be ε, and it is given:

$$\varepsilon \sigma T^4 = M_{bb}$$

The temperature T of the object can be obtained from the above formula. Generally, the lower the specific emissivity ε is, the greater error of the measurement result is.

For example, iron with specific emissivity of 0.8 and aluminum with specific emissivity of 0.1 are heated to 500 °C, and the radiation temperatures of iron and aluminum read by the total radiation thermometer calibrated with blackbody are respectively

$$T_{Fe} = 0.81/4\,(500 + 273) - 273 = 458\,^{\circ}\mathrm{C}$$
$$T_{Al} = 0.11/4\,(500 + 273) - 273 = 162\,^{\circ}\mathrm{C}$$

The measured temperature is T, the temperature of the object being measured is T_o, and the relationship is

$$T_0 = \frac{T + 273}{\varepsilon^{1/4}} - 273$$

4.1.6 The Electromagnetic Spectrum of Radiation

Electromagnetic radiation is distributed over a wide range. According to the division of radio spectrum by International Telecommunication Union (ITU), extremely low frequency (ELF) is 3 to 30 Hz and wavelengths are 10^7 to 10^8 m. The high frequency of electromagnetic radiation is cosmic ray frequency $>3 \times 10^{24}$ Hz, wavelength $<10^{-16}$ m. The most commonly used segment in optical remote sensing is visible spectrum, as defined by the International Commission on Illumination (CIE), ranges from 0.38 μm wavelength for violet light at 789 THz frequency to 0.78 μm wavelength for red light at 385 THz frequency. The spectral representation of electromagnetic radiation is shown in Fig. 4.14.

The electromagnetic spectrum division method of the International Telecommunication Union is basically 10 times the frequency to divide the electromagnetic wave band.

Very low frequency (VLF) and low frequency (LF) are located in the longwave band and are used for transoceanic long-distance communication and navigation.

Intermediate frequency (MF) is located in the middle wave band and is used for AM broadcasting (535–1700 kHz).

High frequency (HF) is located in the shortwave band and is used for broadcasting, navigation, amateur use and civil radio telegraphy.

Very high frequency (VHF) is located in the ultra-shortwave band (meter wave), used for television, remote control, FM broadcasting (88–108 MHz), walkie-talkies and radio navigation.

Ultra-high frequency (UHF) is located in the microwave band (decimeter wave) and is used for cell phones, pagers, cordless phones, GPS, UHF TV channels and microwave ovens.

Super-high frequency (SHF) is located in the microwave band (centimeter wave) and is used for radar and satellite communications.

Extremely high frequency (EHF) is located in the microwave band (millimeter wave) and is used for satellite communications.

Tremendously high frequency (THF) is in the submillimeter wave band, 300–3000 GHz, 100–1000 μm, overlaps with the far infrared band (see Fig. 4.14) and also overlaps with the terahertz wave 0.1–10 THz, 30–3000 μm.

Infrared, in the spectral range, 3–300 THz, 1–100 μm.

Visible light is divided into:

Red, 780–620 nm, 385–484 THz;
Orange, 620–600 nm, 484–500 THz;
Yellow, 600–580 nm, 500–517 THz;
Green, 580–490 nm, 517–612 THz;
Blue, 490–450 nm, 612–667 THz;
Violet, 450–380 nm, 667–789 THZ;
Ultraviolet, 400–10 nm, 750 THz–30,000 THz;
X-ray, 10 nm–0.1 nm, 30,000 THz–3,000,000 THz;

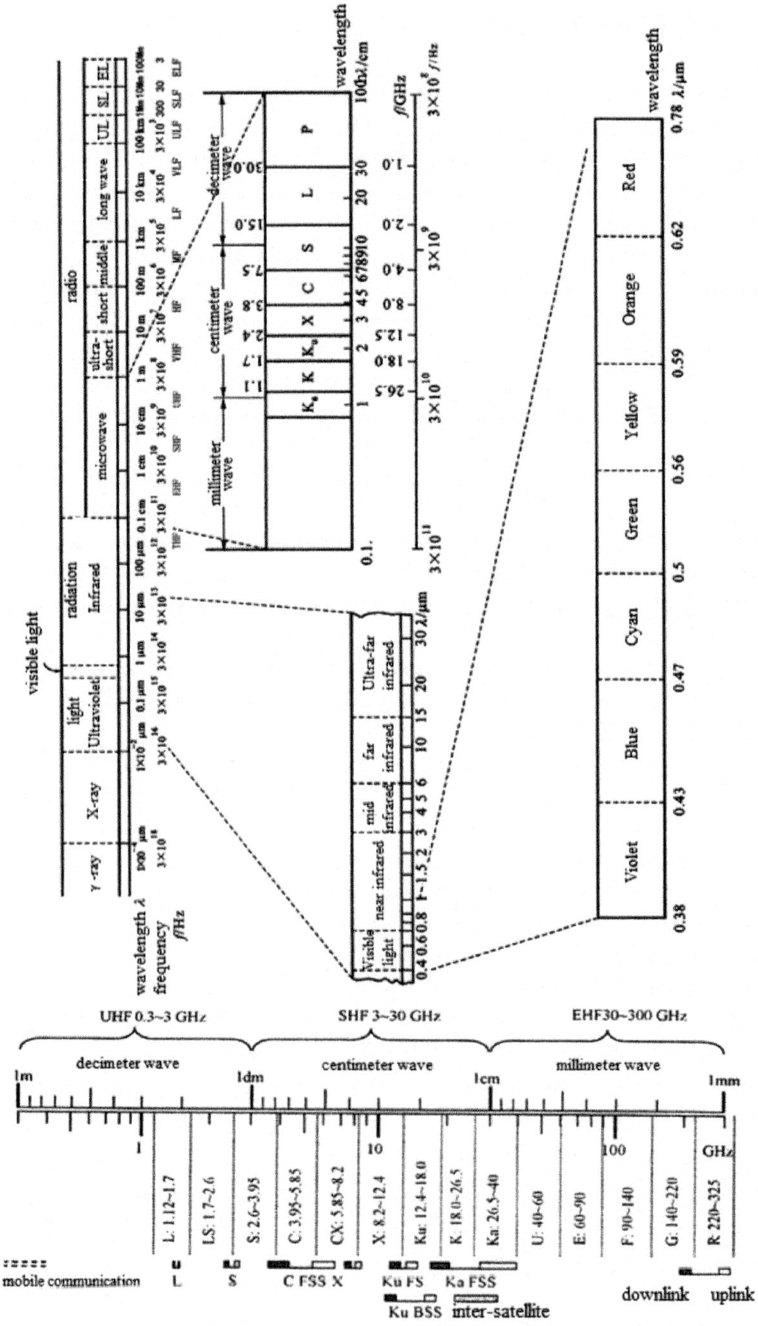

Fig. 4.14 Electromagnetic radiation spectrum

Y-ray, <0.1 nm, >3,000,000 THz.

It is commonly to expand ultraviolet to X-ray range, infrared to terahertz wavelength range and put them into the optical radiation scope. The definition of optical radiation frequency range is mainly on the basis of the interaction of electromagnetic radiation and transmission media, because the medium with different spectrum has very different response, and the processing of the electromagnetic wave method in this frequency range is basically the same as the visible light. The range of light radiation can be defined in the extreme ultraviolet wavelength of 10 nm and frequency of 30,000 THz, to the extreme infrared wavelength of 1 mm and frequency of 300 GHz.

4.2 Solar Radiation

4.2.1 Solar Profile

The sun is the most important light source and energy source in space optical remote sensing. The radius of the sun observed at the mean sun–earth distance is about 16', or 4.6524 mrad. The earth's angle of extension to the sun is only 8.79418″, or 42.6354 μrad. Since the angle between the sun and the earth is very small, direct sunlight can be considered as a parallel beam, but this is only a simple intuitive impression. When studying optical remote sensing, its resolution can reach 0.1″. At this time, the parallelism of sunlight rays cannot be considered as a parallel beam compared with the parallel rays required for camera testing.

The total power of solar radiation is 3.846×10^{26} W. Its enormous power comes from thermonuclear reactions on the sun where hydrogen atoms fuse together to form helium. The sun consists of 75% hydrogen, 25% helium, 0.8% oxygen and a few other elements. The composition of the sun suggests that it will not run out of energy for quite some time. It is estimated that the central temperature of the sun is 15×10^6 K, and the color temperature is about 6000 K. The hierarchical structure of the sun is shown in Fig. 4.15.

From earth, the sun has a clear, spherical shape, which is represented by the photosphere. The photosphere is a very thin layer of the solar atmosphere, about 300 km thick. The photosphere is opaque. All the radiation inside the sun is absorbed by the solar atmosphere in the photosphere, so it is impossible to see what is going on inside the sun. What can be seen is only radiation from the photosphere itself. The temperature of the photosphere ranges from 7500 to 4300 K from the bottom to up.

Above the photosphere to a height of 7000–8000 km is the chromosphere of the solar atmosphere. The thickness of the chromosphere is 2000 km, and the temperature is 4000–6000 K [6]. Due to the influence of the bright photosphere, people cannot see the chromosphere normally. Only in total eclipses does this layer of gas appear as a beautiful rose-colored border of the sun's wheel, called a chromosphere. The serrated edges of the chromosphere are caused by strong updrafts. Sometimes the

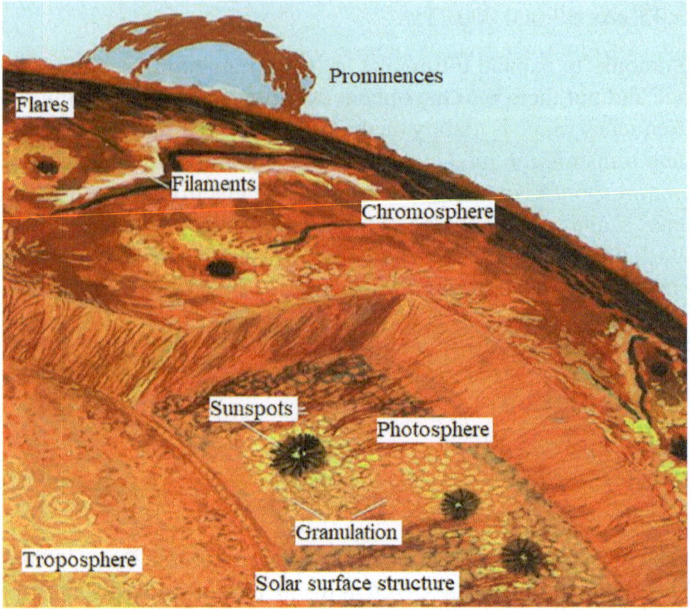

Fig. 4.15 Sun spots on the solar sphere

updrafts are so strong that they soar into the space as columns of gas tens or even millions of kilometers in the space. Such columns are called prominences. For a period of time, prominences hang like earrings from the rim of the sun's wheel. The temperature of the chromosphere increases with the altitude, and it increases sharply above 2000 km and reaches 10^6 K order of magnitude at the top of the chromosphere.

The corona is the outermost layer of the sun's atmosphere. Its brightness is only one thousandth of that of the chromosphere and one millionth of that of the photosphere. Therefore, the corona cannot be seen at ordinary times. The yellowish or white corona can only be seen during a total eclipse or with special viewing tools. The corona varies in shape and thickness, reaching 7–8 million kilometers at its thickest, equivalent to a dozen solar radiuses. The corona typically extends to 4 to 5 solar radii. The temperature of the corona is 10^6 K, which is similar to the temperature on the top of the chromosphere. The matter in the corona moves at a very high speed, such as the average speed of hydrogen nuclei (H^+) can reach 220 km/s. Therefore, some particles in the corona can escape the binding of the sun's gravity and enter into the space. So the sun not only radiates electromagnetic waves, it also spews out a constant stream of high-speed particles known as the solar wind. The solar wind travels at 320 to 770 km/s, and it takes only five to six days to reach earth orbit from the sun. Measured near earth, these particles have an average temperature of 150,000 °C and a particle density of about five per cm^3.

The sun's matter is in constant motion and change. Its activity shows spots and facular on the photosphere, prominences and flares on the chromosphere and changes in the shape of the corona. The sun at low tide is called the quiet sun, and the sun at high tide is called the disturbance sun. An important aspect of solar activity is the activity of sunspots, but a more important indicator of solar activity is the frequent occurrence of flares.

The brightness of different parts of the photosphere is not the same. Sunspots often appear on the photosphere. The temperature of the sunspots is slightly lower than that of the normal photosphere, so they appear black on the background of the bright photosphere. The average cycle of sunspots is about 11 years. When sunspots increase, it means that solar activity is strong, which interferes with radio communication and space flight.

In small areas above sunspot groups, the brightness sometimes suddenly increases, and this phenomenon is called a flare. Flare is an eruption on the sun that lasts only a few minutes, sometimes a few hours. During this time, the sun emits as much energy as 10 billion megaton hydrogen bombs. During a flare, the radio intensity of the sun can be suddenly increased by millions of times, and it also emits a large amount of ultraviolet rays, X-rays and a large number of high-energy particles, as shown in Fig. 4.16.

Solar radiation is mainly shortwave, and the energy is huge. The sun loses 4 million tons of mass every second, becoming energy into space. Although the energy captured by the earth is only their 2,200,000,000th of one, but it can still get the equivalent of 400 million tons of coal per minute quantity of heat, so the influence of solar radiation on earth and humans is very big. The sun has consumed only 0.03% of its mass in its five billion years and is now in its stable, exuberant middle age.

Fig. 4.16 Solar radiation storm

4.2.2 Solar Radiation Spectrum

The solar constant is a physical quantity describing the density of solar radiant energy flux. It refers to the total radiant energy projected by the sun on the unit area perpendicular to the ray direction outside the earth's atmosphere in unit time at the average distance between the earth and the sun (AU = 1.49597870 × 10^8 km [7]). The value of the solar constant published by the World Meteorological Organization (WMO) in 1981 is $E_0 = 1367$ W/m^2. The solar constant measured by China's space flight in 2002 is 1366.7 W/m^2 with an uncertainty of 0.2%. During the nearly 30 years from 1975 to 2002, the uncertainty of the solar constant published is 0.2%, which reflects the stability and reliability of the numerical value. The solar constant is measured with the spectral range 0.1 μm ~ 40 μm, and the measured radiation energy accounts for 99.998% of the full spectrum radiation energy of the sun. From the solar constant, it can be calculated that the total solar radiation power is 3.8444 × 10^{26} W, and the solar surface radiation emission M_e is 6.316 × 10^7 W/m^2. The effective temperature of the sun is calculated to be 5777 K using Stephan-Boltzmann's law. The spectrum of solar radiation is continuous, as shown in Fig. 4.17. In the visible and infrared range, the sun has a color temperature of 6000 K.

The continuous spectrum is emitted by the sun's photosphere. Since the photosphere absorbs all the high-temperature (10^6 K) radiation in the sun, that is, the absorption rate $\alpha = 1$, so the radiation characteristics of the photosphere are consistent with the absolute blackbody, and its color temperature is about 6000 K. When the spectrum of the sun is carefully examined with a high-resolution spectrometer, it is found that there are many discrete dark lines on a bright continuous background, which are called Fraunhofer absorption line. This is due to the absorption of atoms in the photosphere. The sun's atmosphere contains a large number of atoms of various elements, which are scattered when radiation streams collide with them. When the photon energy $h\nu$ is equal to the energy difference $\Delta E = h\nu$ of some two energy levels in the atom, the atom will absorb the photon energy and be excited to the excited state with higher energy, which is called resonance absorption. Atoms in an excited state will radiate photons of the same energy $h\nu$, as they revert back to their original state. When absorption and radiation are in dynamic equilibrium, the total number of absorbed photons is equal to the total number of radiated photons, but the radiated photons are in the direction of 4π, while the original radiation flow is directional radiation, so the number of photons with energy of $h\nu$ decreases when observed against the direction of the original radiation flow, resulting in absorption lines. More than twenty thousand absorption lines are said to have been found in the visible region of the solar spectrum that shows the Fraunhofer absorption spectrum line formed by the absorption of elements at the bottom of the chromosphere toward continuous spectral radiation of the photosphere, and the details of the high-resolution solar spectrum within the spectral width of 1 nm are shown in the figure[6] (Figs. 4.18 and 4.19).

Since the wavelength of the absorption line is same as that of the emission line, the composition of the solar atmosphere can be learned from the Fraunhofer absorption

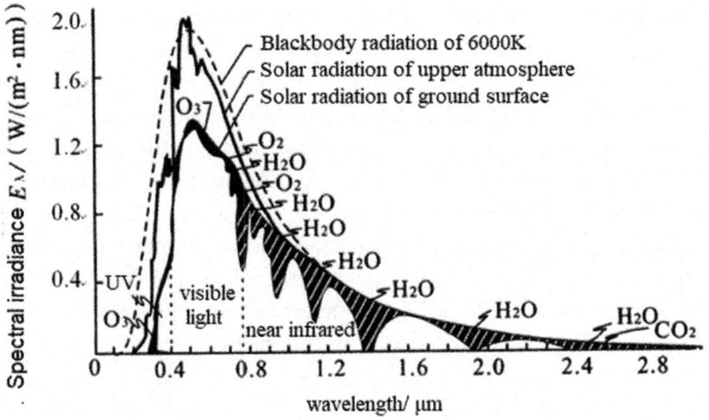

Fig. 4.17 Solar radiation spectrum of upper atmosphere and ground surface

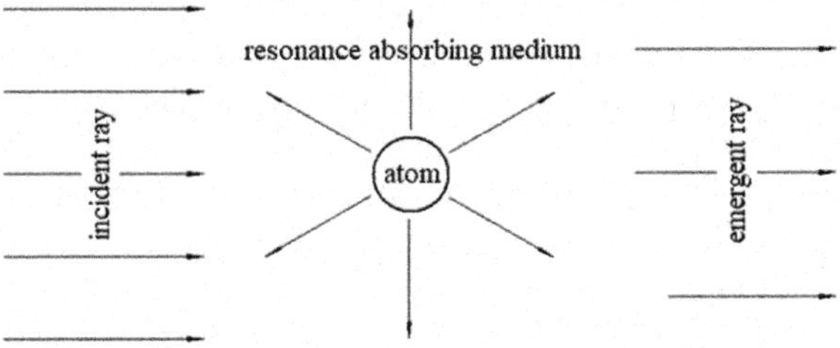

Fig. 4.18 Generation of absorption spectra

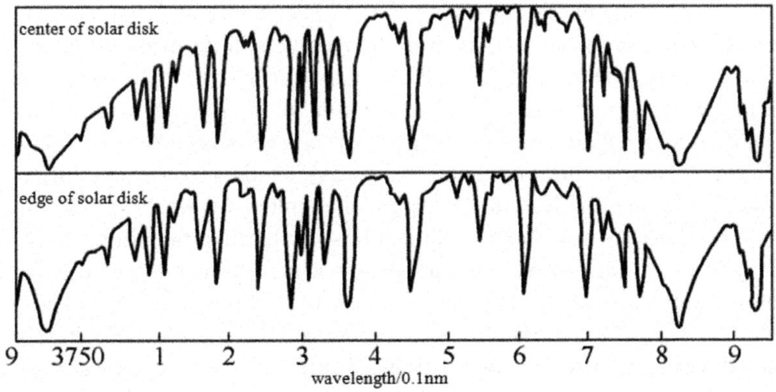

Fig. 4.19 Fraunhofer absorption lines of solar radiation

line. The chromosphere's solar atmosphere is transparent and radiates a line spectrum, many of which are the lines of ionized atoms. The corona radiates a continuous spectrum similar to that of the photosphere, but only a millionth of the intensity. It is caused by free electrons in the corona scattering radiation from the photosphere. The chromosphere and corona do not contribute much to the solar radiation, and it can be thought that the solar radiation in the visible and infrared spectrum almost all comes from the photosphere.

Table 4.4 [2] shows the spectral solar irradiation parameters in the outer atmosphere of the earth and directly gives the spectral width and the corresponding irradiance. The $E_0(\lambda)$ in the Table 4.4 is the average spectral irradiation at the mean distance from the earth to the sun (1 AU), within a small $\Delta\lambda$, in unit W cm^{-2} μm^{-1}, on the IPTS1968 scale, and the center is in wavelength λ, micron /μm. In the Table 4.4, in 0.2–0.3 μm spectrum band, $\Delta\lambda = 0.01$ μm; in 0.6–1.0 μm spectrum band, $\Delta\lambda = 0.01$ μm; in 1.0–5.0 μm spectrum band, $\Delta\lambda = 0.1$ μm; in 5.0–20 μm spectrum band, $\Delta\lambda = 1$ μm. In the Table 4.4, the total area under spectral irradiation curve is 1359.784 W/m^2. For spectral brightness or irradiance of any spectral band, please refer to Table 4.5 parameters of solar spectral irradiance and integral irradiance of outer atmosphere [7]. However, no matter which table is used, the solar constant outside the atmosphere is less than 1367 W/m^2, which needs to be corrected for accurate solution. In the Table 4.5, $E_0(\lambda)$ is the spectral brightness of the sun at the mean distance from the earth to the sun (1AU) outside the atmosphere, in unit Wm^{-2} μm^{-1}, $\int_0^{\lambda} E_0(\lambda)\mathrm{d}\lambda$ is the integral irradiance from $\lambda = 0$ to λ, from the mean distance between the sun and the earth (1AU) outside the atmosphere, in unit Wm^{-2} and the total area under spectral irradiation curve (solar constant) is 1353 W/m^2.

The percentage of solar radiant energy is shown in the table. As can be seen from Table 4.6, most of the energy of solar radiation is concentrated in the near UV-mid-infrared region, and the solar radiation changes very little in this spectral region, which can be considered as a very stable radiation source. The solar constant is related to the activity of sunspots. When the number of sunspots increases, the solar constant increases slightly, but the intensity of the change is not more than 1– 2%.

In X-ray, Y-ray, far ultraviolet and microwave bands, as shown in Table 4.6, the solar radiation energy is less than 1%, but their intensity varies greatly under the influence of sunspots and flares, as does the solar particle radiation. For example, the X-ray intensity of 2–8 Å can vary by three orders of magnitude between the minimum and maximum sunspots and can vary by 100 times before and after a flare in Table 4.7. These X-ray changes can have a big impact on the earth's ionosphere and can even disrupt radio communications on earth. The solar microwave and meter-wave radiation intensity, when quiet and disturbance can also differ several orders of magnitude, especially the meter wave can differ 10^3–10^6 times, but it does not affect the general communication, because the sun's angular width is very small, only 32′, and only the large radio antenna pointing to the sun can be affected.

The brightness temperature T_b of solar radiation at different wavelengths is shown in Table 4.8, in which the brightness temperature of meter wave reaches 10^6 K, because the meter-wave radiation of the sun originates from the corona.

Table 4.4 Exo-atmospheric solar spectral irradiance

λ	$E_0(λ)$	λ	$E_0(λ)$	λ	$E_0(λ)$	λ	$E_0(λ)$	$Λ$	$E_0(λ)$
0.120	0.000010	0.400	0.1440	0.585	0.1776	0.940	0.0842	4.100	0.00087
0.140	0.000003	0.405	0.1696	0.590	0.1756	0.950	0.0828	4.200	0.00078
0.150	0.000007	0.410	0.1736	0.595	0.1736	0.960	0.0810	4.300	0.00071
0.160	0.000023	0.415	0.1755	0.600	0.1723	0.970	0.0790	4.400	0.00065
0.170	0.000063	0.420	0.1736	0.610	0.1709	0.980	0.0775	4.500	0.00059
0.180	0.000125	0.425	0.1691	0.620	0.1673	0.990	0.0756	4.600	0.00053
0.190	0.000271	0.430	0.1628	0.630	0.1639	1.000	0.0741	4.700	0.00048
0.200	0.00133	0.435	0.1658	0.640	0.1612	1.100	0.0592	4.800	0.00045
0.210	0.00244	0.440	0.1787	0.650	0.1567	1.200	0.0484	4.900	0.00041
0.220	0.00517	0.445	0.1907	0.660	0.154	1.300	0.0396	5.000	0.0003830
0.230	0.00606	0.450	0.1968	0.670	0.1517	1.400	0.0336	6.000	0.0001750
0.240	0.00600	0.455	0.1995	0.680	0.1487	1.500	0.0287	7.000	0.0000990
0.250	0.00717	0.460	0.1991	0.690	0.1454	1.600	0.0244	8.000	0.0000600
0.260	0.0126	0.465	0.1982	0.700	0.1419	1.700	0.0202	9.000	0.0000380
0.270	0.0200	0.470	0.1977	0.710	0.1389	1.800	0.0159	10.000	0.0000250
0.280	0.0223	0.475	0.1995	0.720	0.1359	1.900	0.0126	11.000	0.0000170
0.290	0.0435	0.480	0.1992	0.730	0.1331	2.000	0.0103	12.000	0.0000120
0.300	0.0526	0.485	0.1905	0.740	0.1297	2.100	0.0090	13.000	0.0000087
0.305	0.0578	0.490	0.1905	0.750	0.1271	2.200	0.0079	14.000	0.0000055
0.310	0.0653	0.495	0.1943	0.760	0.1245	2.300	0.0068	15.000	0.0000049
0.315	0.0715	0.500	0.1928	0.770	0.1217	2.400	0.0064	16.000	0.0000038
0.320	0.0774	0.505	0.1908	0.780	0.1191	2.500	0.0054	17.000	0.0000031
0.325	0.0876	0.510	0.1871	0.790	0.1163	2.600	0.0048	18.000	0.0000024
0.330	0.0951	0.515	0.1831	0.800	0.1137	2.700	0.0043	19.000	0.0000020
0.335	0.0965	0.520	0.1841	0.810	0.1111	2.800	0.00390	20.000	0.0000016
0.340	0.0965	0.525	0.1856	0.820	0.1083	2.900	0.00350	25.000	0.000000610
0.345	0.0976	0.530	0.1872	0.830	0.1049	3.000	0.00310	30.000	0.000000300
0.350	0.1010	0.535	0.1884	0.840	0.103	3.100	0.00260	35.000	0.000000160
0.355	0.1029	0.540	0.1851	0.850	0.1005	3.200	0.00226	40.000	0.000000094
0.360	0.1038	0.545	0.1822	0.860	0.0985	3.300	0.00192	50.000	0.000000038
0.365	0.1093	0.550	0.1807	0.870	0.0965	3.400	0.00166	60.000	0.000000019
0.370	0.1128	0.555	0.1800	0.880	0.0944	3.500	0.00146	80.000	0.000000007
0.375	0.1111	0.560	0.1787	0.890	0.0926	3.600	0.00135	100.000	0.000000003
0.380	0.1090	0.565	0.1787	0.900	0.0908	3.700	0.00123	1000.000	0.000000000
0.385	0.1040	0.570	0.1791	0.910	0.089	3.800	0.00111		
0.390	0.1075	0.575	0.1795	0.920	0.0875	3.900	0.00103		

(continued)

Table 4.4 (continued)

λ	$E_0(\lambda)$	λ	$E_0(\lambda)$	λ	$E_0(\lambda)$	λ	$E_0(\lambda)$	Λ	$E_0(\lambda)$
0.395	0.1140	0.580	0.1785	0.930	0.0859	4.000	0.00095		

The solar irradiance received by the ground is related to the zenith angle θ of the sun, as shown in Fig. 4.20. In the case of ignoring atmospheric loss, the ground irradiance can be considered to be approximately proportional to cosθ, i.e.,

$$E = \cos\theta\, E_0/D^2$$

where E_0 is the solar constant, D is the distance between the sun and the earth in the unit of the mean distance between the sun and the earth, $D \approx 1$, $D^2 \in [1.034, 0.967]$, and θ is the zenith angle of the sun. Solar zenith angle varies with latitude, season, local time and other factors and can be calculated by the following formula:

$$\cos\,\theta = \sin\phi\,\sin\delta + \cos\phi\,\cos\delta\,\cos t \tag{4.33}$$

where φ is geographical latitude, δ is solar declination, and t is solar hour angle.

Solar declination refers to the angle between the sun's rays and the earth's equatorial plane. It varies between $\pm 23°\ 26'$ in a year. The sun is directly at the earth's equator at spring and autumnal equinoxes, so $\delta = 0$, as well as $\delta = 23°\ 26'$ at summer solstice and $-23°\ 26'$ at winter solstice. The solar hour angle is defined as $t = 0$ at local 12 o'clock, t is $-\pi/2$ at 6 o'clock, and t is $+\pi/2$ at 18 o'clock. The first term sinφ sinδ is the effect of seasonal change on solar zenith angle, while the second term cosφcosδcos$t$ is the change of zenith angle with time during a day. At high latitude, sinφ is large, so the zenith angle varies greatly with the seasons, but little with the days. At low latitudes, the opposite is true, sinφ is small, cosφ is large, and zenith angle changes little with the season, but greatly with the day.

4.3 Earth Radiation

The space optical remote sensing must pass through the atmosphere, so the earth radiation must be mentioned in the earth and air system radiation.

The earth's own radiation is reflected in changes of the earth's albedo, which varies with the local surface of the earth. Cloud reflectance is 0.10–0.80. The albedo of land ranges from 0.05 to 0.45. The reflectance of water surface is 0.03–0.20. The distribution of clouds has the greatest influence on the albedo of the earth. Therefore, although the distribution of land and water in the northern and southern hemispheres is very different, the albedo of the earth is basically symmetric with respect to the equator, as shown in Fig. 4.21. The albedo of the earth varies between 0.25 and 0.6 from the equator to the poles, with a global average of 0.30 ± 0.02. The changes of

Table 4.5 Parameters of exo-atmospheric solar spectral irradiance and integral irradiance

$\lambda/\mu m$	$E_0(\lambda)/Wm^{-2}\ \mu m^{-1}$	$\int_0^\lambda E_0(\lambda)d\lambda$	$\lambda/\mu m$	$E_0(\lambda)/Wm^{-2}\ \mu m^{-1}$	$\int_0^\lambda E_0(\lambda)d\lambda$
0.120	0.10	0.0059993	0.375	1157	89.0641
0.140	0.03	0.0073	0.380	1120	94.7566
0.150	0.07	0.0078	0.385	1098	100.302
0.160	0.23	0.0093	0.390	1098	105.792
0.170	0.63	0.0135	0.395	1189	111.509
0.180	1.25	0.023	0.400	1429	118.054
0.190	2.71	0.0428	0.405	1644	125.737
0.200	10.7	0.10985	0.410	1751	134.224
0.210	22.9	0.27785	0.415	1774	143.037
0.220	57.5	0.67985	0.420	1747	151.839
0.225	64.9	0.98585	0.425	1693	160.439
0.230	66.7	1.31485	0.430	1639	168.769
0.235	59.3	1.62985	0.435	1663	177.024
0.240	63.0	1.9356	0.440	1810	185.707
0.245	72.3	2.27385	0.445	1922	195.037
0.250	70.4	2.6306	0.450	2006	204.857
0.255	104	3.0666	0.455	2057	215.014
0.260	130	3.6516	0.460	2066	225.322
0.265	185	4.4391	0.465	2048	235.607
0.270	237	5.4816	0.470	2033	245.809
0.275	204	6.5716	0.475	2044	256.002
0.280	222	7.6366	0.480	2074	266.297
0.285	315	8.9791	0.485	1976	276.422
0.290	482	10.9716	0.490	1950	286.237
0.295	584	13.6366	0.495	1960	296.012
0.300	514	16.3816	0.500	1942	305.767
0.305	603	19.1741	0.505	1920	315.422
0.310	689	22.4041	0.510	1882	324.927
0.315	764	24.0366	0.515	1833	334.214
0.320	830	30.0216	0.520	1833	343.379
0.325	975	34.534I	0.525	1852	352.592
0.330	1059	39.6191	0.530	1842	361.827
0.335	1081	44.9691	0.535	1818	370.977
0.340	1074	50.3566	0.540	1783	379.979
0.345	1069	55.7141	0.545	1754	388.822

(continued)

Table 4.5 (continued)

$\lambda/\mu m$	$E_0(\lambda)/Wm^{-2}\ \mu m^{-1}$	$\int_0^\lambda E_0(\lambda)d\lambda$	$\lambda/\mu m$	$E_0(\lambda)/Wm^{-2}\ \mu m^{-1}$	$\int_0^\lambda E_0(\lambda)d\lambda$
0.350	1093	61.1191	0.550	1725	397.519
0.355	1083	66.5591	0.555	1720	406.132
0.360	1068	71.9366	0.560	1695	414.669
0.365	1132	77.4366	0.565	1705	423.169
0.370	1181	83.2191	0. 570	1712	431.712
0.575	1719	440.289	2.600	48	1307.96
0.580	1715	448.874	2.700	43	1312.51
0.585	1712	457.442	2.800	39	1316.61
0.590	1700	465.972	2.900	35	1320.31
0.595	1682	474.427	3.000	31	1323.61
0.600	1666	482.797	3.100	26	1326.46
0.605	1647	491.079	3.200	22.6	1328.89
0.610	1635	499.284	3.300	19.2	1330.98
0.620	1602	515.469	3.400	16.6	1332.77
0.630	1570	531.329	3.500	14.6	1334.33
0.640	1544	546.899	3.600	13.5	1335.73
0.650	1511	562.174	3.700	12.3	1337.02
0.660	1486	577. 159	3.800	11.1	1338.19
0.670	1456	591.869	3.900	10.3	1339.26
0.680	1427	606.284	4.000	9.5	1340.25
0.690	1402	620.429	4.100	8.7	1341.16
0.700	1369	634.284	4.200	7.8	1341.99
0.710	1344	647.849	4.300	7.1	1342.73
0.720	1314	661.139	4.400	6.5	1343.41
0.730	1290	674.159	4.500	5.9	1344.03
0.740	1260	686.909	4.600	5.3	1344.59
0.750	1235	699.384	4.700	4.8	1345.1
0.800	1107	757.934	4.800	4.5	1345.56
0.850	988	810.309	4.900	4.1	1345.99
0.900	889	857.234	5.000	3.83	1346.3906
0.950	835	900.334	6.000	1.75	1349.1806
1.000	746	939.859	7.000	0.99	1350.5506
1.100	592	1006.76	8.000	0.6	1351.3456
1.200	484	1060.56	9.000	0.38	1351.8356
1.300	396	1104.56	10.000	0.25	1352.1506

(continued)

Table 4.5 (continued)

$\lambda/\mu m$	$E_0(\lambda)/Wm^{-2}\,\mu m^{-1}$	$\int_0^\lambda E_0(\lambda)d\lambda$	$\lambda/\mu m$	$E_0(\lambda)/Wm^{-2}\,\mu m^{-1}$	$\int_0^\lambda E_0(\lambda)d\lambda$
1.400	336	1141.16	11.000	0.17	1352.3606
1.500	287	1172.31	12.000	0.12	1352.5056
1.600	244	1198.86	13.000	0.087	1352.6091
1.700	202	1221.16	14.000	0.055	1352.6801
1.800	159	1239.21	15.000	0.049	1352.7321
1.900	126	1253.46	16.000	0.038	1352.7756
2.000	103	1264.91	17.000	0.031	1352.8101
2.100	90	1274.56	18.000	0.024	1352.8376
2.200	79	1283.01	19.000	0.020	1352.8596
2.300	68	1290.36	20.000	0.016	1352.8776
2.400	64	1296.96	25.000	0.00610	1352.9328
2.500	54	1302.86	30.000	0.00300	1352.9556
35.000	0.0016	1352.9671	80.000	0.00007	1352.9855
40.000	0.00094	1352.9735	100.000	0.00003	1352.9865
50.000	0.00038	1352.9801	1000.000	0.00000	1353.0000
60.000	0.00019	1352.9829			

Table 4.6 Energy ratio of solar radiation spectrum [8]

Wavelength/μm	Spectrum band	%
<0^{-3}	x, γ rays	} 0.02
10^{-3}–0.20	Ultraviolet	
0.20–0.31	Mid ultraviolet	1.95
0.31–0.38	Near ultraviolet	5.32
0.38–0.76	Visible light	5.32
0.76–1.5	Near infrared	43.50
1.5–5.6	Mid-infrared	36.80
5.6–1000	Far infrared	} 0.41
>1000	Microwave	

the earth's albedo over time are listed in Table 4.10. Within a short time, the earth's albedo varies greatly, up to 100%, and the diurnal and seasonal variations are about ±10% (Table 4.9).

Specific emissivity of various surfaces is shown in the table below.

The mean emission of the earth–air system is 237 ± 7 W/m^2, and the short term and seasonal variation is about ±30%, as shown in Table 4.10. As shown in Fig. 4.22, the emittance of the earth–air system varies with latitude in different seasons. It can be seen that the emittance from the equator to the poles decreases from 250 to 140

Table 4.7 Relative variation of solar radiation during 11-year cycle [7]

Spectrum band	Wavelength	Irradiance/ $Wm^{-2}\ \mu m^{-1}$	Relative variation
X-ray area	<0.01 μm	10^{-7}–10^{-1}	10–10,000 times
Extreme ultraviolet	0.01–0.12 μm	0.1	1–2 times
Ultraviolet	0.12–0.3 μm	0.1–1000	1%–2 times
Visible light, infrared region	0.3–10 μm	1000–0.1	<0.2%
Far infrared region	10 μm ~ 1000 μm	10^{-1}–10^{-8}	<0.1%
Microwave, radio region	>1000 μm①	10^{-8}–10^{-17}	100 times

Note ① The original text is 1 nm

Table 4.8 Brightness temperature of solar radiation [2]

Wavelength/μm	0.44	0.55	5	10	20	50	100	1000	10,000	10^6
Brightness temperature/ T_b(K)	5850	5850	5500	5050	4740	4500	4370	5500	8200	10^6

Table 4.9 Specific emissivity of various surfaces (infrared) [6]

Type	Surface feature	Specific emissivity ε
Soil	Black soil	0.85–0.95
	Wet dust	0.80–0.90
	Dry dust	0.65–0.80
	White sands	0.60–0.70
	Dust	0.73–0.82
Tree	Coniferous forests, shrubbery, etc.	0.80–0.90
Vegetation	Grassland	0.70–0.85
	The desert	0.65–0.75
Crops		0.75–0.90
Water	Natural water	0.90–0.94
	The sea	0.80–0.90
Ice and snow	The ice	0.65–0.85
	Dry clean new snow	0.05–0.20
	Clean wet snow	0.30–0.40
	Dirty snow	0.50–0.60

W/m^2. Due to the absorption of H_2O vapor, CO_2, O_3 and other gases in the earth's atmosphere, the t emission of the earth–air system has an obvious absorption band, and its spectral characteristics are shown in Fig. 4.23 [7].

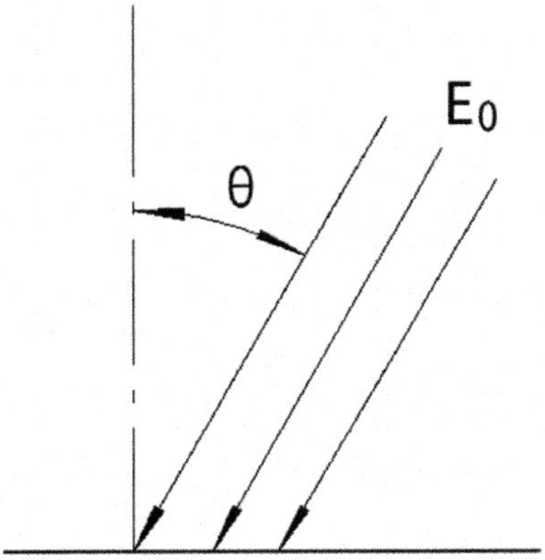

Fig. 4.20 Ground irradiance of E_0

Table 4.10 Variation of the earth's albedo over time

Time interval	Earth albedo		Radiant emission from earth–air system W/m^2	
	Average	Range	Average	Range
<0.3 h	0.3	0.15–0.60	237	140–265
0.3–3 h	0.3	0.20–0.40	237	189–261
3–24 h	0.3	0.25–0.35	237	216–258
Seasonal variation	0.3	0.25–0.35	237	166–308

4.3.1 Shortwave Radiation

The earth's radiation can be divided into shortwave (0.3–2.5 μm) and longwave (above 6 μm). As can be seen from Fig. 4.24, the shortwave part of the earth's radiation mainly comes from the reflection for the solar irradiance, while the earth's own thermal radiation is negligible. In the longwave part, only the thermal radiation of the surface object itself is considered because of the minimal influence of solar irradiation in this region. In the mid-infrared spectrum from 2.5 to 6 μm, the effects of solar radiation and earth thermal radiation cannot be ignored.

For reflected radiation, there are two factors affecting the radiation brightness: one is the solar irradiance *E*, and the other is the reflectance ρ of the ground object. Irradiance E consists of two parts. One is the direct light of the sun, which is a good parallel beam. The intensity of direct light, the variation of zenith angle and the spectral distribution have been discussed in detail in the above, and it has strong directionality. There are exact azimuth angle φ_i and altitude angle θ_i at every moment, and the irradiance $E_s(\theta_i, \varphi_i)$ is a function of φ_i and θ_i. The other part is the diffuse light in the sky, which is mainly caused by the scattering of sunlight by the atmosphere. There is no exact incident direction, but it is shot to the ground from all directions

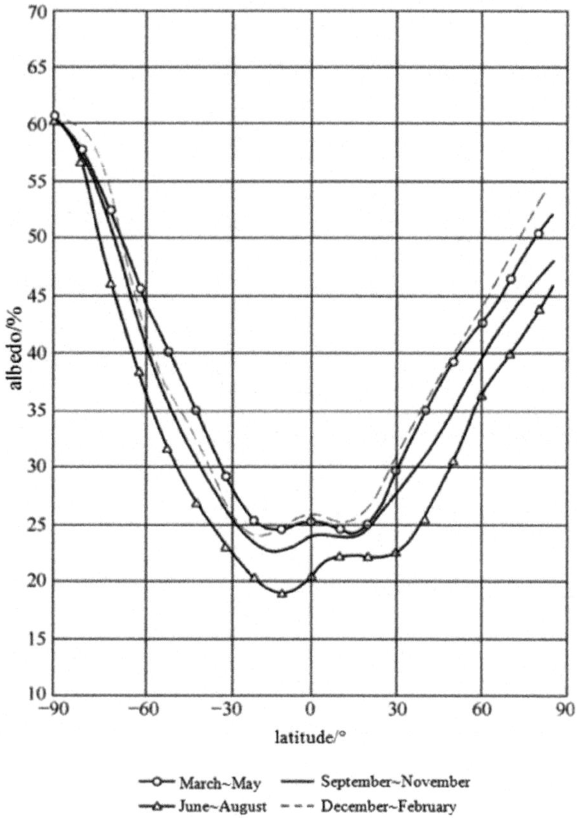

Fig. 4.21 Variation of the earth's albedo with latitude in different seasons

in the air, and the illumination caused by it is E_D. The total ground irradiance E is the sum of direct solar light and diffuse sky light:

$$E = E_S + E_D$$

There are great differences in the reflectivity ρ of different ground objects, and people rely on these differences to judge various objects.

The reflectance spectrum of some plants and soils is shown in Figs. 4.25, 4.26, 4.27 and 4.28.

Plants usually have a small reflection peak at 0.55 μm, which is responsible for the "greenness" of plants. In the near infrared region above 0.7 μm, the plant reflectance is higher, up to 50%. There were two water absorption peaks near 1.4 and 1.9 μm and two weak water absorption peaks near 0.94 and 1.13 μm.

For diseased plants, the peaks and valleys near 0.6 μm gradually disappeared, the infrared reflectance decreased, and the absorption peaks near 1.4 and 1.9 μm also weakened.

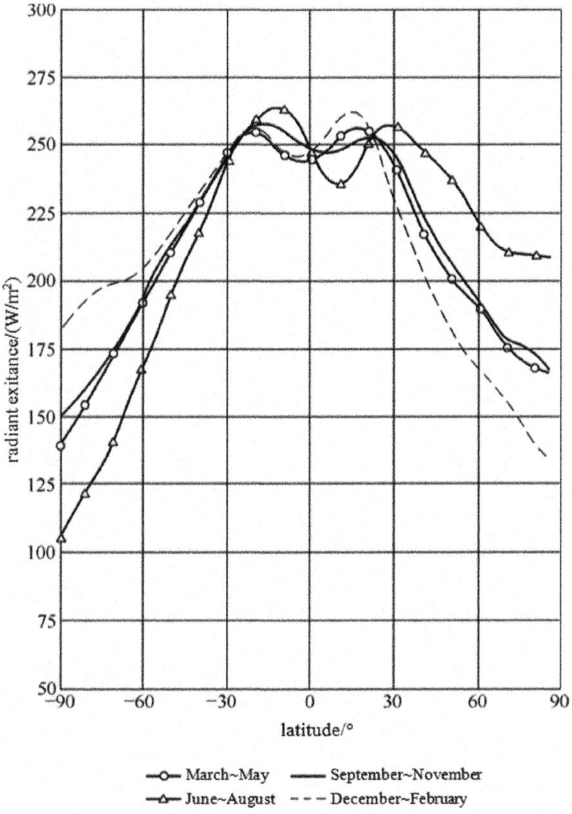

Fig. 4.22 Variation of radiant existence of earth–air system with latitude in different seasons

Soil reflectance is closely related to soil water content. The higher the water content, the lower the reflectance.

The reflectance spectra of objects can be used to distinguish different objects and indicate various information such as soil moisture and the degree of plant diseases.

In the selection of remote sensing spectral segment, the spectral characteristics of various ground objects should be measured first to find the region with large spectral difference. As can be seen from Fig. 4.25, snow can be easily distinguished from other objects by using 0.4–0.5 μm spectrums. By using 0.5–0.6 μm spectrum, desert can be separated from wheat and wet areas. Wheat and wetland can be easily separated by using 0.7–0.9 μm spectrum. For different research purposes and objects, it is necessary to select the best spectrum segment and the best measurement season and time according to their respective spectral characteristics.

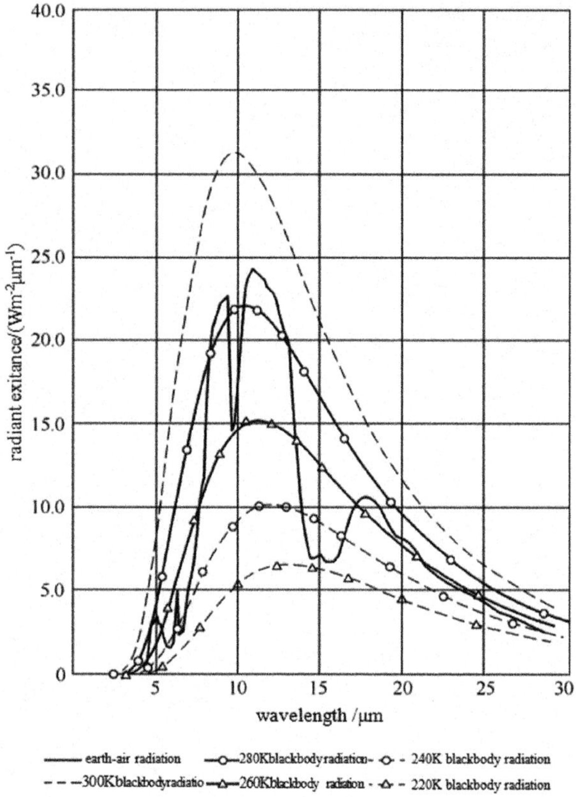

Fig. 4.23 Spectral characteristics of emittance of earth–air system

4.3.2 Longwave Radiation [2]

The earth's longwave radiation comes from the earth's thermal radiation, and the radiance of the ground object is

$$L(\lambda T) = \varepsilon(\lambda T)\, L_b(\lambda T)$$

Radiance of ground object is related to the specific emissivity ε of the ground object, the ground object temperature T and the ground object radiation wavelength λ. In the formula, temperature T refers to the surface temperature of the ground, not the temperature above the ground or let alone the temperature below the surface.

Figure 4.29 shows the spectrum curves of specific emissivity of various magmatic rocks in the vertical direction, indicating that the specific emissivity is different even for the same type of ground objects. It can be seen from the figure that the specific emissivity of various rocks has a valley value between 9 and 11 μm. This valley value is related to the SiO_2 content in the rock. With the increase of SiO_2 content, the valley floor moves to shortwave direction.

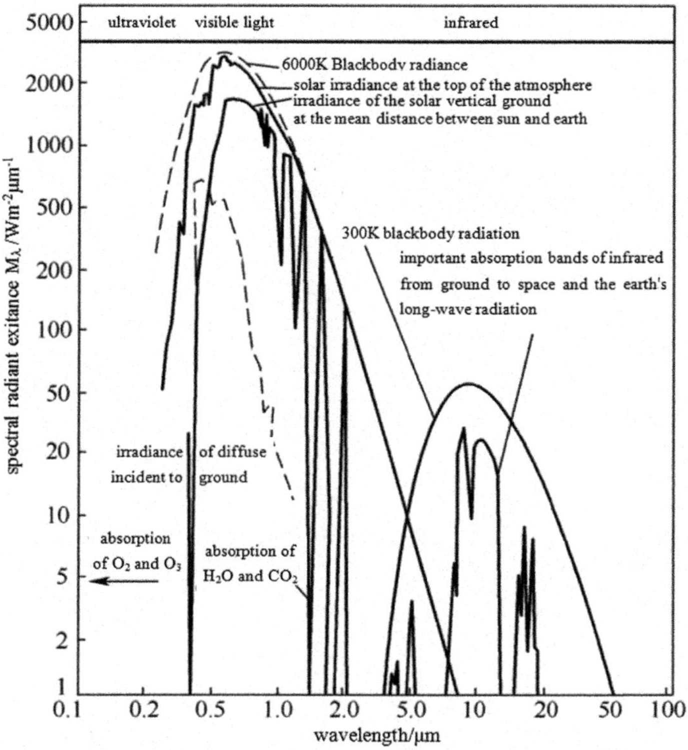

Fig. 4.24 Electromagnetic spectrum of solar and earth surface radiation

Fig. 4.25 Spectral reflectance of typical ground objects

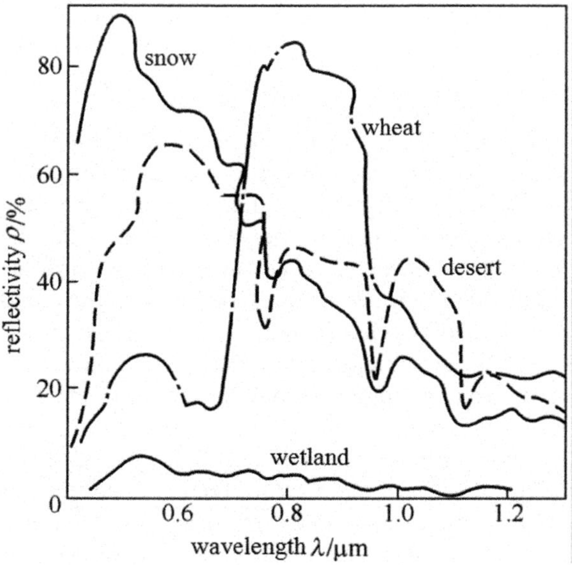

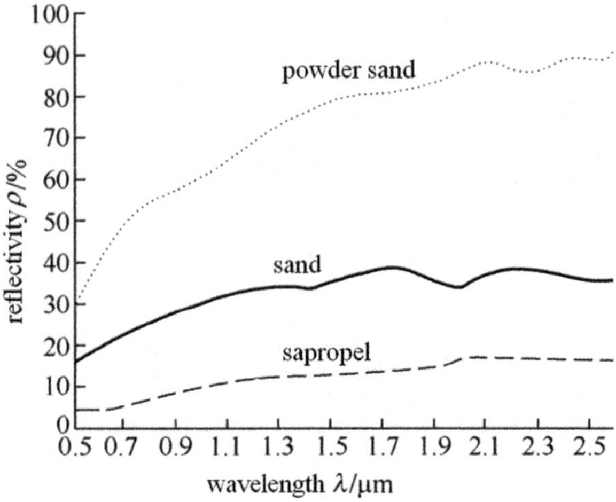

Fig. 4.26 Spectra of three soils

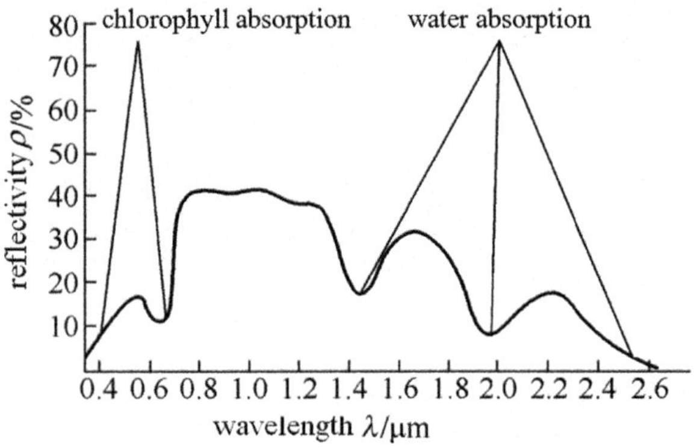

Fig. 4.27 Spectral reflectance of green plants

Figure 4.30 shows the distribution of atmospheric and underground temperatures at different times in a day. It can be seen from the figure that the temperature of the surface is different from that of the atmosphere. At noon, the temperature of the surface is higher than the air, while at midnight, the temperature of the surface is lower than air, and the range of the surface temperature is larger than that of the air temperature. Surface temperature changes periodically, with a daily cycle of 24 h and an annual cycle.

As the surface temperature changes periodically, so does the underground temperature, which propagates from the surface to the ground in the form of temperature waves. A simplified model can be used to estimate temperature wave propagation. Let the coordinate perpendicular to the ground be the X-axis, take the ground as the origin $x = 0$, and the X-axis direction is positive downward and negative upward. Suppose there is no transverse heat conduction at the ground, so the problem is reduced to heat conduction at a one-dimensional semi-infinite boundary. According to the heat conduction equation

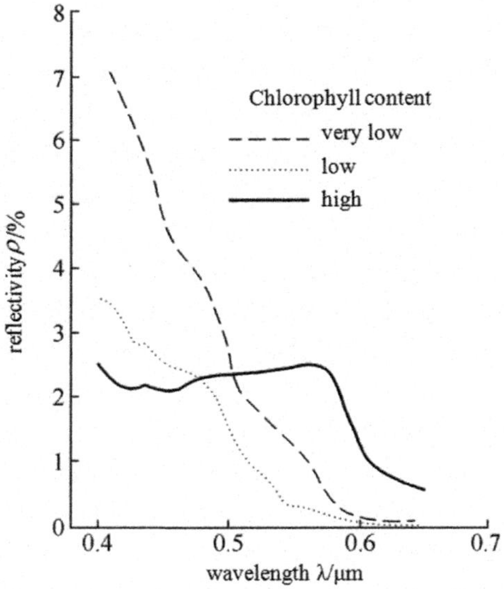

Fig. 4.28 Seawater spectra of different chlorophyll concentrations

$$\frac{\partial T}{\partial t} = k\frac{\partial^2 T}{\partial x^2} \tag{4.34}$$

where T is temperature, t is time, and k is the thermal diffusion coefficient (m^2/s) of the object.

The boundary condition at $x = 0$ is

$$T(0, t) = T_0 \cos\frac{2\pi t}{t_0}$$

where t_0 is the period of temperature wave and T_0 is the amplitude of temperature change at the surface ($x = 0$).

T can be written by variable separation approach

$$T(x, t) = \varphi(t)\psi(x)$$

It can be inferred from physics that the temperature of an object at a certain depth also changes periodically, so the solution of $\varphi(t)$ is

$$\varphi(t) = e^{ipt} = \cos(pt) + i\ \sin(pt)$$

And make $p = kq^2$

So

$$T(x, t) = e^{ipt}\ \psi(x) \tag{4.35}$$

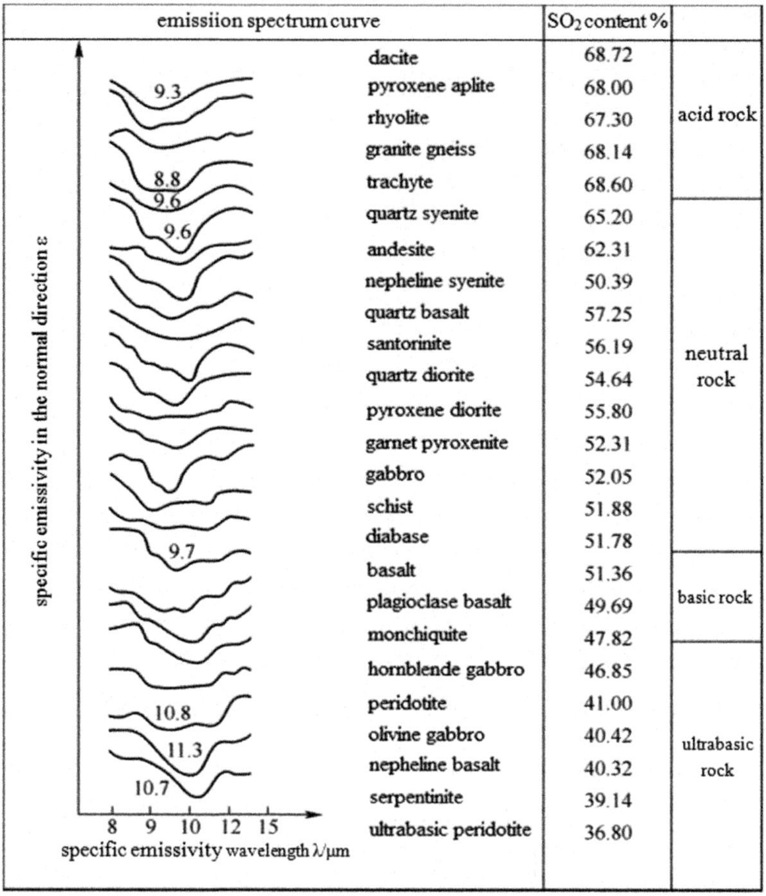

Fig. 4.29 Specific emissivity of various types of magmatic rocks

Fig. 4.30 Temperature distribution near earth surface

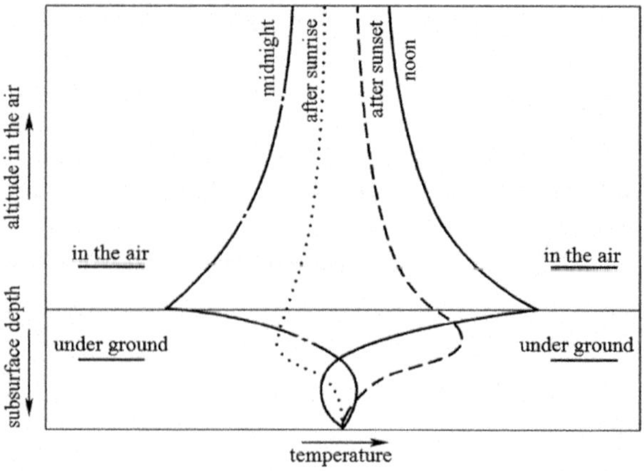

Substitute 4.35, into 4.34

$$\frac{\partial^2 \psi}{\partial x^2} - iq^2\psi = 0$$

And

$$\sqrt{-i} = \pm(1-i)\sqrt{\frac{1}{2}}$$

Then

$$\frac{\partial^2 \psi}{\partial x^2} + \left[\pm(1-i)\sqrt{\frac{1}{2}q}\right]^2 \psi = 0$$

Result of solution

$$\psi(x) = Ce^{\pm i[(1-i)\sqrt{\frac{1}{2}q}]x}$$

where C is a constant.

Substituting $\psi(x)$ into Eq. 4.35

$$\begin{aligned}
T(x \cdot t) &= Ce^{ikq^2t} \cdot e^{\pm[(1+i)\sqrt{\frac{1}{2}q}]x} \\
&= Ce^{\pm\sqrt{\frac{1}{2}qx}} \cdot e^{i\left(kq^2t \pm \sqrt{\frac{1}{2}qx}\right)} \\
&= Ce^{\pm\sqrt{\frac{1}{2}qx}}\left[\cos\left(kq^2 \cdot t \pm \sqrt{\frac{1}{2}qx}\right) + i\sin\left(kq^2 \cdot t \pm \sqrt{\frac{1}{2}qx}\right)\right]
\end{aligned}$$

Since the amplitude of temperature decays with the increase of distance x, the solution of e^+ exponential is not reasonable and only e^- exponential solution can be taken.

So

$$T(x \cdot t) = C_1 e^{-\sqrt{\frac{1}{2}qx}} \cos\left(kq^2 \cdot t - \sqrt{\frac{1}{2}qx}\right) + iC_2 e^{-\sqrt{\frac{1}{2}qx}} \sin\left(kq^2 \cdot t - \sqrt{\frac{1}{2}qx}\right)$$

According to boundary conditions

$$\begin{aligned}
T(0, t) &= T_0 \cos\frac{2\pi t}{t_0} \\
&= C_1 \cos(kq^2 t) + i\, C_2 \sin(kq^2 t)
\end{aligned}$$

where

$$C_1 = T_0$$
$$C_2 = 0$$
$$kq^2 = \frac{2\pi}{t_0}$$

So the solution of temperature field is

$$T(x \cdot t) = T_0 e^{-\sqrt{\frac{\pi}{kt_0}}x} \cos\left(\frac{2\pi}{t_0} \cdot t - \sqrt{\frac{\pi}{kt_0}} \cdot x\right) \tag{4.36}$$

This formula indicates that the wavelength of temperature wave λ is

$$\lambda\sqrt{\frac{\pi}{kt_0}} = 2\pi$$
$$\lambda = 2\sqrt{\pi k t_0} \tag{4.37}$$

It can be seen that the amplitude of temperature wave $T(x{\cdot}t)$ decays exponentially with depth x. At $x = \lambda$, that is, at the depth of one wavelength, the amplitude of T is

$$\begin{aligned} |T| &= T_0 e^{-2\sqrt{\pi k t_0}\cdot\sqrt{\frac{\pi}{kt_0}}} \\ &= T_0 e^{-2\pi} \\ &= \frac{T_0}{540} \end{aligned}$$

That is, the amplitude is 1/540 of the surface amplitude, so the temperature at $x = \lambda$ can be considered to be stable.

Therefore, the depth of temperature wave propagation can be estimated by using Eq. 4.37. Take sand for example. Its thermal diffusion coefficient is $k = 0.3 \times 10^{-6}$ m^2/s, and the temperature wave is $t_0 = 24\ h = 24 \times 3600$ s, then

$$\begin{aligned} \lambda_d &= 2\sqrt{0.3 \times 10^{-6} \times 24 \times 3600 \times \pi} \\ &= 0.55\ \text{m} \end{aligned}$$

For the annual period of the temperature wave λ_y

$$\begin{aligned} t_0 &= 365 \times 24 \times 3600\ \text{s} \\ \lambda_y &= \sqrt{365}\lambda_d \\ &= 10.5\ \text{m} \end{aligned}$$

The results of λ_d and λ_y show that temperature change within a day in sandy land only occurs within 0.55 m, and temperature change below 0.55 m is very small, which can only be observed by year. The ground temperature below 10.5 m is constant and not affected by the ground temperature.

The temperature variation at depth of λ_d in a day is not the same, but there is a certain phase difference. For example, noon is the time in a day of the highest surface temperature and 0.5 λ_d underground, and it is the lowest time of the day. It can also be seen from Fig. 4.31 that there is a large temperature gradient at the surface. The ground temperature measured by conventional methods is the average temperature at a certain depth, which is quite different from the real surface temperature. The temperature of the surface can be measured with a radiometer, but the radiation thermometer is measured by the brightness temperature of the ground T_b, so it is necessary to measure the specific emissivity of the surface to calculate the actual temperature of the surface $T = T_b/\varepsilon$.

The change of surface temperature is related to the energy budget of surface and the thermal properties of surface.

Energy on earth comes from solar radiation and includes both direct solar energy and diffuse energy from the sky. But some of the incoming energy is reflected by the ground, and the actual absorbed energy is

$$\alpha Q_i = (1 - \rho) Q_i$$

In the formula, Q_i is the energy of solar radiation, and α and ρ are the total absorption rate and reflectance of the whole spectrum of the surface.

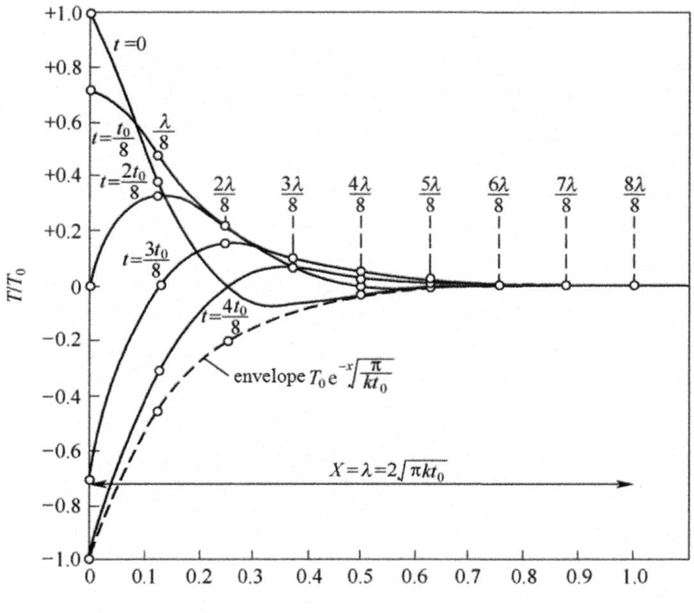

Fig. 4.31 Temperature wave propagation

While absorbing solar radiation energy, the earth's surface also radiates energy to outside in the form of radiation Q_{TO}, and the atmospheric longwave radiation Q_{T} also contributes to the earth's surface. Therefore, the surface radiation net income Q_{n} is

$$Q_n = (1 - \rho)\, Q_i + Q_T - Q_{TO}$$

Generally, daytime income is greater than expenditure, net income Q_{n} is positive, and ground temperature keeps rising. The size of the net income is related to the reflectivity ρ, net income is big when ρ is small, and temperature rise is also high. Blackbody $\rho = 0$, net income is the largest, and temperature rise is also the highest. In the night without the sun radiation, $Q_{\mathrm{i}} = 0$, generally $|Q_{\mathrm{TO}}| > |Q_{\mathrm{T}}||$ surface energy loss, temperature falling. Due to the different reflectance ρ, the diurnal variation of surface temperature is that the solar radiation energy absorbed by objects with small ρ has a higher solar radiation energy and temperature rise, while the solar radiation energy absorbed by objects with large ρ has a lower solar radiation energy and temperature rise.

Radiation net income is not all used for heating the earth's surface, one of them heating the atmosphere in the form of conduction and convection, another part of the energy supply of water in the transformation of state (such as evaporation, condensation, distillation, melting and ice) which makes latent heat, the energy value is related to microclimate factors such as wind speed, humidity and temperature conditions. If there is vegetation on the ground, the energy absorbed by photosynthesis is deducted. As noted above, subtracting this amount of energy from the net income is what really changes the surface temperature.

The change of surface temperature is not only related to the ground energy budget, but also to the thermal properties of the object itself. According to Eq. 4.37, the depth of temperature wave propagation and the depth of energy propagation are proportional to the square root of the diffusion coefficient $k^{1/2}$. In addition, the temperature rise of object is related to the specific heat C (W kg^{-1} °C^{-1}) and the specific gravity ρ (kg m^{-3}) of the object, that is, related to the thermal inertia P of the object.

$$P = C \cdot \rho \cdot \sqrt{k}$$

The unit of P is J m^{-2} s$^{-1/2}$.

In the soil, heat is transferred in accordance with Newton's law of cooling.

$$\frac{\partial Q(x)}{\partial t} = -\mu \frac{\partial T}{\partial x}$$

In the formula, μ is thermal conductivity of the object. It refers to the thermal power passed through unit distance by unit planes when the temperature difference is 1 K. The unit of μ is W m^{-1} K^{-1}

$$\mu = C \cdot \rho \cdot k$$

The heat flows through the surface at $x = 0$

$$\frac{\partial Q(0)}{\partial t} = -\mu \frac{\partial T}{\partial x}|_{x=0} \tag{4.38}$$

Take the partial derivative of Eq. 4.36

$$\begin{aligned}
\frac{\partial T(x)}{\partial x}|_{x=0} &= \left\{-T_0\sqrt{\frac{\pi}{kt_0}}e^{-\sqrt{\frac{\pi}{kt_0}}x}\left[\cos\left(\frac{2\pi}{t_0}t - \sqrt{\frac{\pi}{kt_0}}x\right) - \sin\left(\frac{2\pi}{t_0}t - \sqrt{\frac{\pi}{kt_0}}x\right)\right]\right\}_{x=0} \\
&= -T_0\sqrt{\frac{\pi}{kt_0}}\left(\cos\frac{2\pi}{t_0}t - \sin\frac{2\pi}{t_0}t\right) \\
&= -T_0\sqrt{\frac{2\pi}{kt_0}}\left(\cos\frac{2\pi}{t_0}t + \frac{\pi}{4}\right)
\end{aligned}$$

Substitute it into Formula 4.38

$$\begin{aligned}
\frac{\partial Q(0)}{\partial t} &= T_0\sqrt{2\pi}\frac{\mu}{\sqrt{kt_0}}\cos\left(\frac{2\pi}{t_0}t + \frac{\pi}{4}\right) \\
&= P\sqrt{2\pi}\frac{T_0}{\sqrt{t_0}}\cos\left(\frac{2\pi}{t_0}t + \frac{\pi}{4}\right)
\end{aligned}$$

Integrate the above equation to get the total energy through $x = 0$, but since the heating is periodic, the integral is 0 over one period. In terms of daily cycles, the energy budget of the earth's surface is in balance during one day. But the absolute value of absorption due to different reflectance ρ can have a great difference. The one absorbs more energy expends more, which the absolute value of the integral over half a period from 0 to $t_0/2$ is the same as the absolute value of the integral over half a period from $t_0/2$ to t_0, so

$$\begin{aligned}
Q(0)|_{0\to\frac{t_0}{2}} &= \int_0^{t_0/2}\sqrt{2\pi P\frac{T_0}{\sqrt{t_0}}\cos\left(\frac{2\pi t}{t_0} + \frac{\pi}{4}\right)dt} \\
&= \sqrt{2\pi}P\frac{T_0}{\sqrt{t_0}}\frac{t_0}{2\pi}\sin\left(\frac{2\pi t}{t_0} + \frac{\pi}{4}\right)|_0^{t_0/2} \\
&= -\sqrt{\frac{1}{\pi}}P\sqrt{t_0}T_0
\end{aligned}$$

The above equation shows that if bodies absorb or lose the same amount of energy, the amplitude of their temperature change T_0 is inversely proportional to their thermal inertia P. A body with high thermal inertia P has small temperature change amplitude, while a body with low thermal inertia P has large temperature change amplitude.

4.4 The Effect of Atmosphere on Radiation

Twenty-six percent of the sun's radiant energy is absorbed by atmosphere and heating the atmosphere directly. Thirty-one percent of the energy is reflected back into space by the earth–air system. About 43% of the energy is absorbed by the surface, and 14% of this energy in the form of infrared radiation, 23% in the form of latent heat and 6% in the form of sensible heat transport are returned to the atmosphere through the earth–air boundary [6].

4.4.1 Atmospheric Composition

There is no definite limit to the atmosphere around the earth, but it gets thinner as it gets higher above the earth, so that it almost empties into space. From the highest spots of meteors and aurora borealis, there is a little air left 800 km above the ground. The thickness of the atmosphere is about 1000 km, or about one-twelfth of the diameter of the earth.

Ninety-nine percent by volume of the dry and clean atmosphere is nitrogen and oxygen, plus argon makes up about 99.9% of the total atmosphere by volume. The atmospheric pressure is caused by the weight of the column of air. In the atmosphere, the pressure difference between the upper and lower sides $\mathrm{d}h$ is equal to the weight trapped in the layer.

$$Dp = -\rho g \mathrm{d}h = -nmg\mathrm{d}h$$

where ρ is the density of air, g is the acceleration of gravity, n is the number of air molecules per unit volume, and m is the average mass of air molecules.

From the ideal gas equation PV = RT = NkT

$$P = nkT$$

where k is the Boltzmann constant and T is absolute temperature, and the formula above can be obtained by substituting it into formula $\mathrm{d}p$

$$\frac{\mathrm{d}p}{p} = -\frac{mg}{kT}\mathrm{d}h$$

Integration of the above formula with h

$$p = p_0 e^{-\int_0^h \frac{mg}{kT}\mathrm{d}h}$$

where P_0 is the pressure at height $h = 0$ and P is the pressure at height h. If m, g and T do not vary with h, the formula above can be written as

$$P = P_0 e^{-h/H} \tag{4.39}$$

And

$$H = \frac{kT}{mg}$$

Lead $P = nkT$ into the formula above

$$n = n_0 e^{-h/H} \tag{4.40}$$

According to Eqs. 4.39 and 4.40, atmospheric pressure or the number of molecules per unit volume decays exponentially with height h. H is called the elevation of the atmosphere, and it is the height that corresponds to e times the pressure change.

For $T = 273$ K, $H = 7.99$ km, $P = P_0 e^{-h/H}$ shows that half of the whole atmosphere is below 5.5 km above the ground, 1/4 is concentrated between 5.5 and 11 km, and 99% of the total mass is concentrated within 30 km.

In the vertical direction, the atmosphere can be divided into troposphere, stratosphere, mesosphere and thermosphere according to its thermodynamic properties (Fig. 4.32).

(1) Troposphere

Troposphere is atmosphere from ground up to where the first temperature minimum occurs. The lowest layer of atmosphere is troposphere, and the further decomposition of the troposphere has a 2 m layer at the bottom. The height of its upward extension is related to latitude, which is only 7–8 km above the poles, but 16–18 km above the equator. The atmospheric temperature of the polar and equatorial troposphere top can drop to 220 K and 190 K, respectively [7]. The air in the troposphere has significant vertical activity, and the main atmospheric phenomena are almost concentrated in this layer. In the troposphere, the temperature decreases 6.5 K for every 1 km increase. The upper limit of the troposphere is called the tropopause, and its height varies according to latitude, season and even from day to day in the same place.

(2) Stratosphere

Stratosphere is atmosphere from the tropopause up to the height at which the temperature maximum occurs. This is an atmosphere where ozone is concentrated, because ozone absorbs ultraviolet radiation from the sun, and the temperature of the stratosphere increases slowly with altitude, up to a height of 50 km, and the average temperature is about 273 K, which is slightly lower than the temperature at sea level. The stratosphere is different with the troposphere, which has no significant mixing up and down. There is little weather in the stratosphere. The lower stratosphere has a distinct isothermal layer extending about 20 km from the tropopause.

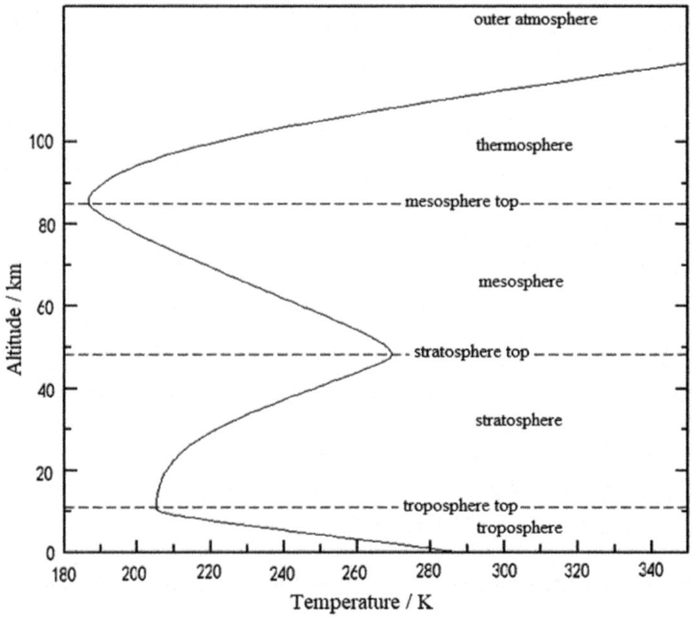

Fig. 4.32 Atmospheric vertical temperature profile

Because the amount of ozone is unevenly distributed across the globe and varies seasonally, the temperature in the stratosphere varies considerably, both spatially and temporally. Within this layer, there is a seasonal circulation of strong winds.

(3) Mesosphere

Mesosphere is the atmosphere from the stratosphere up to where the second temperature minimum occurs. The temperature of mesosphere decreases with the increase of altitude and reaches the lowest point of 190 K at about 85 km. In high latitudes, the temperature of mesosphere has a strong seasonal change and can drop to 160 K in summer. The mesosphere is the alternate layer between the two energy absorption layers, and the stratosphere below it is also hotter than the mesosphere due to the concentration of O_3. The temperature in the mesosphere decreases with height, decreasing by an average of 3 K per 1 km, so there is also some vertical air movement in this layer.

(4) Thermosphere

Thermosphere is the atmosphere from the mesosphere up to an altitude of 400–700 km. The temperature of the thermosphere is between 500 and 700 K, depending on solar activity. At a height of about 90–200 km, the atmosphere absorbs far ultraviolet solar radiation with a wavelength less than 200 nm, resulting in photochemization and ionization of atmospheric molecules accompanied by an exothermic process, which rapidly increases the temperature of atmosphere. Above 90 km, the temperature rises because O_2 and N_2 absorb excess energy as they decompose and dissociate, heating up the air. Above 200 km, the heat of the atmosphere gradually decreases with the increase of altitude. If there is no energy input from the earth's magnetosphere, the thermospheric atmosphere gradually tends to isothermal state.

(5) Outer atmosphere

The isothermal atmosphere above the thermosphere top is called outer atmosphere. Because H and He atoms have lightweight and certain energy, sometimes they can escape from the earth's gravitational field to the outer space environment, so the outer atmosphere is also called escape layer. The outer atmosphere consists of O, He and H atoms, from lowest to highest. Solar activity and magnetic storms have great influence on the outer atmosphere.

(6) The properties of the atmosphere

In terms of atmospheric properties, from the ground to 90 km is about below the mesosphere, which is the homogeneous layer of the atmosphere. The atmosphere is evenly mixed through turbulence, the composition is basically uniform, and the average molar mass is constant and follows the hydrostatic pressure equation and the ideal gas state equation. Above the homogeneous layer, the average molar mass of the heterogeneous layer decreases with the increase of the height, and the atmospheric turbulence of the heterogeneous layer below 105 km is the main mixing effect. The decrease of the average molar mass is the result of oxygen decomposition, and the hydrostatic pressure equation and ideal gas state equation are still satisfied in this range. Molecular diffusion in the heterogeneous layer above 105 km plays a major role of mixing and is in diffusion equilibrium state. The distribution of various components follows their own diffusion equation, and the atmospheric pressure and density decrease exponentially with the height.

Temperature is a description of intensity of molecular thermal motion from the knowledge of molecular physics. Look at it statistically, each particle has $kT/2$ energy per degree of freedom. If an atom in the upper atmosphere is taken as a particle, the motion of this particle in space has three degrees of freedom, so the average kinetic energy of each atom E_{k} is

$$E_{\mathrm{k}} = \frac{3}{2}kT$$

$$T = \frac{2E_{\mathrm{k}}}{3k}$$

where k is the Boltzmann constant. T formula shows that the temperature is proportional to the kinetic energy of the molecule. The object with high temperature has more intense molecular movement, and the average kinetic energy of each molecule is large. The temperature in thermosphere can reach 1500 K, which means that each gas atom in the layer has a large kinetic energy. However, since the gas density in the layer is very small, only 10^{-10}–10^{-12} of that at sea level, the total energy obtained by the objects in the thermosphere (such as the earth satellite) due to gas collisions is still insignificant. And the heat inside the layer does not burn anything in the layer.

In the lower atmosphere below 80 km, all kinds of gases are well mixed. Except for a few gases such as H_2O and O_3, the proportion of each gas is almost unchanged, so it is called uniform layer. In the process of radiation transmission, the influence of atmosphere mainly comes from uniform layer, because the pressure at 80 km is only 10^{-5} of sea level, which has little effect on radiation transmission. The composition of the lower atmosphere is shown in Table 4.11. The distribution of most gases in the atmosphere, such as N_2, O_2 and CO_2, conforms to Eq. 4.40, they decay exponentially with height, and the number of molecules decreases about 10 times with each increase of 15 km. The water vapor in the atmosphere is mainly distributed below 5 km, and there is almost no water vapor above 12 km. Water is also the leading role in weather phenomena, appearing in various forms such as gas, cloud, fog, rain, snow and ice.

Table 4.11 Atmospheric composition [2]

Gas	Molecular weight	Content of dry atmosphere		Total/atm cm	Note
		Volume × 10^{-6}	Mass × 10^{-6}		
N_2	28.0143	780,840	755,230	624,000	
O_2	31.999	209,470	231,420	167,400	
H_2O	18.015	1000–28,000	600–17,000	800–22,000	b, d
Ar	39.948	9340	12,900	7450	
CO^2	44.010	320	500	260	a
Ne	20.179	18.2	12.7	14.6	
He	4.003	5.24	0.72	4.2	
CH_4	16.043	1.8	1.0	1.4	
Kr	83.80	1.14	3.3	0.91	
CO	28.010	0.06–1	0.06–1	0.05–0.8	a
SO_2	64.06	1	2	1	a
H_2	2.016	0.5	0.04	0.4	
N_2O	44.012	0.27	0.5	0.2	
O_3	47.998	0.01–0.1	0.02–0.2	0.25	b, c
Xe	131.30	0.087	0.39	0.07	
NO_2	46.006	0.0005–0.02	0.0008–0.03	0.0004–0.02	a
Rn	222	$0.0^{13}6$	$0.0^{12}5$	5×10^{-14}	
NO	30.006	Trace	Trace	Trace	a

[a]More in industrial areas;
[b]It varies with meteorological and geographical conditions;
[c]Increased in the ozone layer;
[d]Decreases with increasing height;
[e]Atmospheric pressure cm = simplifies to thickness of atmosphere in centimeters at standard temperature and pressure

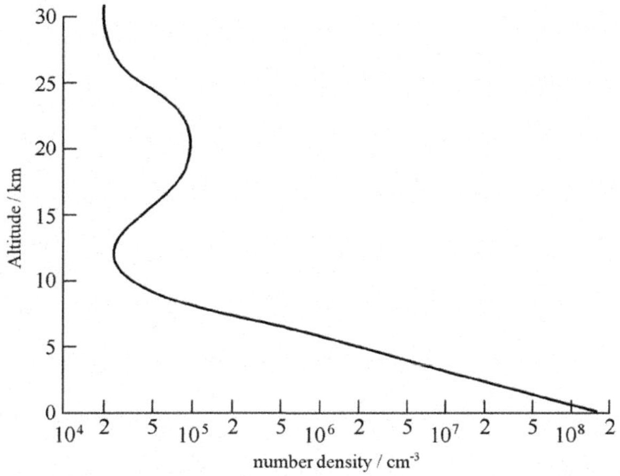

Fig. 4.33 Height distribution of aerosol particle concentration

The amount of water in the atmosphere varies greatly with climate and location. For example, the water vapor content in sea surface, basin area and rainy season is larger, while the water vapor content in desert area and dry season is less. The total ozone content in the atmosphere is very small, and the distribution is not uniform: the ozone content is very small at the ground and above 60 km, but concentrated near 25 km altitude.

There is also a kind of solid and liquid suspended substance called aerosol in the atmosphere. Generally, there is a solid core covered by a liquid outer layer, with different refractive index and shape. The core of aerosols can be wind-blown dust, pollen, microorganisms, embers of meteors, salt grains from the sea, volcanic ash, etc., with diameters of 0.01–30 μm. Aerosols are mainly distributed below 5 km, in Fig. 4.33.

4.4.2 Atmospheric Absorption and Atmospheric Windows

In the ultraviolet, visible and infrared regions, the main absorbing gases in the atmosphere are H_2O, CO_2, O_3 and O_2, followed by CO, CH_4 and N_2O. The atmospheric absorption spectra at medium resolution are shown in Fig. 4.34, and the ordinate in the figure is transmittance. The absorption spectra of each component of H_2O, CO_2, O_3, O_2, N_2O and CH_4 are shown in Fig. 4.35. The ordinate in the figure is the absorptivity, and the last row of the spectrum is synthetic absorption spectrum of the absorptivity of the above-mentioned components. Absorption spectra of main gas components in the atmosphere are shown in Table 4.12 [6].

Absorption of Light Waves by N and O

About 99% of the components in the earth's atmosphere are N_2 and O_2, which are diatomic molecules without fixed dipole moments. Therefore, there is no vibration

Table 4.12 Absorption spectra of main gases in the atmosphere [6]

Composition	Center wavelength/μm	Absorption band/μm
O_2	0.69; 0.76	0.1752–0.1926 (S-R band)
		0.2420–0.2600 (Herzberg band)
O_3	0.255	0.2–0.3 (Hartley band)
	~0.23	0.3–0.36 (Huggin band)
	~0.60	0.44–0.80 (Chappuis band)
	9.6	
H_2O	0.718	0.70–0.74 (α band)
	0.810	0.79–0.84 (β band)
	0.935	0.926–0.978 ($p\sigma$ band)
	1.13	1.095–1.165 (ψ band)
	1.395	1.319–1.418 (Ψ band)
	1.87	1.762–1.977 (Ω band)
	2.68	2.520–2.845 (X band)
	6.3	4–8
		13.0–1000
CO_2	2.7	2.3–3.0
	4.3	3.9–4.7
	14.7	12–18
CH_4	3.3	
	3.8	
	7.66	
N_2O	4.5	
	7.8	
CO	4.7	

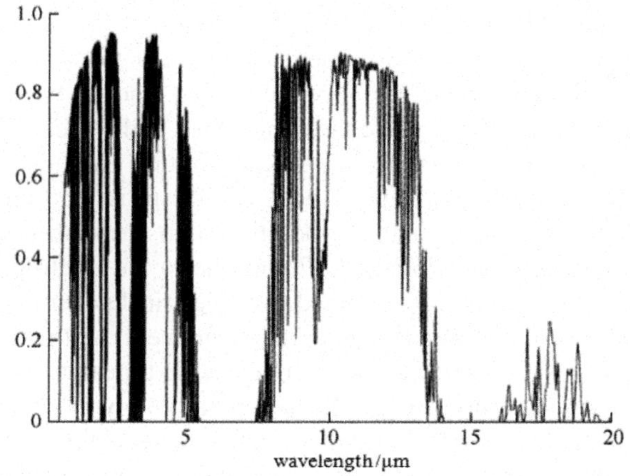

Fig. 4.34 Atmospheric absorption spectra at medium resolution

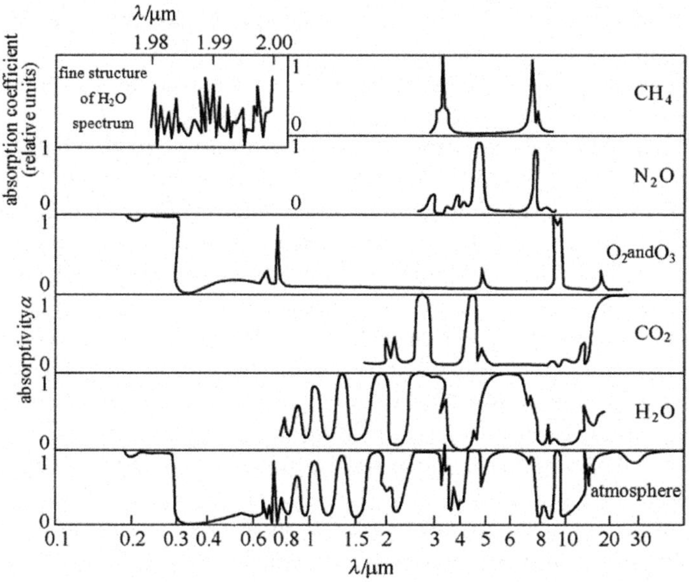

Fig. 4.35 Absorption spectra of major atmospheric components

and rotation spectrum in the absorption spectrum, but there can be a strong electron transition spectrum. Due to the absorption of N_2 and O_2, the atmosphere becomes opaque when the wavelength is less than 0.24 μm.

O_2 has a strong absorption band in ultraviolet, which is located in 0.1752–0.1925 μm, called Suman-Roogen band, which is caused by the transition of O_2 electronic energy level. The Suman-Roogen band can be extended to 0.1300 μm, and $\lambda = 0.1752$ μm wavelength causes decomposition of O_2 into O atom, and the excess energy of photon increases the kinetic energy of O, that is, the temperature of O increases. The absorption coefficient of the Suman-Roogen band is so large that only a very thin layer of O_2 can make the atmosphere in this band opaque.

At 0.242–0.260 μm, there is also a weak absorption band due to the electron transition, called the Hertzberg band. About 0.242 μm is the longwave limit for the decomposition of O_2 molecules. In addition, two weak absorption bands are generated in the vicinity of 0.76 and 1.27 μm due to the electron transition. The above three absorption bands are prohibited from electric dipole moment transitions. Their transitions are caused by electric quadrupole moments, so the absorption is very weak and only causes a slight decrease in atmospheric transmittance. The radiation intensity of electric quadrupole moment is generally only 10^{-8} times that of the electric dipole moment and 10^{-3} times that of the magnetic dipole moment.

There are strong absorption bands of nitrogen and oxygen in the range of 0.020–0.100 μm. From 0.030–0.100 μm, a strong absorption band is composed of a series of N_2 and O_2 absorption bands. From about 0.020–0.090 μm, N and O free transitions also cause strong continuous absorption bands.

Continuous absorption at wavelengths less than 0.020 μm is due to the inner electron excitation of N and O atoms.

Due to the magnetic dipole moment, O_2 also has two strong absorption peaks at 5 and 2.5 mm in the microwave region.

The content of O_3 in the atmosphere is very small, accounting for only 0.01–0.1%, which is formed by the combination of O_2 and O atoms after the decomposition of O_2. O_3 has a strong absorption band in ultraviolet, ranging from 0.22 to 0.32 μm, known as Hartley zone. Adjacent to the Hartley zone is the Huggin zone, located in the spectral region of 0.30–0.345 μm, and there is a much weaker Chappuis zone in the visible region of 0.6 μm or so. O_3 is a molecule with an inherent electric dipole moment and has a strong absorption peak in the infrared absorption spectrum, among which the most important infrared absorption peak is located at 9.6 μm. There are also two weak absorption peaks for O_3 at 4.75 and 14 μm. O_3 is mainly distributed near the altitude of 30 km, so it has little influence on the absorption peak of airborne sensors with a flight altitude less than 10 km, but has a great influence on spaceborne sensors.

Due to the absorption of O_3 and O_2, the shortwave part of the solar spectrum with a wavelength less than 0.30 μm is almost completely absorbed.

Absorption of Light Waves by CO_2

The content of CO_2 in the atmosphere only accounts for 0.032%. It is also a molecule with an inherent electric dipole moment, so it has a strong absorption peak in the infrared spectrum, which is located at 2.7, 4.3 and 14.5 μm. At 1.43, 1.54, 1.57, 1.60, 1.64, 1.96, 2.01, 2.06, 9.4, 10.4 and 12.0 μm, there are some weak absorption bands.

Absorption of Light Waves by H_2O

H_2O absorption has a wide impact on the atmosphere, from visible, infrared to microwave, and H_2O absorption peaks can be found everywhere. The amount of

H_2O in the atmosphere varies widely over time and location, from about 0.1–3%. The vibration-rotational absorption peaks of H_2O have two extremely strong absorption bands with centers of 2.66 and 2.74 μm distributed in the range of 2.5–3.0 μm and with centers of 6.27 μm distributed in the range of 5.0–7.5 μm. There were some moderate intensity absorption bands at 0.94, 1.13, 1.38 and 3.2 μm, and some weak absorption peaks at 0.72, 0.82, 2.01 and 2.05 μm. Due to absorption by pure rotational-level transitions of H_2O, the atmosphere with a wavelength greater than 24 μm becomes almost completely opaque and absorbs most of the energy in the 16–24 μm wavelength range.

In the microwave band, H_2O also has three absorption peaks at 0.94, 1.63 mm and 1.35 cm.

Absorption of Light Wave by N_2O and CH_4

N_2O has absorption peaks at 3.90, 4.06, 4.50 and 7.78 μm. CH_4 has absorption peaks at 3.3 and 7.6 μm. NH_3 has absorption peaks at 2.0 and 10.5 μm.

Considering the influence of various gas absorption, only some spectral segments of the atmosphere have relatively high transmittance, which is called atmospheric window. For the non-atmospheric window, the atmospheric absorption rate can be considered as 1, which can be regarded as a blackbody.

In the visible and infrared regions, the commonly used atmospheric windows are: 0.3–1.4 μm, 1.4–1.9 μm, 1.9–2.7 μm, 2.7–4.3 μm, 4.3–5.9 μm and 8–14 μm. There is also a translucent window between 16 μm ~ 24 μm.

In the microwave region, the available atmospheric windows are 1.4, 3.3, 8 mm and 1.6 cm to 15 m.

Whether the electromagnetic radiation of 15–30 m can pass through the atmosphere is determined by the specific conditions of the ionosphere. Radio waves larger than 30 m are reflected back to the universe by the ionosphere and cannot pass through the atmosphere.

When using remote sensing technology (such as earth resources satellite and reconnaissance satellite) to study the condition of the earth surface and communication work, the working spectrum of electromagnetic wave must be selected in the atmospheric window. However, some special applications are deliberately selected in the non-transparent spectrum, such as some meteorological satellites select some absorption areas of H_2O, CO_2 and O_3 and measure their content and temperature.

4.4.3 Atmospheric Scattering and Radiation

1. **Atmospheric scattering**

When electromagnetic waves encounter small particles such as atoms, molecules or aerosols on their propagation path, they will change the direction and spread out in all directions. This phenomenon is called scattering. In the visible spectrum, absorption has little effect on the light wave, and scattering is the main cause of extinction. The effects of scattering can be expressed in the following ways:

(1) The radiation reaching the ground is weakened, or the intensity of ground radiation to outside is weakened.
(2) It changes the direction of solar radiation. Most of the energy of solar radiation still travels in its original direction, while scattered sunlight is directed in different directions of 4π space depending on the pattern of scattering.
(3) Part of the scattered light is radiated downward to increase the irradiation of the ground, and part of the radiated upward enters the sensor. The intensity of downward radiation is generally different from that of upward radiation, and the upward radiation entering the sensor is not good for space optical remote sensing, because it reduces the contrast of the target and reduces MTF.

The scattering of particles in the atmosphere can be divided into three cases:

(1) Rayleigh scattering: it is the scattering of particles (<0.1 times) much smaller than wavelength, namely air molecules, and the scattering intensity is proportional to λ^{-4} [3].
(2) Meter scattering: it is the scattering of particles with similar size with wavelength (0.1–10 times), such as smoke, aerosol and other small particles [3].
(3) Non-selective scattering: it is the scattering of particles whose size is much larger than the wavelength, such as the scattering of clouds and fog on visible light.

In Fig. 4.36, (a) is the angular distribution of scattering intensity of Rayleigh scattering. Then, (b) and (c) are the scattering intensity angle distribution of meter scattering. With the increase of particle scale, the amplitude of scattering light intensity fluctuation gradually decreases, and the scattered light is mainly concentrated in the incident direction.

In cloudless sky with excellent visibility, the radiation attenuation is almost entirely caused by Rayleigh scattering. The scattering material is a gas molecule with a molecular size of approximately 1 nm, which is much smaller than the shortwave limit of the atmospheric window 0.3 μm. Molecular scattering is dipole scattering, the scattering light intensity is proportional to λ^{-4}, there is a directional factor $1 + \cos^2\varphi$, (the normalized Rayleigh scattering phase function $3(1 + \cos^2\varphi)/4$ [9]), and scattering light is partially polarized light. The extinction coefficient $k_R(\lambda)$ caused by Rayleigh scattering is

$$k_R(\lambda) = 32\pi^3 \frac{(n_0 - 1)^2}{3N_1\lambda^4}$$

where N_1 is the number of molecules per unit volume and n_0 is the refractive index of air.

The value of N_1 and n_0 in the atmosphere is quite stable and does not change much with climate, so the attenuation amount caused by Rayleigh scattering is also relatively stable. Because Rayleigh scattering is proportional to λ^{-4}, the scattering

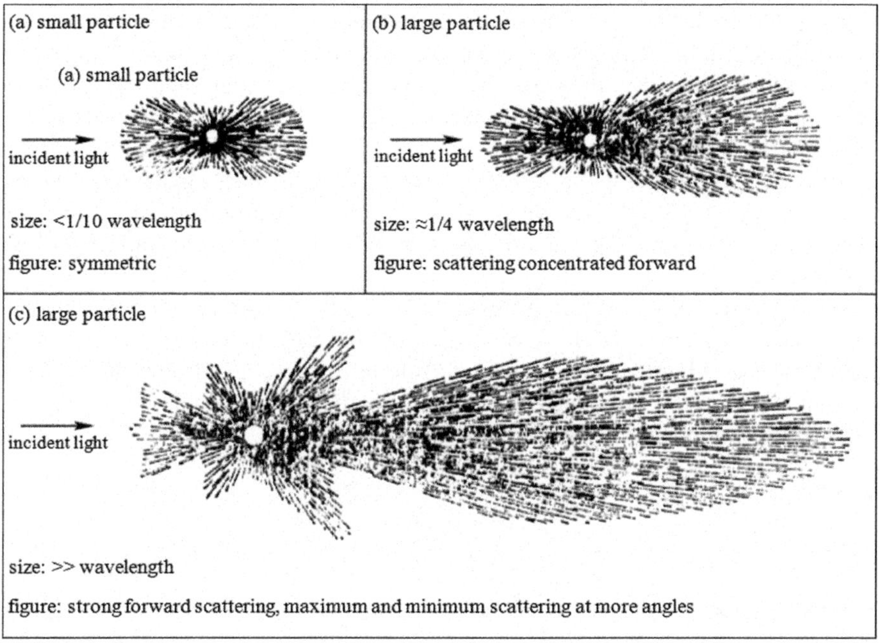

Fig. 4.36 Angular distribution of scattering intensity of particle size relative to wavelength

Fig. 4.37 Illustration of atmospheric refraction

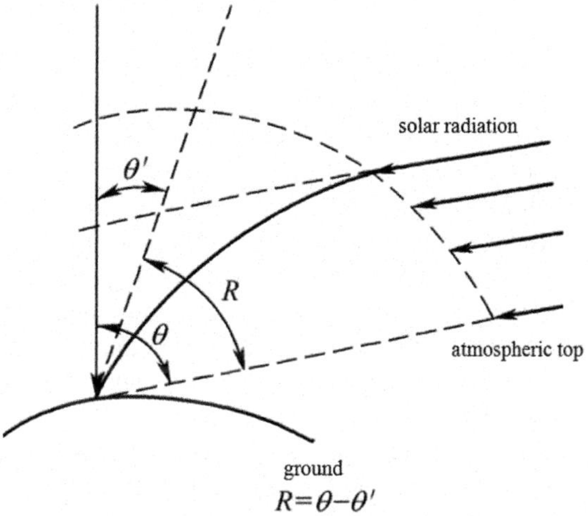

intensity decreases rapidly with the increase of wavelength. For example, for infrared light $\lambda = 4$ μm, its scattering intensity is only 1/10,000 of blue light $\lambda = 0.4$ μm. Therefore, Rayleigh scattering has a large effect on visible light, but a small effect on infrared radiation and even more negligible effect on microwave.

The particles that accord with the Mie scattering are all kinds of dust, aerosols and water droplets in the atmosphere. When the particle size is close to the wavelength of electromagnetic wave, the scattering is not dipole scattering, and the coherence between the scattering elements must be considered. The scattering intensity of Mie scattering is proportional to λ^{-2}, that is, $I \propto \lambda^{-2}$ [8], and it is mainly forward scattering, with smaller polarization of scattering light than Rayleigh scattering. For example, the particle size of cloud is close to the wavelength of infrared ray (0.76–15 μm), so the Mie scattering in wet weather has a great impact on light transmission.

When the size of the particle is much larger than the wave length, the scattering intensity is independent of the wavelength, and the reflection and refraction of the particle to the radiation play a dominant role in macroscopic scattering, which is called non-selective scattering. The scattering coefficient β is equal to the sum of the cross sections of N particles with radius r_i per unit volume

$$\beta = \pi \sum_{i=1}^{N} r_i^2$$

The radius of fog droplets is much larger than visible light waves. Fog scatters all wavelengths of visible light in the same way, so fog appears white. For rain, the scattering coefficient in the infrared spectrum is

$$\beta_{rain} = 0.248 V^{0.67}$$

where V is the rainfall rate (mm/h).

Clouds have a great influence on the sun's radiation. These large droplets absorb very little radiation, and absorption becomes constant when the thickness exceeds 500 m, while reflection plays a major role. When the cloud thickness exceeds 1200 m, the solar radiation has been completely blocked.

In the atmospheric window, radiation attenuation is lost mainly by scattering. As in the visible region, the absorbed energy accounts for only 3% of the decay energy. The type and intensity of scattering attenuation are closely related to wavelength. For example, the scattering of small raindrops in clouds is non-selective scattering to visible light, while in microwave band, the diameter of raindrops is smaller than microwave wavelength, so it belongs to Rayleigh scattering. According to the law of Rayleigh scattering, the scattering light intensity is proportional to λ^{-4}, and the microwave wavelength is more than 1000 times longer than the visible light wave length, and the scattering intensity is 10^{-12} times weaker, so microwave has a strong ability to penetrate clouds. Infrared radiation is less capable of penetrating clouds than microwaves, but more than 10 times more powerful than visible light.

2. **Atmospheric radiation**

Shortwave radiation of the atmosphere is mainly caused by the scattering of solar radiation by the atmosphere, which has been discussed previously.

Another form of atmospheric radiation is air glow, a luminous phenomenon in the upper atmosphere. After solar ultraviolet radiation enters the earth's atmosphere, it can directly cause the excitation of atoms and molecules in the atmosphere or indirectly through the secondary process. When the excited atoms and molecules return to normal state, the light emitted is air glow. Air glow exists day and night, respectively, called solar air glow, night air glow of dawn and dusk.

When the satellite operates on one side of the earth's sunlight, the solar air glow is inundated by solar radiation, reflected light and scattered light from the earth. At night, however, without earth's reflection, gas glow becomes an important optical phenomenon.

There is a continuous spectrum in the night air glow, extending from about 0.4 μm to the infrared region, superimposed on the continuous spectrum of atomic lines and molecular bands. In the ultraviolet, there is O_2 Herzberg band from 0.24 to 0.50 μm and H L_α line 0.1215 μm. There are 0.5577 μm and 0.6300 μm lines for O and 0.5893 μm lines for Na in the visible region. There is a 0.8645 μm line of O_2 in the infrared region, and the Maynell band of OH ranges from 0.38 to 4.5 μm, in which the main energy is concentrated in 1.0–4.5 μm.

Air glow is only produced in a certain layer of the atmosphere, at an altitude of 77–100 km, with a thickness of 24 ± 3 km. The intensity of air glow is very weak, and its ground illumination is equivalent to brightness of a candle at a distance of 100 m. The intensity of air glow is affected by season, geomagnetic activity, latitude and solar activity.

There is another luminous phenomenon that is often seen in polar regions, called the aurora. The aurora that fills the sky has the intensity of a full moon.

Atmospheric longwave radiation is related to atmospheric absorption, and atmospheric absorption is a function of wavelength. For the non-transparent spectrum, the atmospheric absorption is 1, and the longwave radiation of ground is all absorbed by atmosphere, equivalent to a blackbody, so the radiation of atmosphere in non-transparent region is the same as the blackbody radiation spectrum. The substances with the largest absorption coefficients in infrared spectrum are H_2O and CO_2, which are mainly concentrated in troposphere. H_2O, in particular, drops to zero at a few kilometers above tropopause. It can be seen from Fig. 4.35 that other major absorption peaks except for water absorption are few, which means that the atmosphere above the stratosphere is quite transparent to infrared radiation. The isothermal layer at the bottom of the stratosphere can also be seen as a blackbody, with a temperature equivalent to 218 K. Therefore, from the outer space, the longwave radiation in the opaque region of the atmosphere is equivalent to the blackbody radiation of 218 K, and the radiation peak is about 13 μm. In the transparent spectral region of the atmosphere—the atmospheric window, the atmospheric specific emissivity $\varepsilon(\lambda T) = \alpha(\lambda T)$ is very small because the atmospheric absorption is very small, so the atmospheric longwave radiation has little effect on the atmospheric radiation in atmospheric window.

4.4.4 Atmospheric Turbulence

When the sky is looked on a clear night, stars blinking can be seen. This phenomenon is called scintillation. The phenomenon of scintillation is caused by atmospheric turbulence, especially small-size turbulence, which leads to local inhomogeneity of atmospheric temperature and pressure, also leads to local variation of atmospheric refractive index and convergence or divergence of plane wavefront. While this fluctuation is random, so the beam drifts and intensity fluctuates. Scintillation exists not only in visible region, but also in infrared and microwave region. It is equivalent to a kind of noise, which affects the improvement of resolution of instrument and changes the properties of optical transfer function.

Atmospheric turbulence reflected in space optical remote sensing is the drift, jitter and extension of imaging beam, which is the turbulence effect of atmospheric scintillation, phase fluctuation and beam extension. The influence and form of atmospheric turbulence on beam characteristics are related to the ratio of beam diameter d to turbulence scale l. When $d/l \ll 1$, the propagation direction of beam drifts randomly on the detector receiving surface, which is beam drift. When $d/l \approx 1$, the beam section is deflected randomly, and the arrival angle is fluctuated, which leads to jitter of image points on focal plane. When $d/l \gg 1$, the turbulence diffuses the beam, resulting in random distribution of beam intensity and phase in time and space, leading to the expansion of the beam section.

Turbulence leads to changes in atmospheric refractive index. The air refractive index n determined by air density is a function of temperature t/°C, atmospheric pressure p/kPa and relative humidity RH/% that is shown as follows [3]:

$$n = 1 + 7.86 \times 10^{-4}\frac{p}{273 + t} - 1.5 \times 10^{-11}\,\mathrm{RH}\left(t^2 + 160\right) \tag{4.41}$$

It is generally believed that the refractive index fluctuation of dry air is exactly the same as temperature. The fluctuation characteristics of atmospheric refractive index can be studied by measuring the fluctuation characteristics of atmospheric temperature field.

The turbulence intensity of the near-earth atmosphere has significant diurnal variation. During the day, turbulence can fully develop with large intensity, and it can reach the maximum value at noon. Night turbulence is not easy to develop, and the average intensity is relatively low. At 1 h after sunrise and 1 h before sunset, solar radiation and ground radiation are in equilibrium and turbulence is weakest. Turbulence in the near-earth atmosphere is also influenced by geographical features.

If the specific optical system is not considered, and the angle between normal direction of the wavefront and tangent direction of the light propagation can be called arrival angle. For interferometers, arrival angle refers to the phase difference between two points, so the distance between these two points affects the measurement of arrival angle. For a telescopic system, arrival angle corresponds to the tilt direction of the wavefront at the receiving aperture or the direction of the focus away from the optical axis.

For the whole atmosphere, a typical RMS of arrival angle in a telescope of 400 mm diameter is about 9 μrad (about 2″) [3]. For high spatial resolution space optical remote sensing, the effect of atmospheric turbulence on remote sensing resolution must be paid more attention.

4.4.5 Atmospheric Refraction and Its Effect on Signal

1. **Atmospheric refraction**

In optical band, the air refractive index n determined by air density as a function of temperature t/°C, atmospheric pressure p/kPa and relative humidity RH/% is shown in Eq. 4.41. While from Eq. 4.39, it can be seen that pressure P attenuates exponentially with the increase of height H. Therefore, the refractive index of atmosphere also decreases gradually with height, and $n = 1$ at the top of atmosphere.

Due to the constant change of atmospheric refractive index, the propagation path of electromagnetic wave in the atmosphere is a curve. If the electromagnetic wave passes through the whole atmosphere, there is a certain difference R between the actual zenith angle θ and the measured apparent zenith angle θ' due to the influence of atmospheric refraction, which is called the atmospheric refraction value

$$R = \theta - \theta'$$

This can be proved at normal temperatures

$$R = 58.3'' \, tg\theta' - 0.067'' tg^3\theta' \tag{4.42}$$

When $\theta' = 0$, the atmospheric refractive index has no effect. With the increase of θ', the influence of atmospheric refractive index becomes more significant. As seen from Table 4.13, the object can be seen when it is actually below the geometric horizon, because $\theta' = 90°$, $\theta = 90\,°\,35'$. Similarly, in photogrammetry, due to the influence of atmospheric refraction, the angular distance on the image is not consistent with the angular distance of the actual object. In addition, the higher the flight altitude, the larger the angular distance from the zenith, the larger the error and the larger the amount of correction.

Atmospheric refractive index is also a function of wavelength. However, the effect of dispersion is very small and can be ignored in general. Only in high spatial resolution and high orbit optical remote sensing, it should be paid attention to this effect.

2. **Atmospheric attenuation of radiation signals**

Atmospheric attenuation of radiation is called extinction. It can be expressed by the extinction coefficient $k(\lambda, h)$, k in unit m^{-1}, which represents the proportion of radiation attenuation across a unit distance, i.e.,

Table 4.13 Atmospheric refraction values and atmospheric mass values

θ'	θ	R	Sec θ'	Atmospheric mass	θ'	θ	R	Sec θ'	Atmospheric mass
0°	0°	0″	1.000	1.000	80	80° 5″	319″	5.76	5.60
10°	10° 0′	10″	1.015	1.015	81°	81° 6′	353″	6.39	6.18
20°	20° 0′	21″	1.064	1.064	82°	82° 7′	394″	7.19	6.88
30°	30° 1′	34″	1.155	1.154	83°	83° 7′	444″	8.21	7.77
40°	40° 1′	49″	1. 305	1. 304	84°	84° 8′	509″	9.57	8.90
45°	45° 1′	59″	1.414	1.413	85°	85° 10′	593″	11.47	10.40
50°	50° 1′	70″	1.556	1.553	86°	86° 12′	706″	14.34	12.44
55°	55° 1′	84″	1.743	1.740	87°	87° 14′	865″	19.11	15.36
60°	60° 2′	101″	2.000	1.995	88°	88° 18′	1163″	28.65	19.8
65°	65° 2′	125″	2.366	2.359	89°	89° 25′	1481″	57.3	27.0
70°	70° 3′	159″	2.924	2.904	89°51′	90° 0′	1760″	116	32
75°	75° 4′	215″	3.864	3.816	90°	90° 35′	2123″	∞	38

Table 4.14 Apparent magnitude and color temperature of the planet when it is brightest

Name	Moon	Venus	Mars	Jupiter	Mercury	Saturn
Apparent magnitude	−12.2	−4.28	−2.25	−2.25	−1.8	−0.93
Color temperature/K	5900	5900	5900	5900	5900	5900

$$\frac{\mathrm{d}E(\lambda)}{E(\lambda)} = -k(\lambda, h)\mathrm{d}h$$

The radiation intensity after passing through the vertical atmosphere of thickness l is

$$\int_{E_0}^{E} \frac{dE(\lambda)}{E(\lambda)} = -\int_{0}^{1} k(\lambda, h)\mathrm{d}h$$

$$E(\lambda) = E_0(\lambda)e^{-\int_0^l k(\lambda,h)\mathrm{d}h}$$

In the formula, $E_0(\lambda)$ is the radiation intensity at $l = 0$. When considering the radiation from ground to outer space, $l = 0$ is taken as the ground; when considering the attenuation of solar radiation through the atmosphere, $l = 0$ is taken as the top of atmosphere. $E(\lambda)$ is the intensity of radiation attenuated through the atmosphere at vertical height l.

If the radiation is incident oblique at a certain altitude angle θ, then

$$\mathrm{d}h' = \sec\theta\mathrm{d}h$$

The intensity of radiation is

$$\begin{aligned} E'(\lambda) &= E_0(\lambda)e^{-\sec\theta \int_0^l k(\lambda,h)\mathrm{d}h} \\ &= E_0(\lambda)e^{-m(\theta)\tau(\lambda)} \end{aligned} \tag{4.43}$$

where $\tau(\lambda) = \int_0^l k(\lambda, h)\mathrm{d}h$ is called vertical optical thickness. $m(\theta) = \sec(\theta)$ is called atmospheric mass. It represents the ratio of equivalent optical path of atmosphere at oblique incidence to atmospheric optical path at vertical incidence. Therefore, the atmospheric mass at vertical incidence is 1. $m(\theta)$ increases rapidly as θ increases. $m(\theta) = \sec(\theta)$ is derived by treating the atmosphere as a flat layer, ignoring its curvature and refraction. When $\theta < 60°$, the use of $m(\theta)$ as atmospheric mass is fairly accurate. When $60° < \theta < 80°$, the curvature of atmosphere becomes important, and $m(\theta)$ is less than the value calculated using $\sec(\theta)$. The relationship between $m(\theta)$ and θ can be found in Table 4.13.

Theory and experiment show that the extinction effect caused by different atmospheric components and different physical processes has the property of linear superposition, and the total extinction characteristic quantity can be written as the sum of components [6].

The vertical optical thickness of atmosphere is related to absorption and scattering of atmosphere

$$\tau(\lambda) = \tau_\alpha(\lambda) + \tau_s(\lambda)$$

where $\tau_\alpha(\lambda)$ is vertical optical thickness caused by absorption

$$\tau_\alpha(\lambda) = \int_0^l k_\alpha(\lambda, h)\mathrm{d}h$$

where $k_\alpha(\lambda, h)$ is extinction coefficient of absorption

$\tau_s(\lambda)$ is vertical optical thickness caused by scattering

$$\tau_s(\lambda) = \int_0^l k_s(\lambda, h)\mathrm{d}h$$

where $k_s(\lambda, h)$ is extinction coefficient of scattering action.

$k_s(\lambda, h)$ can also be expressed as

$$k_s(\lambda, h) = k_R(\lambda, h) + k_M(\lambda, h) + k_L(\lambda, h)$$

$k_R(\lambda, h)$, $k_M(\lambda, h)$ and $k_L(\lambda, h)$ are extinction coefficients caused by Rayleigh scattering, Mie scattering and large particle scattering, respectively, and their corresponding vertical optical thickness is

$$\tau_s(\lambda, h) = \tau_R(\lambda, h) + \tau_M(\lambda, h) + \tau_L(\lambda, h)$$

Similarly, $k_\alpha(\lambda, h)$ can be expressed as

$$k_\alpha(\lambda, h) = k_{O3}(\lambda, h) + k_{CO_2}(\lambda, h) + k_{H_2O}(\lambda, h) + k_A(\lambda, h)$$

In the formula, $k_{O_3}(\lambda, h)$, $k_{CO_2}(\lambda, h)$, $k_{H_2O}(\lambda, h)$ and $k_A(\lambda, h)$ are extinction coefficients caused by absorption of O_3, CO_2, H_2O and other gases, respectively.

Their respective vertical optical thickness is

$$\tau_\alpha(\lambda, h) = \tau_{O_3}(\lambda, h) + \tau_{CO_2}(\lambda, h) + \tau_{H_2O}(\lambda, h) + \tau_A(\lambda, h)$$

Substituting τ_s and τ_α into Eq. 4.43

$$E'(\lambda) = E_0(\lambda)e^{-m(\theta)(\tau_{O_3}+\tau_{CO_2}+\tau_{H_2O}+\tau_A+\tau_R+\tau_M+\tau_L)}$$

Atmospheric attenuation can also be expressed by transmittance T

$$T = \frac{E'}{E_0} = e^{-m(\theta)\tau}$$

In the formula, $m(\theta)\cdot\tau$ is also called optical thickness. T and τ are both functions of wavelength and other meteorological factors and vary greatly in different spectral band, time and place.

3. Atmospheric attenuation of signal contrast

The influence of atmosphere on radiation signal is not only attenuation, but also great influence on signals. On the platform of height H, the radiance of object observed on the ground L is

$$L = L_G T(h) + L_P$$

where L_G is radiance of objects on the ground, $T(h)$ is atmospheric transmittance from the ground to altitude h, L_P is radiation brightness of atmospheric column upward from ground to altitude h, and these quantities are all functions of wavelength, atmospheric thickness and observation angle.

The shortwave region dominated by solar radiation, if the ground is Lambert, can be expressed as

$$L_G = \frac{E \cdot \rho}{\pi}$$

ρ is hemispheric reflectance of ground, E is irradiance of ground, and E includes the sum of irradiance of direct solar E_s and sky diffuse incident E_D

$$E = E_s + E_D$$

According to $E = \cos\theta \cdot E_0/D^2$ and considering the effect of atmospheric transmittance, E_s can be written as

$$E_s = \frac{\cos\theta T \cdot E_0}{D^2}$$

where E_0 is spectral solar constant, D is sun–earth distance in unit of the mean sun–earth distance, T is transmittance of the whole atmosphere, and its relationship with transmittance T_0 of the vertical whole atmosphere is

$$T = e^{-m(\theta)\tau}$$

At vertical, the whole atmosphere, $m(\theta) = 1$ and $T_0 = e^{-\tau}$

$$T = e^{-m(\theta)\tau} = (e^{-\tau})^{m(\theta)} = T_0^{m(\theta)}$$

where $m(\theta)$ is atmospheric mass.

Substitute the above four formulas into $L = L_G T(h) + L_P$

$$L = \left(\frac{E_0}{D^2}\cos\theta \cdot T_0^{m(\theta)} + E_D\right) \cdot \frac{\rho}{\pi} \cdot T(h) + L_P$$

In the formula, E_0, E_D, T, T_0, ρ and L_P are all functions of wavelength. E_0, T, T_0 and L_P are also related to the absorption and scattering of atmosphere, as well as are functions of time and space. They are also related to factors such as ground conditions and observation angles. Therefore, it is very complicated to correct the influence of atmospheric transmission accurately.

Because L not only includes target brightness E_0 and E_D, but also includes the atmosphere backscatter intensity L_P. L_P occupies a considerable proportion in L and brings adverse effect on the quality of space optical remote sensing imaging. Moreover, the influence of solar radiation and earth–air system on optical radiation calculation generally adopts the method of LOWTRAN software to calculate, MODTRAN can also be used when spectral resolution is high.

4.5 The Moon and Planetary Radiation

The setting of apparent magnitude is to divide stars visible to naked eye into magnitude 6, the stars just visible to naked eye are defined as magnitude 6, and the stars slightly brighter than magnitude 6 are defined as magnitude 5.

The apparent magnitude of two stars is defined as m_1 and m_2, and their radiance intensity is E_1 and E_2. If $m_2 - m_1 = 5$, then $E_1 = 100\, E_2$.

According to the definition:

$$lg\frac{E_1}{E_2} = 0.4(m_2 - m_1)$$

The irradiance of zero apparent magnitude is 2.65×10^{-6} lx, which translates into radiometric is

$$2.65 \times 10^{-6}/683 = 3.88 \times 10^{-9}\,\mathrm{W/m^2}$$

According to above, the ground irradiance or atmospheric top irradiance of each magnitude can be calculated. And its results can be used as reference for radiometric calibration of space optical remote sensing.

Planet

Since the moon and planets reflect sun's radiation, their spectra are close to sun with 5900 K blackbody radiation, but much weaker. The following table lists visual magnitudes and color temperatures of moon (full moon) and planets (when they are brightest).

The photometric stability of lunar surface can reach 10^{-9}/year, making it a stable natural reference source. China's FY-2 (03 batch) satellite adopted the internal blackbody calibration method based on lunar radiometric correction, and the test showed that the calibration accuracy of infrared spectrum could reach 0.5 K @ 300 K [10].

References

1. McCartney EJ. In: Naixian P, Jietai M, Yongsheng W, translators. Optics of the atmosphere. Beijing: Science Press; 1988.
2. Sihua L. Physics fundamentals of remote sensing. Beijing: The Commercial Press; 1981.
3. Jingzheng L. Hand book of optics. Xian: Shaanxi Science and Technology Press; 2010.
4. Zongfan W, Meilin L, Shaojv Z, Yuze Z, Hanbin L. Infrared and low light technology. Beijing: National Defense Industry Press; 1998.
5. Jingxian L, Yudan L, Weiqi J. Low light level and infrared imaging technology. Beijing: Beijing Institute of Technology Press; 1995.
6. Xiuji Z, Shancang T, Keya Y. Advanced atmospheric physics. Beijing: China Meteorological Press; 1991.
7. Heng D, Zonghai Y, editors. Space environment manual for low orbit spacecraft. Beijing: National Defense Industry Press; 1996.
8. Anxin M. An introduction to remote sensing. Beijing: Higher Education Press; 1989.
9. Nanrong X, Nanhua B. Infrared radiation and guidance. Beijing: National Defense Industry Press; 1997.
10. Qiang G, Boyang C, Yong Z, Na X, Xuan F, Changjun Y, Xin W, Xiuzhen H, Yuan L, Xiuqing H, Zhixin Z. Progress in on-orbit radiometric calibration technology for Fengyun-2 Satellite. Adv Meteorol Sci Technol. 2002;3(6):6–12.

Chapter 5
Photoelectronic Detectors

5.1 Overview

Whether it is passive or active optical remote sensing system, the selection of detector plays a key role in the energy requirements of optical remote sensing system. With the development of photoelectric technology, the types of detectors are increasing, and the corresponding performance is also improving. Traditional photosensitive film is gradually fading out of the historical stage, and photodetectors have sprung up, not only in military, but also in civil. However, as a main method of traditional optical remote sensing, some properties of film still need to be mentioned, because understanding the basic principles of this field will also be beneficial to the research and parameter selection of optical remote sensor.

Film introduction

Introduction to film characteristics. The photosensitive layer of photosensitive emulsion is composed of silver bromide micrograins suspended in emulsion solution. When the emulsion is exposed, the halogen ion in the photosensitive layer absorbs photons and excites an electron. The electron is absorbed by the nearby silver ion to reduce a silver ion to a silver atom:

$$\mathrm{Ag}^{+} + \mathrm{Br}^{-} + \mathrm{h}\nu \rightarrow \mathrm{Ag} + \mathrm{Br}$$

Or it can be expressed in two steps

$$\mathrm{Br}^{-} + \mathrm{h}\nu \rightarrow \mathrm{Br} + \mathrm{e}^{-}$$
$$\mathrm{Ag}^{+} + \mathrm{e}^{-} \rightarrow \mathrm{Ag}$$

There are few silver atoms produced by the action of photons, which are called latent image centers. The latent image center expands continuously under the chemical action of the developer, resulting in a large number of silver atoms. After development, dissolve some residual silver bromide that has not been decomposed in the

J. Tao, *Space Optical Remote Sensing*, Advances in Optics and Optoelectronics,
https://doi.org/10.1007/978-981-99-3318-1_5

photosensitive layer with fixing solution. This enables a stable and dense layer of silver atoms to be formed in the photosensitive layer. The silver atom layer precipitated more in the place with strong light looks black, and the silver atom layer precipitated less in the place with weak light looks transparent.

Characteristic curve of latex. The degree of blackening of latex is expressed by density D.

$$D = \log \frac{i_0}{i}$$

where i_0 is incident light intensity, I is output light intensity through negative film.

$$i = i_0,\ D = 0$$
$$i = 0,\ D = \infty$$

It is related to illuminance, exposure time, composition of emulsion, composition of developer and development time.

Exposure E

$$E = I \cdot t$$

I is the irradiance of the emulsion and t is the exposure time.

The relationship between D and logE is called the characteristic curve of latex.

As can be seen from Fig. 5.1, the characteristic curve of emulsion between curves B–C is linear. The greater the exposure, the greater the density of the negative is, that is, the darker the negative.

The background density D_0 value displayed by the unexposed latex is between 0.01–0.02, and the D_0 value of the expired latex will increase.

The resolution of film. It refers to the ability of emulsion to distinguish small details in photogrammetry, which is usually expressed by the number of line pairs that can be distinguished per mm. The resolution is not only related to the particle size of the emulsion itself, but also related to the contrast of the photographed object. Objects with high contrast have high resolution, and objects with low contrast have low resolution. The resolution of common latex is tens to hundreds of lines / mm, and holographic negative can reach to more than 1000 lines/mm.

At present, expect for special needs, most photosensitive elements use photodetectors, so this chapter mainly introduces photodetectors. As an important part of space optical remote sensing system, its functional principle and performance index are introduced.

Photodetector is a device that converts optical radiation into electronic quantity. In principle, there are two categories: photon effect photodetector and photothermal effect photodetector.

Photon effect photodetector is a photoelectric conversion method in which photons $h\nu$ excite electrons e^-. The common feature of photon detectors is that they have a working wavelength threshold, that is, the long wavelength limit

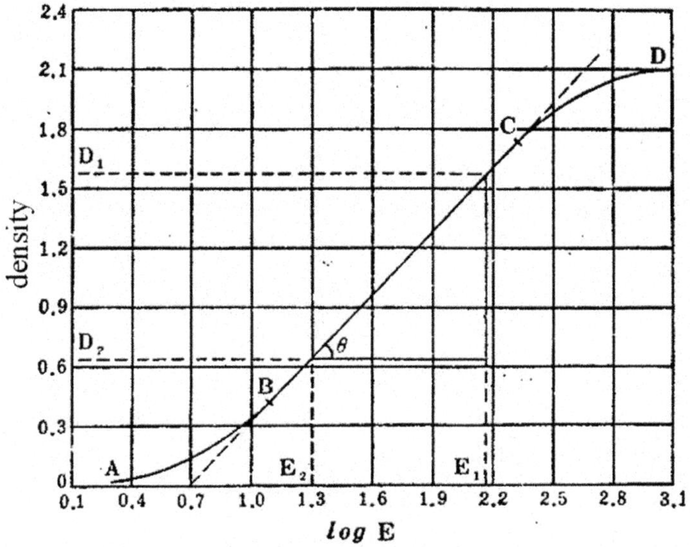

Fig. 5.1 Characteristic curve of photographic emulsion

$\lambda_f, \lambda_f = 1.24/E_g$, E_g the threshold energy determined by the material of photon detector, unit: $\lambda_f/\mu\text{m}$, E_g/eV.

Photothermal effect detector is the heat energy that is converted into lattice vibration after the device absorbs light energy, which causes the temperature of the element to rise, resulting in the changes of electronic or other physical properties of the device. These changes are finally converted into electronic output and detected. There is no direct relationship between photothermal effect and photon energy $h\nu$, and no choice of radiation frequency. Such devices usually have slow response speed and are easily affected by ambient temperature. A pyroelectric effect detector based on the temperature change rate of materials has fast response speed and has application prospects.

5.2 Characteristic Parameters of Photoelectronic Detectors

The performance parameters of photodetectors generally include spectral sensitivity, integral sensitivity, quantum efficiency, noise equivalent power and normalized detection rate. Sensitivity is often called response, which is a measure of the photoelectric conversion characteristics of the detector.

1. Spectral sensitivity R_i

The spectral power density of incident light is p_λ, the response of the detector is i_λ or u_λ, then the spectral sensitivity of the detector is

$$R_{i\lambda} = \frac{di_\lambda}{dp_\lambda}(A/W)$$
$$R_{u\lambda} = \frac{du_\lambda}{dp_\lambda}(V/W)$$

They are called current spectral sensitivity $R_{i\lambda}$ and voltage spectral sensitivity $R_{u\lambda}$.

If R_λ is a constant, then the detector is a non-selective detector, such as a photoconductive detector. Photovoltaic detector is a selective detector, and its relative spectral sensitivity is S_λ by

$$S_\lambda = \frac{R_\lambda}{R_{\lambda m}}$$

$R_{\lambda m}$ is the peak response, S_λ is a dimensionless quantity, the curve of $S_\lambda(\lambda)$ is the spectral sensitivity curve of the detector.

2. The integral sensitivity of the detector R

If the incident light power is p and the response of the detector is i or u, the integral sensitivity of the detector is

$$R_i = \frac{di}{dp} = \frac{i}{p}(\text{In the linear region})(A/W)$$
$$R_u = \frac{du}{dp} = \frac{u}{p}(\text{In the linear region})(V/W)$$

They are called current sensitivity R_i and voltage sensitivity R_u, respectively. P is the total optical power within a certain incident wavelength range.

3. Quantum efficiency η

The ratio of the number of photoelectrons produced to the number of incident photons is usually a function of wavelength [1].

4. Detector noise

Noise is a kind of irregular electricity output, which is the average value in the statistical sense.

(1) Thermal noise

Any material with resistance will produce thermal noise as long as its temperature is higher than absolute zero. It is formed by the random fluctuation of voltage caused by the collision of the thermal motion of electrons at both ends of the resistance.

$$U_n^2 = 4kTR\Delta f$$
$$i_n^2 = \frac{4kT\Delta f}{R}$$

where U_n noise voltage (V), in noise current (A), K Boltzmann constant, T thermodynamic temperature (K), R resistance (Ω), Δf detector bandwidth (Hz).

(2) Photon shot noise

Originally found in vacuum diodes, it is caused by irregular fluctuations in the speed of cathode electrons reaching the anode. A similar phenomenon occurs when carriers pass through the transition region of the semiconductor PN node. They are like fluctuations in the number of bullets in the target, so they are named. This phenomenon exists in the number of photons radiated by the object, the number of electrons emitted by the cathode, the number of carriers in the semiconductor and the multiplication coefficient of the photomultiplier tube.

$$i_n^2 = 2\mathrm{e}I\Delta f$$

where e^- is electron charge, $\mathrm{e} = 1.6 \times 10^{-19}$(c); I is current flowing through the detector, Δf is band.

When I and Δf are not easy to determine, the photon shot noise can be expressed in the form of photon number, i.e.,

$$N_n = \sqrt{N_p}$$

where N_n is the number of photons fluctuates with shot noise, and N_p is the number of photons of incident radiation.

(3) Generation recombination noise

The number of carriers in a semiconductor fluctuates, which is caused by the randomness of generation recombination. The resulting noise is called generation recombination noise.

$$i_n^2 = 4eI_0\Delta f\frac{\tau_c}{\tau_d}\frac{1}{1+(2\pi f\tau_c)^2}$$

where I_0 is average current in detector; τ_c is carrier lifetime; τ_d is carrier transit time; f is modulation frequency of incident light. When $f << (1/2\pi\tau_c)$, $i_n{}^2 = 4eI_0\Delta f\tau_c/\tau_d$, Δf is frequency band.

(4) 1/f noise

It is caused by the small fluctuation of the current flowing through the detector, which mainly occurs in the low frequency domain of about 1 kHz and is inversely proportional to the modulation frequency of optical radiation. The noise can be suppressed when the modulation frequency is higher than 1 kHz [2].

$$i_n^2 = \frac{k_1 I^\alpha \Delta f}{f^\beta} = \frac{C_1 I^\alpha \Delta f}{LAf^\beta}$$

where l is detector length; A is detector sectional area; α is a constant ≈ 2; β is a constant ≈ 1; k_1 is a proportional constant; C_1 is related to the material itself and independent of material size; I is current of detector; Δf is frequency band; f is frequency.

$$u_n^2 = i_n^2 R^2$$

(5) Background noise

Noise caused by random fluctuation of background radiation photon number. The power of photon noise is independent of frequency, and the mean square noise power within Δf bandwidth is

$$p^2 = 8\varepsilon\sigma k T_b^5 A_b$$

where ε is background specific emissivity; σ is Stefan–Boltzmann constant; k is Boltzmann constant; T_b is background temperature; A_b is background area.

In space optical remote sensing, the whole earth atmosphere system is taken as the scenery, so the target and background are not distinguished. The whole radiation is taken as the target radiation, so this noise is taken into account in the photon shot noise.

(6) Noise of photovoltaic detector and photoconductive detector

The noise of photoconductive detector is mainly thermal noise (1); generation compound noise (3) and $1/f4$) have

$$i_n^2 = \left(\frac{4kT}{R} + \frac{4eI_0\tau_c}{\tau_d} + \frac{C_1 I^\alpha}{LAf^\beta}\right)\Delta f$$

The noise of photovoltaic detector is mainly thermal noise and shot noise, and the generated compound noise is very small, have

$$i_n^2 = \left(2eI + \frac{4kT}{R}\right)\Delta f$$

When the photovoltaic detector is in the reverse bias working state, the resistance R is very large, and the thermal noise can be ignored.

5. Noise equivalent power NEP

The incident optical power required to generate the noise level is the noise equivalent power NEP.

6. Normalized Detectivity D^*

When describing the sensitivity of an infrared detector, it can be expressed by the noise equivalent power NEP, which is related to the material, process, pixel area A_d

of the detector and the bandwidth of the measuring circuit Δf. Proof of theory and practice proved that $(A_d \Delta f)^{1/2}$/NEP = constant, $D = 1$ / NEP, called the detection rate W^{-1}.

Let $D^* = D(A_d \Delta f)^{1/2}$ = constant, called normalized detectivity, unit is $cmHz^{1/2}W^{-1}$. It can be understood as the signal-to-noise ratio (SNR) measured on the $1cm^2$ photosensitive surface, the illumination of 1 Wcm^{-2}, and the measuring circuit of 1 Hz bandwidth.

5.3 Solid Image Components (CCD & CMOS)

Solid-state imaging device is a kind of photoelectric imaging device, which can realize photoelectric conversion, storage, electric charge transfer and readout functions on an integrated semiconductor chip and work under the drive of a specific external circuit. CCD and CMOS are widely used. Charge coupled devices (CCD) is a large-scale integrated diode array imaging device, which has two categories: area array and linear array. It integrates photoelectric transfer, charge storage, charge transmission and pickup. It is an all solid-state photoelectric imaging device. Complementary metal oxide semiconductor (CMOS) is a photoelectric imaging device that integrates paired metal oxide semiconductor field effect transistors (MOSFETs) on a chip.

CCD device is characterized by low noise, large dynamic range, time delay integration (TDI) imaging, high effective photosensitive area coefficient of pixel, small pixel size, and easy to obtain large-scale pixel quantity. The disadvantages are that it is difficult to integrate with CMOS circuits, the readout speed is limited, the power consumption is large, the blue light efficiency is poor, and the information is difficult to be stored and retrieved immediately.

CMOS devices are characterized by high circuit integration, fast readout speed, low power consumption, wide response spectrum, real time memory and access, and easy to integrate pixel array with analog and digital functions on one chip. The disadvantages are large noise, small dynamic range and difficult to realize time delay integral imaging, mainly area array photoelectric imaging devices.

5.3.1 CCD Image Components

5.3.1.1 Working Principle of CCD

Figure 5.2 is a schematic diagram of the structure and bias voltage of a Photoelectric MOS (metal oxide semiconductor). SiO_2 is deposited on p-type Si, a metal film is plated on SiO_2, and then a circuit (electrode) is photolithographed. Different voltages are applied to the electrode, as shown in the figure. Three electrodes correspond to one pixel. The bottom surface of the pixel p-type Si acts as a negative electrode,

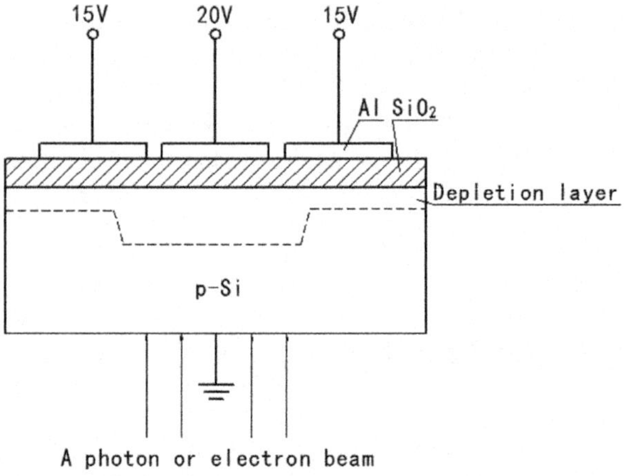

Fig. 5.2 Pixel structure of a CCD

and a reverse bias voltage is formed on the pixel. The photoelectrons formed by the illumination of the pixel are gathered under the high-voltage electrode. The pixels of a linear array CCD imaging device are composed of a plurality of the same pixels arranged in a row. A plurality of linear arrays are arranged in a column, which forms the pixel composition of the area array device (area array CCD). The amount of charge in a pixel represents the intensity of the pixel's illumination, and the pixel's description of the intensity of illumination is the image.

1. Photoelectric conversion and charge storage

When the pixels of CCD are illuminated, holes will be generated in p-type Si. According to semiconductor physics, under the action of reverse bias voltage, most carriers (holes) in p-Si will be far away from SiO_2 to form a "potential well". When illuminated, the electrons produced by the pixel will be stored in a deep "potential well" to form a charge packet.

2. Charge transfer

The charge transfer is shown in Fig. 5.3. By applying the reverse bias voltage on the pixels in sequence, the charge packets can be driven along the surface. Electrode 2 in Fig. 5.3a is 2, the electrodes 2 and 5 in Fig. 5.3b are 20 V, and the charge package is under these two high-voltage electrodes. When the voltage of 2 and 5 is converted to 15 V and the voltage of 3 and 6 (not shown in the figure) is converted to 20 V, the state of the charge packet is shown in Fig. 5.3c. After the electrode voltage is stabilized and the charge packet transfer is completed, the state is shown in Fig. 5.3d. This is a charge packet transfer process.

3. Charge pickup

Figure 5.4 shows the basic composition of linear CCD imaging device. 1 is the optical integration region, 2 is the shift register, 3 is the output diode, 4 is the output grid, 5 is the transverse transfer electrode, 6 is the storage electrode, and Φ_1, Φ_2, Φ_3 is the

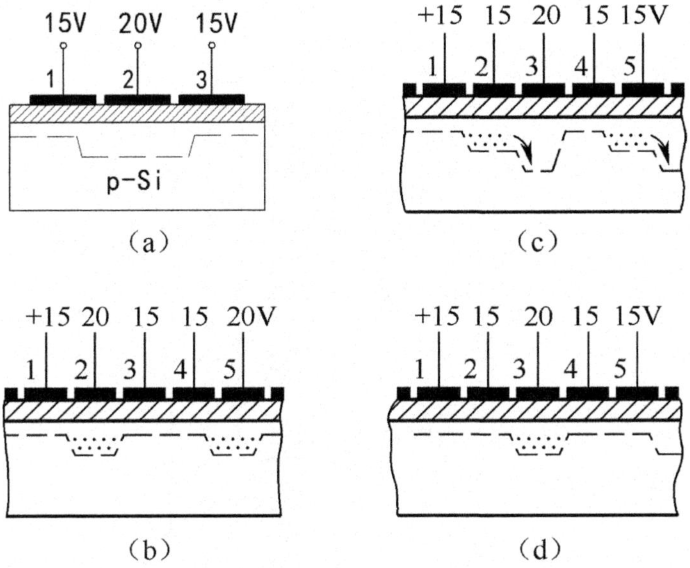

Fig. 5.3 Transfer of CCD charge

longitudinal transfer electrode. When the charge transfer pulse arrives, the charge packets under each pixel in photosensitive region 1 are driven by each electrode of transverse transfer electrode 5 and output to shift register 2 within the integration time of a pixel imaging. Then, driven by a longitudinal transfer electrode Φ_1, Φ_2, Φ_3, the charge packets of each pixel are output in turn by output diode 3.

4. Charge packets to voltage conversion

As shown in Fig. 5.5, (a) is the output circuit structure of CCD, and (b) is the output voltage waveform of CCD unit. When the charge package enters electrode Φ_3, the voltage is high. The output grid Φ_{OG} voltage V_{OG} is lower than Φ_3 voltage, reset pulse Φ_R is positive, The MOSFETs T_1 is on, and the output diode D is in the state of backbias. Its nodule capacity C_s is charged to a fixed DC voltage V_{cc}, and the output voltage of source follower T_2 is duplicated at the reset level V_{OS} and slightly less than V_{cc}. When reset pulse Φ_R positive voltage ends, T_1 tube cut off. As there is leakage current in T_1 tube, the output voltage has a T_1 tube voltage drop and a small drop, which is called feed-through voltage. Meanwhile, Φ_3 voltage becomes lower than V_{OG}. Therefore, the charge packet under Φ_3 electrode is transferred to C_s, resulting in a voltage drop to the voltage proportional to the charge packet quantity, the output voltage of source pole follower T_2 also has a corresponding reduction of the output voltage of a TV signal level, namely

$$V_{OS} = G\Delta V_{OS} = G\Delta Q_S / C_s$$

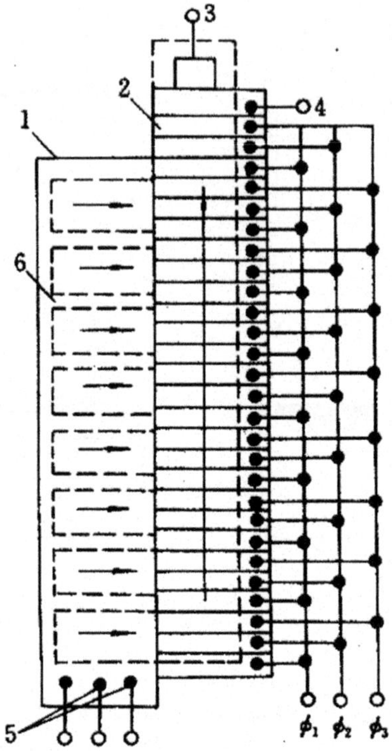

Fig. 5.4 Linear array CCD component diagram

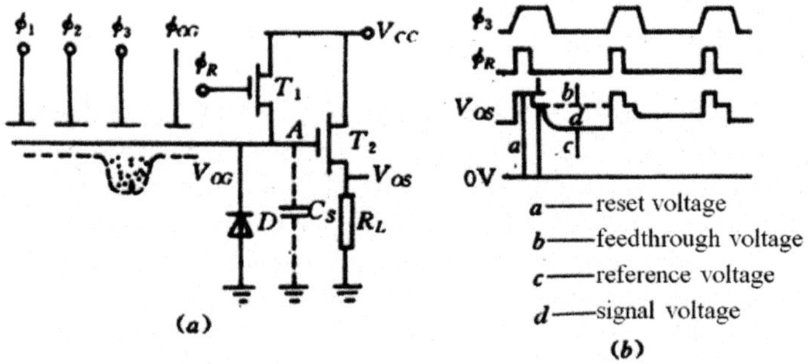

Fig. 5.5 CCD output circuit structure and voltage conversion

where G is the gain of the output amplifier, ΔQ_S is the charge pack quantity, ΔV_{OS} is the drop amplitude of the output voltage.

For array CCD, in order to improve the speed and efficiency of charge picking, frame transfer CCD charge picking mode is usually adopted. There are two modes of frame transfer: the full frame transfer mode from photosensitive area A to readout storage area B as shown in Fig. 5.6, and the interline transfer mode as shown in Fig. 5.7.

The photosensitive region *A* of the full frame transfer area array CCD corresponds to the pixel of the readout memory region *B*, which is not photosensitive. When the signal is picked up, the charge packet in the sensitive area *A* is transferred to the corresponding pixel in the storage area *B* through the sequence Φ_{11}, Φ_{21}, Φ_{31} of frame transfer in the blanking time of each frame. A row of pixel charge packets is output by output diode through timing Φ_{12}, Φ_{22}, Φ_{32} and Φ_{13}, Φ_{23}, Φ_{33} drivers in a row blanking time.

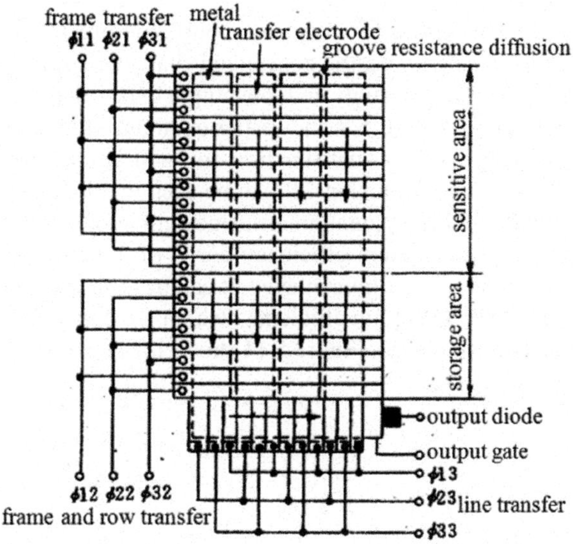

Fig. 5.6 Full frame transfer mode of area array CCD

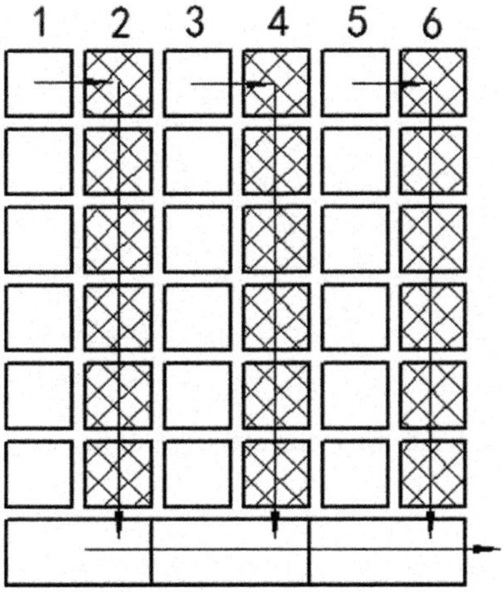

Fig. 5.7 Interline transfer mode of area array CCD

If there is a column of photosensitive pixels in the inter-line transfer area array CCD, a column of storage pixels is set on its space. In Fig. 5.7, the odd number of frames is listed as photosensitive pixels, while the even number is listed as storage pixels. When the signal is picked up, it outputs all the charge packets of each photosensitive pixel within a time of line blanking. Therefore, the frame speed of inter-line transfer CCD is fast, but the disadvantage is the existence of photosensitive blind area, while the full frame transfer CCD is just the opposite.

5.3.1.2 Performance Metrics of CCD

The relationship between the input exposure $H(lx{\cdot}s)$ and the output voltage V is called the photoelectric conversion characteristic of CCD. A good CCD should have high light responsivity and low output dark voltage. This is embodied in its quantum efficiency. Figure 5.8 shows the quantum efficiency curve of a linear array CCD and the relative visual function curve of human eyes (dashed line dark vision and solid line bright vision). It can be seen that its quantum efficiency can reach 30%. Figure 5.9 is the quantum efficiency curve of a color linear array CCD. Figure 5.10 shows the quantum efficiency curve of a area array CCD and the relative visual function curve of human eyes. It can be seen that the average quantum efficiency is only 15%. Quantum efficiency, which is a function of wavelength, also depends on the spectrum used. Figure 5.11 shows the quantum efficiency curve of the color area array detector.

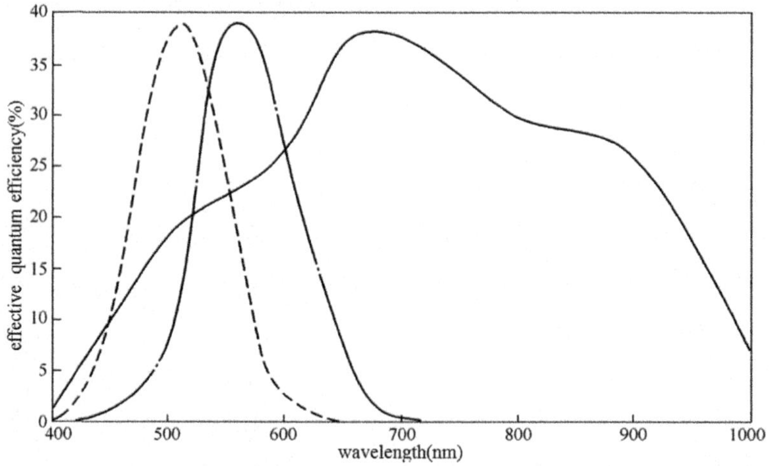

Fig. 5.8 Quantum efficiency of linear array CCD and the relative visual function curve of human eye

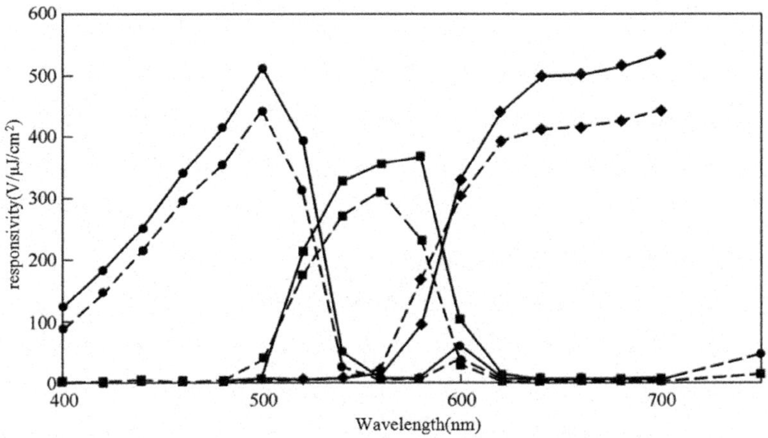

Fig. 5.9 Quantum efficiency curve of color linear array CCD

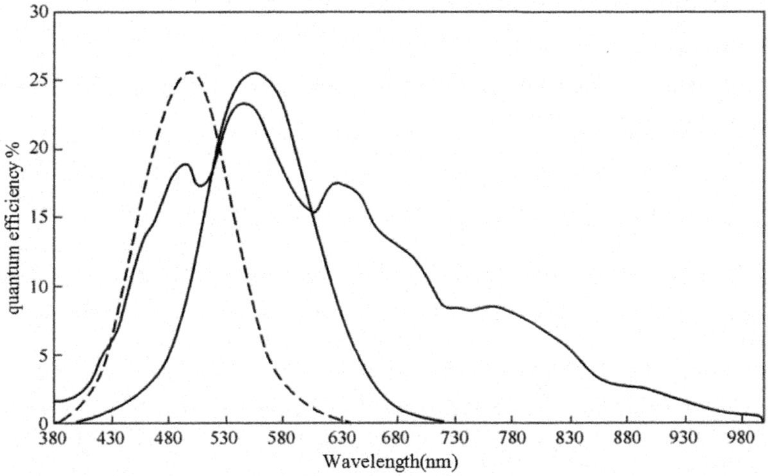

Fig. 5.10 Quantum efficiency of area array CCD and the relative visual function curve of human eye

1. Spectral response

CCD is mostly made of Si material with photoelectric diode and photoelectric MOS array structure, and the spectral response curve is shown in Fig. 5.8. The spectral curve of standard visual function is also given. It can be seen from the figure that the peak wavelength λ_{max} of Si-CCD spectral response is in the visible range and the wave limit $\lambda_f \approx 1.1$ μm, which corresponds to the intrinsic light absorption cutoff wavelength of Si semiconductor band gap width 1.12 eV. The short wavelength of

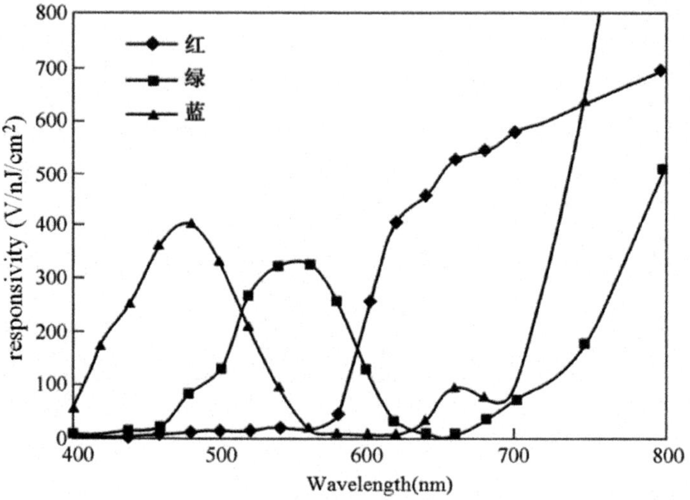

Fig. 5.11 Quantum efficiency curves of area color area array CCD

spectral response depends on the shortwave absorption characteristics of the input surface material and Si material.

2. MTF and resolution

The essence of CCD imaging is discrete pixels sampling incident light discretely. The spatial sampling frequency f_0 is expressed in pixels per mm. According to the sampling theorem, the highest sampling frequency is Nyquist limit sampling frequency $f_N = f_0/2$. Like other imaging devices, CCD has different modulation transmission characteristics for sinusoidal waves with different spatial frequencies, and the higher the spatial frequency, the worse the transmission capability. After normalization (MTF(0) = 1), the measured MTF curve is the MTF of the device. There are three factors affecting the high-frequency MTF of CCD and other array devices:

(1) Pixel size;
(2) Crosstalk caused by transverse diffusion of carriers. Because the CCD potential wells are not completely isolated, the photoelectrons of the pixels may diffuse into neighboring pixels, and the crosstalk becomes more serious with the increase of the input wavelength;
(3) Carrier transfer loss, the higher the frequency, the greater the loss.

In the actual CCD imaging system, the MTF value at Nyquist limit sampling frequency f_N is about 0.5 due to the influence of optical lens and stray light.

At present, the pixel size of CCD device in large-scale array has reached 140 M.

3. Moire effect

If the CCD array is 512 × 320, then $R_H = 512, R_V = 320, f_{0H} = 512, f_{0V} = 320, f_N = f_0/2$. The signal of frequency >f_N will be reflected back to the fundamental frequency domain. This effect interferes with landscape images in the basic frequency domain. So the entertainment TV system wants to reduce the high frequency MTF response; television systems intended for military and industrial applications want to retain the Moire pattern because it contains more information about the object being viewed and the background. It can be collected and interpreted by a trained interpreter.

4. Full well charge, dark charge, dynamic range

Each CCD pixel is equivalent to a capacitive C_p. The maximum number of electrons that can be held in a potential well is called a full well charge and is typically 10^6 e^-/pixel. The charge number is:

$$N_s = \frac{ES_pA_p \cdot t}{e^-}(e^- = 1.602 \times 10^{-19}\,C)$$

where E is illumination absorbed on the detector in W/cm^2, S_p is sensitivity in A/W, A_p is pixel areas in cm^2, t is pixel integration time in s.

Therefore, the pixel's full well charge N_{smax}, average current I_s and average voltage V_{os}:

$$N_{s\,\max} = \frac{C_p \Delta V_{\max}}{e^-}$$

$$I_s = N_s f_H e^-$$

$$V_{OS} = G\Delta V_S = \frac{G\Delta Q_S}{C_p}$$

where $\Delta V_{\max}$ is maximum voltage change during pixel integration time, f_H is pixel sampling frequency, usually video scanning frequency, ΔV_S is pixel signal voltage, V_{OS} is output signal voltage after pixel signal voltage gain G times, ΔQ_S is pixel charge.

5. Dark charge

The charge output of CCD pixels in the absence of light is determined by three kinds of noise:

(1) Shot noise, resulting from the random fluctuation of photon number and photo-electron number, follows the Poisson distribution and is characterized by uniform power distribution over a wide frequency range (white noise). The noise $N_P^{1/2}$ caused by the N_p fluctuation of the number of signal particles is a kind of photoinduced noise, which is also the lowest detectable noise limit level of low light photography. At present, the CCD can reach 10 e^-/ pixel after cooling.

(2) Dark current noise, one is depletion layer thermal excitation, it is random noise, Poisson distribution; The other is caused by the non-uniform distribution of the production–recombination center in the device, especially the dark current peak

formed by the dense defect on the pixel, which will form a fixed graph noise related to the defect location.

The noise reduction measures are cooling, from 25 to −196 °C, dark current drop three orders of magnitude. CCD can reach 1 nA/cm^2 at room temperature, equivalent to 6.25×10^3 e$^-$/ pixel (10 × 10 μm). After the liquid nitrogen cooled, it is about 6.25 e$^-$/ cm^2, the dark current is no longer an important factor.

(3) Transfer noise is caused by transfer loss, interfacial state capture and bulk state capture. The measures and effects to suppress transfer noise are as follows:

Measures effects (e$^-$/pixel).
Same chip amplifier, $C_P \leq 0.03$ pF reset noise reduction 69.
Floating gate amplifier, no reset noise < 69.
Distributed floating amplifier (DFGA) 10 –20.
DFGA modified "Surface/Float well Detector" 1.2
Cooling −196 °C liquid nitrogen, reduce dark current noise < 6.5
After the dark current and transfer noise are reduced, the photon noise is restricted.

Another widely used noise reduction method is to use correlation double sampling signal readout method. The correlation double sampling hold circuit is realized by sampling voltage signals at both ends of photodiode twice during pixel readout. Once before the integration, the pixel is reset, and the sampled signal is noise signal. The other is before the end of the pixel signal integration; sampling is the response of the light change of the electrical signal. Finally, the two signals are subtracted by difference to reduce signal noise.

6. Dynamic Range

Ratio of pixel saturation voltage to RMS noise voltage. The same type of CCD due to the structure and working mode of different dynamic range will be different, generally 70 dB.

The working temperature can reach 60 °C.

5.3.2 CMOS Image Components

CMOS image detector is widely used for its high circuit integration, small size, low power consumption, flexible readout mode, low price and simple drive. Figure 5.12 shows a wide-angle image taken by a 1 × 1 K CMOS camera with a field angle of more than 100 degrees and a weight of only 150 g.

5.3.2.1 CMOS Detector Working Principle

CMOS image detectors have been studied since 1960s. With the continuous progress and improvement of CMOS integrated circuit technology, passive pixel image detectors (PPS), active pixel image detectors (APS) and digital image detectors (DPS)

Fig. 5.12 CMOS wide-angle camera images

appear in CMOS image sensors. Their pixel and chip structures are shown in Figs. 5.13 and 5.14.

Figure 5.13a is the schematic diagram of passive pixels, and Fig 5.13b is the schematic diagram of active pixels, which appeared in 1990s. As the pixel unit of CMOS APS (Active Pixel Sensor) is equipped with an amplifier for each pixel, the performance of the detector has been significantly improved. At present, its performance in terms of SNR and dynamic range is similar to that of CCD. A typical APS active pixel structure has three transistors, which are used for pixel reset, selection, amplification and buffering. In general, a correlation double sampling hold circuit is used to obtain the pixel voltage signals before and after the integration. The noise signal is obtained by sampling before integration, and the noise + optical change signal is obtained by sampling after integration. Finally, the two signals are subtracted by difference. Thus, the circuit requires two sample-hold signals and two identical sample-hold circuits, which increases the chip area and the complexity of the drive circuit. Some researches on novel CMOS image detector readout circuit try

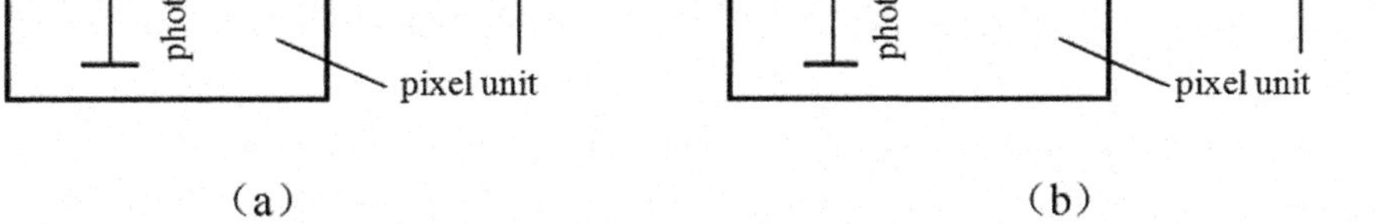

Fig. 5.13 CMOS image detector pixel schematic

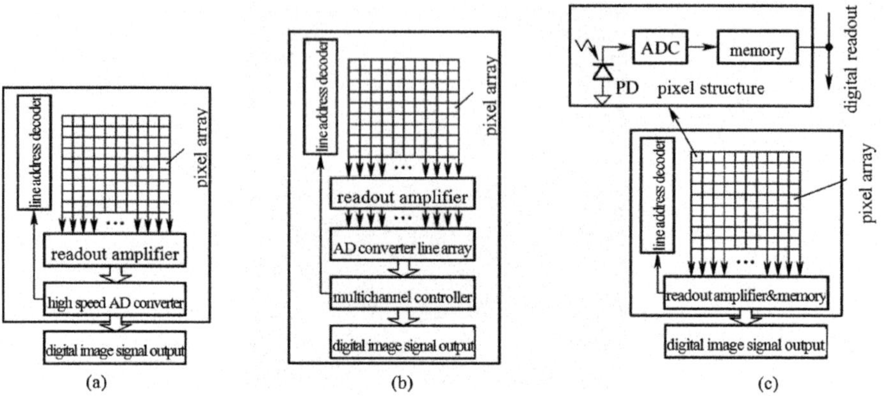

Fig. 5.14 Imaging principle diagram of CMOS image detector on chip

to simplify the related double sampling circuit and make the CMOS image detector circuit more simplified.

Figure 5.14 is the schematic diagram of three CMOS digital image detectors. Figure 5.14a is the digital image signal output mode at the chip level, Fig. 5.14b is the digital image signal output mode at the column level, and Fig. 5.14c is the digital image signal output mode at the pixel level. The flexibility of three kinds of digital image signal output mode is improving, and the pixel level digital image signal output mode is the most flexible and can realize functions such as window opening.

5.3.2.2 CMOS Detector Performance

1. Spectral response of CMOS detector

The spectral response of CMOS detector is shown in Fig. 5.15a. The peak value of its spectral response is located in the visible spectral region, and its spectral response characteristic is similar to that of planar CCD detector (See Fig. 5.10). The quantum efficiency is greater than 60% at the peak of spectral response. Figure 5.16 shows the quantum efficiency curve of area array color CMOS detector.

2. Noise

The noise of CMOS detector rises rapidly with temperature, as shown in Fig. 5.15b. At 40 °C, the noise is about 150 e^-, while the CCD of large array is about 15 e^-, and the difference between them is in the order of magnitude.

3. Resolution

The main application form of CMOS detector is the area array, and the size of the area array (pixel number) is its important index. The pixel scale of the spectral response

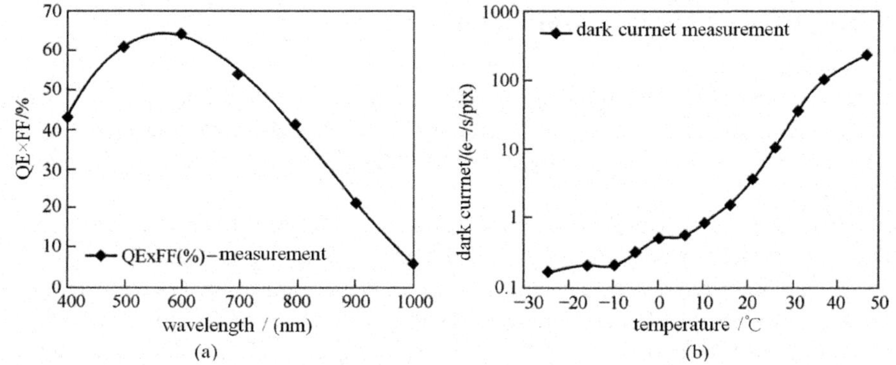

Fig. 5.15 Spectral response and noise of CMOS detectors

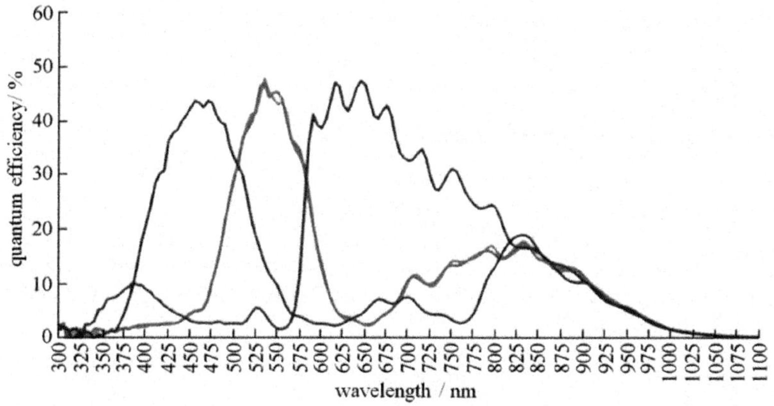

Fig. 5.16 Quantum efficiency curves of color CMOS detectors

of the two detectors shown above is 2 × 2 K, and the pixel of the large-scale CMOS detector has reached 100 M.

4. Dynamic range

The dynamic range of CMOS detectors is comparable to that of CCDs of the same size. The aforementioned 2 × 2 K pixel device has a dynamic range of about 70 dB, and the larger 70 M pixel device has a dynamic range of about 60 dB.

The working temperature can reach 60 °C.

5.4 Infrared Image Components

Objects whose absolute temperature is not zero radiate electromagnetic waves. Low temperature results in long peak wavelength. Infrared detector is used to detect infrared wave; it is divided into two categories, thermal detector and infrared photon detector.

Thermal detectors include thermal resistance, thermocouple and pyroelectric detectors, which absorb infrared rays and heat up, resulting in changes in physical parameters. Infrared photon detector is based on the principle of photoconductance or photogenerating volt effect generated by the interaction between photon and electron.

At present, infrared photon detector is mainly used, which has high sensitivity and fast response speed. The response speed of photon detector is on the order of microsecond, and that of thermal detector is on the order of millisecond.

The infrared detector at room temperature has a large noise and needs cooling. It needs scanning imaging in not area array staring imaging.

5.4.1 Principle of IR Components

1. Principle of photoconductive infrared detector

The phenomenon that the resistance of a semiconductor is changed by light is called photoconductivity effect, as shown in Fig. 5.17. There are three conditions for the photoconductance effect:

a. The photon energy $hv \geq Eg$ (band gap)
b. The irradiation intensity makes the device in a linear operating range
c. It has a certain external bias *E*.

2. Principle of photovoltaic single-pixel infrared detector

As shown in Fig. 5.18, there is a thin *p–n* nodule on the transparent space. In the absence of light, the majority carrier holes (h^+) in p-Si in *p–n* spread toward the side of n-Si. Majority carrier electrons (e^-) in n-Si spread out to the side of p-Si. Depletion layers form near the *p–n* junction, building the self-built electric field *E*. The electric field is in the direction of $n \rightarrow p$. The size of the self-built electric

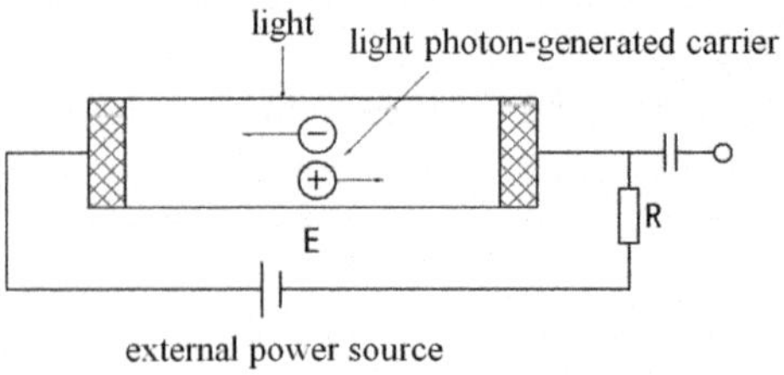

Fig. 5.17 Principle of photoconductive infrared detector

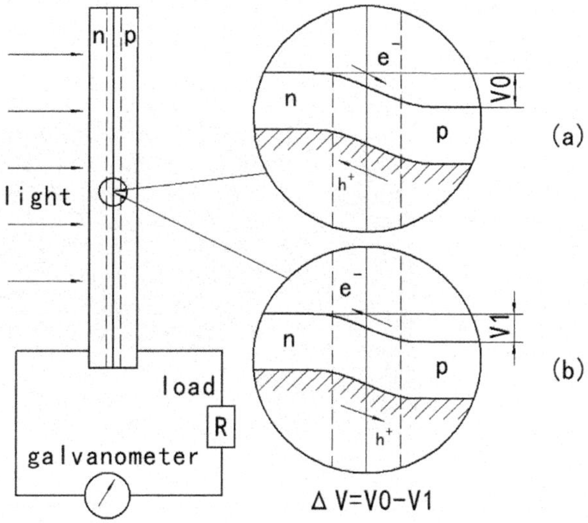

Fig. 5.18 Principle of photovoltaic infrared detector

field is V_0. When light is present, minority carriers are excited in the depletion layer, holes (h^+) are excited in n-Si, electrons (e^-) are excited in p-Si. A small number of carriers h^+ diffuse toward the p-Si space, e^- diffuses to the n-Si space, a new voltage is formed at the p–n junction $V_1 = V_0 - \Delta V$. If the external circuit is closed, ΔV provides the electromotive force to the load.

Infrared detector material. It determines the spectral response range. 3–5 μm InSb, InAs, Si:Ni, Si:S, Si:Ti; 8–14 μm HgCdTe, PbSnTe, Si:Mg, Si:Sc.

Sprite infrared detector. It was developed by British company Mullard Sprite—Signal Processing in the Element. By applying bias voltage in the length direction, the excited carrier moves synchronously with the image and finally accumulates the output in the readout region at the end of the device with the function of delay integration. Figure 5.19 shows the main structural parameters and working principle. From the working principle, it is similar to the existing time delay integration device.

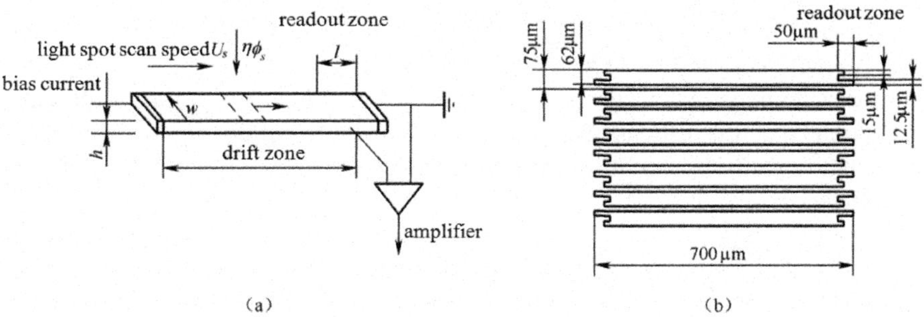

Fig. 5.19 Principle diagram of the sprite infrared detector structure

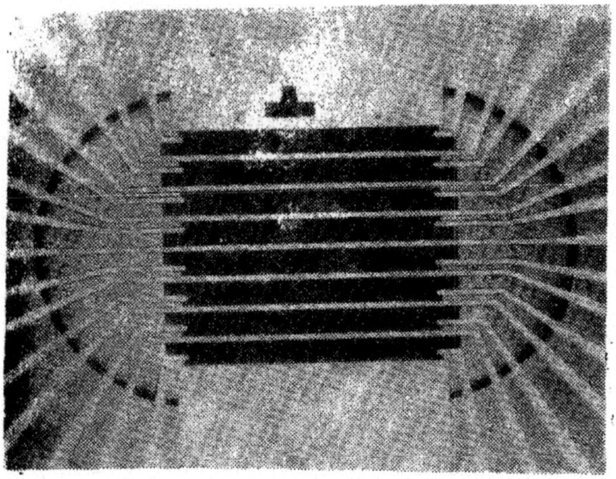

Fig. 5.20 Sweep-accumulated infrared detector

Figure 5.20 shows a 8—pixel Sprite (sweep–accumulated) infrared detector, which is equivalent to about 96 time delay integral pixels. The working spectrum is 8–14 μm. At 77 K absolute temperature, D* can reach 1×10^{11} cm·Hz$^{1/2}$·W^{-1}, and responsiveness can reach 6×10^4 V/W.

3. Principle of area array infrared imaging device

For the single-pixel infrared detector to complete a scene with m pixels, scanning imaging is needed. If the frame period is T_p, the dwell time of each pixel is T_p/m. Bandwidth $\Delta f_d = m/T_p$. If a "row, column" pixel array $n_H \times n_V = m$ detector is used to image the scene, the bandwidth $\Delta f_A = \Delta f_d/m = 1/T_p$. Since the noise is proportional to $\Delta f^{1/2}$, the noise ratio of the single pixel to the area array imaging is:

$$\sqrt{\frac{\Delta f_A}{\Delta f_d}} = \frac{1}{\sqrt{m}} = \frac{1}{\sqrt{n_H \times n_V}};$$

In other words, the SNR of the area array detector is increased by $m^{1/2}$ times than that of the single-pixel scanning imaging.

There are four types of area array infrared imaging devices.

(1) A single-chip infrared CCD integrates an infrared array detector with a CCD on a semiconductor substrate. Intrinsic semiconductor materials with appropriate spectral response should be selected, such as PtSi, InSb and HgCdTe. Photosensitive elements and charge readout structures are fabricated on these materials.
(2) Hybrid infrared CCD. Photosensitive element array and CCD scan reader are made of two kinds of semiconductor materials respectively. In this way, the high quantum efficiency of the infrared array detector and the technology of mature

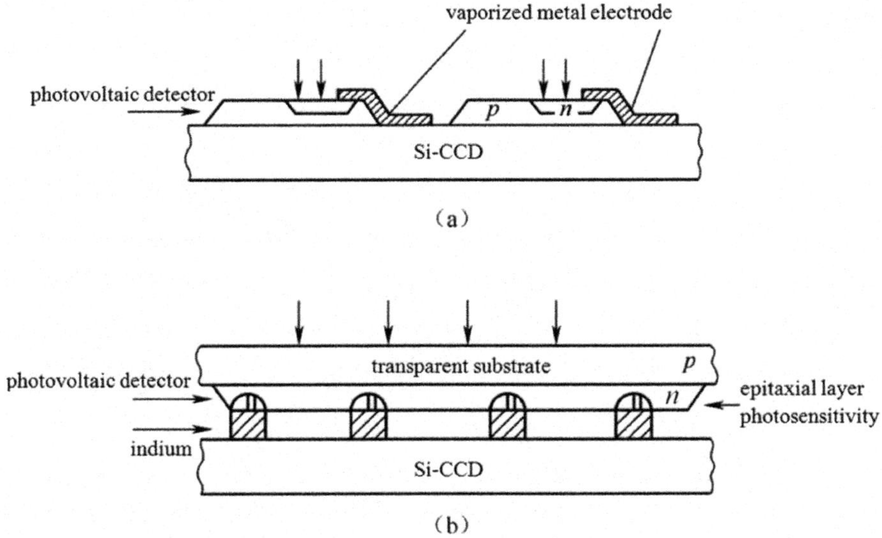

Fig. 5.21 Structure of hybrid infrared CCD

silicon CCD can be coupled together to make a high-performance infrared area array detector.

The key technology is the interconnection of photosensitive elements and CCD, including thermal matching and electrical contact. Indium column interconnection technology and ring hole technology have high yield and become the main technology form of hybrid infrared CCD. Infrared array detector structure mainly consists of "front illumination" and "back illumination", as shown in Fig. 5.21.

The electric signal of the infrared detector with the front illumination structure is elicited in the illumination surface, and the electric lead must pass the detector edge to the multiplexer, as shown in Fig. 5.21a. For this reason, the detector must be made very thin, and the interconnection takes up part of the area, resulting in the reduction of the photographic surface.

The backlight structure, as shown in Fig. 5.21b, requires the embedded detector to have a thin photosensitive layer, and the photoelectron diffuses from the back to the front to be detected by the *p–n* junction to obtain the signal. At present, most of them are back-illuminated structures. A thin single crystal layer (10–20 μm) is generated on the transparent substrate. *p–n* junction is formed by ion implantation to make high-performance HgCdTe photovoltaic array, coupled with SiCCD into a high density array.

(3) Schottky barrier infrared focal plane array (SB-IRFPA) is also a highly integrated infrared area array detector which is expected to be practical. There is 512 × 512 pixels infrared area array detector.
(4) 3D infrared area array detector, as shown in Fig. 5.22, combines chips with signal readout and processing functions (including impedance transform amplifier,

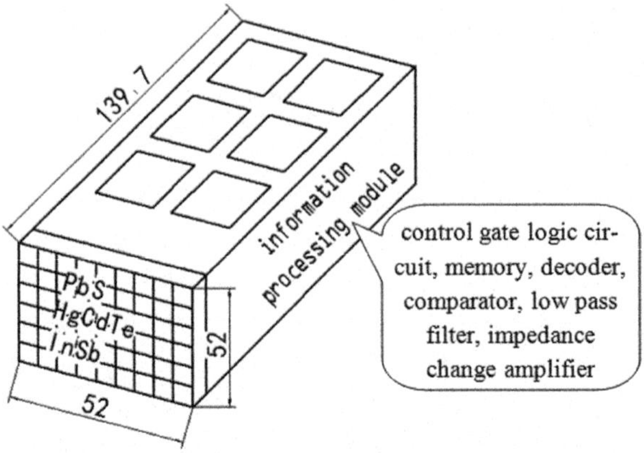

Fig. 5.22 3D infrared area array detectors

low-pass filter, memory, decoder and control gate logic circuit, etc.) in a layered way to form a signal processing module. Then the module is directly connected with the detector to complete photoelectric conversion, signal storage and scanning output functions in the infrared thermal imaging process. At present, it has been made into mid-infrared InSb and HgCdTe components. The detector adopts indium welding technology or binds conductive epoxy resin on the edge of the chip. Pixels from 64 × 64 to 256 × 256.

The biggest advantage of 3D infrared area array detector is that all the technology is based on the existing semiconductor mature technology, which can be mass-produced, easy to modularize, convenient to maintain, and has the ability of data pretreatment. It is of great significance to suppress noise, improve sensitivity and reduce the volume of the whole machine, especially suitable for multitarget recognition and imaging tracking system.

5.4.2 Characteristic Parameters of IR Components

The characteristic parameters of infrared detector include: responsivity, noise equivalent power, detection rate, specific detectivity and spectral response.

1. Responsivity R. Ratio of the detector output voltage V_s or current I_s to the radiation flux φ incident to the photosensitive surface of the detector.

$$R_u = V_s/\varphi \mathrm{V/W}$$
$$R_i = I_s/\varphi A/W$$

2. Noise equivalent power NEP. The radiated power incident to a detector when the output signal power is equal to the noise power. That is:

$$NEP = \frac{EA_d}{V_s/V_n} W \tag{5.1}$$

where E is the irradiance on the detector, A_d is the pixel area of the detector, V_s/V_n is the signal-to-noise ratio.

3. Detection rate D and normalized detectivity D^*.

$$D = \frac{1}{NEP} W^{-1} \tag{5.2}$$

The definition of NEP does not consider the influence of A_d and Δf on V_n, which is generally proportional to $(A_d \Delta f)^{1/2}$.

$$NEP \propto \sqrt{A_d \Delta f} \tag{5.3}$$

Formula (5.2) times formula (5.3), so:

$$D\sqrt{A_d \Delta f} = \text{const.} = D^* \tag{5.4}$$

D^* is called normalized detection rate, which is detection rate per unit area per unit bandwidth. The unit of D^* is cm $Hz^{1/2}$ W^{-1}. Since D^* has more practical application significance in the performance comparison of various infrared detectors, the normalized detection rate D^* is simply referred to as detection rate.

D^* relates to test conditions and should be indicated as (500, 800, 1). That is, the blackbody temperature is 500 K, the modulation frequency is 800 Hz, and the bandwidth is 1 Hz.

Substitute formulas (5.1) and (5.2) into formula (5.4), so:

$$\begin{aligned} D^* &= \frac{1}{V_n} \frac{V_s}{EA_d} \sqrt{A_d \Delta f} \\ &= \frac{R_u}{V_n} \sqrt{A_d \Delta f} \end{aligned} \tag{5.5}$$

where R_u is the response of the detector.

In Formula (5.5), D^* and R_u should be the mean of the spectral response or spectral response at the same time, and E should also be the mean of the spectral response or spectral irradiance.

$$
\begin{aligned}
SNR &= \frac{V_s}{V_n} = \frac{D^* E A_d}{\sqrt{A_d \Delta f}} \\
&= D E A_d \\
&= \frac{E A_d}{NEP} \\
&= \frac{E A_d R_u}{V_n} \\
&= \frac{V_s D^*}{R_u \sqrt{A_d \Delta f}}
\end{aligned}
\tag{5.6}
$$

Formula 5.6 reflects the relationship between SNR and D^*, D, NEP, R_u, as well as the relationship between D^* and R_u.

$$
R_u = \frac{V_n D^*}{\sqrt{A_d \Delta f}} \tag{5.7}
$$

4. Time constant τ. When a rectangular radiation pulse irradiates a detector, the time response of the detector generally rises or falls according to the law of $V = V_0(1-\mathrm{e}^{-t/\tau})$, where τ is a time constant. τ refers to the time that the signal voltage increases from zero to $V = V_0(1-\mathrm{e}^{-1})$. The photon detection τ can reach microseconds or even nanoseconds. The thermal imaging τ should be less than the dwell time. V_0 is the highest voltage the detector.
5. Spectral response. The relationship between the signal voltage generated by unit power of incident single frequency radiation on the detector and the radiation wavelength is usually expressed by single frequency response R_λ or spectral ratio spectral specific detection rate relative to wavelength mapping, as shown in Fig. 5.23. The response of the heat detector is wavelength independent.

The photon detector has a limiting wavelength of λ_c, $\nu_c = c/\lambda_c$, where C is the speed of light in a vacuum. The response characteristics can be expressed as:

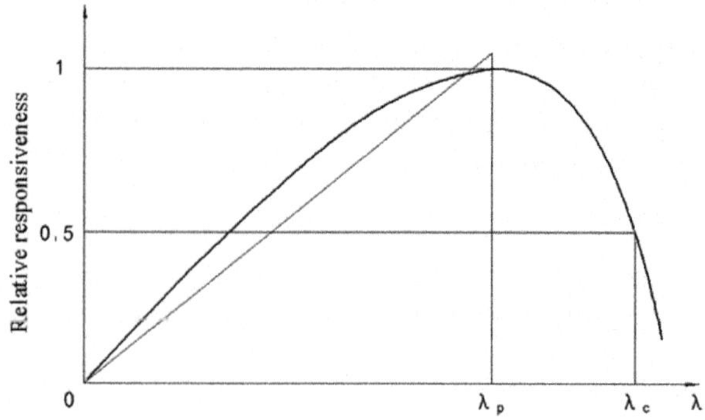

Fig. 5.23 Linear characteristics of infrared detectors

$$D_\lambda^* = \frac{\lambda D_{\lambda c}^*}{\lambda_c} \quad \lambda \le \lambda_c$$
$$= 0 \quad \lambda > \lambda_c$$

The response of the actual photon detector at λ_c does not suddenly go to zero; instead, the wavelength at which the response drops to 50% is defined as the cutoff wavelength.

6. Background limit detection rate. The noise of detector itself is ignored, and its performance is only limited by the noise of background photon number fluctuation.

At a certain wavelength, the theoretical maximum value of the background light-limited detector D^* is:

$$D_\lambda^* = \frac{\lambda}{2hc}\sqrt{\frac{\eta}{Q_b}} = 2.52 \times 10^{18}\lambda\sqrt{\frac{\eta}{Q_b}}$$

where η is the quantum efficiency of the detector, Q_b is the hemispherical background irradiance of the detector, h is Planck constant, and c is the speed of light.

For photovoltaic detectors, because there is no recombination mechanism, D^*_λ should be multiplied by $\sqrt{2}$, so:

$$D_\lambda^* = 3.56 \times 10^{18}\lambda\sqrt{\frac{\eta}{Q_b}}$$

Many photon detectors are approaching the background noise limit.

Figure 5.24 shows the spectral detection rates of various infrared photon detectors. Figure 5.25 shows the relationship between the detection rate and temperature of a typical infrared detector.

Note: PbS 300 K of the maximum spectral detection rate curve of PbS in the figure should be PbS 193 K, for reference.

5.5 Low Light Level Image Intensifiers

Low light level image intensifier, also known as low light level image tube (LLL tube) or image tube, is a kind of imaging device that can reproduce the scene under weak light into visible image through photoelectron conversion of photocathode, enhancement of electron multiplier and electric-light conversion of fluorescent screen. Broadly speaking, they are a class of multispectral and multifunctional photoelectronic imaging devices, which can be used to detect, enhance and image scenes irradiated by ultraviolet, visible, near-infrared, X-rays and γ-rays. It is used in the fields of low light level night vision and astronomical observation. It can be divided into three broad categories.

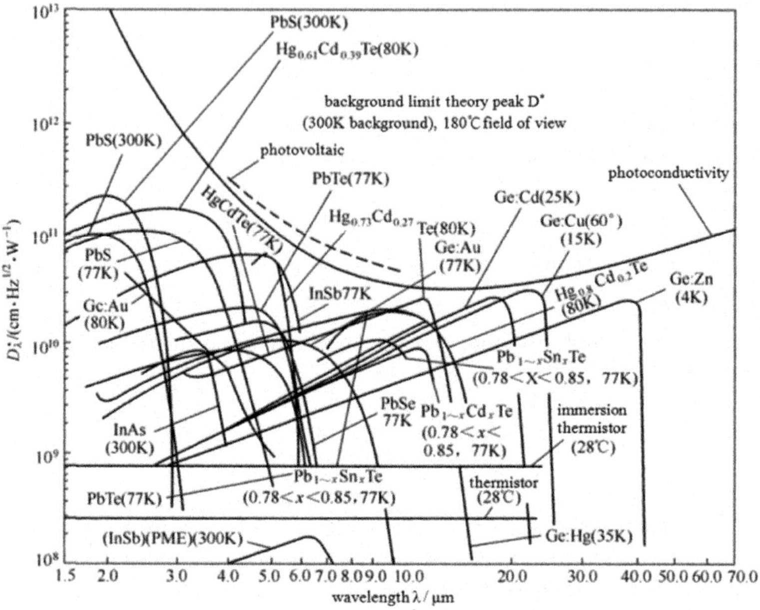

Fig. 5.24 Spectral detectability of various infrared photon detectors

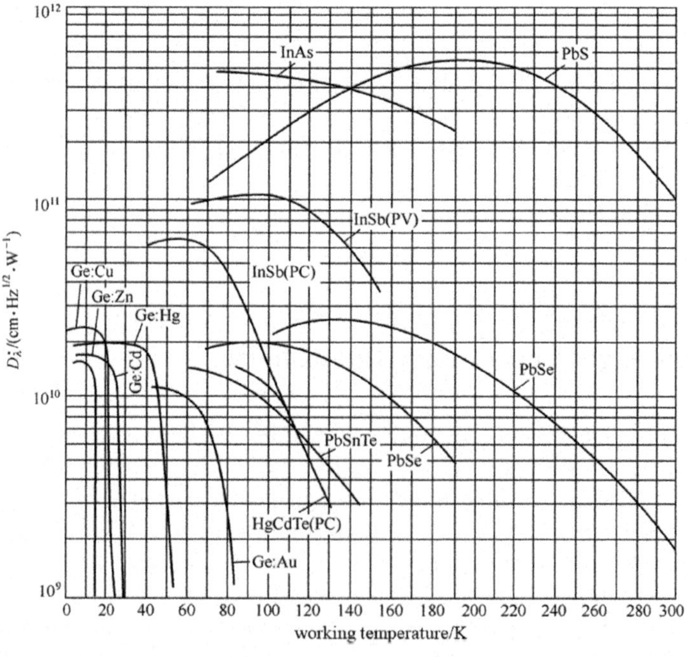

Fig. 5.25 Relationship between the detection rate of typical infrared detectors and temperature

Proximity focusing class. There are 10 kinds of TV camera tubes, such as first-generation proximity focusing tube, second-generation proximity focusing tube, third-generation proximity focusing tube, photon counting image tube, fast photomultiplier tube, ICCD, biplanar photodiode, triplanar phototransistor and MCP enhanced solid-state array television camera tube.

Electrostatic focusing class. There are seven kinds of TV camera tubes including zero-generation invert tube, first-generation invert tube, second-generation invert tube, enhanced CCD (ICCD), electron bombardment CCD (EBCCD), MCP electron multiplication solid array imaging tube, etc.

Electromagnetic focusing class. There are six kinds of image tubes including electromagnetic focusing type the first-generation image tube, MCP electron multiplier image tube, solid array television imaging tube, high-speed photography stripe tube, TV shunt tube and video image storage tube.

5.5.1 Principle and Function of Image Intensifiers

Generally, LLL tube consists of a photocathode, an electron lens, an electron multiplier, and a fluorescent screen, as shown in Fig. 5.26. The photocathode generates photoelectron images under the excitation of the input photons of the scene. In the ultra-high vacuum tube, these electrons obtain energy from the external high-voltage power supply and are focused (deflected) by the electron lens to bombard the fluorescent screen with high energy to emit light. Thus, the corresponding photon image visible to human eyes is generated, and the brightness is enhanced.

Figure 5.26 shows the working principle of an electrostatic focusing image intensifier. The object is projected onto a cathode fiber optic panel (MCP) through an objective lens. The photon excites the photoelectrons, which are excited by electron multiplication in the cathode MCP. Under the action of the electric field composed of

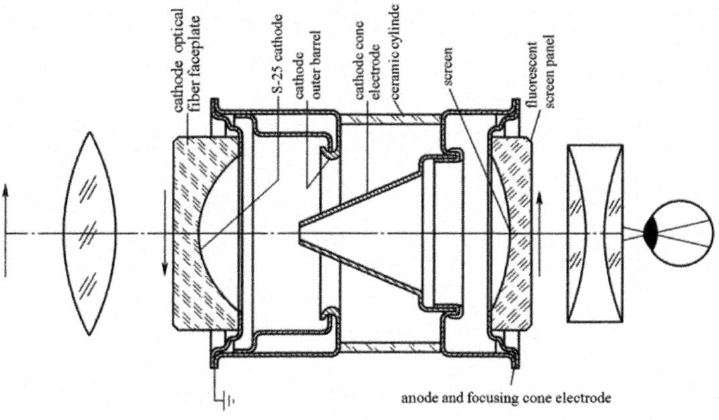

Fig. 5.26 Image intensifier principle

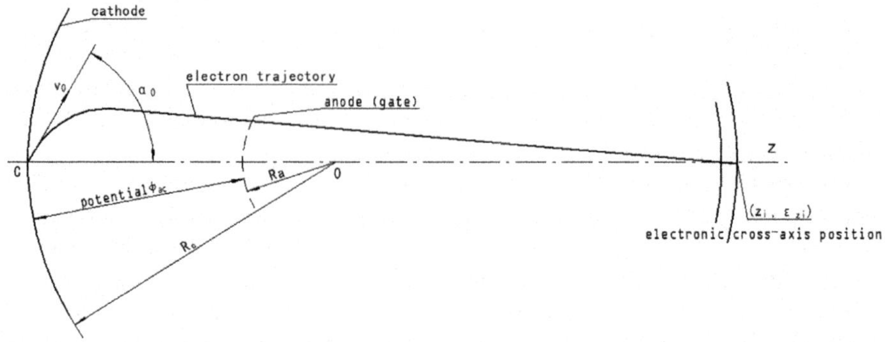

Fig. 5.27 Principle of electron lens

the photocathode and the anode, the photoelectrons bombast the fluorescent screen, which lead to glowing of the fluorescent screen and producing the enhanced image visible to the human eyes.

The principle of the so-called electron lens is that two concentric spherical electrodes are used as the anode and cathode, and the electric line moves along the radial direction. Under the action of the electric field force, electrons move along the electric line and produce the focusing effect (Fig. 5.27).

The MCP is used as the intensifier, as shown in Fig. 5.28. Under the action of electric field force, the electrons incident to MCP will act on the space wall of the microchannel to generate secondary electrons, thus producing electron multiplication, which will be output to the fluorescent screen to generate optical signals for observation. MCP has gain saturation and is suitable for strong light applications. Figure 5.29 shows the microchannel plate structure diagram.

The main functions of the image enhancer are:

Brightness increases. All kinds of LLL tubes have sufficient brightness gain (10^4), which enables people to observe distant objects under low illumination at night (10^{-1}–10^{-4} lx), or to take high-speed photography of fast luminescence phenomena requiring high exposure. Low light level image photon counter for photoelectric

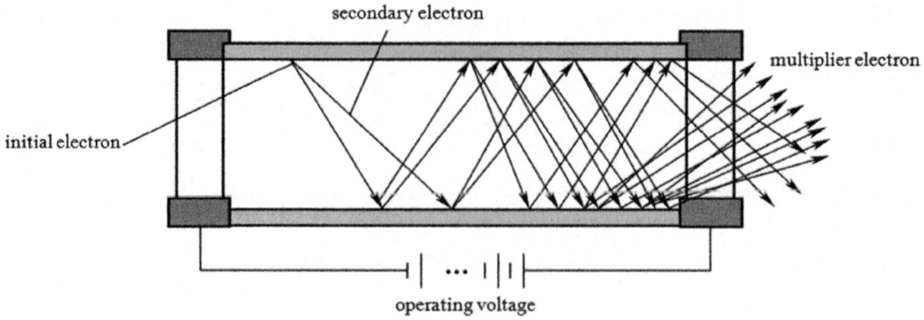

Fig. 5.28 MCP working principle diagram

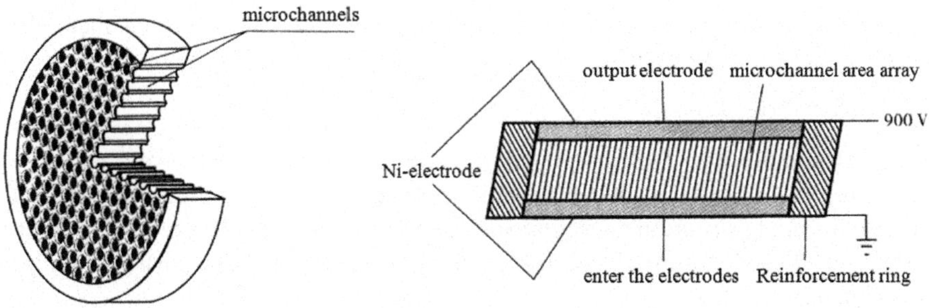

Fig. 5.29 Microchannel plate structure diagram

warning of remote ($\geq$ 100 km) targets with tube gain up to 10^6–10^7 are also included in this category.

Spectral conversion. It can convert the invisible X-ray, ultraviolet ray and infrared ray into the corresponding photoelectronic image, and reappear into the enhanced visual image through the same MCP electron multiplication and fluorescent screen electro-optical conversion.

High-speed photography. Due to the photoelectron in the vacuum cavity by the electron lens focus, deflection, accelerating, and multiply the whole process in the MCP transit time is very short, and the field-assisted photocathode structure can be used to shorten the time of photogenerated electron transport to the backward interface and escape into vacuum, the fast response characteristics can be used to achieve picosecond even femtoseconds transient scenery of high-speed photography. An electronic shutter device belonging to this class of applications is called a fringe image tube.

5.5.2 Major Characteristic Parameters of LLL Tubes

1. Main Parameters:

Geometric dimensions. Photocathode effective diameter, screen effective diameter, device size.

Photoelectric conversion gain characteristics. Photocathode integral sensitivity, spectral sensitivity, brightness gain, maximum output brightness, brightness automatic gain control range.

Contrast transfer characteristics. Identification, MTF, screen afterglow, equivalent spatial frequency bandwidth.

Optical properties. Magnification, distortion, image alignment error, field of view defect.

Power supply conditions. Input voltage, current, power consumption.

Environmental adaptability. High temperature, low temperature, shock, vibration, humidity adaptability.

2. Photocathode sensitivity

Integral sensitivity. The integrated sensitivity of the device is defined as the photocurrent generated by the photocathode per receiving unit of light flux, and under the condition of saturation operating voltage at the color temperature of 2856 K standard tungsten lamp illumination of the cathode. $S = I$ (photocurrent)/Q_{υ} (luminous flux), mA/lm.

Spectral sensitivity. The photocurrent generated by the photocathode per receiving unit of radiation flux under monochromatic illumination of a particular wavelength is defined as the spectral sensitivity of the device at this particular wavelength. $S_{\lambda} = I$ (photocurrent)/Q_{λ} (monochromatic radiation flux), mA/W.

3. Luminance gain

Under standard light source illumination and rated operating voltage, the ratio of luminance of fluorescent screen output L_p to luminance of photocathode input E_k is called luminance gain of LLL Tube. $G_B = L_p/E_k$, cd m^{-2} lx^{-1}. Or, its number of photons Gain G_n is the number of output photons/the number of input photons.

4. Background equivalent input illuminance

When there is no light input, the dark background of the fluorescent screen caused by the dark emission of the photocathode, the field emission or the noise of other components is L_d, divided by the brightness gain of the device, namely L_d/G_B, which is equivalent to the background equivalent input illumination (E_d, l_X) on the photocathode input surface.

5. Contrast transfer characteristic

Modulation transfer function MTF. MTF of imaging devices = output modulation/input modulation.

Limit discrimination. Spatial frequency at MTF = 0.02.

Low contrast, low illumination resolution. Low contrast and low illuminance resolution of the device are measured under the conditions of different contrast resolution test cards at 10^{-3} lx illuminance.

Angular resolution of the imaging system α (rad). $\alpha = 1/(f\ R_k) f$ is the focal length of the imaging objective lens mm, and R_k is the limiting discrimination rate of the imaging device lp/mm.

6. Signal-to-noise ratio transmission characteristics

Signal-to-noise ratio(SNR). The ratio of the average brightness measured from the output end of the LLL tube fluorescent screen to the RMS value of noise. The main noise factors include input photon noise, photocathode quantum conversion noise, dark emission noise and fluorescent screen particle noise.

Number of flickers. The number of flickering points per square centimeter on a fluorescent screen in the absence of light. The scintillation seen on the fluorescent screen of LLL tube includes electron scintillation, ion scintillation and field to emission spot scintillation. The scintillation point caused by electron number fluctuation is small and dark. The scintillation point caused by the fluctuation of ion number is large and sparse. It is caused by the ions of the residual gas of device and residual gas molecules produced in the process of electron multiplication of MCP bombards the photocathode after ionization.

5.6 Photon Counters for Low Light Level Imaging

The image photon counter can be understood as a two-dimensional image sensor array composed of multiple photomultiplier channels each working in the single photon counting mode. All kinds of noise are the main factors affecting the photoelectric detection threshold. By the photomultiplier process output pulse height distribution test and analysis, looking for all kinds of noise distribution and sources of energy, with high and low threshold circuit to suppress and cut off the thermal noise and high-power flashnoise, allow only the photons with single peak distribution pass subsequent amplification, processing and display system, can reduce the noise of several orders and greatly improving the system signal-to-noise ratio.

Weak signal working capacity of photomultiplier tube:

DC operation, 10^{-14} W (10^4–10^5 photons/s).

Phase-locked amplifier, 10^{-16} W (10^2–10^3 photons/s).

Photon counting, 10^{-19} W (10^0–10^1 photons/s).

The main characteristics of photomultiplier tubes:

Photon number gain (10^6– 10^7).

Dynamic range (10^6).

Output brightness ($\geq$ 400 cd/m^2).

Response speed($\leq$ 1 ms).

Noise Source and Pulse Height Distribution (PHD) in MCP LLL Tube.

Figure 5.30 shows the source of noise and its energy distribution region in MCP LLL tube. According to the noise energy from low to high distribution of four regions *A*, *B*, *C*, *D*. The signal comes from the scene. Noise sources are: MCP heat and its ionic noise A, the photocathode heat emits noise *B*, photon noise *B*, end screen light feedback noise *B*, flying electron ion feedback noise *C*, intrinsic ion feedback noise *C*, cosmic ray radiation noise *D*, MCP material emission noise *D*, fixed scintillation point noise *C*, MCP fixed image noise, total 11 kinds noises. Among them, the noise of fixed scintillation point is a kind of visual interference in nature, which comes from the field emission between the internal devices of light tube. MCP fixed image noise is caused by non-uniform geometrical size and electronic characteristics of MCP channel, and has no fixed energy range.

These signals and noises were analyzed according to the frequency and intensity of their occurrence in Fig. 5.31a–b. Figure 5.31a shows the quantity and intensity of various noises. Figure 5.31b is made after the statistics of frequency and intensity of signal in Fig. 5.31a, where the abscissa is the relative intensity and the ordinate is

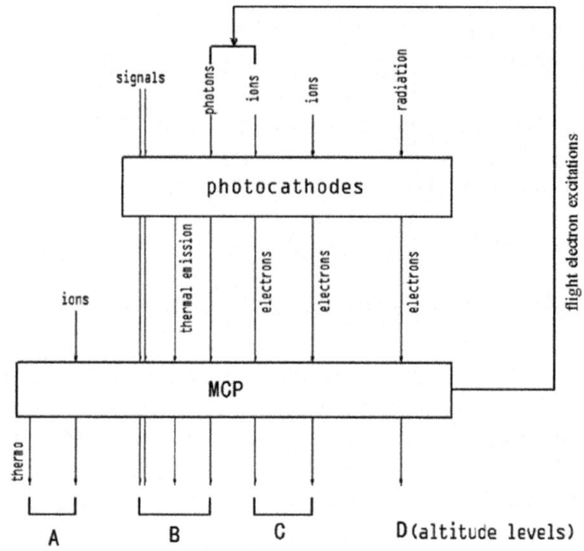

Fig. 5.30 Noise source and its distribution in MCP LLL tube

the frequency of signal occurrence, and the signal distribution in regions *A*, *B*, *C* and *D* is calculated according to the noise energy from low to high. This is pulse height differentiation (PHD). It can be seen that the peak values of noise and signal statistics exist in region *B*, which is exactly the region where the signal is located. According to this feature, the noise can be greatly reduced by suppressing and cutting the *A* region with low signal strength and the *C* and *D* region with high signal strength, and only allowing the signal from the *B* region containing the signal to pass through. It is the PHD distribution of low light signal that provides the basis and direction for the system noise analysis, suppression and elimination of devices, and the realization of photon counter.

5.6.1 Operation Principle of Photon Counters

According to the PHD distribution diagram for analyzing microlight signals, the main signal amplification component of the photon counter developed is MCP. MCP has two working modes, the multiphoton DC working mode is shown in Fig. 5.32a, and the single-photon pulse counting mode is shown in Fig. 5.32b. Among them, the multiphoton DC working mode is the common working mode of microlight devices. After the same amplification of signal and noise, the minimum detectable threshold is impossible to detect the signal smaller than noise due to the interference of more than 10 kinds of noise. The low light level detector working in the single-photon pulse counting mode can detect signals 2–3 orders of magnitude lower than noise and reach the signal level of detecting single photon irradiation only after electronic removal the gas of MCP and vacuum chamber and the use of PHD signal peak screening.

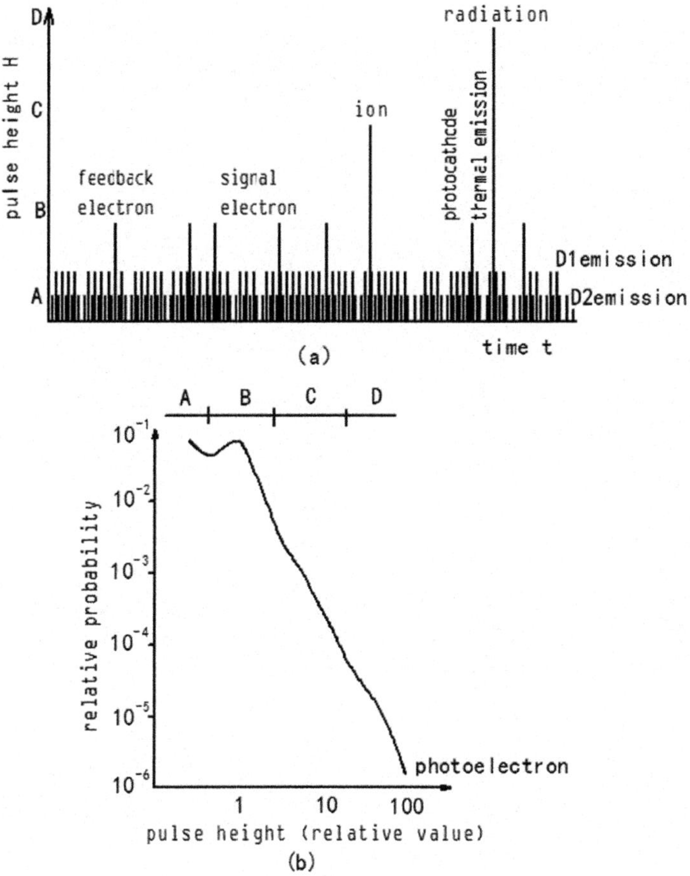

Fig. 5.31 Output waveform and pulse height distribution of MCP LLL tube

Figure 5.33 schematic diagram of a photon counter, the imaging objective will project the photon stream from the scene to the photocathode and convert it into photoelectrons. After the MCP electron multiplication, the photon stream will be projected to the fluorescent screen and converted into enhanced light signals, which will reproduce into enhanced scene images. It consists of imaging objective, photon counter image tube, gating pulse generator, image tube power controller, relay optics, high-speed camera, high-speed CCD camera, video processor, display, cooling system and so on.

The photon counter image tube has very low dark count number rate and single-photon PHD characteristics. After coupling to a high frame rate CCD camera through a relay lens, the output image is a two-dimensional image amplified by nearly one million microchannels on the MCP plane array. The amplified electronic signal contains signal and noise, which is processed by the amplitude discrimination processor in the gated pulse generator and video image processor to eliminate

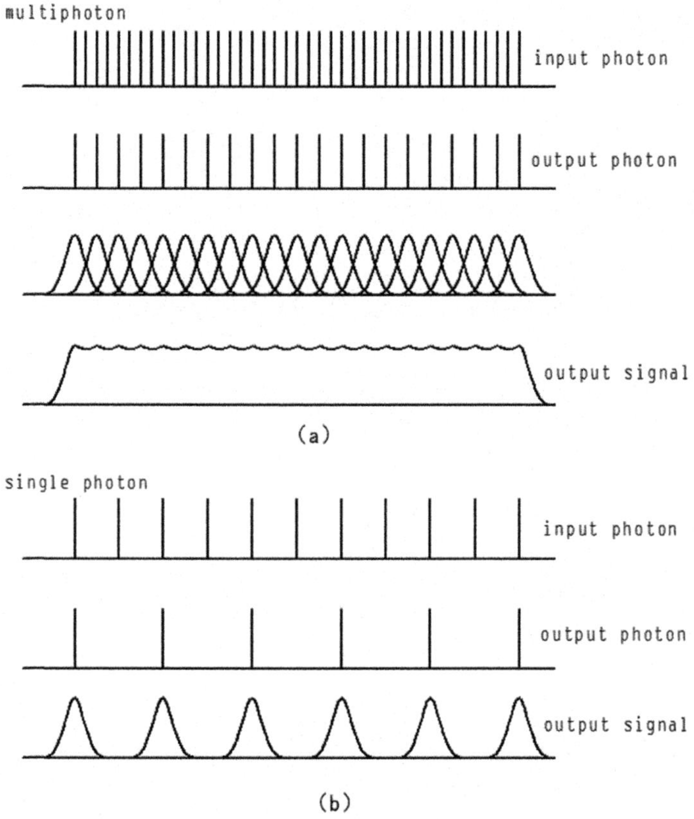

Fig. 5.32 Two working modes of photon counter

high-energy particle scintillation noise and low energy MCP thermal noise, and only the target information and the thermal noise of the photocathode pass through. This composite signal is analyzed and processed by computer and sent to the terminal display to reproduce a scene image with a greatly improved signal-to-noise ratio. This is how the photon counter works.

5.6.2 Characteristic Parameters of Photon Counters for Low Light Level Imaging

The parameters can be classified into eight categories:

1. Quantum detection efficiency. Including photocathode quantum efficiency, MCP on the input electron collection efficiency, fluorescent screen electro-optic conversion efficiency, CCD photoelectric detection efficiency;

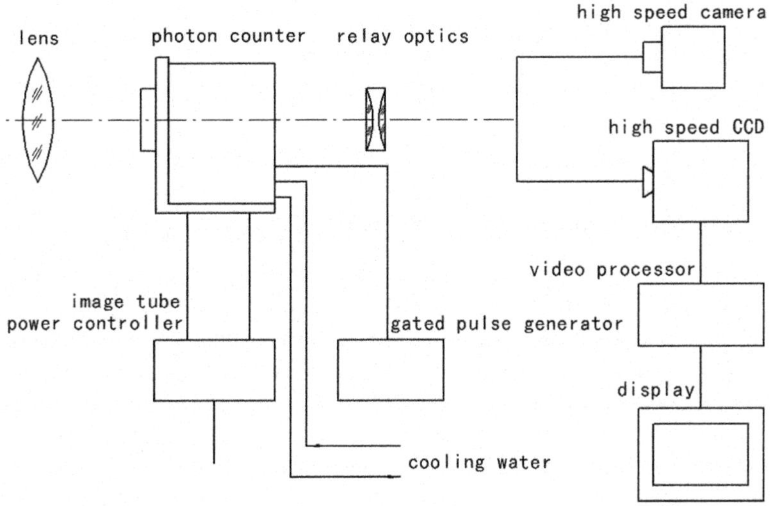

Fig. 5.33 Photon counter principle block diagram

2. Photoelectric conversion characteristics. Photon number gain, maximum output brightness, linear dynamic range of photoelectric conversion;
3. Signal-to-noise ratio characteristics. PHD, peak-to-trough ratio, energy resolution, dark counting, signal-induced noise;
4. Noise factor. Device input SNR/device output SNR;
5. Spatial contrast transfer characteristics. Dynamic and static spatial resolution, MTF, distortion, field of view uniformity;
6. Time response characteristics. Afterglow, response time;
7. The others. Geometric dimensions, environmental adaptability, cooling conditions, shielding.

To be clear, in general, the shorter the afterglow, the better. However, the existence of short afterglow of LLL tube is one of the necessary conditions for successful coupling to high frame rate CCD.

5.7 Television Camera Tubes

Television camera tube is a kind of photoelectronic imaging device which converts spatial optical image of scene into sequential television video signal. At present can be divided into two categories: television vidicon and solid imaging device (CCD). CCD class has been introduced, here only television vidicon.

Any television vidicon has three functions: photoelectric conversion, charge storage and scanning-reading. The target plane in the vacuum television vidicon performs the first and second functions, and the electron gun performs the third function through scanning. For CCD detector, MOS array plays the first two roles of the target plane of the imaging tube, and the specific array conductive wiring and peripheral drive line assume the functions of signal picking and output timing video signal.

5.7.1 Operation Principle of Television Camera Tubes

Figure 5.34 shows the working principle of a simple television camera tubes. The imaging tube is composed of lens, target plane, electron gun and electron beam controller, as shown in Fig. 5.34a. The electron gun consists of a filament, a cathode, a control grid, an accelerator pole and a focusing pole. The voltage of the focusing pole is adjustable, and it forms an electronic lens with the accelerating pole to assist the focusing. The focusing coil, deflection coil and correction coil out space the tube control the electron beam. The net electrode on the right space of the target forms a uniform electric field with the target so that the electron beam is perpendicular to the target plane.

Figure 5.34b shows the photoelectric conversion principle of the target surface, and the base of the target is a glass window. The target surface material is a highly photosensitive material such as Sb_2S_3, which is evaporated on a transparent conductive substrate such as SnO_2 by vacuum deposition method. The target surface is facing the electron gun. In order to meet the requirements of signal charge storage and small inertia, the photoconductive target should meet:

(1) The resistivity of the photoconductive layer $\geq 10^{12}$ Ωcm;
(2) The electrostatic capacitance of the target ranges from 600 to 3000 pF;
(3) The band gap of photoconductive materials is 1.7 eV $\leq E_0 <$ 2 eV.

Photoelectric conversion process

The highly photosensitive photoconductor forms the target surface of the imaging tube, the conductivity of the photoconductor is proportional with the illuminance. There are 10^4–10^5 pixels on it, and the volume resistance and capacitance of each pixel are combined into an equivalent RC loop, which consists of the bias voltage V_T, scanning electron beam, photoelectric target surface and *R*, and reads the video signal through the capacitor *C*.

Charge storage process

When the target surface is not scanned by electron beam, the photoelectrons in the pixel are stored in the volume capacitance. The higher the illuminance, the more photoelectrons in the pixel, the higher the potential of the corresponding point, and the stronger the electrical signal output when scanned by electron beam. Therefore, the amount of charge stored is the brightness of the corresponding point in the scene.

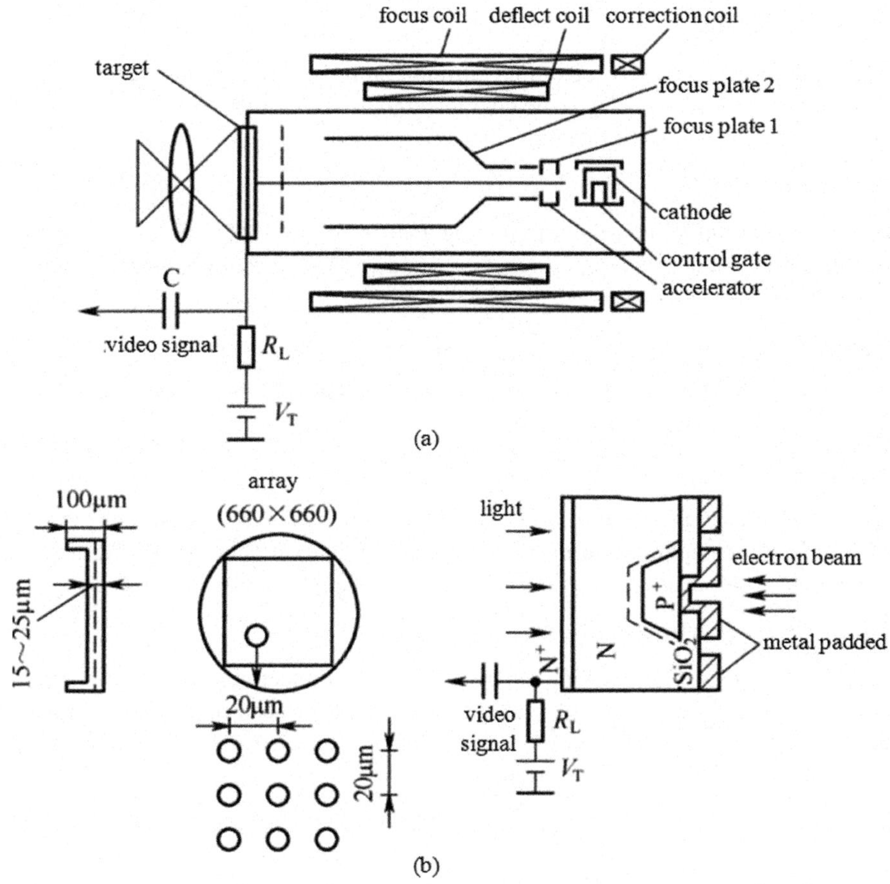

Fig. 5.34 Structure principle of television camera tube

Scan and pick up process

Under the action of the deflection coil, the electron beam emitted by the electron gun scans different pixels of the target plane, and the photoelectrons are released to form a current, which flows through *R* to generate potential of each pixel, and then the video signal of the scene is obtained.

5.7.2 Characteristic Parameters of Television Camera Tubes

Sensitivity

Output signal current per unit luminous flux at 2856 K color temperature standard light source, mA/lm, mA/W.

Gray Coefficient

The function relation between output current i and input illuminance E:

$i = \sigma E^g$, where g is gray scale coefficient, σ is scale coefficient.

g is close to 1 for most imaging devices.

Resolution

Optical resolution R_p is usually lp/mm, and TV resolution RTV (line). The resolution of TV device has horizontal resolution and vertical resolution, usually refers to horizontal resolution.

Video SNR

The ratio of the output video signal current to the RMS of the noise current in the same frequency band, in dB.

Hysteresis

The ratio between the remaining signal and its initial value, the remaining signal is measured at the beginning of the third scan after the input optical signal turned off. Commercial broadcasting requires a hysteresis value of no more than 10%.

Example: reverse beam vidicon

10,000 lines of resolution, high capacity of photoconductivity target, no granular structure, area of 50 $\times$ 50 mm, high gain. Signal pickup can be continuous pickup mode, single frame storage mode and fast multiframe scan display mode. The image can achieve continuous electron doubling.

5.8 X-Ray Imaging Devices

There are four kinds of sensors that can realize X-ray detection and imaging: X-ray fluorescence conversion screen, photocathode, MCP cathode and special semiconductor photoelectric element.

5.8.1 Principles of X-Ray Imaging Devices

In 1895, the German physicist Roentgen discovered the ability of X-ray exposure through objects and realized the non-real-time imaging of X-ray.

X-ray fluorescence conversion screen is composed of input window base/reflective metal film/X-ray phosphor/leaded transparent glass and other layers. The phosphor atoms are excited by X-ray photons to produce fluorescence sensitive to human eyes or photographic plates, and the fluorescence brightness is proportional to the irradiation intensity. X-rays through the object due to the different parts of the object transmittance to form a detection of the object image. Two important characteristics of X-ray fluorescent screen are fluorescence brightness conversion efficiency and resolution. In order to improve the conversion efficiency, it is necessary to adopt large particle size, thickening of powder layer, powder material with high content of heavy elements and Al film plating. This reduces the resolution, so a practical tradeoff is needed.

The monocrystal array fluorescent screen grown by vacuum deposition method achieves both high photoelectric conversion efficiency and high resolution.

The main characteristic parameters of several common X-ray and γ phosphors are given in the following table (Table 5.1).

X-ray photocathodes

CsI is one of the most common and efficient X-ray photocathode materials. After being irradiated by X-ray, it produces primary photoelectrons, but its spillovers probability is very small, and most of them exchange energy with neighboring electrons at higher energy levels continuously to produce secondary electrons. The loose X-ray material is favorable for the secondary electrons to double and overflow under high voltage, forming photocurrent. The spillover electrons spill out in the way of reflection and transmission to form the following two photocathode structures: CsI/MCP-reflected photocathode and window glass /CsI transmitted photocathode.

CsI/MCP-reflected photocathode

It is formed by vapor deposit CsI onto the surface of MCP input end surface, as shown in Fig. 5.35. Figure 5.35a shows the structure of CsI/MCP-reflected photocathode, and Fig. 5.35b shows the electron generation process of CsI/MCP-reflected photocathode. There are three possibilities for an X-ray photon of energy E(keV) or wavelength λ to strike at point A of CsI layer: X-ray reflection, which is very few; X-ray folding into CsI layer produces one and two secondary electrons as signals, which is the source of X-ray photocathode electronic signal. X-rays enter adjacent channels through the CsI layer to generate a crosstalk signal.

Table 5.1 Main characteristic parameters of X-ray and γ-ray phosphors

Name	Mass component %	Density/g cm^{-3}	μ/ρ		Note
			20 keV	40 keV	
ZnS:Ag	0.67Zn0.33S	4.1	24.39	4.39	Xscreen
ZnS/CdS:Ag	0.341Zn0.382S0.271Cd	4.46	20.18	10.76	Xscreen
$CaWO_4$:W	0.142Ca0.634W0.224O	6.06	49.50	9.57	Xscreen
CsI:TI	0.51Cs0.49I	4.51	23.52	23.32	Xscreen
CsI:TI Single crystal array	0.51Cs0.49I	4.51	23.52	23.32	Xscreen
CsI:TI	0.51Cs0.49I	4.51	23.52	23.32	γscreen
CsI:Na	0.51Cs0.49I	4.51	23.52	23.32	γscreen

Note μ is linear absorption coefficient (1/cm), ρ is density (g/cm^3), μ/ρ is linear absorption coefficient per unit density (cm^2/g) 0.382S. The original for 0.382 *Cd*

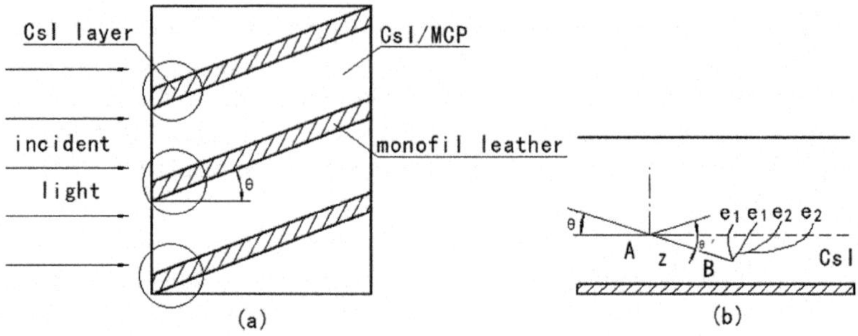

Fig. 5.35 Structure and principle of CsI/MCP-reflected photocathode

Quantum efficiency of CsI/MCP-reflected photocathode

The quantum efficiency of CsI/MCP-reflected photocathode is denoted by the symbol η. The number of photoelectrons generated by an X-ray photon incident on the photocathode depends on the photon absorption efficiency and the photoelectron overflow efficiency of the photocathode material.

$$\eta_{\text{reflection}}(E,\theta) \propto P_{\text{absorption}}(E,\theta)P_{\text{effusion}}(E,\theta),\ P \text{ is the probability}$$
$$P_{\text{absorption}}(E,\theta') = 1-\exp(-\mu\ \csc\theta' z)$$
$$P_{\text{effusion}}(E,\theta') = P_s(0)\exp(-z/L_s)$$
$$\eta_{\text{reflection}}(E,\theta') = \sigma(E,\theta')P_s(0)\frac{\mu L_s}{\mu L_s+\sin\theta'}[1+\exp(-(\mu\ \csc\theta' + 1/Lz)z)$$

where θ' is X-ray grazing angle, z is the oblique distance at which X-rays enter the cathode layer, σ (E, θ') is the energy of secondary electron produced, μ is linear absorption coefficient of cathode material, L_s is the distance from photoelectron produces to recombination.

Window glass/CsI transmission photocathode

It is formed by vacuum deposition on the X-ray transparent window glass, as shown in Fig. 5.36. The principle is that the X-ray is incident to the photocathode material CsI through the window glass, and the secondary electrons excited by the X photon travel along the x photon direction and escape into the vacuum from the back interface. A voltage of several thousand volts behind the photocathode allows for the multiplication of secondary electrons. CsI photocathode was evaporated into loose structure under argon filling condition to improve the escape length and efficiency of secondary electrons.

Quantum efficiency of CsI transmission photocathode:

$$\eta_{\text{transmission}}(E,0) = \sigma(E,0)P_s(E,0)\frac{\mu L_s}{\mu L_s-1}\left(e^{-d/Ls}-e^{-\mu d}\right)$$

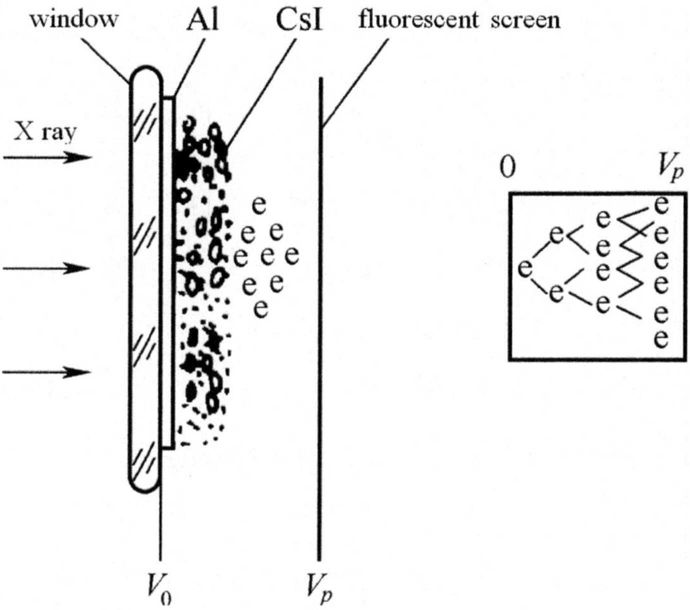

Fig. 5.36 Windows/CsI transmission photocathode schematic diagrams

where d is the thickness of the fluorescent material layer of the cathode fluorescent screen, others are the same as above. Where $\theta' = 0$ means that x photons are incident vertically.

The material with higher absorption coefficient μ and larger overflow depth L_s has higher quantum efficiency of transmitting photocathode.

Principle of X-ray image intensifiers

X-ray imaging devices include photographic plate, fluorescence conversion screen and image intensifier. Proper combination can form direct vision device and TV imaging device. X-ray image intensifier can increase the image brightness by several orders of magnitude, greatly reduce the radiation dose of the subject, simplify the device structure, easy to operate, easy to protect, and has a wide range of practical value. Figure 5.37 shows the principle of X-ray image enhancement device. Figure 5.37a shows the principle of a single CsI/MCP photocathode and a single proximity focusing X-ray image intensifier. Figure 5.37b shows the principle of a glass window /CsI/MCP dual photocathode, dual proximity focusing X-ray image intensifier.

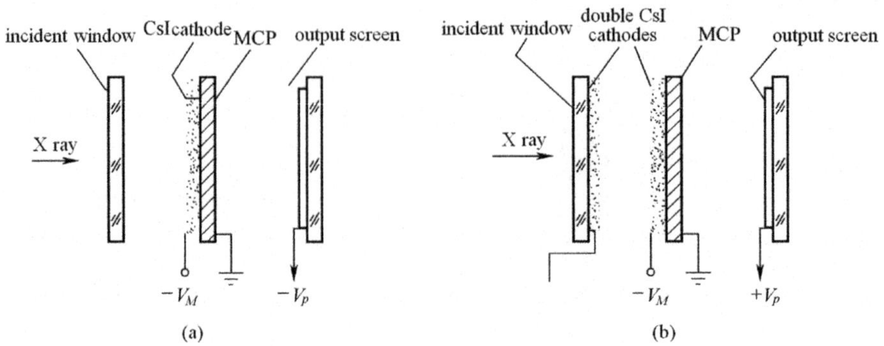

Fig. 5.37 X-ray image intensifier principle

5.8.2 Characteristics of X-Ray Imaging Devices

Image contrast of X-ray imaging device

The transmittance of the object to monochromatic X-rays

$$\tau = \frac{I}{I_0} = \exp(-\frac{\mu}{\rho}\rho t)$$

The components of the examined materials (μ/ρ), the density (ρ), the thickness(t), and the X-ray transmittance are different, that formed the two-dimensional original image of the X-ray, and its contrast is defined as:

$$C_T = \frac{L_1 - L_2}{L_1}$$

where L is the image brightness.

1. Luminance conversion factor

$$C = \frac{L}{P_0}$$

where P_0 is the X-ray input dose rate, mrad /s; L is the output brightness of X-ray imaging device, cd/m^2.

2. Derivation of luminance conversion factor for X-ray image intensifier

Items	Parameters	Unit
Input dose rate	P_0	mrad/s
Input irradiation intensity	$I_0 = 3.12 \times 10^9\ P_0$	keV/cm^2 s

(continued)

(continued)

Items	Parameters	Unit
Input × photon flow density	$N_0 = 3.12 \times 10^9\ P_0/E$	Number of photons/cm^2 s
After the input window is attenuated	$N_\tau = N_0\tau$	Number of photons/cm^2 s
Photocathode electron density	$n_e = N_0\tau\eta_k$	Number of optoelectronics/cm^2 s
After MCP electron multiplication	$n_m = N_0\tau\eta_k\ G_m$	Number of electronics/cm^2 s
Output screen brightness	$L = N_0\tau\eta_k\ G_m\ eV_P\eta_P$	cd/cm^2

Finally, the X-ray image intensifier brightness conversion factor is:

$$C = \frac{L}{P_0} = 8.63 \times 10^{12}\tau\eta_k G_m e V_P \eta_P / E$$

where η_k is X-ray photocathode quantum efficiency(%), η_P is screen conversion efficiency (lm/W), τ is incident window X-ray transmittance (%), $e^-$is electronic charge, G_m is MCP electron gain, V_p is fluorescent screen voltage (V).

The luminance conversion factor under non-monochromatic X-ray irradiation can be obtained by numerical integration.

Resolution and MTF

The resolution of X-ray imaging device is low, which is represented by lp/cm;

X-ray penetration is strong, using the test card with thick lead plated lines;

The MTF of X-ray imaging devices is difficult to measure, so image contrast and contrast sensitivity are used to characterize the contrast transfer characteristics of devices.

The parameters and performance of typical X-ray imaging devices are shown in the following table (Table 5.2).

5.9 Other Photoelectronic Detectors

1. Position sensitive detector

It is a Position-Sensitive Detector (PSD) for detecting the position of incident light on a detector surface. It is a kind of continuous simulator which is sensitive to the center of gravity of non-uniform incident light energy based on the transverse photoelectric effect of PN junction of semiconductor. Compared with CCD and CMOS, it has no pixel size, wide spectral response, fast response and simple signal processing (Fig. 5.38).

The incident position of the light spot is (x, y), and the charge product components of the four corners are A, B, C and D respectively, then the position of the light spot is:

Table 5.2 X-ray image intensifier performance

Items	Parameters	Unit
Input/output effective diameter	Φ50	mm
The whole long pipe	65	mm
Maximum diameter of the whole pipe	Φ85	mm
The power supply voltage	3 ± 0.4	V
X-ray tube voltage	40	keV
X-ray tube current	≤ 80	μA
Focal spot to skin distance	135	mm
Fluorescent screen output brightness	≥ 15	cd/m^2
Resolution	≥ 4	lp/mm
The working life	≥ 1000	h
The environment temperature	0–45	°C
Environmental humidity	≤ 80	%

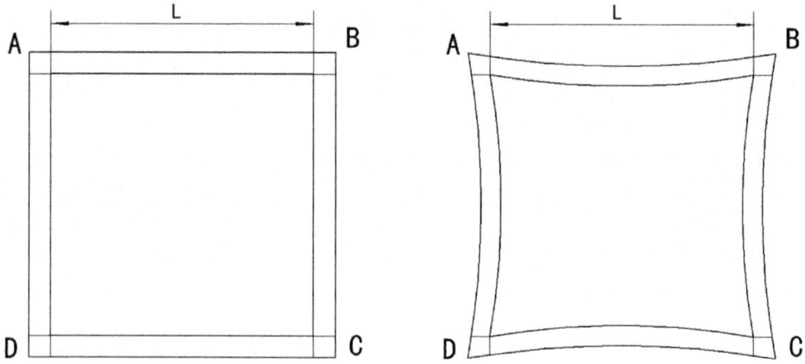

Fig. 5.38 Schematic diagram of PSD

$$x = L\frac{A + D}{A + B + C + D}$$
$$y = L\frac{C + D}{A + B + C + D}$$

2. Photoresistor

It is a photodetector that works by using the effect of photoconductivity. Photoconductivity effect is a bulk effect of semiconductor materials, the stronger the light is, the smaller the resistance of the device itself. Photoresistors have no polarity, and only sense changes in light by measuring changes in resistance.

3. Photovoltaic cell

It is a photoelectric conversion device that senses the output electric energy of illumination.

4. Photodiode

It is a junction photodetector working in photoconductance mode, which has very important applications in weak and fast light signal detection. Specifically, there are avalanche photodiodes, color-sensitive photodiodes and other types.

In addition, there are thermistors, pyroelectric detectors, etc.

Reference

1. DALSA INC. 1998–1999 DATABOOK.
2. Zongfan W, Meilin L, Shaojv Z, Yuze Z, Hanbin L. Infrared and low light technology. Beijing: National Defense Industry Press; 1998.

Chapter 6
Optical System Selection of Remote Sensor

6.1 Introduction

The optical system of space optical remote sensor is a basic condition that determines the performance of space optical remote sensing system. Large field of view means short revisit cycle, while long focal length remote sensors can be on high orbit. For optical systems with a small ratio of tube length to focal length, the optical remote sensor is light, and the platform can be small, and the optical system with large tolerance has stable performance. In traditional optics system, the processing method is simple, and general cost is low. For an optical system, one of the most important evaluation indexes is the optical modulation transfer function (MTF). MTF is a basic indicator to evaluate the performance of imaging camera, infrared camera, imaging spectrometer or stereo mapping camera. The optical system $F^{\#}$ is most closely related to MTF. Generally, small $F^{\#}$ means high optical performance and difficult optical design. Theoretically, MTF at characteristic frequency is high. If $F^{\#}$ is large, the difficulty of optical design is low. Theoretically, MTF at the characteristic frequency is low. Different types of optical systems have different characteristics. They can have different transmittance, different blocking ratio, different modulation transfer function and different field size. To study the space optical remote sensor, it is an effective way to analyze its characteristics according to the type of optical system. Here, different types of optical system will be simply analyzed and compared.

J. Tao, *Space Optical Remote Sensing*, Advances in Optics and Optoelectronics,
https://doi.org/10.1007/978-981-99-3318-1_6

6.2 Basic Concepts of Optical Systems

6.2.1 Basic Laws

1. **The law of straight-line propagation of light**

In an isotropic uniform medium, light travels in straight lines, which is the straight line propagation law of light. When there is an opaque object, a small hole or a slit in the path, the light travels away from the straight line. The law of straight line propagation of light holds only if there is no barrier in a uniform medium.

2. **The law of independent propagation of light**

When the light emitted from different light sources passes a certain point in space in different directions, each light does not affect each other and spreads independently, which is the law of independent propagation of light. Different light sources here refer to incoherent light; if it is coherent, the light will produce interference.

3. **Laws of reflection and refraction**

As shown in Fig. 6.1, incident ray AO, normal NO, reflected ray OB and refracted ray OC. The incidence angle ∠AON = I, the reflection angle ∠NOB = −I ′ and the refraction angle ∠N′OC = I ′.

The rays AO, OB and OC are coplanar with the normal ON.

Angle of incidence ∠AON = Angle of incidence ∠NOB (Laws of reflection).

Refractive index n of incident medium; n' of the refracting medium. The speed of light in vacuum c_0, the speed of light v in medium n and the speed of light v' in medium n', then

$$n = \frac{c_0}{v}$$
$$n' = \frac{c_0}{v'}$$

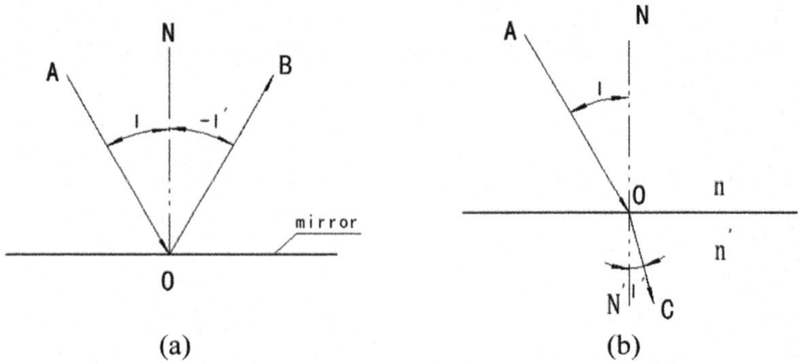

Fig. 6.1 Reflection and refraction of light

The refractive index of light in vacuum is $n_0 = 1$, and the refractive index of light in air at standard pressure 760 mm Hg and 20 °C temperature is $n = 1.00028 \approx 1$ [1].

$$n \sin I = n\prime \sin I\prime \text{ (Laws} \cdot \text{of} \cdot \text{refraction)}$$

4. **The light path is reversible**

The propagation of light can travel back along the original path.

The forward propagation path in the figure is $A \to O \to B$, $A \to O \to C$, then the reverse propagation path $A \leftarrow O \leftarrow B$, $A \leftarrow O \leftarrow C$ must be established.

6.2.2 Concepts of Imaging Optical System

The optical system imaging mentioned here refers to geometric optical imaging. Geometric optics is a subject that studies the imaging problem of light propagation in transparent media, only based on the straightness of light, which does not consider the wavy nature of light. **Light** is a geometric line without diameter and volume, which represents the direction of light propagation. The **luminous point** is a geometric point without volume or size but with energy. It is clear that the point and line of light in geometric optics are a simplified abstraction; however, it can convert the energy transmission and imaging problems of complex to simple geometric operations and guide the optical system design.

Light wave is electromagnetic wave, and electromagnetic wave is shear wave; that is, the vibration direction is perpendicular to the propagation direction, and the surface formed by each point with the same vibration phase at the same time is called the **wave surface**, and the normal direction of the wave surface is the propagation direction of the light. Plane waves correspond to parallel beams, and spherical waves correspond to converging or diverging beams. Converging or diverging beams are collectively called concentric beams. When the rays neither intersect at any point nor are unaligned, such beams are called astigmatic beams.

An **object point** can be understood as a point on the visible surface of the imaged target or, more generally, as an intersection where the incident beam is extended in reverse. **Object space** is the set of object points. An **image point** can be understood as a point on the image plane, or more generally as the intersection of imaging beams. **Image space** is the set of all image points. Object space is not only the geometric space in front of the optical system, but also extends behind the optical system. Image space is not only the geometric space behind the optical system, it may also extend in front of the optical system, and there may also be overlap. Whether a point belongs to object space or image space depends mainly on whether it is the center of the incident beam or the center of the outgoing beam, rather than whether it is in front of or after the optical system [2].

Imaging is to present each object point of the target scene on a plane called the image plane through optical components for human eyes to observe or detectors to receive. The combination of these optical parts is an optical system. Optical parts are usually composed of curved surfaces, and the curvature centers of optical parts are usually on a straight line. Such an optical system is called a coaxial system, and this straight line is called the optical axis. Most optical systems are coaxial systems. Objects and images are both formed by the intersection of light rays. The condition for an optical system to form perfect images is that both the incident and the outgoing are spherical waves.

Ideal optical system

An ideal optical system is an optical system that can form a perfect image in an arbitrarily large space with an arbitrarily wide beam. It has the following properties:

1. A point in the object space has one and only one point in the image square corresponding to it, and the corresponding two points are the conjugate points of the object image space.
2. A straight line in the object space has one and only one straight line in the image space corresponding to it, and the corresponding two straight lines are the conjugate lines of the object image space.
3. If the object space is located at a point on a straight line, the conjugate point in the image space must also be on the conjugate line of the straight line in the image space.
4. For any plane in the object space, there is only one plane in the image space corresponding to it, and the two corresponding planes are the conjugate planes of the object image space.

It can be deduced from this that any concentric beam in the object space corresponds to a conjugate concentric beam in the image space. The above definition is called the collinear imaging theory, and the imaging conforming to the collinear imaging theory must be a perfect image. The ideal optical system was established by Gauss in 1841, so it is also called Gaussian optics. The paraxial region of the actual imaging system can satisfy the collinear imaging theory, and the imaging results in the paraxial region are often used as the standard to measure the imaging quality of the actual optical system.

The base points and planes of an ideal optical system

In geometric optics, the points and planes representing the imaging characteristics of optical systems are called base points and base planes, which are the focal points and focal planes, principal point and principal plane and node and node planes of ideal optical systems.

1. Focal points and focal planes

According to the conjugate imaging cha(a)racteristics of ideal optical system, as shown in Fig. 6.2, incident ray A_1E_1 with height h and parallel to optical axis FO_1 must have a conjugated ray parallel or not parallel (intersecting with optical axis) to it in the image square. Parallelism is a telescopic system and will not be conspacered here. If they intersect, the image space ray G_kF' intersects with the optical axis at F'. The light FO_1 along the optical axis of the object space still propagates in the original direction on the image space after passing through the optical system; that is, O_kF', G_kF' and O_kF' intersect at F'. The image point F' corresponds to the on-axis object point at infinity in the object space, that is, parallel light, and the point F′ is the image focus of the optical system. The image point F' corresponds to the on-axis object point at infinity in the object space, that is, the parallel light, and the point F' is the imaging focus of the optical system. The plane passing through the point F' and perpendicular to the optical axis is the image space focal plane (image plane) as shown in Fig. 6.3a, which is conjugate with the object space infinity vertical axis plane. After passing through the optical system, the parallel beam of the infinity off-axis object point on the object space must be imaged on this image plane.

Similarly, according to the conjugate imaging theory, the incident ray A_kE_k with height h and parallel to the optical axis FO_1 must have a conjugate G_lF on the object space, which intersects with the optical axis at F. The ray $F'O_k$ along the optical axis

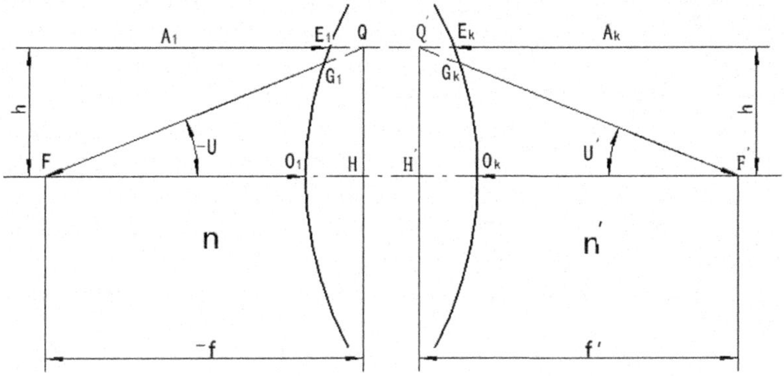

Fig. 6.2 Schematic diagram of the formation of focus

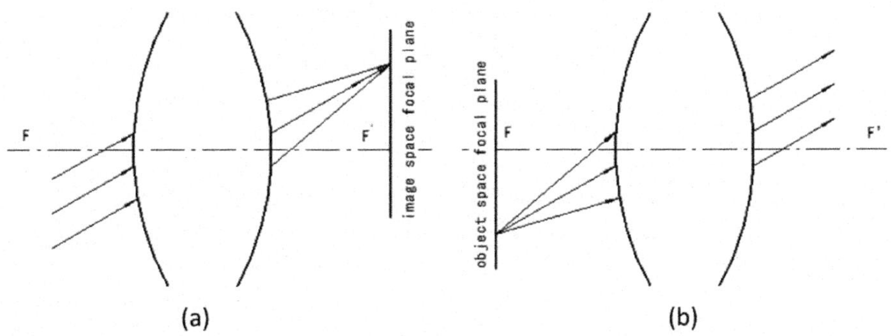

Fig. 6.3 Focal points and focal planes

of the image space and the conjugate ray O_1F intersect at F, and point F is the object focus of the optical system. The plane passing through point F and perpendicular to the optical axis is the focal plane of the object space as shown in Fig. 6.3b, which is conjugate with the infinite vertical axis plane of the image space. The light emitted by the point on the focal plane of the object must be emitted as a parallel beam after passing through the optical system.

2. Principal point, principal plane and focal length

As shown in Fig. 6.2, the object space ray A_1E_1 is conjugated with the image space ray G_kF'; the object space ray FG_1 is conjugated with the image space ray E_kA_k. The extended object space rays A_1E_1, FG_1 intersect at point Q; the extended image space rays G_kF', E_kA_k intersect at point Q'. Obviously, Q and Q' are a pair of conjugate points. Make the vertical line of the optical axis O_1O_k through Q and Q', and H and H' are the foot point. $QH = Q'H = h$, and they are on the same space of the optical axis, so their lateral magnification is 1, that is, $\beta = +1$. The pair of conjugate planes QH and $Q'H'$ with a lateral magnification of 1 is the main planes of the optical system. Point H is called the principal point of the object space, point H′ is called the principal point of the image space, and H and H' are a pair of conjugate points. QH is the principal plane of the object space, and $Q'H'$ is the principal plane of the image space.

The distance from the principal point of the object space to the focal point of the object space is the focal length of the object space. In this example, the direction is opposite to the direction of light propagation, and the focal length is negative, which is $-f$. The distance from the principal point on the image space to the focal point on the image space is the focal length on the image space. In this example, the direction is the same as the direction of light propagation, which is positive f'.

If the object space index of refraction is n, while the image space index of refraction is n', then

$$\frac{f}{f'} = -\frac{n}{n'}$$

If the optical system is in the same medium, that is, $n = n'$, then

$$-f = f\prime$$

3. Node and node plane

In an optical system, there is also a pair of conjugate points with angular magnification $\gamma = +1$, which is similar to the principal plane with lateral magnification $\beta = +1$, called nodes. The plane passing through the node and perpendicular to the optical axis is the nodal plane. The node of object space is called the object space node J, and the node of image space is called the image space node J', as shown in Fig. 6.4a. The object space ray BN passes through the object space node J. Since the angular magnification of the light passing through the object space node is $+1$, its conjugate ray must pass through the image space node J' and in the same direction as BN, that is, $BJ \| J'B'$.

$$\because \Delta FHQ \cong \Delta J'F'B'$$
$$\therefore FH = J'F' = -X'_J = -f, \text{ that is, } X'_J = f.$$

The same $HJ = H'J'$, that is, $X_J = f'$.
If the optical system is in the same medium, that is, $n = n'$, then

$$-f = f'$$

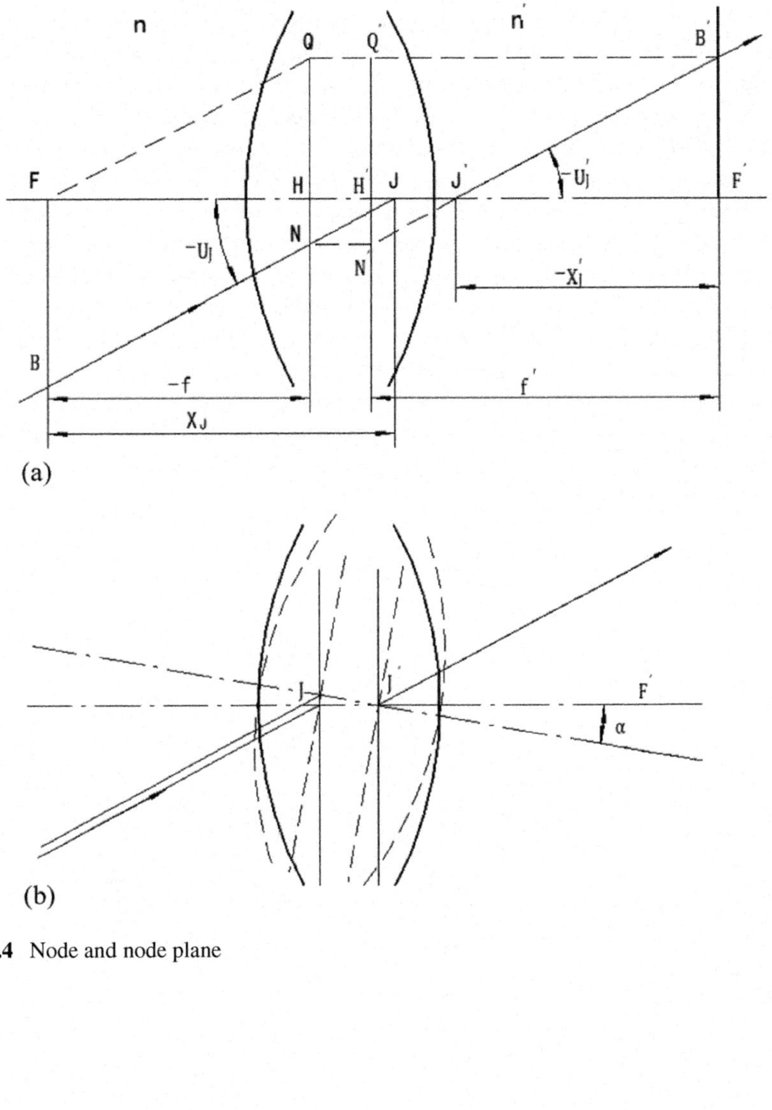

Fig. 6.4 Node and node plane

so

$$-X'_J = f' = -f = X_J.$$

That is, in an optical system in the same medium, the object space principal point coincides with the object space node, and the image space principal point coincides with the image space node. Therefore, as shown in Fig. 6.4b, rotate the optical system at the angle of α with the image space node J' as the axis, and the conjugate rays of the parallel incident rays passing through the object space node before and after the rotation are the same on the image space.

Aperture stop, entrance pupil and exit pupil of an actual optical system

An **aperture stop** is a stop that limits the aperture of an on-axis spot imaging beam. As shown in Fig. 6.5, the aperture $Q_1Q_2Q_3$ limits the on-axis spot beam, which is the aperture stop of the system. An optical system has a series of apertures, how to determine which one is the aperture stop. In the figure, through the optical aperture $Q_1Q_2Q_3$ is imaged by the front lens L_1 to the object space $P_1P_2P_3$, then among the aperture $P_1P_2P_3$, the frame of L_1 and the frame of L_2, the opening angle of $P_1P_2P_3$ to the center of the object surface is the smallest, so the optical aperture $Q_1Q_2Q_3$ corresponding to the image $P_1P_2P_3$ is the aperture stop. The image $P_1P_2P_3$ in the object space is the entrance pupil of the optical system, referred to as the entrance pupil. It is the public entrance of the imaging beam at each point on the object surface, and it determines the size of the imaging beam entering the optical system. Through the optical aperture $Q_1Q_2Q_3$, $P_1'P_2'P_3'$ is imaged by the rear lens L_2 to the image space, then among the aperture $P_1'P_2'P_3'$, the frame of L_1 and the frame of L_2, the opening angle of $P_1'P_2'P_3'$ to the center of the image surface is the smallest, so the image $P_1'P_2'P_3'$ is the exit pupil of the optical system. It is the public exit of the imaging beam at each point on the image surface, and it determines the size of the imaging beam exiting the optical system.

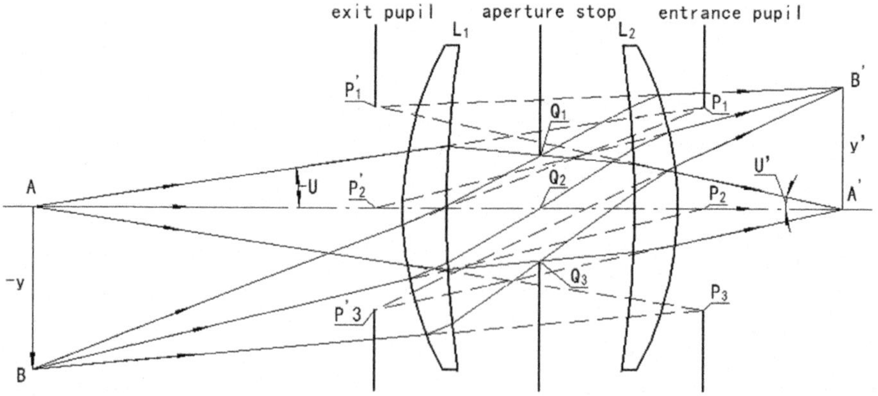

Fig. 6.5 Aperture stop, entrance pupil and exit pupil

The angle between the ray from the center of the object plane to the edge of the entrance pupil and the optical axis is the object space aperture angle U, and the angle between the ray from the center of the image plane to the edge of the exit pupil and the optical axis is the image space aperture angle U'. The exit pupil and the entrance pupil are conjugate to the entire optical system. The ray passing through the center of the entrance pupil is called the chief ray. For an ideal optical system, the chief ray must also pass through the center of the aperture stop and the exit pupil. The chief ray is the axis of the imaging beam from each point on the object plane that fills the entrance pupil.

The position of the aperture stop is very important to the optical system. Reasonable selection of the position of the aperture stop can reduce the lateral size of the optical system, make the structure symmetrical and improve the imaging quality of the optical system.

In addition to aperture stops, there are apertures placed in the object plane and the image plane that limit the imaging range called field stops. A diaphragm in an optical system that limits the imaging beam at an off-axis point, which may cause vignetting on the image plane, but can improve image quality, is called a vignetting diaphragm (such as a mirror frame). In order to eliminate the influence of non-imaging light on imaging contrast, there is also a stray light diaphragm.

The plane formed by the chief ray and the lens axis is the meridian plane, and the plane perpendicular to the meridional plane through the chief ray is the sagittal surface.

Relative aperture or *F* number

The relative aperture is the ratio of the diameter of the incident pupil D to the focal length f', that is, D/f'. The number F is generally expressed as $F^{\#}$, $F^{\#} = f'/D$. $F^{\#}$ is also called aperture.

MTF of an ideal optical system

The optical modulation transfer function of an ideal diffraction limited circular pupil optical system is

$$MTF = \frac{2}{\pi}[\arccos\frac{f}{f_m} - \frac{f}{f_m}\sqrt{1-\left(\frac{f}{f_m}\right)^2}]$$

where

$$f_m = \frac{1}{\lambda F^{\#}} f < f_m$$

$$x = \frac{f}{f_m}, \text{ then}$$

$$MTF = \frac{2}{\pi}(\arccos x - x\sqrt{1-x^2})$$

6.3 Aberrations of Optical Systems

The paraxial region of the optical system has the properties of an ideal optical system, and the imaging (Gaussian image) of the paraxial region is considered to be an ideal image. The difference between the image formed by the actual optical system and the image formed by the paraxial area is the aberration. There are six categories of aberrations: spherical aberration, coma, astigmatism, field curvature, distortion and chromatic aberration, all of which belong to the concept of geometrical optical aberrations. Light itself is a highly abstract concept, and the light itself as electromagnetic wave has wave properties. Therefore, in addition to these six types of geometric aberrations, there are also wave aberrations.

Geometric aberration of optical systems

1. Spherical aberration

The phenomenon that concentric monochromatic rays passing through a sphere at different heights do not intersect at the same point on the axis is called spherical aberration (mrad). As shown in Fig. 6.6, Figure 6.6a positive lens causes negative spherical aberration, Fig. 6.6b negative lens causes positive spherical aberration, and positive and negative lens combination can be used for spherical aberration correction. Spherical aberration is the only monochromatic aberration at the point on the axis. The spherical aberration can only be corrected to zero for one aperture, usually for the edge aperture. Some optical systems can correct the spherical aberration for two aperture bands, but generally cannot make all aperture bands zero. When the spherical aberration of the edge aperture is not zero, the optical system with negative spherical aberration is called undercorrection, and the optical system with positive spherical aberration is called overcorrection

For spherical reflectors, the minimum dispersion circle diameter due to spherical aberration is $\delta_s = 15.6/F^{\#3}$, milliradian (mrad). For a single lens with a given focal

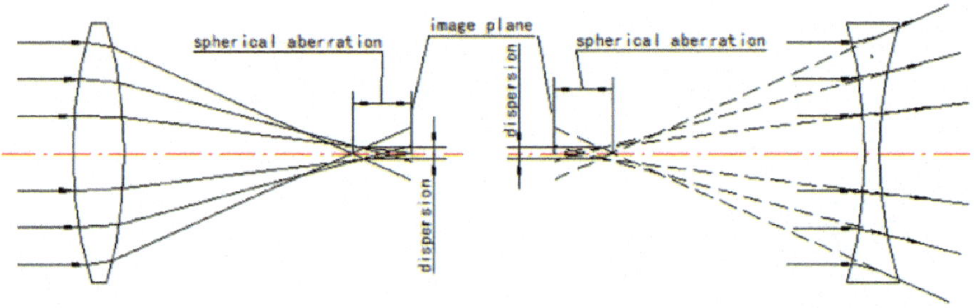

(a) (b)

Fig. 6.6 Spherical aberration

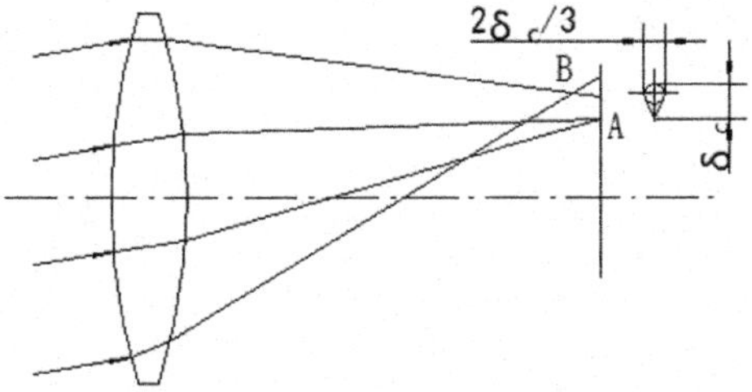

Fig. 6.7 Coma

length, when $r_1/r_2 = -(4 + n - 2n^2)/(n + 2n^2)$, the spherical aberration is minimal [3], while n is for lens refractive index.

2. Coma

A series of light spots of different sizes and non-confocal planes caused by the different ring band of infinity beams at off-axis points, which are shaped like comet tails, so they are called coma aberrations, as shown in Fig. 6.7. Coma is a type of monochromatic vertical axis aberration related to the field of view and aperture.

$$\delta_{\mathrm{cs}} = \omega/\left(16\, F^{\#2}\right).$$

where:

δ_{cs} unit rad;
ω half field of view;
$F^{\#}$ the ratio of focal length to entrance pupil diameter [4]

For parabolic mirrors, meridional coma $\delta_c = 0.1875\, u/F^{\#2}$, u is aperture angle, and δ_c is consistent with u for unit. When the coma is equal to the spherical aberration, the aperture angle $u = 4.8/F^{\#}$. For infrared systems with a small field of view, other monochromatic aberrations except spherical aberration and coma have little effect [3].

3. Astigmatism

The off-axis point beamlets form two focal lines in the meridional focal plane and the sagittal focal plane. The light spot in the middle of the two focal lines is the smallest, and the distance between the two focal lines is astigmatism X_{ts} in Fig. 6.8. After the off-axis point beam passes through the optical system, the wavefront of the beam becomes an aspheric surface, and its curvature in the two main sections of the meridian and sagittal is different, so it is focused into the meridional image point and the sagittal image point, forming astigmatism. The correction of astigmatism is to make the astigmatism of a certain field of view to zero, while other fields of view still have residual astigmatism.

For the imaging of a single thin lens at an infinite object point, the diameter of the astigmatic dispersion circle is $\delta_a = \omega^2/(2F^{\#})$, and the astigmatism has nothing to do with the shape of the lens.

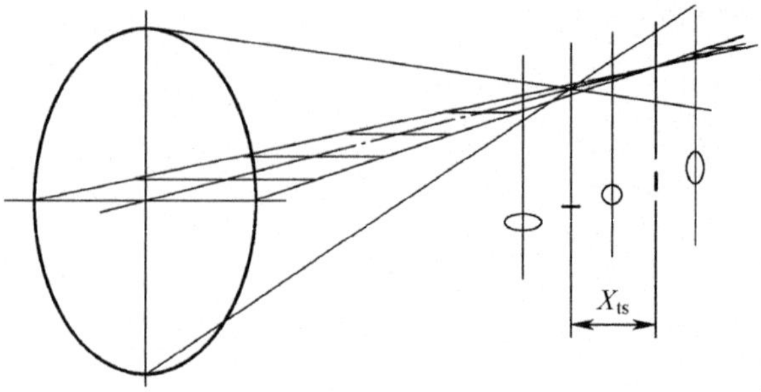

Fig. 6.8 Astigmatism

When the parabolic mirror itself is the aperture, the diameter of the astigmatism is $\delta_a = \omega^2/(2F^\#)$, rad.

4. Curvature of the field

In general, the image plane that the actual optical system can clearly image is a curved surface, and the axial distance between it and the image plane of the ideal optical system is the field curve, as shown in Fig. 6.9. Due to the existence of the astigmatism of the beamlet at the off-axis point, the meridional image point and the sagittal image point form the meridional image plane and the sagittal image plane, both of which are surfaces of revolution symmetric to the optical axis. Since the on-axis point has no astigmatism, the on-axis object point forms an ideal image, and its image is located on the Gaussian image plane, so the meridional image plane and the sagittal image plane must be tangent to the Gaussian image plane at the focus of the ideal imaging.

Even if there is no astigmatism, through the imaging analysis of a single sphere, a single sphere can only form a perfect spherical image concentric with a single sphere

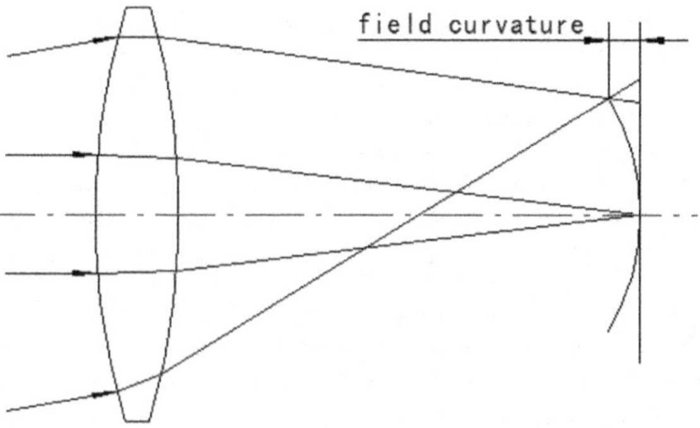

Fig. 6.9 Curvature of the field

for the object space sphere, while a perfect planar image for a flat object with a single sphere cannot be formed, and a field curvature still exists, detailed in reference [5].

5. Distortion

The transverse magnification of the image varies with the field of view, resulting in distortion, as shown in Fig. 6.10. Figure 6.10a is ideal image without distortion, Fig. 6.10b is pillow-shaped positive distortion, and Fig. 6.10c is barrel-shaped negative distortion. For a single thin lens or group of thin lenses, no distortion occurs when the diaphragm coincides with it. Distortion only distorts the image relative to the object and does not affect the resolution of the image.

6. Chromatic aberration

Since the optical material has different refractive index for the light of different wavelengths, the light of different wavelengths has different axial imaging positions and different imaging magnification after passing through the optical system, which is called chromatic aberration. Figure 6.11 illustrates the chromatic aberration of position

Wave aberrations of optical systems

In geometric optics, the ray direction is the normal line of the wave surface, and the concentric beam emitted by point light source or object point is a spherical wave, which changes the curvature after passing through the optical system. If the optical system is an ideal system, a new spherical wave with the ideal image point as the center is formed. In fact, even in an ideal optical system, the diffraction produced by limited optical aperture is not a point image, but a certain size of the diffuse spot. Moreover, the actual optical system always has residual aberration, which deforms the wavefront more or less. The deviation between the deformed wavefront and the ideal wavefront is the wavefront aberration.

(a) (b) (c)

Fig. 6.10 Distortion

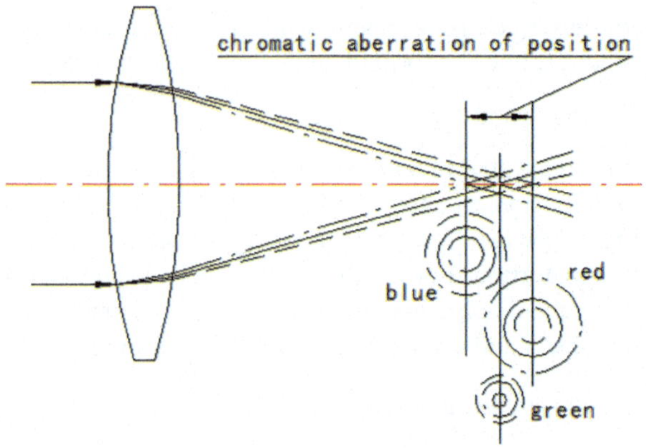

Fig. 6.11 Chromatic aberration of position

The central point luminance (Strelby), Rayleigh criterion, resolution and optical transfer function used in evaluating the image quality of optical system are all concepts established from wave optics. For example, the tolerance of wave aberrations according to the famous Rayleigh criterion is λ/4.

6.4 Introduction to Several Main Optical Systems

The optical system used in space optical remote sensing is mainly photographic optical system, whose performance is mainly determined by three parameters, namely focal length, relative aperture and field of view. A basic feature of space optical remote sensing for optical system is long focal length, because the orbit of spacecraft at least 200 km above, focal length is above meters. In order to shorten the regression period, it is generally required to have a large field of view, which is also related to the orbit height. The lower the orbit is, the larger the field of view is generally expected. Due to the characteristics of space applications, it is expected that optical remote sensors are light in weight to save launch costs. As the focal length is long, the *F* number is generally large. It is hoped that the optical system is not sensitive to the position error of the optical mirror and the vacuum degree of the optical system, because of the poor thermal environment of the spacecraft and the influence of microgravity and vacuum.

At present, almost all space optical remote sensors use photodetectors. In order to improve the signal-to-noise ratio, it is hoped that the incident signal is strong; that is, the transmittance of optical system is high.

The above requirements of space optical remote sensing for optical system can be summarized as follows:

1. Small weight;
2. High transmittance;
3. High stability.

So, the optical system is divided into refraction type, reflection type and catadioptric type according to the form of light transmission. Space cameras are generally larger, and cameras using reflective systems are generally light in weight, high in transmittance, but difficult in manufacturing process; the camera with refraction system is generally simple in manufacturing process, but the weight is generally larger, and the transmittance is generally lower. The catadioptric system is somewhere in between.

6.4.1 Transmission Optical System

In transmission system, the field of view can be tens of degrees or dozens of degrees, transmittance between 0.6 and 0.7.

Transmission optical systems are generally suitable for light and small cameras or aerial cameras. An aerial camera optical system is shown in Fig. 6.12, which shows the optical path of full field of view imaging rather than that of a specific field of view.

This kind of optical system is easy to process, install and adjust and has low manufacturing costs, but the transmittance is low at generally about 0.6–0.7. It is sensitive to temperature, has strict requirements for thermal control and has a relatively large tube long to focal length ratio, so it is generally used for aerial imaging remote sensing. The optical system shown in Fig. 6.12 is flexible in aerial imaging and can be used as both panoramic camera and tilting camera.

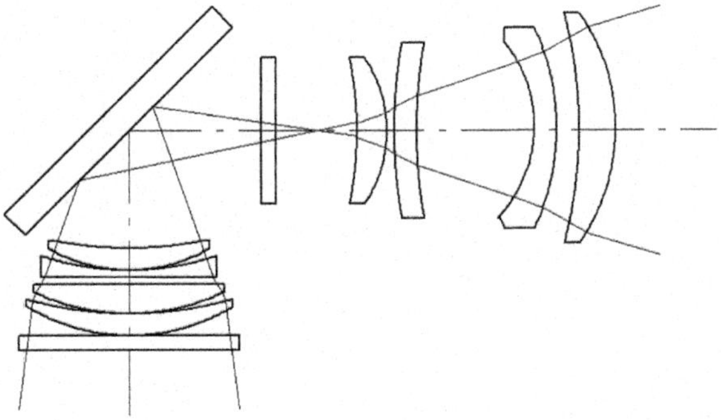

Fig. 6.12 Transmission optical system

6.4.2 Catadioptric Optical Systems

A catadioptric system with a field of view of 2–4° can be used for larger optical remote sensors.

1. A single-mirror catadioptric optical system

Chromatic aberration is unavoidable in refractive systems, and both spherical aberration and off-axis aberration are relatively serious. Pure reflection system can eliminate chromatic aberration, but in general, the off-axis aberration is large, and it is difficult to obtain good image quality in a large field of view. The aspherical mirror is not easy to process, inspection is also troublesome, and the manufacturing cost is high, so the early aerospace optical remote sensing cameras adopted a compromise method, using a catadioptric-type optical system.

The space optical remote sensing camera of SPOT satellite, for example, uses such an optical system, as shown in Fig. 6.13. Its main optical element is a spherical mirror with a set of corrective lenses without focal degrees placed at its center. Its focal length is 1082 mm, $F^{\#} = 3.24$, and $2\omega = \pm 2.07$. In order to shorten the length of the tube, the method of folding optical path is adopted due to large correction lens aperture, the large ratio of the tube length and focal length and larger lens size and weight.

2. Two-mirror catadioptric optical system

A single-mirror catadioptric optical system has a relatively large ratio of the tube length and focal length, while two-mirror catadioptric optical systems can obtain a smaller ratio of the tube length and focal, as shown in Fig. 6.14.

A set of correction lenses in front of the system is used to correct the on-axis aberrations and determine the relative aperture of the system. A set of field correction mirrors is added in front of the image plane to correct the off-axis aberration of the system and increase the field of view of the system. The reflector can be a spherical mirror. Such a system is suitable for long focal length and large aperture systems.

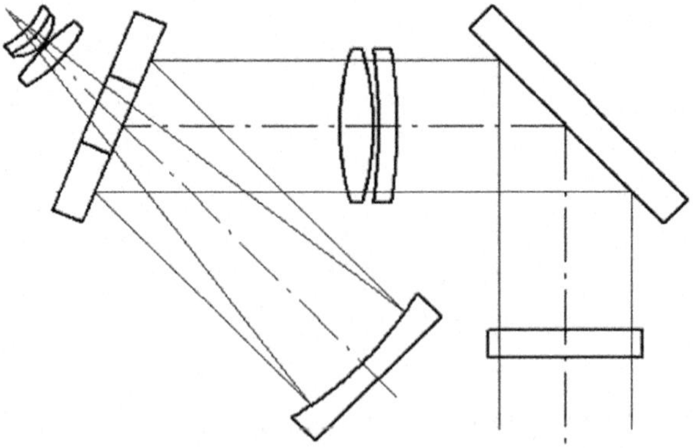

Fig. 6.13 Single-mirror catadioptric optical system

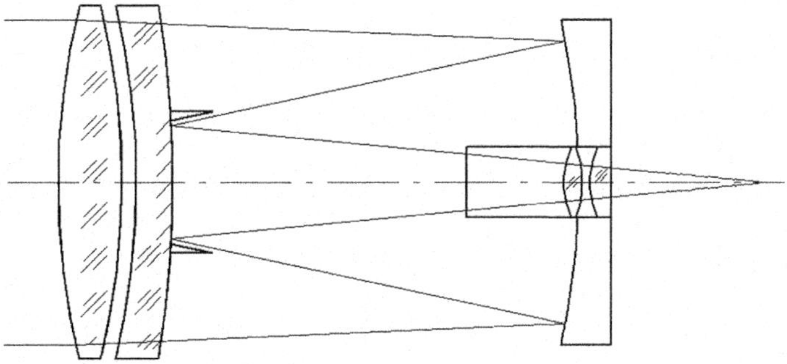

Fig. 6.14 Two-mirror catadioptric optical system

The disadvantage of this obstructing coaxial catadioptric hybrid optical system is that when the aperture of the front group correction lenses is required to be large, the size of each lens is large, so the weight of the system is large, and the uniformity of materials is required to be high. Since the former group of correction lenses and the rear lens group are both heavier, a lens tube with greater rigidity is required, which further increases the weight of the optical remote sensing camera.

6.4.3 Reflective Optical System

Reflective systems include coaxial reflection systems, partial field of view reflection systems and off-axis reflection systems. The field of view of the coaxial reflection system and the partial field of view of reflection system are about 2°, and the field of view (2ω) of the off-axis reflection system can reach 10–20°. It can be used for large optical remote sensors, with wide working spectrum and high transmittance.

In order to improve the resolution, the aerospace optical remote sensing camera has a large aperture and a long focal length. When the aperture reaches the order of 1 m, the high-quality lens material required by the optical system is difficult to manufacture, the system weight is also large, and the large-diameter lens is difficult to support, so aerospace optical remote sensing cameras mostly use reflective optical systems.

Reflective optical systems have the following advantages:

1. It is easier to make a large-diameter lens, and optical materials are easier to obtain. Glass materials or other materials can be used. Only the surface of the glass material is used instead of the transmission performance of the glass, so the material of the mirror is not high relative to the lens.
2. Compared with the transmission optical system, the reflective optical system generally has less light energy loss. After the surface of the reflective optical mirror is coated with a reflective film, its single surface reflectivity can reach more than 97%.

Fig. 6.15 R–C system

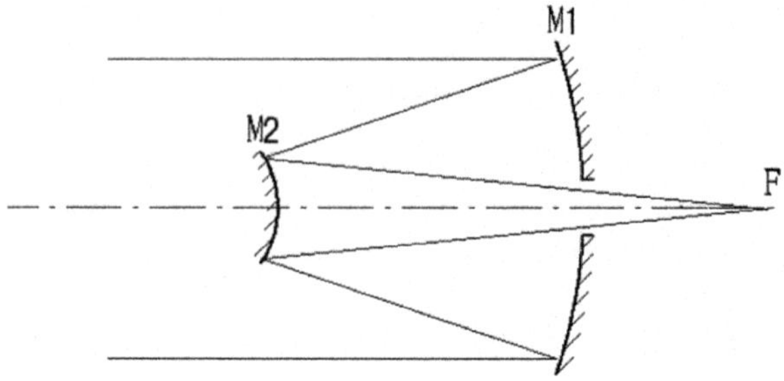

3. The spectral range is wide, and the reflector does not produce chromatic aberration. The reflectivity of each wavelength is basically equal in terms of the visible band.

Reflective optics also have significant disadvantages:

Since the reflective optical system uses an aspherical mirror, it is difficult to manufacture, highly cost and difficult to assemble. However, these difficulties have been gradually overcome with the development of computer numerical control machining technology and computer-aided assembly and adjustment technology.

1. Central obscured reflective optical systems

(a) **Coaxial two-mirror systems**

The common type of reflective optical system with central obscuration is a two-mirror system, as shown in Fig. 6.15. When the primary mirror is a paraboloid and the secondary mirror is a hyperboloid, there is no aberration at the point on the infinity axis. When both primary and secondary mirrors are hyperboloids, the system can eliminate both spherical and coma aberrations. When the primary reflector is ellipsoid and the secondary reflector is spherical, the system can eliminate spherical aberration.

The advantages of this system are that the ratio of the length of the tube and the focal length is small, the structure is compact, and the condensing light path can be designed to pass through the central hole of the primary mirror, so that the focal plane is located behind the primary mirror, which is convenient for the adjustment and installation of the detector assembly.

However, this type of system can only eliminate one or two types of aberrations, and the field of view cannot be made too large.

(b) **Coaxial three-mirror system**

Figure 6.16 shows a three-mirror reflective aspheric optical system that simultaneously corrects spherical aberration, coma, field curvature and astigmatism and is also free of chromatic aberration as a reflective system. The optical path is folded, the structure is compact, and the image quality of the entire field of view is relatively uniform.

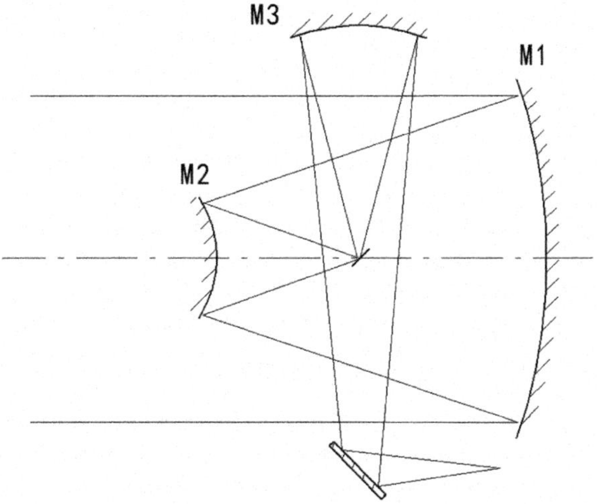

Fig. 6.16 Coaxial three-mirror aspheric system

(c) **Partial field of view three-mirror system**

Figure 6.17 shows a three-mirror optical system with a coaxial partial field of view, where the small deflection mirror is located outside the optical path to avoid secondary obscuration.

However, the field of view of such an optical system is small, and the light hole in the center of the primary mirror is also prone to stray light.

2. Central unobscured reflective optical systems

For a central obscured optical system, the obscuration reduces the effective aperture of the optical system and causes the reduction and redistribution of the modulation transfer function of the optical system. These changes will also cause the resolution of optical remote sensing camera to decrease, so the off-axis aspheric three-mirror optical system is mostly studied.

The advantage of this optical system is that it avoids the obscuration of the coaxial reflection optical system, improves the effective aperture of the optical system and

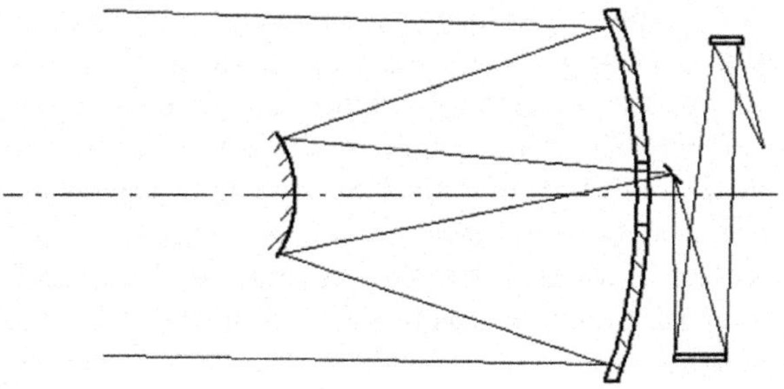

Fig. 6.17 Partial wide field of view three-mirror system

then improves the modulation transfer function of the optical system, which also provides the possibility to improve the resolution of the system. It eliminates spherical aberration, coma, field curvature and astigmatism and has no chromatic aberration as a reflective system.

The realization of this off-axis aspherical three-mirror optical system is crucial from optical design, processing and inspection of aspherical surfaces, support of optical mirrors and high-rigidity structure design of the whole machine to system installation and inspection. In particular, the support of high-precision optical mirrors is a key technology.

Three off-axis aspheric three-mirror optical systems are presented here.

(a) **Once imaging off-axis three-mirror system**

Figure 6.18 shows an off-axis aspherical three-mirror optical system without intermediate images. Its primary mirror is a concave off-axis aspherical mirror, its secondary mirror is a convex one, and its tertiary mirror is a concave off-axis aspherical mirror. Figure 6.18a is the *y–z* coordinates view of the system, and Fig. 6.18b is the *x–y* coordinates view of the system. An *x–y* coordinate view shows that the field of view of such optical systems is very large. Figure 6.19 is the MTF of the system. It can be seen that its MTF is very close to the diffraction limit in such a large field of view.

In addition to the common characteristics of unobstructed reflective optical systems, this optical system has its own characteristics:

1. It is suitable for optical remote sensing cameras with a wide field of view, and the field of view can be close to 20°.
2. Its ratio of the tube length and the focal length is relatively small, which is beneficial to the high rigidity and lightweight design of the entire structure of the optical remote sensing camera.
3. Pay full attention to the design of eliminating stray light.
4. The supporting structure of its primary mirror and tertiary mirror is difficult to design. They are all elongated mirrors with irregular shapes, which can be clearly seen from Figures (a) and (b).

(b) **Secondary reimaging off-axis three-mirror system**

Figure 6.20 shows an off-axis aspheric three-mirror optical system with an intermediate image. Its primary mirror is a concave off-axis aspheric mirror, and the secondary mirror is a convex off-axis aspheric mirror. The third mirror is a concave off-axis aspherical mirror. Figure 6.20a is a *y–z* coordinate system view of the system, and Fig. 6.20b is an *x–y* coordinate system view of the system. The intermediate image plane of this type of optical system can be seen from Fig. 6.20a. From Fig. 6.20b, it can be seen that the field of view of this type of optical system is small, only a few degrees. It can be seen from Fig. 6.20a and b that this type of optical system is very compact. Figure 6.21 is the MTF of the system, and it can be seen that the MTF is close to the diffraction limit. When the field of view is made 1–2°, it can be very close to the diffraction limit.

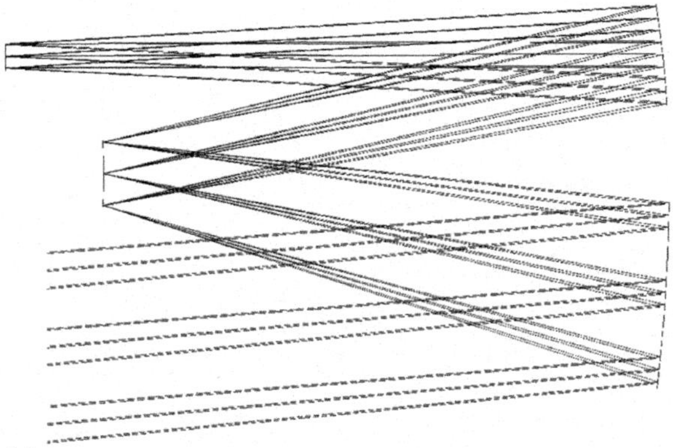

(a)

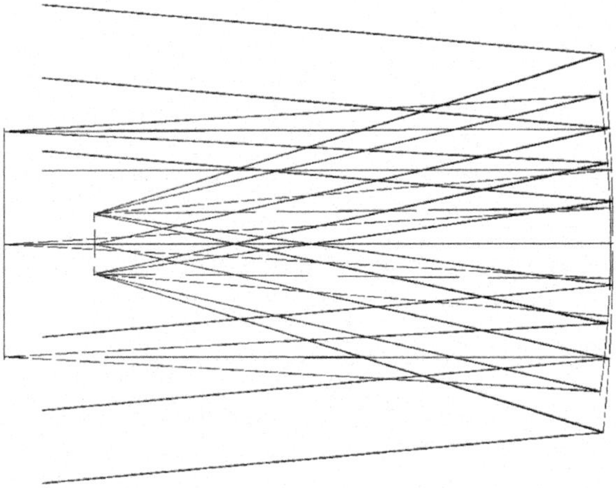

(b)

Fig. 6.18 Once imaging off-axis three-mirror system

In addition to the common characteristics of unobstructed reflective optical systems, this optical system has the following characteristics:

1. Its ratio of the tube length and the focal length is small, which is beneficial to the high rigidity and lightweight design of the entire structure of the optical remote sensing camera.
2. Its field of view is smaller than that of the previous once imaging off-axis three-mirror system, which is suitable for designing an optical remote sensing camera with a small field of view and a long focal length.
3. The design of eliminating stray light is easier than the previous once imaging off-axis three-mirror system.
4. The shape of its primary mirror is regular, the length to width ratio of the secondary mirror is relatively close to 1, and the shape of the third mirror is not regular. The overall supporting structure is easier to design than the once imaging off-axis three-mirror system introduced above, but the shapes of the

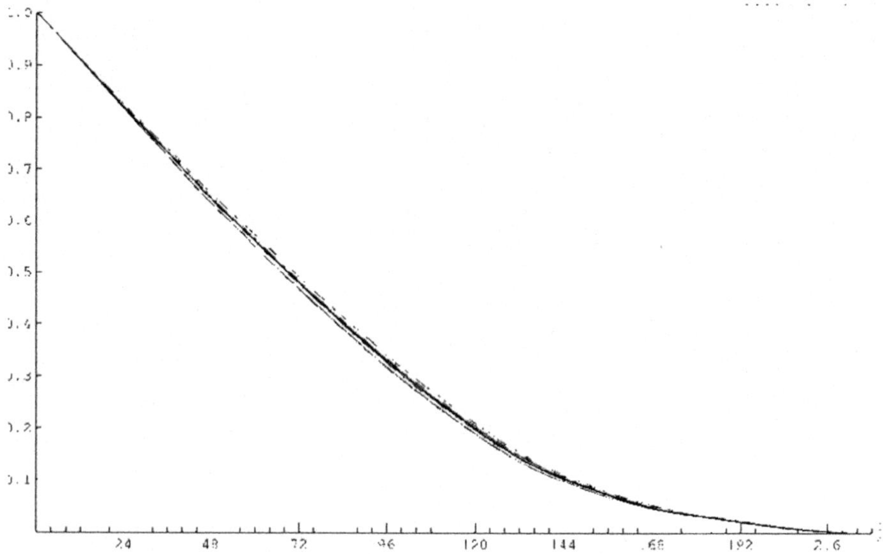

Fig. 6.19 MTF of once imaging off-axis three-mirror system

secondary mirror and the tertiary are also irregular, which adds difficulty to the structural design.

(c) **Large field of view off-axis three-mirror system**

Figure 6.22 is a large field of view off-axis aspherical three-mirror optical system. The first mirror of this system has a convex spherical surface, the second mirror has a concave off-axis ellipsoid, and the tertiary mirror has a concave spherical surface. The first reflecting mirror is designed as a convex spherical surface, which is conducive to realize a large two-dimensional field of view.

In addition to the common characteristics of unobstructed reflective optical systems, this optical system has the following characteristics:

1. It can realize a large two-dimensional field of view, and its field of view can reach 20° × 30°, which is the largest field of view in the current unobstructed reflective optical system.
2. It has two spherical mirrors, which reduces the difficulty of processing and manufacturing the optical system to a certain extent.
3. From the optical path of the system, it can be seen that the design of eliminating stray light of the optical remote sensing camera of this optical system needs to be given considerable attention.
4. From the design results of the optical system, it needs to be further improved, and its image plane distortion is relatively large. This is intolerable for remote sensing imaging surveying and mapping, but it can still be considered for applications in imaging reconnaissance and infrared surveillance systems, and most of the

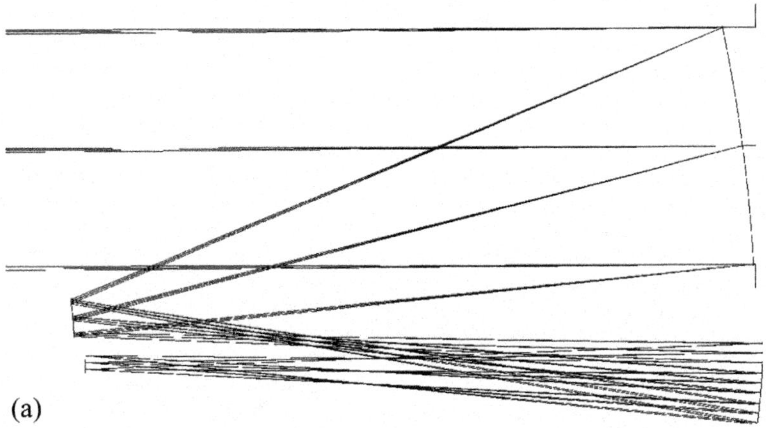

(a)

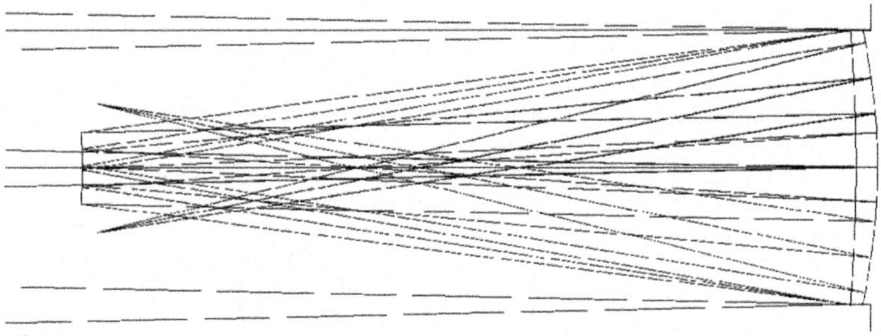

(b)

Fig. 6.20 Secondary reimaging off-axis three-mirror system

current aerospace optical imaging cameras are digital photoelectric transmission remote sensing cameras, so image surface distortion can be considered using image processing to correct it.

5. The structure and layout of the optical remote sensing camera of this optical system is relatively bulky, because it requires a large light shield, and the optical mirror configuration is not compact enough.

6.5 Analysis and Comparison of Several Typical Optical Systems

As the refraction optical system is sensitive to temperature and pressure and has strict requirements on thermal control, it is difficult to obtain high-quality lens materials with large size. Moreover, due to the development of computer-aided processing technology and digital interferometry technology, the processing technology of aspheric

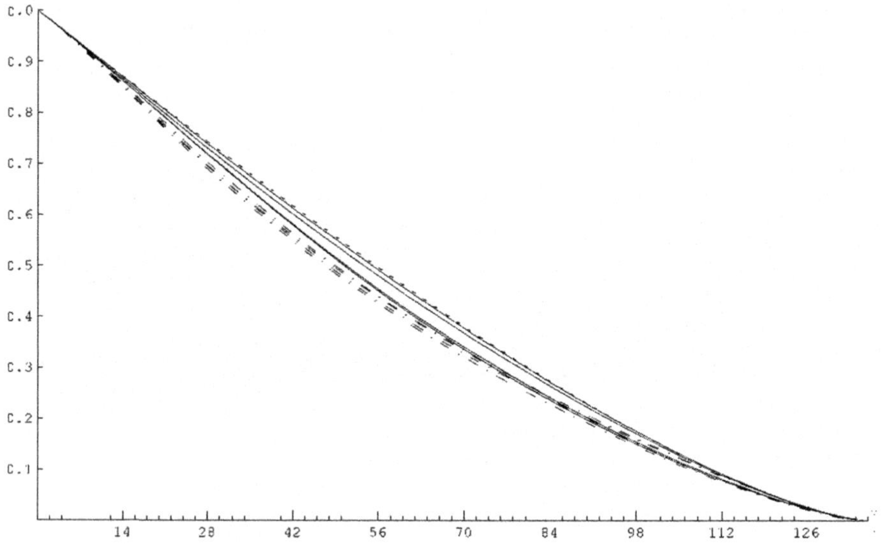

Fig. 6.21 MTF of secondary reimaging off-axis three-mirror systems

Fig. 6.22 Large field of view off-axis three-mirror system

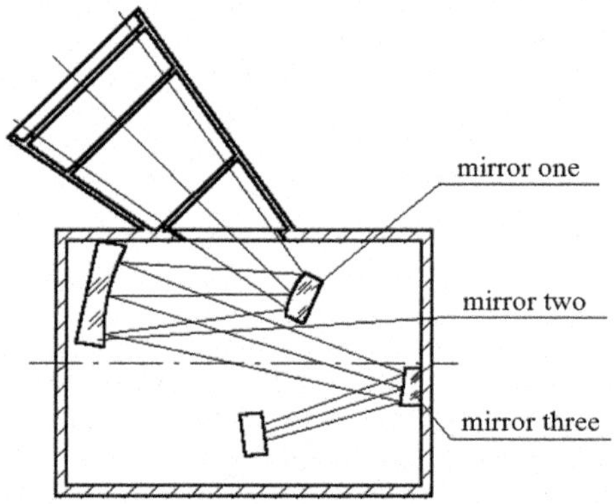

surface has been breakthrough, so the optical system of space optical remote sensing camera is preferred to reflect optical system.

The reflective optical system is divided into obstructed optical system, such as coaxial system, coaxial partial field of view system and coaxial three reflection system, and unobstructed optical system, such as once imaging three reflection system, secondary reimaging three reflection system and large field of view three reflection system.

The advantages of obstructed coaxial reflection optical system are that the accuracy of installation and structure system is lower, and the sensitivity to heat is less. The disadvantage is that its field of view is small and the image quality uniformity in the whole field of view is lower than that of the latter two off-axis aspheric three-mirror optical systems. From the perspective of development trend, the off-axis aspherical three-mirror optical system has more development advantages.

Especially for the last one, the off-axis aspherical three-mirror optical system, its large field of view will be more and more advantageous with the development of small pixel and large area array CCD detector. This imaging mode using the area array detector is also a favorable factor to improve the resolution of the on-orbit image of the space optical remote sensing camera. But its optical design has yet to be perfected.

In recent years, three kinds of optical systems are mainly used in the optical system of the space optical remote sensing camera: the coaxial system of double mirrors or this kind of optical system supplemented by small aberration correction lens. This system has the advantage of the coaxial system, but fewer lenses; off-axis aspheric three-mirror optical system without intermediate image; and off-axis aspherical three-mirror optical system with an intermediate image, no shading and high transfer function of optical modulation.

In a reflective optical system with central obstructed, its optical modulation transfer function is lower than that of an unobstructed optical system, and the distribution of its optical modulation transfer function is quite different from that of an unshading optical system. The on-orbit image resolution of an aerospace optical remote sensing camera is directly related to the optical modulation transfer function of its optical system.

References

1. Fenglin Z, Xuezhu S, Engineering optics, 3rd ed. Tianjin: Tianjin University Press; 1990.
2. Kecheng F, Jingsheng L. Infrared optical system. Beijing: Ordnance Industry Press; 1994.
3. Nanrong X, Nanhua B. Infrared radiation and guidance. Beijing: National Defense Industry Press; 1997.
4. Boruo C. Infrared system. Beijing: Ordnance Industry Press; 1995.
5. Yimo Z, editor. Applied optics. Beijing: Publishing House of Electronics Industry; 2010.

Chapter 7
Main Types of Optical Remote Sensors

7.1 Introduction

There are many types of earth observation optical remote sensors, and common ones are space imaging cameras, space mapping cameras, space infrared cameras, space multispectral cameras and imaging spectrometers. There are also professional optical instruments such as radiometers, LiDARs and high-speed photographic cameras, which will not be introduced here.

Space imaging cameras are the most important remote sensors among aerospace remote sensors. In terms of civilian use, it can be used in mineral resources survey, urban planning, land use monitoring, resource census, agricultural survey, environmental monitoring, disaster monitoring and geographic information services.

Optical remote sensing technology has been applied to all aspects of human life, and the dynamic changes of global climate change, El Niño phenomenon and impact, global desertification, vegetation shift, marine iceberg drifting, etc., have attracted wide attention; remote sensing has become an important role in the study of marine fisheries, maritime transportation, marine ecology and other aspects. Mineral resources, land resources, forest and grassland resources, wildlife resources, water resources surveys and crop yield estimation are all with the application of optical remote sensing methods; remote sensing has its own unique role in solving various environmental changes, such as urbanization, desertification, land degradation, salinization, environmental pollution and so on. In addition, optical remote sensing plays a major role in disaster monitoring, such as the prediction, forecasting and assessment of multiple meteorological hazards and crop pests and diseases.

In various project constructions, various scales and types of optical remote sensing play a role at different levels, such as large-scale water conservancy projects, port projects, high-speed railway construction, highway construction, nuclear power plants, airport construction and urban planning all obtain important data from remote sensing images. Several high-tech local wars that have taken place in modern times have comprehensively used optical remote sensing technology to obtain important information.

J. Tao, *Space Optical Remote Sensing*, Advances in Optics and Optoelectronics,
https://doi.org/10.1007/978-981-99-3318-1_7

7.2 Optical Imaging Cameras

There are two main types of space cameras for the purpose of obtaining figure images, namely film photography cameras and CCD real-time transmission. Compared with the two photographic methods, the resolution of the film photography method is higher than the resolution of the CCD real-time transmission method, and the format size of the film photography camera is large. However, CCD real-time transmission cameras have real-time transmission capabilities, especially for space earth observation. Also widely used later than CCDs are CMOS detector cameras. Compared with CCD cameras, CMOS cameras have the advantages of small power consumption, simple electronics and easy integration of video processing circuits, but their responsive performance and image quality are inferior to CCD cameras. The main indicator of camera performance is the resolution of the camera, followed by the width of the image coverage.

7.2.1 Main Parameters of the Imaging Cameras

The parameters that determine the resolution of the camera and the width of the imaging coverage are the camera's focal length f, field of view angle 2ω and light aperture D. Related parameters are also the pixel size and quantum efficiency of the detector used.

1. The imaging coverage width of the camera

The camera's image coverage width is directly determined by the camera's field of view angle and the orbital height, such as Fig. 7.1. The Camera coverage width B, orbit height H, focal length f, image plane length L, field of view angle 2ω, pixel size a, ground pixel resolution r_g of a camera. The imaging coverage width of the camera

$$B = 2H \tan \omega = \frac{LH}{f} \tag{7.1}$$

2. The camera imaging resolution

Image resolution refers to the smallest scale that a camera can resolve ground scene, i.e., r_g can be seen from the Fig. 7.1.

$$r_g = \frac{aH}{f} \tag{7.2}$$

This is the most basic geometric relationship to which the resolution of the imaging camera is to be observed. In order to truly obtain such a resolution, it is necessary

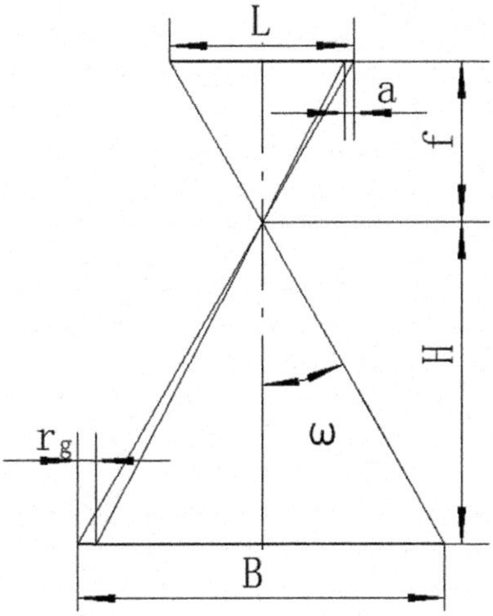

Fig. 7.1 Imaging coverage width

to meet the requirements of optical diffraction resolution, according to the Rayleigh resolution angle of the diffraction theory object space [1, 2].

$$\alpha = 1.22\frac{\lambda}{D} \tag{7.3}$$

When the opening angle of the diameter of the Airy spot relative to the principal point of the image space is taken as the angle resolution

$$\delta = 2.44\frac{\lambda}{D}$$

where

λ wavelength of incident light;
D entrance pupil diameter;
α, δ unit radian.

In addition, the resolution of the imaging camera ultimately requires the observation of the human eye, so it is also necessary to meet the conditions of the human eye observation threshold, that is, the requirements of contrast, which is expressed in brightness L,

$$\frac{\Delta L}{L} = 2\%$$

3. Signal-to-noise ratio of photoelectric camera

Space optical remote sensing development to today, space cameras are almost all photoelectric transmission cameras, so the main technical parameters of the camera, not only optical parameters requirements, but also electronic index requirements, as a camera electronic parameters of the signal-to-noise ratio is also a key technical indicator.

The reflected energy from the ground reaches the CCD through the optical system and is photoelectrically converted to the output voltage

$$V_{\text{CCD}} = \frac{\pi}{4F^2}\Delta t A(1-\xi)\cos^4\omega \int_{\lambda_1}^{\lambda_2} R(\lambda)\tau_a(\lambda)\tau_o(\lambda)L(\lambda)\text{d}\lambda \tag{7.4}$$

where

V_{CCD}	voltage value, V;
F	$F^{\#}$;
Δt	integration time, s;
A	CCD pixel area, m^2;
ξ	occlusion coefficient;
ω	camera half-field of view;
$R(\lambda)$	CCD spectral response, $\text{V J}^{-1}\ \text{m}^{-2}\ \mu\text{m}^{-1}$;
$\tau_{\text{a}}(\lambda)$	atmospheric spectral transmittance;
$\tau_{\text{o}}(\lambda)$	optical system spectral transmittance;
$L(\lambda)$	the brightness of the spectral radiation of object, $\frac{\text{W}}{\text{m}^2\text{Sr}\,\mu\text{m}}$;
$\xi = \left(\frac{D_1}{D}\right)^2$, D_1	obscuration diameter.

Signal-to-noise ratio

$$\frac{S}{N} = \frac{V_{\text{CCD}}}{V_N}$$

The V_N is the RMS noise voltage of the CCD photodetector, generally a few millivolts.

From the perspective of optical remote sensor function, the imaging camera is the simplest of these major types of optical remote sensors, but it is also the technical basis of other types of optical remote sensors, and the other major optical remote sensors are evolved and developed on the basis of imaging cameras.

7.2.2 Space Application of Imaging Cameras

Imaging cameras have two major categories: film type and CCD real-time transmission type, film-type cameras develop earlier and cameras with CCD as image receivers appear later than film-type cameras. Traditionally, the photographic spatial scale resolution ability of a film-type camera is called resolution, which is expressed as the number of line pairs of the black and white strip that can be resolved within

1 mm on the photographic film, which is recorded as lp/mm, and the corresponding ground size of a line pair is called the photographic ground resolution. CCD real-time transmission cameras often use pixel resolution or instantaneous field of view (IFOV) to measure resolution, and the corresponding ground dimension of a pixel on the receiver is called ground pixel resolution.

The focal length f of satellite reconnaissance film-type cameras is related to the satellite orbital altitude h, camera photographic resolution N and photographic ground resolution R as follows: $f = h/(\mathrm{RN})$. For example, $h = 300$ km, $R = 1$ m, $N = 100$ lp/mm, $f = 3$ m. At this time, the meaning of the photographic ground resolution of the film-type camera is that there are two strips of one black and one white distributed in this spatial scale of 1 m, and the width of each band is only 0.5 m.

If a CCD camera has a ground pixel resolution of 1 m, it means that its pixel size is projected to the ground corresponding to the ground scale of 1 m, which corresponds to one black strip or one white strip on the ground with a width of 1 m.

It can be seen that the traditional film-type camera with lm ground resolution and the CCD real-time transmission camera with a ground pixel resolution of 0.5 m have the same imaging resolution capability at the same spatial scale. It should be paid attention to the meaning of distinguishing the spatial resolution of these two cameras, or it can be agreed that for the traditional film-type camera, it is still called its resolution, and for the CCD camera, it is called resolution capability (resolvability), because the film camera has become a history, the current space camera is basically CCD camera, and the newly revised national optical standard recommends the use of resolvability, which is exactly the same as the resolution expression of the existing CCD camera mentioned here.

It is not enough to achieve a certain camera resolution only with the corresponding focal length but also needs to have a certain size of optical aperture D, and the size of the optical aperture is determined by the following two conditions. First of all, the requirements for angular resolution of an ideal optical system must be met. According to the diffraction theory of light, the ideal angular resolution α of an optical system with a wavelength of λ and an aperture of D is $\alpha = 1.22\lambda/D$. Second, it is necessary to meet the requirements of the image surface illumination. The image surface illuminance E is proportional to the relative aperture $(D/f)^2$. Therefore, according to the sensitivity of the receiver, the aperture D is appropriately selected to meet the requirements of light energy.

In short, in order to improve the resolution, the aperture of the space camera is getting bigger and bigger, such as the American KH-12 reconnaissance camera has adopted the aperture $D = 3.8$ m, reaching the ground resolution of 0.1 m.

Imaging optical cameras for military applications, that is, space reconnaissance cameras, can be divided into two categories: census cameras and detailed inspection cameras, the ground resolution of census cameras is 1–3 m, and the ground resolution of detailed inspection cameras is generally better than 0.5 m.

The field of view 2ω of the optical system determines the ground coverage area. The field of view of the camera is several degrees, and it can cover the ground more than ten kilometers, and the census camera has a larger field of view. When the field of view is large, due to the decrease in the illumination of the image plane outside the axis field, coupled with the off-axis vignetting, it is not easy to achieve the uniformity of the image plane illumination.

On the other hand, due to the image motion compensation requirements of space cameras, the lens distortion requirements are very strict (such as less than 0.1–1%), so it is difficult to correct the distortion when the field angle is large. Therefore, it is difficult to increase the field angle of space reconnaissance cameras.

The key technologies include overall optimal design of camera, development of long focal length and high stability optical lens, focal plane technology of large array CCD and imaging stability control technology.

One of the main features of space cameras is their high reliability. Unlike ground equipment, space cameras are not so much about maintainability, because once launched, they are almost impossible to repair, with the exception of the Hubble telescope (Table 7.1).

Since the Soviet Union and the United States began to launch photographic reconnaissance satellites in the 1960s, the United States has gone through the development of six generations of space optical photographic reconnaissance cameras from the KH-1 to the current KH-12 and completed the process from film return remote sensing camera to photoelectric imaging transmission remote sensing camera, and the ground resolution has also increased from the original 7.5–0.1 m. The former Soviet Union also began to develop military camera of reconnaissance satellites in the 1960s and developed from the film transmission remote sensing camera to the photoelectric transmission remote sensing camera. Its ground resolution developed from the initial 20–0.4 m at the end of the twentieth century (Table 7.2).

7.3 Multiband Optical Camera

Space multiband camera is an optical remote sensor between imaging camera and imaging spectrometer. Generally, its resolution on the ground is lower than that of imaging camera but higher than that of imaging spectrometer. Its few spectral segments are more than those of panchromatic cameras but less than those of imaging spectrometers.

7.3.1 Bands Allocation

The configuration of cameras can be divided into multiple lenses and single lenses. Multiple lenses use different lenses to shoot different spectral segments, such as Skylab in the United States, which uses multiple lenses. However, most space multiband cameras use single lens or filter.

There are many factors affecting the registration accuracy in a multi-lens multiband camera, such as inconsistent focal length, CCD length error, CCD not confocal, optical axis not parallel, distortion, stiffness and stability of the structure, influence

Table 7.1 Application of typical space remote sensors

No.	Countries	Satellite	Sensors	Numbers of spectrum	Spectral range/nm	Spatial resolution/m	Use	Schedule work time
1	America	Landsat-4–5	TM thematic cartographer	7	450–12,500	30, 120	Land use, agriculture, forestry, water resources, cartography, etc.	1982–1997
2	America	Landsat	MSS multispectral scanner	4	500–1100	79	Land use, agriculture, forestry, water resources, cartography, etc.	1988–1997
3	America	NOAA-11	HIRS/21 high-resolution infrared radiation detector	20	690–14,950	20.4	Atmospheric temperature, humidity	1988-
4	America	EOS-AM-1	MODIS medium resolution imaging spectroradiometer	36	400–14,500	250, 500, 100	Geophysical processes, atmosphere, ocean, land	1988-
5	America	EOS-AM-1	ASTER advanced space thermal radiation thermal reflection detector	14	500–900, 1600–2500, 8000–12,000	15, 30, 90	Land surface water and clouds	1996–2001
6	America	Earth Watch	Quick Bird	Multispectral 4 panchromatic	450–900	1, 4	Drawing	1996–2001
7	America	Earth Watch	Early Bird	Multispectral 4 panchromatic	500–890 450–800	3, 15	GIS, drawing	1997–2002
8	America	Orbital Science	OrbView-1	Multispectral 4 panchromatic	450–900 450–900	1, 2, 8	Drawing	1997–
9	America	Space Imaging	Space Imaging	Multispectral 4 panchromatic	450–900	1, 4		1999–
10	America	IKONOS	IKONOS-2 high-resolution imager	Multispectral 4 panchromatic	450–880 450–900	4 1	High-resolution Landsat	2002–

(continued)

Table 7.1 (continued)

No.	Countries	Satellite	Sensors	Numbers of spectrum	Spectral range/nm	Spatial resolution/m	Use	Schedule work time
11	America	EOS	HIRIS high-resolution imager	192	400–2450	30	The biosphere	2000–
12	America	Earth Observing-1	EO-1 hyperspectral imager	220	400–2500	30	Wide ground cover	1995–1998
13	Russia	PRIRODA-1	ISTOK-1 infrared multispectral radiometer system	64	400–16,000	750–3000	Atmospheric radiation	1995–1998
14	Russia	PRIRODA-1	MSU-E infrared multispectral radiometer system	3	500–900	20		1996–1998
15	Russia	ALMAZ-IB	MSU-SK optical multispectral scanner	5	500–1100 400–12,600	80 300		1982–1997

Table 7.2 Visible light detection camera

No.	Country	Type	Age	Orbit altitude km	Orbit inclination°	Working life	Dimensions/m	Ground resolution	Note
1	America	KH-8	1966–1981	400–480	110	7–90 days	8 (1.5)	0.3–0.2	The film camera
2	America	KH-9	1971–1986	260–274	96.4	40–275 days	15 (3)	0.15	The film camera
3	America	KH-11	1976–1982	530,1020	96.9	2–3 years	14–15 (3)	0.15–0.1	Photoelectric detailed investigation
4	America	KH-12	1992–	287–1020	97.8	About 3 years	15 (4.5)	0.1	Photoelectric detailed investigation
5	The former Soviet union	The first generation	1962–1976					18–24	The film camera
6	The former Soviet union	The second generation	1966–1970					6–8, 0.9–1.5	The film camera
7	The former Soviet union	The third generation	1968–1978					1.5–2.4,0.9–1.5	The film camera
8	The former Soviet union	The fourth generation	1975–1982					0.4–0.9	The film camera
9	Russia	The fifth generation	1982–1989			295 days		0.4–0.9	Photoelectric imaging
10	France	Sun God	1995–	600–700		4–5 years			Photoelectric imaging

of satellite attitude, instrument error (parallel light tube, objective lens, microscope, monitor, etc.) and installation error. Therefore, multiband registration is a key technology in multi-lens and multiband cameras. Multi-lens configuration is mostly used in small multiband CCD cameras with low ground resolution and short focal length. The advantage of multi-lens configuration system is that the optical design is easy to realize, but the disadvantage is that the volume is large, and the registration is difficult.

Single-lens splitting which is set up behind the lens spectroscopic system will receive the light decomposition into a number of spectral segments, and this is the current CCD multiband camera more adopted method. The main advantages of single-lens splitting system are small size, compact structure and easy image registration. The disadvantage is that the optical design is difficult due to the spectral band broad. If prism splitting is used, the transmission of the wide spectrum of the optical system will be limited. If multicolor CCD device is used, the spectrum setting will be greatly limited. Figure 7.2 shows a CCD multiband detector which has common the channels of red, green, blue, near infrared and panchromatic.

It can be divided into broadband or narrow-band system according to the spectral bandwidth. The fundamental difference of spectral sensitivity between narrow-band system and wideband system lies in whether the channels overlap or not. Currently, narrow-band system is mostly used in space cameras. The common narrow-band system channel bandwidth $\Delta\lambda$ and channel number n satisfy the following relations

$$S_{\max} - S_{\min} = \Delta\lambda \cdot n$$

where $S_{\max} - S_{\min}$, the spectral response range of multiband camera.

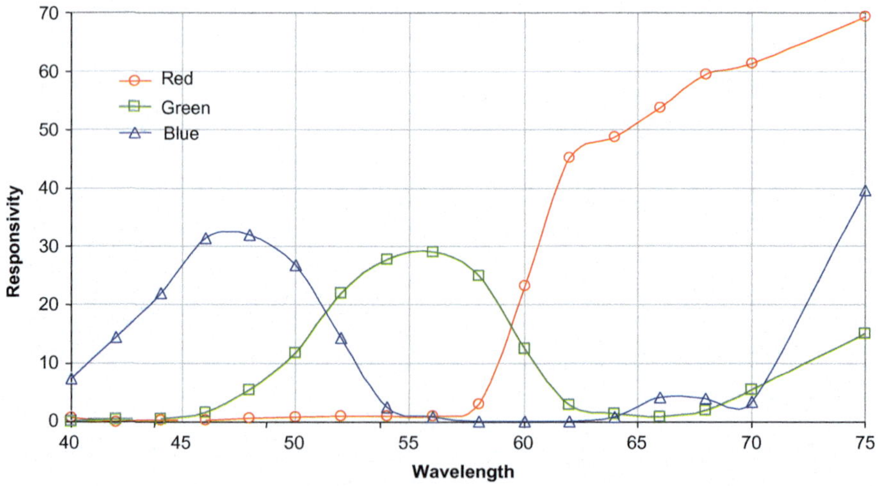

Fig. 7.2 Multiband CCD spectral response curve

Each channel does not overlap in narrow-band system, so there is no crosstalk between different channels. For traditional three-color imaging, signal crosstalk between different channels reduces chroma and is not conducive to the expansion of gamut. This is the theoretical explanation that the narrow-band spectral response sensitivity is almost used in traditional color image reproduction. However, Munsell laboratory of Rochester Institute of Secret Service in New York, America has obtained satisfactory spectral estimation by using broadband mode to obtain multiband images.

According to the type of CCD, it can be divided into monochromatic and multicolor system. An actual multiband camera is typically a combination of monochrome and multicolor CCD. Multicolor CCD can be realized by absorption filter or narrow-band interference filter. There is also a liquid crystal electronic adjustable color filter using polarization principle to achieve channel switching, electronic control to improve the reliability of the acquisition system. Liquid crystal adjustable color filters and interference color filters belong to narrow-band filters. In spectral image acquisition, spectral information is often obtained by equal spectral interval.

Because the number of filters used by multiband cameras is limited by space, the spectral bandwidth is limited, and the spectral transmission energy is small, the spectral transmission wavelength will be offset with the angle of the field of view, and there is an error in image registration, these factors often deteriorate the quality of the obtained spectral images. Multiband camera should pay attention to the compromise of the above parameters and reduce the aliasing of the obtained spectra as much as possible to achieve better spectral reflectance reconstruction accuracy. The essence of multiband imaging is to record and reconstruct the initial spectral properties of objects. Multiband imaging system usually includes several modules such as multichannel spectral image, spectral reflectance reconstruction and spectral analysis pattern recognition.

7.3.2 Applications of Multiband Camera

It is widely used in geological research, water resources utilization, vegetation protection, land monitoring and geographic information system. It has become an important means of obtaining information in many civil fields, such as resource census, disaster reduction and prevention, pollution monitoring and urban planning. It can be used in coastal zone dynamic mapping and ocean water color monitoring to obtain medium resolution images of sea-land interaction areas and serve for marine environmental monitoring and resource development.

Table 7.3 shows the main parameters of civil optical remote sensors launched at home and abroad after 2000. The imaging mode of multiband cameras is push-sweep, while foreign cameras are reflection.

From the current development situation, multiband camera is rarely used alone and is basically used in combination with panchromatic camera or stereo mapping camera.

Table 7.3 Multiband camera

Satellite	EO-1	QuickBird	QuickBird-2	Orbview-2	ZY-2(CBERS)
Imaging method	Push-sweep	Push-sweep	Push-sweep	Push-sweep	Push-sweep
Panchromatic band/m	10	1	0.61	1	20
Multiband/m	30	4	4	4	20
Panchromatic band/μm	0.48–0.69	0.45–0.9	0.45–0.90	–	0.51–0.73
Multiband/μm	0.45–0.52 0.52–0.60 0.63–0.69 0.775–0.805	0.45–0.52 0.52–0.60 0.63–0.69 0.76–0.90	0.45–0.52 0.52–0.60 0.63–0.69 0.76–0.90	0.45–0.52 0.52–0.60 0.63–0.69 0.76–0.90	0.45–0.52 0.52–0.59 0.63–0.69 0.77–0.89

7.4 Mapping Optical Cameras

Mapping camera is basically the same as imaging camera from the form of optical system, but it is very different from imaging camera in the purpose of application. Imaging camera solves the problem of seeing and meets the need of seeing visually. Imaging itself is similar from object to image, as long as this similar relationship cannot feel the difference from the vision. But mapping is to solve the problem of measurement, it not only needs to see, but also requires the accuracy of measurement, in essence, mapping camera is a metering equipment.

7.4.1 Fundamental Theory of Mapping Cameras

In general, the space mapping camera requires a lower ground resolution than the space reconnaissance camera, so the focal length is shorter, such as $F = 300$–400 mm, but the field of view angle is large, such as $2\omega = 60°$–$70°$. However, with the development of technology, the photographic resolution has been gradually improved and is now at the level of several meters. In addition, it is required to calibrate the principal distance, principal point position, distortion and intersection angle strictly. At the same time, through the star camera and other means to the camera's external orientation elements, that is, the center of photography in the ground or geocentric coordinate system position and attitude angle requirements strict determination.

Space photography is quite effective for ground mapping of vast and hard-to-reach areas. The space system to be developed should technically meet the requirements of the compiled maps as far as possible. Any map contains information about three types of ground targets:

1. Plane position;
2. Height;

3. Content (completeness).

The first type of information, plane position, establishes a relationship between the objects shown on the map and a reference point on the earth's surface. However, in areas where there is no ground reference data, the plane position of the target can also be obtained if the position and direction of the camera at the moment of image are known.

The second type of information, height, is usually shown on a mapping map as a horizontal line that gives a cross-section of the terrain. Height data can be obtained from stereoscopic photographs. The initial height position is obtained by relating it to the height of the known punctuation marks. As with plane position, height data can be obtained without reference points, provided that the position of the photographic instrument and its pointing are known; at the moment, the camera is exposed.

The third kind of information and the content of the map, with all the elements and objects of the area being photographed, are clearly and credibly displayed on the map of the appropriate scale. The completeness of the content mainly lies in the resolution of the photos used in object interpretation and the existing supplementary drawing information.

Requirements for resolution of photographs. The basic indexes of quality evaluation for map drawing by photos are the completeness of map content and the probability of image recognition. In order to find the basic targets in the observation area, the resolution of 3.0 m is sufficient, but in order to identify in more detail, it must have the resolution of 1–1.5 m. A resolution of a few tenths of a meter is required to display the characteristics of a geographical target. In any case, the completeness of map content is a probabilistic parameter that depends on the characteristics of the region itself. Statistical studies show that acceptable content completeness can be obtained only when the resolution is 1.5–2.0 m for 1:50,000 scale maps of moderately complex areas and 1.0–1.5 m for 1:25,000 scale maps.

The mapping principle of stereo mapping camera is shown in Fig. 7.3 [3]. Let S_1 be the forward-looking camera, S_2 is the forward-looking, face-looking and rear-looking-looking camera, B be the baseline length, and H be the orbital height. In order to facilitate derivation, the projection positive image is used. $f_1 = S_1O_1$ is the focal length of the forward-looking camera, $f_2 = S_2O_2$ is the focal length of the forward-looking, face-looking and rear-looking-looking camera. In stereoscopic mapping cameras, $f_1 = f_2$ is generally set as f.

A and C are two ground object points with elevation difference h, and their image points on the forward-looking camera and the rear-looking camera are a_1, c_1 and a_2, c_2, respectively. If the coordinate value of a_1 to the principal point o_1 of the forward-looking camera is x_{a1}, then the coordinate value of a_2 to the principal point o_2 of the forward-looking, face-looking and rear-looking-looking camera is $-x_{a2}$. It is something like x_{c1} and $-x_{c2}$. So the parallax between A and C is

$$\begin{aligned} p_a &= x_{a1} - (-x_{a2}) \\ &= x_{a1} + x_{a2} \\ p_c &= x_{c1} - (-x_{c2}) \end{aligned}$$

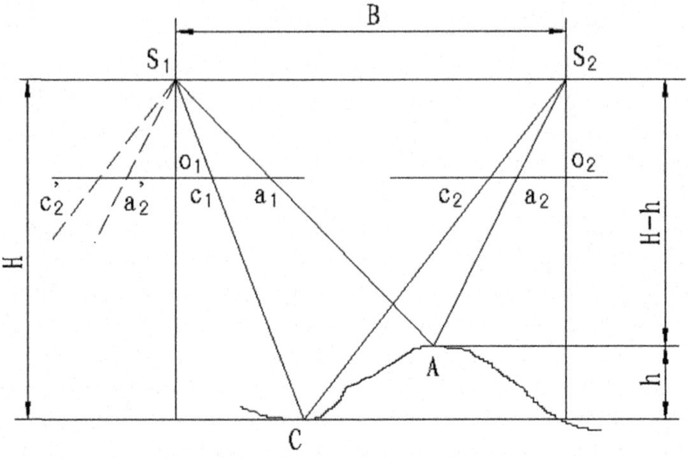

Fig. 7.3 Mapping principle of space optics

$$= x_{c1} + x_{c2}$$

Translate s_2, a_2, c_2 to s_1, a_2', c_2', then

$$\begin{aligned} p_a &= o_1a_1 + o_1a_2' \\ &= a_1a_2' \end{aligned}$$

$$\begin{aligned} p_c &= o_1c_1 + o_1c_2' \\ &= c_1c_2' \end{aligned}$$

$$p_a = \frac{f}{H-h}B$$

$$p_c = \frac{f}{H}B$$

The parallax difference between A and C is

$$\begin{aligned} \Delta p &= p_a - p_c \\ &= \frac{Bf}{H-h} - \frac{Bf}{H} \\ &= \frac{Bf}{H} \cdot \frac{h}{H-h} \end{aligned} \tag{7.5}$$

$$h = \frac{\Delta p \cdot H^2}{\Delta p H + Bf} \tag{7.6}$$

From Eq. 7.6, it can be known that the parallax difference of two points on the image plane is Δp, the orbital height H and the focal length f of the camera are known, and the baseline length $B = 2H \cdot \tan(\alpha/2)$ is known from the intersection angle of the camera. This method of calculating the baseline length B ignores the curvature of the earth. Therefore, the height difference of A and C points is obtained. It can be seen that the elevation difference between them can be known as long as the parallax difference of any two points on the pictures of two cameras is known, and then, the elevation mapping of topographic map can be realized. This is the principle of stereo mapping of optical remote sensing camera. This implies the prerequisite that the camera's viewing axis is perpendicular to the ground and also implies the requirement of high-precision pointing and high stability of satellite attitude.

7.4.2 Stereo Mapping by Satellite Swing

The French SPOT satellite uses the swing to map adjacent orbits in stereo. Of course, this method as shown in Fig. 7.4 has many factors that affect the accuracy of mapping and has great influence.

The American IKONOS satellite achieved stereo mapping by swinging back and forth, with horizontal accuracy of 12 m and elevation accuracy of 10 m, when there was no ground control point. IKNOS2 has a resolution of 1 m, oblique photography resolution of 1.5 m and elevation mapping accuracy of 10 m. But this kind of swing mapping accuracy is low. Our country also realized stereo mapping by using film camera.

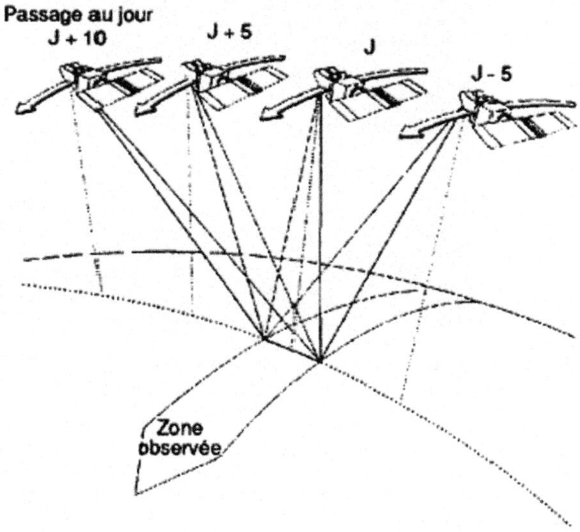

Fig. 7.4 Stereo mapping by satellite swing

7.4.3 Single-Lens Multi-Line Array Stereo Mapping

Since the focal length of space cameras is generally long, and the field of view required by this single-lens multi-line array method is large (about 45°), this lens is difficult to manufacture. Aerial cameras generally have a short focal length, which can be used in this way.

Single-lens multiarray remote sensors are applied to the high-resolution stereo mapping camera (HRSC) and wide-field stereo scanner (WAOSS) for Mars observations made by ESA.

Figure 7.5 is the schematic diagram of stereoscopic and multiband (color) imaging of HRSC, where white lines are used for stereoscopic imaging and color lines for multiband imaging. The resolution of the ground pixel at the orbit altitude of 250 km is 10 m for the face-looking line array, 20 m for the forward-looking line array and 20 m for the rear-looking line array, and the size of the CCD pixel is 7 μm; the focal length is 175 mm (Fig. 7.6).

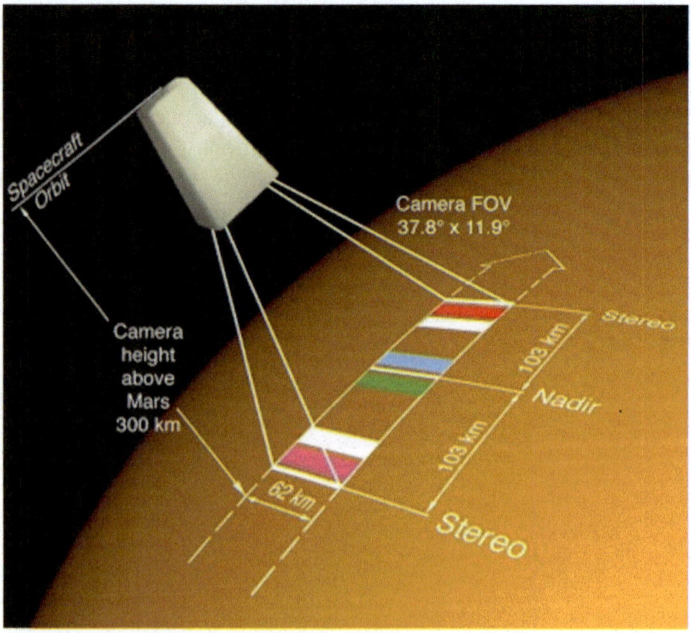

Fig. 7.5 Single-lens multi-line array stereo mapping

(a) (b)

Fig. 7.6 HRSC stereoscopic cameras and multiband (color) maps

7.4.4 Multiple Single-Line Array Lens Combination Mode

Due to the short focal length of single-lens and multi-line array stereo mapping cameras, the mapping accuracy is low when used for space stereo mapping. Therefore, cameras used for space stereo mapping generally combine multiple single-line array lenses into stereo mapping remote sensor for stereo mapping, thus generating multiple single-line array lens combination mode. At present, there are mainly two camera combination and three camera combination.

1. Double camera combination mode

It may be conventionally called a double-line array. India's cosmos-3 M launch vehicle was launched on 27 October 2005. The camera on IRS-P5 (CartoSAT-1) adopted this double-line array stereo mapping method, as shown in Fig. 7.7, with a ground resolution of 2.5 m and an imaging width of 30 km. The two cameras were placed on the satellite platform, and its visual axis is $+ 26°$ and $- 5°$ from the substellar point, respectively, for stereoscopic imaging.

2. Three-line array mode

Three-line array is a combination of three cameras. In this way, three single-line array lenses are used for combined stereo mapping, and symmetrical configuration is generally adopted. A is the forward-looking lens, B is the face-looking lens, and C is the rear-looking lens, as shown in Fig. 7.8. The forward-looking lens and the rear-looking lens are arranged symmetrically with respect to the face-looking lens.

Due to the high orbit of space stereo mapping, it is necessary to use long focal length to obtain high-precision stereo mapping images. In this case, it is not appropriate to use single lens. Multi-lens combination method is mostly used, while three-line array method is the majority. The remote sensors for MAPSAT and STEREOSAT that the United States planned to launch the satellites used this imaging method.

PRISM, a three-line array stereo mapping camera developed by Japan, has a ground pixel resolution of 2.5 m. Figure 7.9a is the image of the camera's appearance

(a) (b)

Fig. 7.7 IRS-P5 double-line stereo mapping

Fig. 7.8 Three-line array stereo mapping

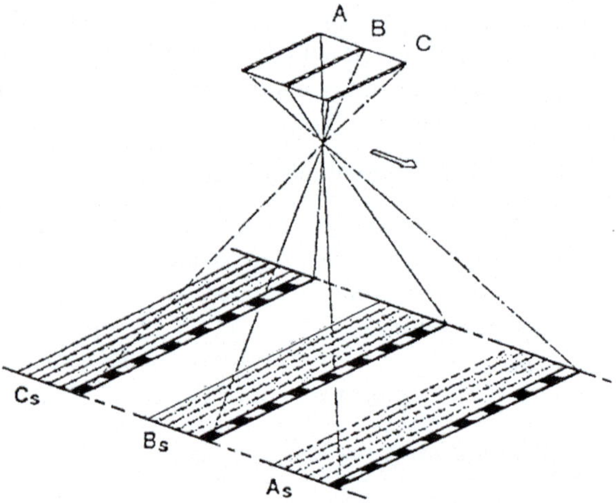

and measurement parameters, Fig. 7.9b is the image of the camera's in-orbit work, and Fig. 7.9c is the image of the camera's in-orbit measurement results. The red and blue colors in Fig. 7.9c represent parallax in stereo mapping.

The general development trend of space stereo mapping is wide cover and high precision. Space stereo mapping is still a relatively new subject, and there are few stereo mapping cameras compared with reconnaissance cameras. In order to seek its development trend, in addition to the way of space stereo mapping and the general direction and principle can be seen some clues, there are many specific problems to be further studied. However, it is good that space reconnaissance has a history of more

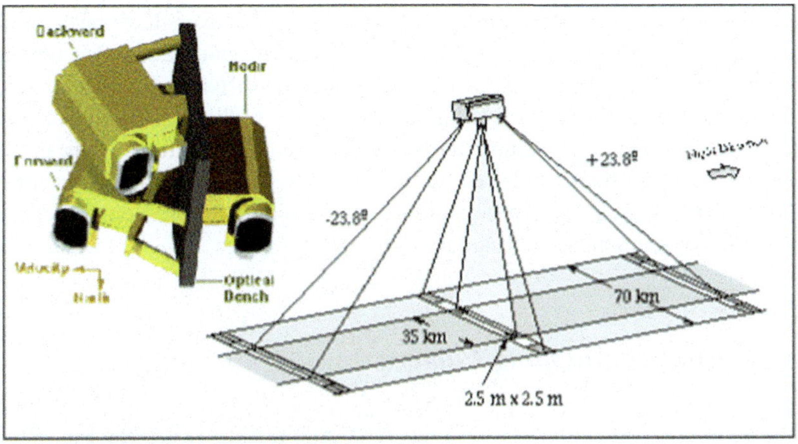

(a)

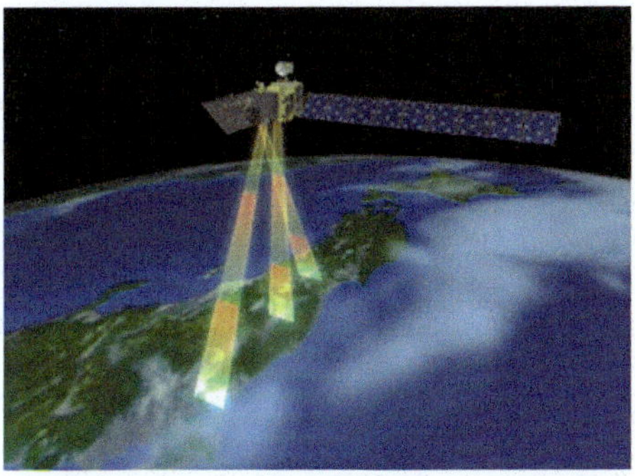

(b)

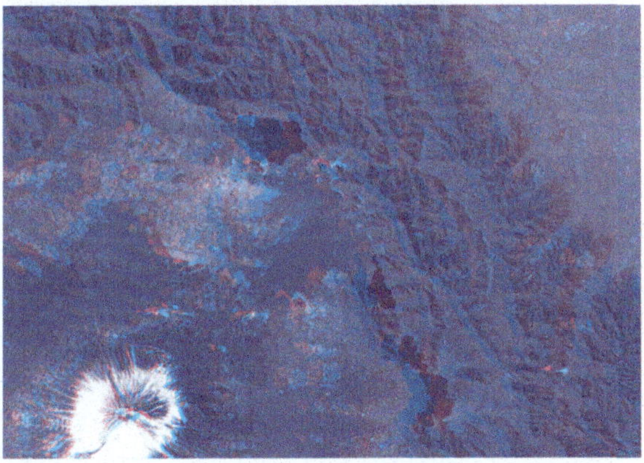

(c)

Fig. 7.9 PRISM three-line array stereo mapping

than half a century, and a large number of space remote sensors have been launched into the space. Although space stereo mapping is different from space reconnaissance, except in the image post-processing, they are different in the manufacture of remote sensors, and space stereo mapping requires higher geometric accuracy of remote sensor image, focal length and several lenses combination. There are many similarities between them, so it can be further studied and solved the special problems in the development of space stereo mapping camera based on some specific technologies of reconnaissance camera.

7.4.5 Testing of Mapping Camera

Inner orientation elements testing Fig. 7.10 and testing principle Fig. 7.11. In Fig. 7.10, 1 is self-collimating parallel light tube, 2 is tested camera, 3 is reading microscope, and 4 is optical goniometer.

Before the test, a grid plate was placed at the focal plane of the camera, and the center point of the grid plate was adjusted to coincide with the master point of auto-collimation. Then, the camera is installed on the bracket of the optical goniometer, the mutual position of the camera and the rotation axis of the optical goniometer are adjusted, and the camera is rotated around the rotation axis of the optical goniometer to make the self-collimating parallel light tubes reach the state of self-collimation. The angle reading corresponding to the dial is recorded as the angle coordinate of the master point of self-collimation. Finally, the precise angle measurement method is used to calculate the data. Test other fields of view by rotating the turntable that holds the camera bracket.

The form of distortion is shown in Fig. 6.10, which can be measured and corrected by grid method, etc.

With the development of remote sensing technology, space stereo mapping as one of the fields of space remote sensing has also been developed, its stereo mapping

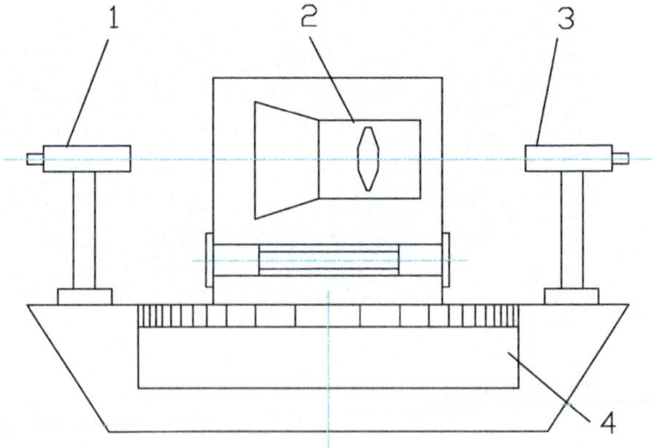

Fig. 7.10 Inner orientation elements testing diagram

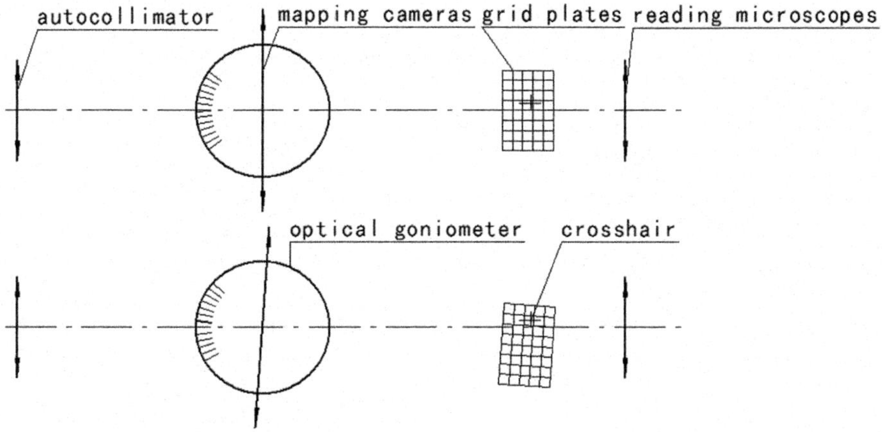

Fig. 7.11 Testing principle of inner orientation elements

accuracy is also improving, as the main way of space stereo mapping three-line array stereo camera, its volume, and weight is also increasing. The focal length of the main parameters of the three-line array stereo mapping camera has developed from tens of millimeters to hundreds of millimeters to several meters today. Cameras are 100 times heavier and 1000 times bigger in volume.

The increase of volume and weight leads to the increase of difficulty in the manufacture of three-line array stereo mapping camera, and the improvement of precision of space stereo mapping camera requires the improvement of design and manufacture precision of three-line array stereo mapping camera. The large three-line array stereo mapping camera is generally composed of three linear array push-sweep cameras, so its design and manufacture have the problems of the position accuracy between the three linear array cameras and its maintenance and accuracy measurement, which is different from the imaging camera, so the mapping camera not only has higher requirements on distortion and stability of optical machine structure but also has higher requirements than the imaging camera.

7.4.6 Space Applications of Mapping Cameras

Table 7.4 lists the space stereo mapping satellites working in orbit, in which the focal length of the mapping camera has developed from 61.8 to 2000 mm, increasing by more than 30 times. China's Ziyuan-3 satellite is the first civilian high spatial resolution stereo mapping satellite. Its three-line array stereo mapping camera has the spatial resolution of 2.1 m in face-looking camera, 3.5 m in forward-looking camera and the rear-looking camera. The spatial resolution of multiband camera is 5.8 m [4]. Table 7.5 lists the main parameters of Ziyuan-3 satellite stereo mapping camera.

Table 7.4 Space stereo mapping satellites in orbit

The name of the instrument/country	MEOSS/German	IRS-P5/India	PRISM/Japan	SPOT-5/France
Orbital altitude/km	400	618	691.65	822
Focal length/mm	61.6	1980	2000	580
Field of view/°	7.6	(2.78)	7.6	4
Ground coverage/km	(53)	27	35	(50)
Ground pixel resolution/m	52 * 80	2.5	2.5	10
Intersection angle/°	23.6	+ 26/− 5	23.8	19.2
Spectral coverage/μm Note: multispectral discrete instrument	0.57–0.72	0.5–0.7	0.52–0.77 0.42–0.50 0.52–0.60 0.61–0.69 0.76–0.89	0.48–0.7 0.5–0.61 0.61–0.68 0.78–0,89 1.58–1.75
Volume/m^3		1.5 × 0.85 × 1	2.0 × 1.0 × 1	1 × 1.3 × 0.4
Launch time	In the 80 s	2005, 5	2006, 1	2000, 5

Table 7.5 Main parameters of Ziyuan-3 satellite stereo mapping camera

Camera	Spectral range/μm	Spatial resolution (m)	Width (km)	Revisit period (day)	Coverage cycle (day)
Forward-looking	0.50–0.80	3.5	52	3–5	59
Rear-looking	0.50–0.80	3.5	52	3–5	59
Face-looking	0.50–0.80	2.1	51	3–5	59
Multiband camera	0.45–0.52 0.52–0.59 0.63–0.69 0.77–0.89	5.8	51	5	59

7.5 Infrared Optical Camera

Military reconnaissance, monitoring, surveying and mapping, industrial material inspection, medical diagnosis and scientific research, etc., often require infrared system to give a detailed and clear thermal image of a field of view range of scenery, and this camera to receive infrared radiation of imaged object is called infrared optical camera. Vehicles, tanks, warships and aircraft that are active on the ground, water and air all have engines, and they often form strong infrared radiation at high temperatures. Even objects at normal temperature, including human bodies, are constantly radiating infrared rays. In short, objects with temperatures higher than absolute zero will emit infrared rays. This is because the atoms and molecules of the constituent substances are constantly moving, and the changes in their motion state form infrared radiation. The intensity and spectral composition of radiation depend on the temperature of the radiator. The higher the temperature is, the shorter the peak wavelength of radiation is. Infrared cameras generate infrared images by receiving infrared radiation.

7.5.1 System Composition of Infrared Camera

With the rapid development of optic-electronic technology and technological level, staring focal plane detectors have been applied more and more in the field of space, but due to cost, technology and other reasons, staring focal plane detectors are often difficult to achieve the required high-resolution and wide-field angle, so most thermal imaging systems have scanning link. Figure 7.12 shows the composition of an infrared imaging system.

There are three types of infrared imaging methods:

1. Optical mechanical scanning imaging;
2. Electron beam scanning imaging;
3. Imaging by plane array detector.

Optic-mechanical scanning imaging is easy to understand, that is, mechanical motion points the optic axis of the optical system at each point of the observed picture, similar to the electron gun scanning of cathode-ray tube of the television.

Electron beam scanning imaging is to image the whole viewing area of the scene on the target surface of the Vidicon, and then, the electron beam is used to scan along the target surface to identify the image signal and display the visible thermal image of the scene through the display device.

Plane array detector imaging is to use the large array detector to image the scene and then read the signals of each pixel serial, also known as staring imaging. Infrared imaging is based on the temperature difference and emissivity difference of various parts of the scene and forms the temperature distribution of the scene, namely the thermal image. It is different from visible light imaging, which depends on the reflection difference imaging of scenery.

Optic-machine scanning imaging has been widely used and is still used today. The content to be discussed here is the working principle of infrared imaging system, scanning mechanism, camera mode and so on.

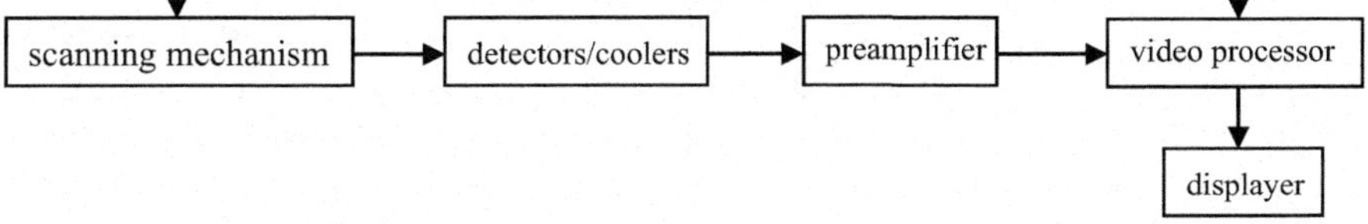

Fig. 7.12 Composition of infrared imaging system

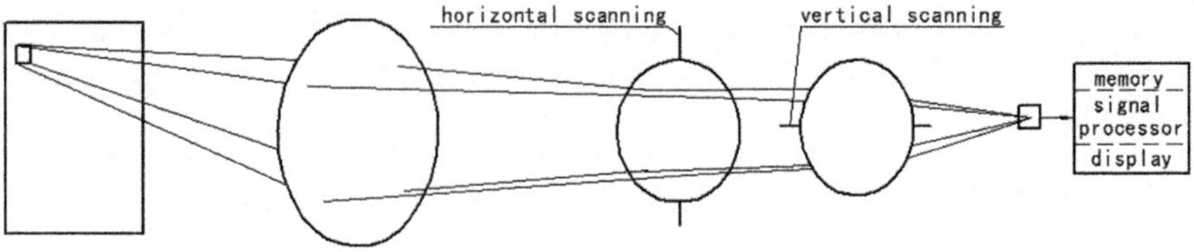

Fig. 7.13 Composition of scanning imaging system

The optical system collects the infrared radiation of the target and background and converges into the focal plane containing the detector. The optical scanning mechanism is placed between the optical system and the detector. The simple scanning mechanism consists of two scanning mirror groups, one for horizontal scanning and the other for vertical scanning, as shown in Fig. 7.13. From left to right is the scenery, instantaneous field of view of the optical system, optical system, horizontal scanning mirror, vertical scanning mirror, detector, storage and display circuit for subsequent processing.

When the scanning mirror moves, the line of sight from the detector to the scene also moves, scanning the scene space into a TV-like grating. The incident radiation of the scene is converted into electrical signals by the detector. Along with the scanning, the two-dimensional scene is converted into one-dimensional analog voltage, which is amplified and processed by electronics, and the visible light image is displayed by the display.

In order to improve efficiency, multidetector parallel scanning is often used. Other systems scan infrared images in real time as quickly as a television. The system was originally used in what is called the forward-looking infrared system carried by the aircraft.

7.5.2 Basic Parameters of Infrared Camera

The basic parameters of the infrared camera are

1. Clear aperture D and focal length f

Clear aperture and focal length are key factors that determine the performance and volume of infrared cameras. Compared with visible cameras, an outstanding feature of infrared cameras is that $F^{\#}$ is small, because infrared radiation is much weaker than visible light. See Chap. 4 for relevant information.

2. Instantaneous field of view ω

Instantaneous field of view is the range of objects that can be observed by an optical system (including a scanning mechanism) when it is stationary. It is determined by the size of the photographic surface (pixel) of the detector and the focal length of the optical system. The size of the instantaneous field of view determines the spatial resolution of the imaging system. The smaller the instantaneous field of view is, the higher the spatial resolution is.

The pixel of the general detector is rectangular, the size is $a \times b$, and the instantaneous field of view plane angle is α, β, then

$$\alpha = \frac{a}{f}$$

$$\beta = \frac{b}{f}$$

$$\omega = \alpha \times \beta$$

The units of α and β are radians or milliradians.

3. The total field of view Ω

$$\Omega = A \times B$$

A and B are mutually perpendicular plane angles of the field of view, in radians. The size of the scene space observed in the general viewing field depends on the focal length of the optical system, the size of the detector and the scanning range.

4. Frame time and frame rate

The time required by the system to scan a complete frame is frame time. The number of frames swept in a second is the frame rate, which is the reciprocal of frame time. In the case of limited scanning speed, interlaced scanning can be adopted, as shown in Fig. 7.14, so that the frame rate is half of the field speed.

5. Scan efficiency η

When the scanning mechanism is scanning the scene, the actual space swept is often larger than the solid angle of the scene. The ratio of the time required by the system to scan the space angle of the complete scene and the time required by the scanning mechanism to scan one cycle is called the scanning efficiency. If the efficiency of horizontal scanning and vertical scanning is η_H and η_V, respectively, then

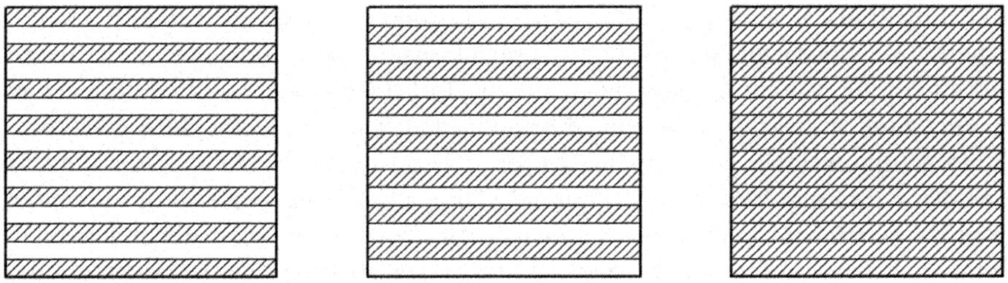

Fig. 7.14 Interlaced scanning imaging

$$\eta = \eta_H \eta_V$$

6. Dwell time t

The time that the instantaneous field of view of the system scans through the single pixel detector.

If the scene space swept by the scanning mechanism is $\Omega = A \times B$, and the instantaneous field angle corresponding to the unit detector is $\omega = \alpha \times \beta$, then the pixel number of the whole scene N

$$N = \frac{\Omega}{\omega} = \frac{AB}{\alpha\beta}$$

Set T for scanning frame time, efficiency η for efficiency and dwell time t_1 for single pixel detector, then

$$t_1 = \frac{T\eta}{N} = \frac{\omega T\eta}{\Omega} = \frac{\alpha\beta T\eta}{AB}$$

If the detector is a n-element parallel detector, the dwell time t is

$$t = nt_1 = \frac{n\alpha\beta T\eta}{AB}$$

The dwell time of the detector should be greater than the time constant of the detector.

7.5.3 Scanning Mode

There are two kinds of scanning methods in infrared system, object scanning and image scanning.

Object scanning is a scanning mechanism placed in front of the focused optical system to scan the radiation from the scene directly. Because the light from the scene is parallel light, also known as parallel light scanning. Figures 7.15 and 7.16 belong to parallel optical scanning. In Fig. 7.15, (a) is one-dimensional scanning imaging, where 1 is a plane swing mirror, 2 is an optical system, and 3 is a detector. (b), (c) and (d) all belong to two-dimensional scanning. In Fig. 7.15b, 1 is a two-dimensional plane swing mirror, 2 is an optical system, and 3 is a detector. In Fig. 7.15c, 1 is a one-dimensional plane swing mirror, 2 is an optical system, 3 is a detector, and 4 is a rotating drum mirror. Figure 7.15d is a scanner composed of two optical wedges, which has a complex two-dimensional scanning trajectory depending on the rotational speed and initial phase of the two optical wedges (see the next section).

When the swing angle of the reflector is large, the aperture of the reflector will be large. To reduce the size of the swing mirror, a telescopic system can be set in

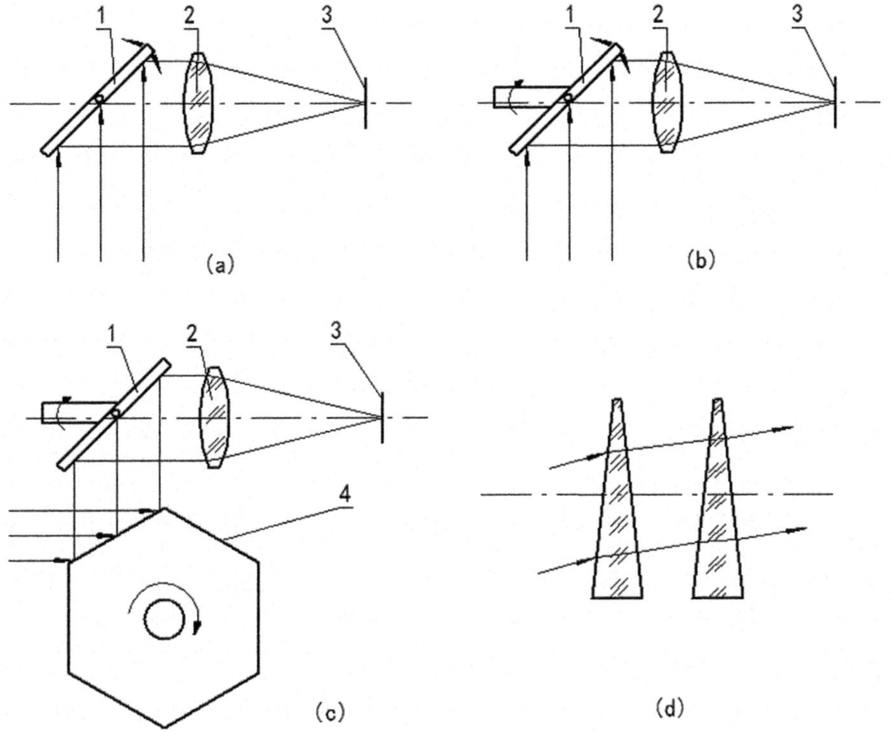

Fig. 7.15 Several parallel light scanning imaging methods

Fig. 7.16 Compression aperture scanning

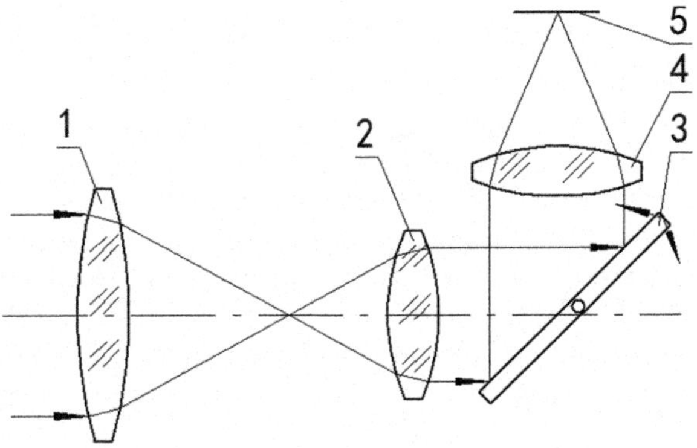

front of the swing mirror, as shown in Fig. 7.16. In the figure, the lens 1 and 2 constitute the telescopic system, 3 is the scanning swing mirror, 4 is the imaging optical system, and 5 is the detector. The telescopic system is added before the object scanning mechanism to form the front-placed telescope system. In this way, the radiation from the scene becomes parallel light after being compressed by the

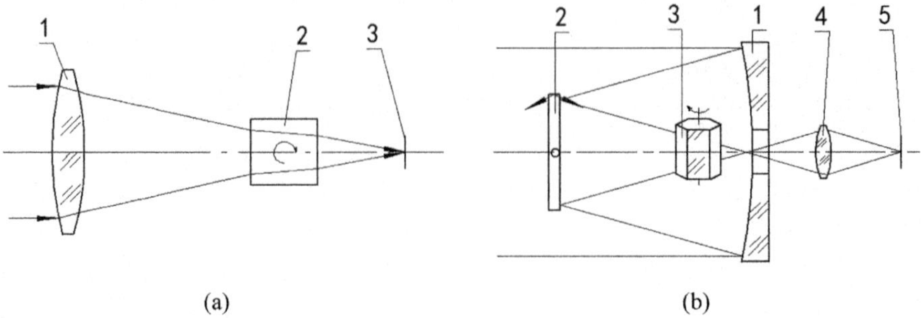

Fig. 7.17 Convergent beam scanning

front-placed telescope system, which is scanned by the scanning mirror and then focused to the detector by the optical system. This scanning mode still belongs to the object scanning.

Image scanning is also called convergent beam scanning, in which the scanning mechanism is placed between the focusing optical system and the detector to scan the image beam, also known as convergent beam scanning. Figure 7.17 is the schematic diagram of two-dimensional image space scanning cases. Figure 7.17a is one-dimensional scanning of rotating prism, 1 is optical system, 2 is prism, and 3 is detector. Figure 7.17b shows the two-dimensional scanning combined with the swing mirror and rotating drum mirror. 1 is the primary mirror, 2 is the swing mirror, 3 is the drum mirror, 4 is the imaging lens, and 5 is the detector.

7.5.4 Common Scanning Mechanism

There are usually several scanning mechanisms as follows: plane swing mirror, rotating drum mirror, rotating refraction prism and rotating optical wedge. The basic requirement of the scanning mechanism is that the angle of the scanning mechanism should be linear with that of the visual axis of object space. When scanning mechanism is scanning, it has little influence on system aberration. The scanning efficiency should be as high as possible, and the scanning parts should be as small and compact as possible in order to reduce the moment of inertia. The following are the main working characteristics of several scanning mechanisms.

1. Plane swing mirror

For a plane swing mirror, if the direction of the incident ray or the outgoing ray remains unchanged, the outgoing ray or the incoming ray will be rotated by 2γ angles when the mirror rotates γ angle about the axis perpendicular to the incident ray and the outgoing ray (Fig. 7.18).

When the plane swing mirror is rotated on an axis parallel to the incident light, the relationship is one times.

Fig. 7.18 Planes wing mirrors canning principle

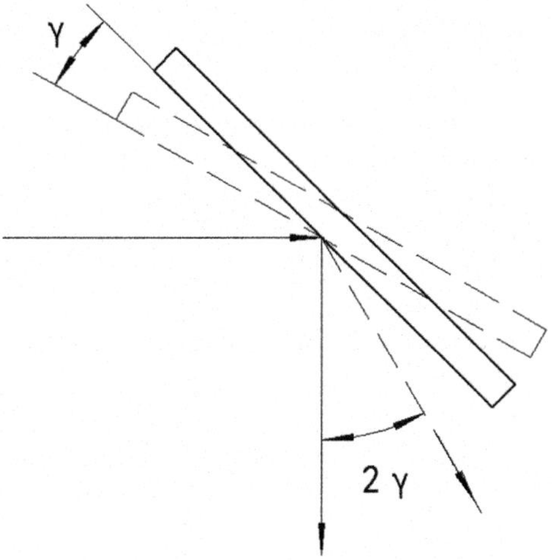

In parallel beam scanning with a planar swing mirror, the light beam passes through the mirror and is emitted in parallel. The linear relationship between the angle of the swing mirror and the angle of the outgoing light beam is realized. As shown in Fig. 7.19, the aperture D of the optical system, γ is the swing angle of the plane swing mirror, and the required mirror length L is changing, and when the scanning angle γ is larger, the mirror length will become larger.

The initial state of the swing mirror is 45° with the optical axis, and the length of the swing mirror is

Fig. 7.19 Scanning characteristics of plane swing mirror mechanism

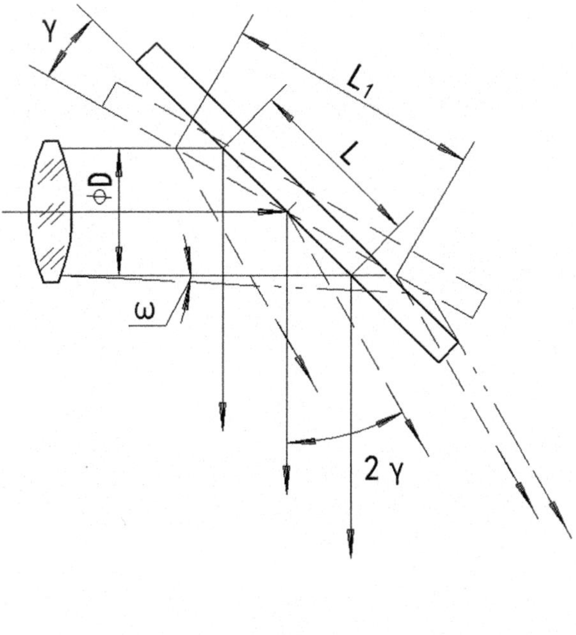

$$L = \frac{D}{\sin 45} = \sqrt{2}D$$

After the γ angle of the swing mirror, the angle of the swing mirror and the optical axis is 45°$- 2\gamma$, the length of the swing mirror is

$$L = \frac{D}{\sin(45 - 2\gamma)}$$

Set $\gamma = 10°$, then the image scanning angle is 20°, and the length of the swing mirror is

$$L = \frac{D}{\sin(45 - 2\gamma)} = \frac{D}{\sin 25} = 2.37D$$

It can be seen that the length of the swing mirror increases rapidly. The optical system always has a certain field of view and should consider the angle of view 2ω, which has a greater impact on the increase of the length of the swing mirror. Figure 7.19 shows the increase in the length of the swing mirror caused by the intersection of the double dot line of ω and the dotted line of the swing mirror. In addition, if the rotation axis of the swing mirror is not on the mirror shown in the figure, the effect of the offset of the mirror caused by the swing of the swing mirror should also be considered. When this effect is combined with the double dotted line of the angle of field ω, the effect on the length of the swing mirror should also be paid attention to, and it will also produce the phenomenon of ray drift from the visual axis. But the amount of movement is not large for space optical remote sensing or can be ignored.

When the plane swing mirror is scanning the convergent beam, the incident beam is convergent through the optical system and images on the detector after passing through the mirror. The relationship between the scanning angle θ and the plane swing mirror before and after the γ angle is shown in Fig. 7.20.

$$\tan\theta = \frac{y}{a+z}$$
$$y = b\sin 2\gamma$$
$$z = b\cos 2\gamma$$

$$\tan\theta = \frac{b\sin 2\gamma}{a + b\cos 2\gamma}$$
$$\theta = \arctan\frac{b\sin 2\gamma}{a + b\cos 2\gamma}$$

When θ and γ are small, approximately

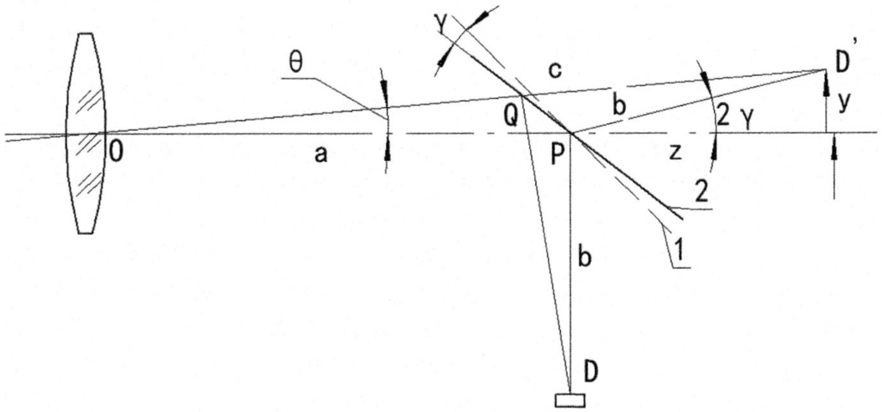

Fig. 7.20 Scanning characteristics of plane swing mirror in image space

$$\theta \approx \frac{2b\gamma}{a+b}$$
$$\gamma \approx \frac{a+b}{2b}\theta$$

There is an approximate linear relationship between the scanning angle θ and the rotation angle γ.

Now let's look at aberrations.

When the swing is at position 1, the ray goes OPD, and when the swing is at position 2, the ray goes OQD

$$c < a + b = f$$

It is shown that when the main light is deviated from the optical axis, the light path from the lens to the detector becomes shorter, defocusing produces dispersion and increases aberration. When the light is deflected θ angle, the optical path length is

$$c^2 = a^2 + b^2 - 2ab\cos(\pi - 2\gamma)$$

That is

$$\begin{aligned} c^2 &= a^2 + b^2 - 2ab\cos\left(\pi - \theta - \arcsin\frac{a\sin\theta}{b}\right) \\ &= a^2 + b^2 + 2ab\cos\left(\theta + \arcsin\frac{a\sin\theta}{b}\right) \end{aligned}$$

In the actual system, when the aberration is not corrected, the image surface is a curved surface, which is called Petzval surface. It can be seen from Fig. 7.20 that the change of c with θ is often inconsistent with the curvature of the system field curvature, thus dispersing and affecting the image quality, as shown in Fig. 7.21. Aberration correction design should be considered in optical design for such systems.

With the help of Fig. 7.22, the scanning defocus of light rays on the axis of a flat focal plane is discussed. Segment A is shared by the deviated light and the undeviated light, the mirror turns γ angle, the ray deviates by 2γ, when γ is less

$$\begin{aligned}\varepsilon &= b' - b \\ &= \frac{b}{\cos 2\gamma} - b \\ &= b(\sec 2\gamma - 1) \\ &\approx 2b\gamma^2\end{aligned}$$

$$\varepsilon \approx \frac{(a+b)^2\theta^2}{2b}$$

The final diameter d of the diffusion circle can be obtained from the triangle shown in the figure.

$$d = (f - c)\frac{D}{f} = \varepsilon\frac{D}{f} = \frac{\varepsilon}{F^{\#}}$$

Curved focal surfaces are designed horizontally to compensate for scanning defocus, and flat focal surfaces are designed vertically to use flat detector arrays.

(1) Close object scanning of the swing mirror

Since the object is not at infinity, it is not scanned by parallel beams. As shown in Fig. 7.23, the swing mirror is rotated from position 1 to position 2 at angle γ

$$s = 2B\gamma$$

Fig. 7.21 Defocus of the image scan in image space

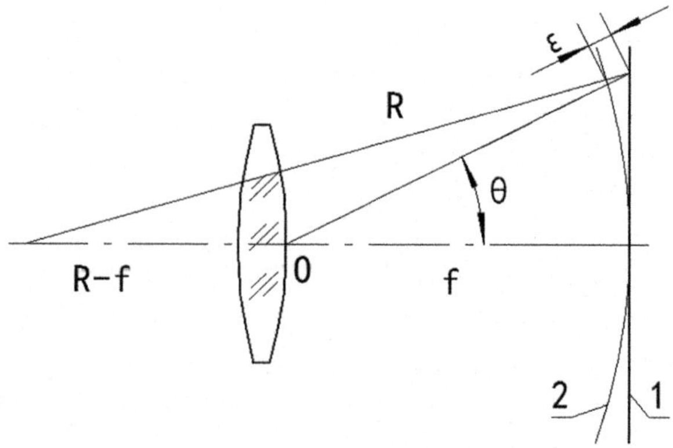

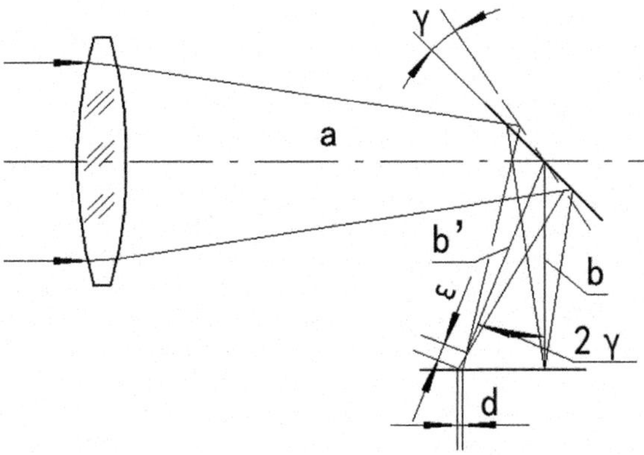

Fig. 7.22 Aberration of the image scan in image space

$$\theta = \frac{s}{OP} = \frac{2B\gamma}{OP} = \frac{2\gamma(OP - A)}{OP} = 2\gamma\left(1 - \frac{A}{OP}\right)$$

(2) Scanning efficiency of the swing mirror

It is determined by the ratio of the observed space angle to the swing angle of the swing mirror. The swing angle of the swing mirror can be designed according to the space angle needed to be observed, so that the empty range is as small as possible. The mirror is generally only used in the forward scanning, so the scanning efficiency is less than 50%, taking into account the inertia, generally only 40%. It can also improve the return speed or phase compensation of the video signal to take advantage of the return scan. Due to inertia, the swing speed of the swing mirror cannot be too large.

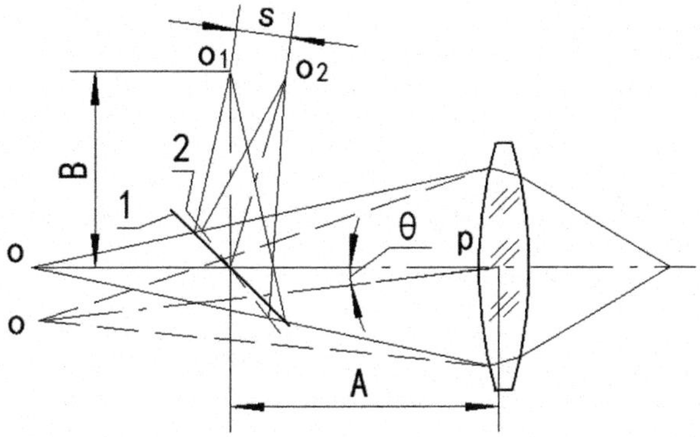

Fig. 7.23 Close object scanning of the swing mirror in object space

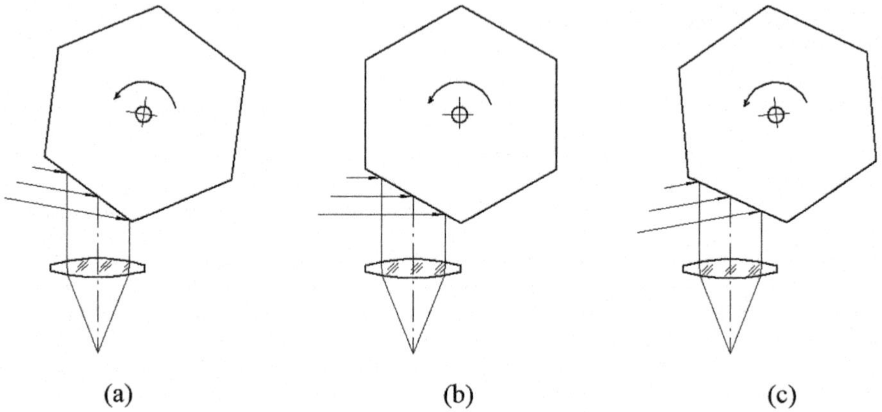

Fig. 7.24 Drum mirror scanning principle

At high speed, the edge of the field of view becomes unstable, and increasing the motor power is required, so it is not suitable for high speed, generally used for frame scanning.

2. Rotating drum mirror

In the case of high-speed scanning, the rotating drum mirror as shown is often used because the movement of the drum mirror is continuous and stable (Fig. 7.24).

The rotating drum mirror and the swing mirror have the same working conditions, and the rotation angle and aberration are basically the same, but the drum mirror rotates around the central axis, and the mirror position has displacement relative to the light. Here are the main parameters:

(1) Mirror width L

As shown in Fig. 7.25, suppose there are n faces

$$\theta_f = \frac{2\pi}{n}$$

$$L = 2r_0 \sin\frac{\theta_f}{2}$$

$$\frac{r_i}{r_0} = \cos\frac{\theta_f}{2}$$

(2) Displacement of mirror when drum mirror rotates

As shown in Fig. 7.26

$$\begin{aligned}\delta &= r_i(1 - \cos\gamma) \\ &= r_0(1 - \cos\gamma)\cos\frac{\theta_f}{2}\end{aligned}$$

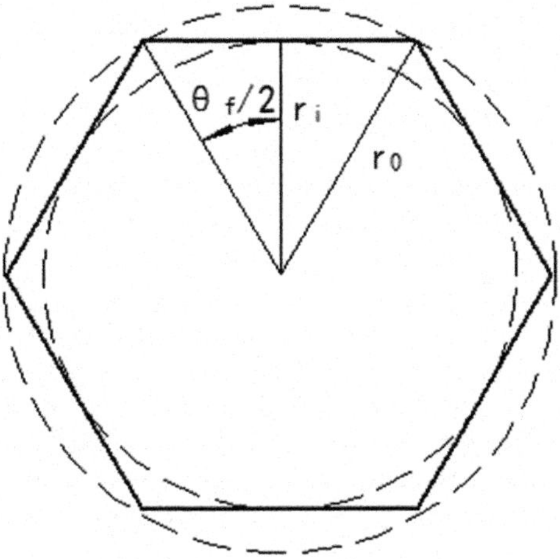

Fig. 7.25 Drum mirror scanning parameters

(3) The minimum radius of drum mirror without vignetting

Because the mirror surface position moves when the drum mirror rotates, the mirror displacement may make the incident light at the edge of the scanning area not all enter the field of view, resulting in vignetting. Therefore, the radius r_o of the mirror drum must be greater than a certain minimum value.

$$r_0 = \frac{D_0}{2\cos\theta \sin\frac{\theta_f - -\gamma}{2}}$$

$$2\gamma = 2w$$

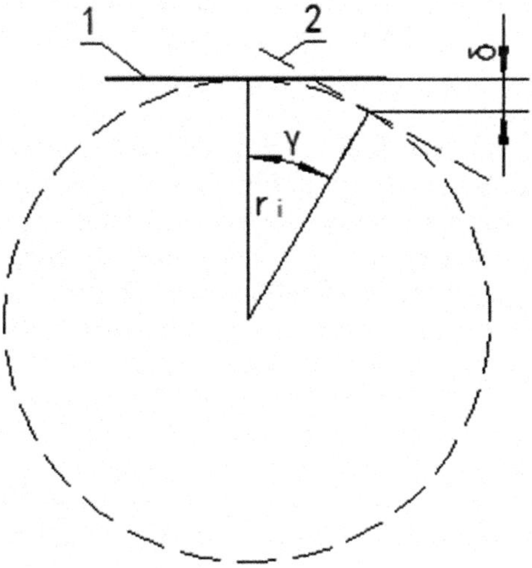

Fig. 7.26 Displacement of the surface of the drum mirror by scanning

D_0 incident beam width, $2w$ field angle, θ incident angle of incident ray to mirror surface.

(4) Scanning efficiency

The scanning efficiency of the rotating drum mirror is directly related to the mirror angle θ_f

$$\eta = \frac{w}{\theta_f} = \frac{\gamma}{\theta_f}$$

It can be seen that the number of mirror surfaces should be increased to improve efficiency, and the drum radius must be increased to ensure the width of incident light. Efficiency, beam and radius are mutually restricted. γ and θ_f cannot be too different; otherwise, it will affect the efficiency.

(5) Limit speed of drum mirror

Limited by the strength of the material cannot be too large

$$\omega_{\max} = \frac{1}{2\pi r_0}\sqrt{\frac{8T}{\rho(3+\eta)}}$$

$\omega_{\max}$ Drum mirror maximum rotation speed
ρ Material density
T Material tensile strength of drum mirror
η Poisson ratio of materials
r_0 Drum mirror radius.

This is only to consider the strength limit of the material, in fact, before reaching the limit, it cannot work because of the mirror deformation, and the analysis of the mirror deformation can use the finite element method.

3. Rotating refraction prism

The prism is a regular prism with N-planes which rotates around the center of the prism, as shown in Fig. 7.27. Rotating refraction prism is only used for scanning convergent beam, and its main characteristics are the relationship between refraction prism rotating angle and ray rotating angle and scanning efficiency.

Longitudinal movement Y and transverse movement Z [5].

Only one chief ray is drawn in Fig. 7.28, refractive index of prism n, rotating angle γ.

$$\frac{a+b}{t} = \tan(\varphi_1 - \gamma)$$

$$a = t\tan(\varphi_2 - \gamma)$$

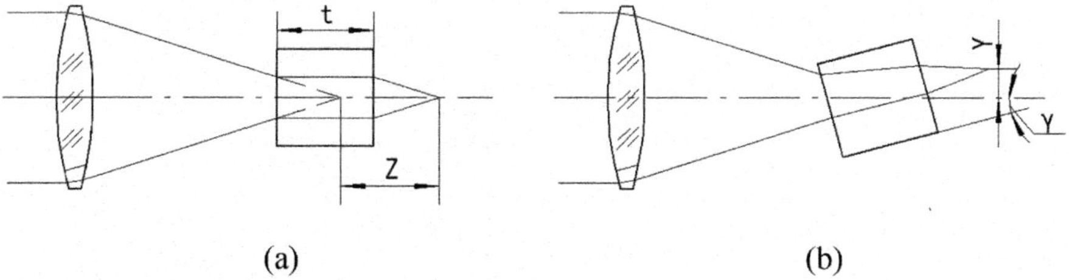

Fig. 7.27 Principle of the prism scanning

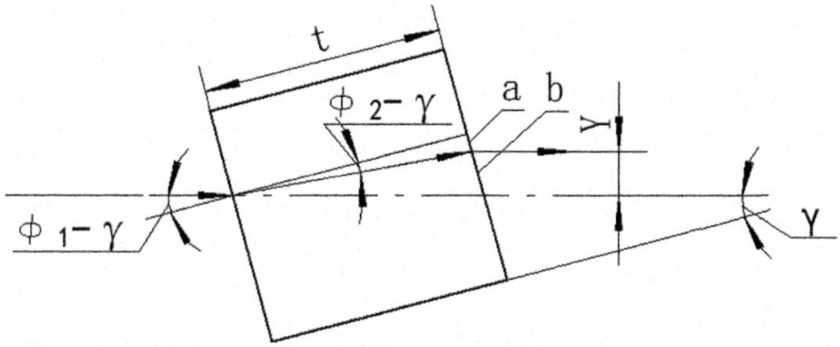

Fig. 7.28 Movement *Y*

$$b = t[\tan(\varphi_1 - \gamma) - \tan(\varphi_2 - \gamma)]$$

According to the law of refraction

$$n \sin(\varphi_2 - \gamma) = \sin(\varphi_1 - \gamma)$$

$$\mathrm{tg}(\varphi_2 - \gamma) = \frac{\sin(\varphi_2 - \gamma)}{\sqrt{1 - \sin^2(\varphi_2 - \gamma)}}$$

$$\begin{aligned} Y &= b\cos\gamma \\ &= t\cos\gamma[\tan(\varphi_1 - \lambda) - \tan(\varphi_2 - \gamma)] \\ &= t\cos\gamma\left[\frac{\sin(\varphi_1 - \lambda)}{\cos(\varphi_1 - \lambda)} - \frac{\sin(\varphi_1 - \lambda)}{\sqrt{n^2 - \sin^2(\varphi_1 - \lambda)}}\right] \end{aligned}$$

When $\varphi_1 - \gamma$ and γ are less

$$\begin{aligned} Y &\approx \left(\varphi_1 - \gamma - \frac{\varphi_1 - \gamma}{n}\right) t\cos\gamma \\ &= t(\varphi_1 - \gamma)\frac{n-1}{n} \end{aligned}$$

Fig. 7.29 Displacement Z

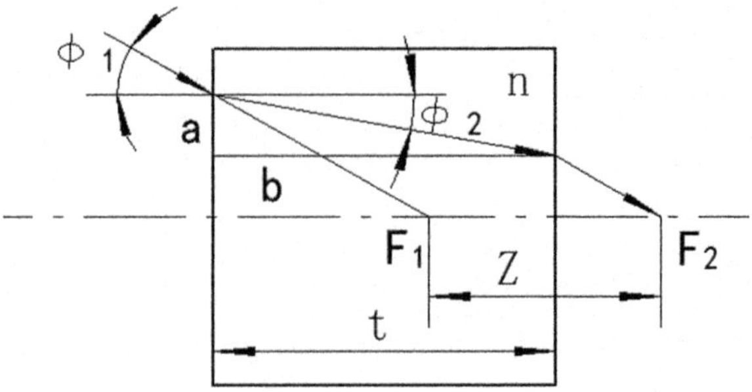

It can be seen that in a small angle range, the rotation of the prism produces approximately linear scanning.

For paraxial rays, $\varphi_1 = 0$

$$Y = -t\left(1 - \frac{\cos\gamma}{\sqrt{n^2 - \sin^2\gamma}}\right)\sin\gamma$$

For paraxial rays, the rate of change of Y with the rotating angle γ of the prism

$$\frac{\mathrm{d}Y}{\mathrm{d}\gamma} = -t\left(\cos\gamma - \frac{\cos 2\gamma}{\sqrt{n^2 - \sin^2\gamma}} - \frac{\sin^2 2\gamma}{4(n^2 - \sin^2\gamma)^{3/2}}\right)$$

It can be seen that when the angle of the prism is general, the relationship between Y and γ is complex and non-linear.

The relationship between Z and γ.

As shown in Fig. 7.29, again only one chief ray is drawn.

As shown in figure,

$$Z = t - b$$

$$b = \frac{a}{\tan\varphi_1}$$

$$a = t\tan\varphi_2$$

$$\sin\varphi_2 = \frac{\sin\varphi_1}{n}$$

Then,

$$\tan\varphi_2 = \frac{\sin\varphi_2}{\sqrt{1 - \sin^2\varphi_2}}$$

$$= \frac{\sin \varphi_1}{\sqrt{n^2 - \sin^2 \varphi_1}}$$

So

$$b = t \frac{\cos \varphi_1}{\sqrt{n^2 - \sin^2 \varphi_1}}$$

$$Z = t\left(1 - \frac{\cos \varphi_1}{\sqrt{n^2 - \sin^2 \varphi_1}}\right)$$

It can be seen that the focal point moves away from the lens axially as the obliquity of the ray increases.

When the prism is rotated at γ angle, the paraxial ray $\varphi_1 \approx 0$, and the displacement Z is

$$Z = t\left(\cos \gamma - \frac{\cos 2\gamma}{\sqrt{n^2 - \sin^2 \gamma}} - \frac{\sin^2 2\gamma}{4(n^2 - \sin^2 \gamma)^{3/2}}\right)$$

When the light passes through the prism, it shifts the focus vertically and horizontally and will cause additional spherical aberration, coma, astigmatism and chromatic aberration. Therefore, special attention must be paid to the problem of aberration in optical system. Rotating refraction prism scanning is used in concentrating light path, the size can be smaller than rotating drum mirror, which is beneficial to improve speed and reduce noise.

The scanning efficiency of the rotating refraction prism is

$$\eta = \frac{2\gamma}{2\pi/n}$$

$\gamma = f\ (2w, f, t)$, prism rotate angle is a function of field of view $2w$, focal length f and prism thickness t. Y can be get by tracing the optical path. The rotation refraction prism can obtain high scanning efficiency.

4. Rotating optical wedge [6]

When the apex angle A of refraction wedge is small, it can be used for parallel light scanning. Serious aberration can be produced if used for convergent path scanning.

When the incident angle i_1 is very small and the optical wedge is in the air, the deviation angle δ of the light is

$$\delta = n\left(i_1' + i_2\right) - A = (n - 1)A$$

Evolve the optical wedge of (a) in Fig. 7.30 into a rectangular optical wedge and establish the coordinate system is with its right-angled edges as X-axis and z-axis and the other side as Y-axis, as shown in Fig. 7.30b. The deflection vector $\boldsymbol{\delta}$ is constructed from the value size and the deviation direction of the deviation angle δ. Rotate the optical wedge of Fig. 7.30b with axis X, the angular velocity ω and the initial phase φ; then, the components of the vector $\boldsymbol{\delta}$ on the y and z axes are

$$\delta_y = (n-1)A\cos(\omega t + \varphi)$$
$$\delta_z = (n-1)A\sin(\omega t + \varphi)$$

From this set of deflection angle equations, the direction of the outgoing ray at any time t can be obtained.

When two identical refraction wedges are scanned in a group, the angular velocity ω_1, ω_2 and the initial phase φ, then

$$\delta_y = (n-1)A[\cos\omega_1 t + \cos(\omega_2 t + \varphi)]$$
$$\delta_z = (n-1)A[\sin\omega_1 t + \sin(\omega_2 t + \varphi)]$$

When the refraction optical wedge scanning system in Fig. 7.31 is used for scanning imaging, the optical axis scanning pattern can be formed by setting ω_1, ω_2 and φ. Setting $\varphi = 0$ and adjusting ω_1 and ω_2 to form the scanning trajectory shown in Fig. 7.32.

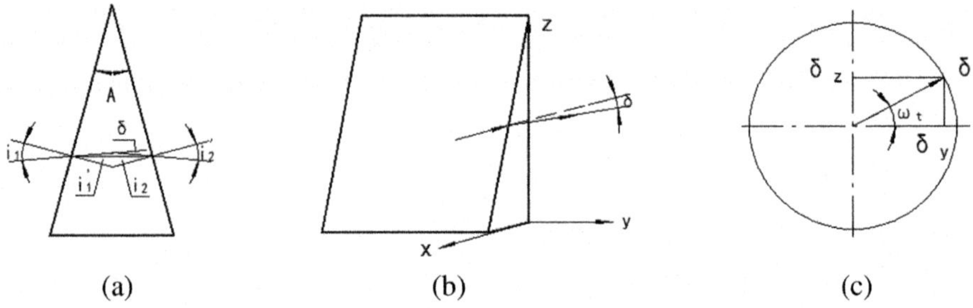

Fig. 7.30 Refracted wedge scanning

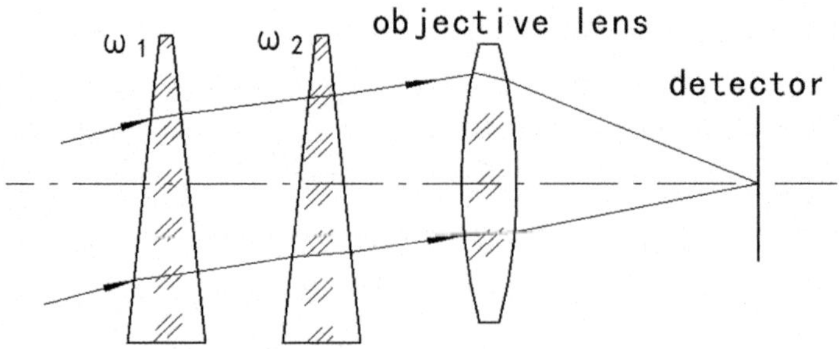

Fig. 7.31 Optical wedge scanning imaging system ($\varphi = 0$)

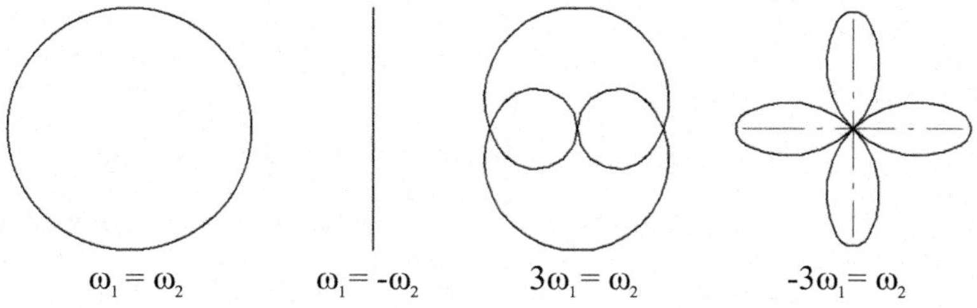

Fig. 7.32 Refracted light wedge scanning trajectory

7.5.5 Several Optical Machine Scanning Schemes

The scanning mechanism is usually composed of a swinging mirror, a drum mirror, a prism, etc. Several optical machine scanning schemes are introduced below.

1. Drum mirror for row scanning and swing mirror for frame scanning

That shown in Fig. 7.33 is such a scanning mechanism for parallel light scanning. Drum mirror 1 is a hexagonal prism, 2 for swing mirrors, 3 for objective mirrors and 4 for detectors. The basic size is determined by the beam width D_0 and the field of view 2ω.

2. Parallel light scanning of the front-placed telescope system

The front-placed telescope system can compress the beam and enlarge the field of view and can also be used as a zoom link to change the field of view.

As shown in Fig. 7.34, there are two types: The objective lens group and the collimating lens group are both positive lenses or one positive and one negative and the focus of the two-mirror groups coincide.

Magnification of telescopic system is

Fig. 7.33 Drum mirror and swing mirror scanning system

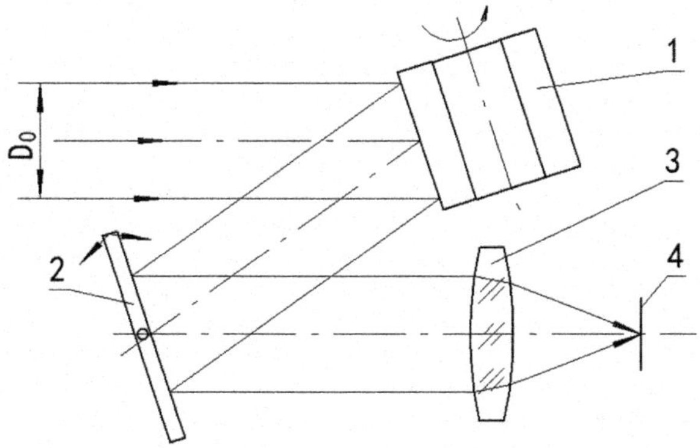

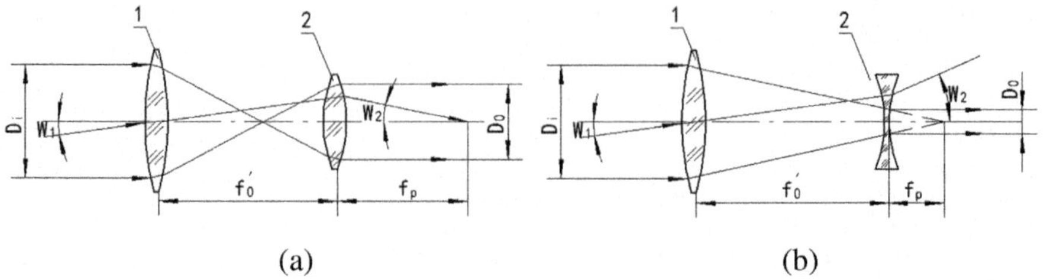

Fig. 7.34 Inverted and positive image front-placed telescope system

$$K = \frac{f_0}{f_p}$$

$$\frac{D_i}{D_0} = \frac{f_0}{f_p} = K$$

$$D_0 = \frac{D_i}{K}$$

$$\frac{W_1}{W_2} = \frac{f_p}{f_0} = \frac{1}{K}$$

$$W_2 = K W_1$$

The imaging beam can be compressed into small beams through the front-placed telescope system, so as to achieve high speed by using small drum mirror and small mechanism.

In Fig. 7.35, (a) is the inverted front-placed telescope system, and (b) is the positive front-placed telescope system. They reduce the size of the subsequent scanning mechanism by compressing the aperture of the through-light, which is conducive to reducing the inertia of the scanning mechanism and improving the scanning speed.

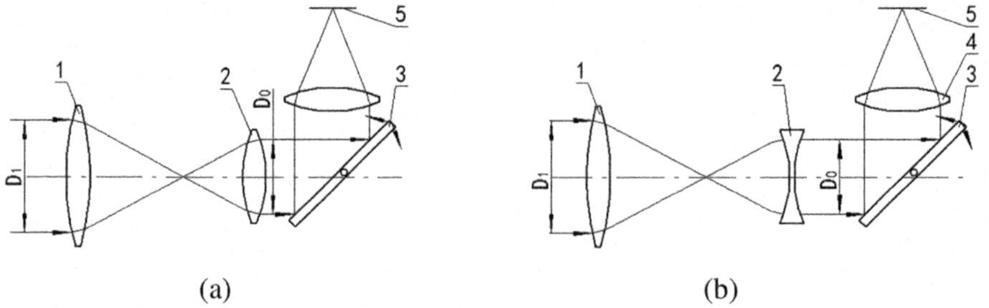

Fig. 7.35 Aperture compression of the front-placed telescope system

3. Swing mirror for frame scanning in parallel light path and drum mirror for row scanning in convergent light path

As shown in Fig. 7.36, the swing mirror is placed between the objective lens group and the collimating lens group for frame scanning and drum mirror for row scanning in the parallel light path. The scanning efficiency of this scheme is the same as that of the parallel beam scanning scheme with a front-placed telescope system. When the field of view is large, the image quality deteriorates, so it is not suitable for multipixel scanning with large field of view.

4. Prism for frame scanning and parallel beam drum mirror for row scanning

As shown in Fig. 7.37, the quadrilateral prism is placed in the middle of the light path of the front-placed optical system for frame scanning, and the drum mirror of the parallel beam is swept in row, so that a relatively stable high speed can be obtained. If the number of prism surfaces is less, the thickness can be reduced.

5. Two-prism scanning

In Fig. 7.38, there is the typical structure of this system, including 1 for objective lens set, 2 for frame scanning refraction prisms, 3 for row scanning refraction prisms, 4 for relay optical system and 5 for detector. The frame scanning is performed before the row scanning. They both use eight-prism make the vertical and horizontal fields of view of the same quality. The beam in the vertical field of view is very close to the optical axis after passing through the frame scanning prism. The row scanning prism can be thinner and smaller in size, which is conductive to improve the scanning speed. The advantages of the system are high scanning efficiency and fast speed, but the image quality design is difficult.

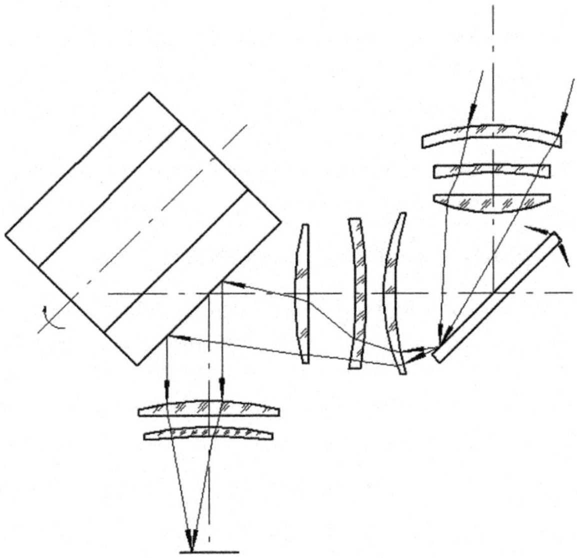

Fig. 7.36 Swing mirror for frame scanning in convergent light path and drum mirror for row scanning in parallel light path

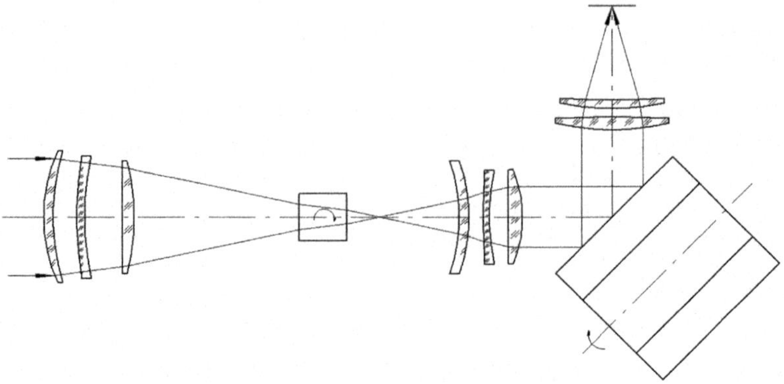

Figure 7.37 Prism for frame scanning and parallel beam drum mirror for row scanning

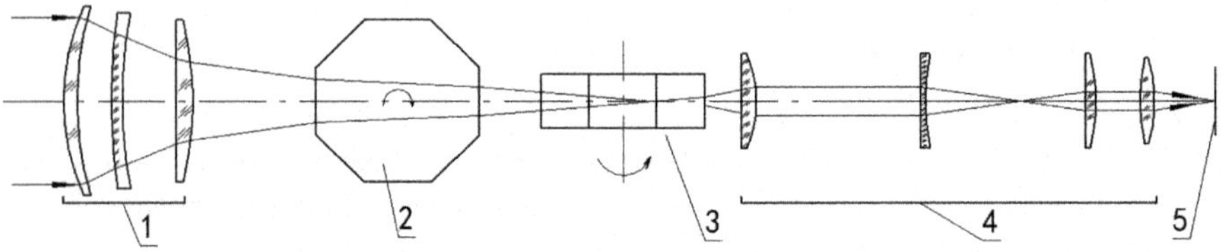

Fig. 7.38 Two-prism scanning

7.5.6 Camera Mode

Since the sensitivity and response speed of single pixel infrared detector are not high, the number of detector pixels can be increased to form an array to improve the SNR, thus forming two scanning imaging modes, serial scanning and parallel scanning. As shown in Fig. 7.39, (a) is the multipixel parallel scanning mode, and (b) is the multipixel serial scanning mode, through which the image (c) is obtained. However, Fig. 7.39c is also an imaging method, which is used to carry out frame imaging on the target with two-dimensional array detector (area array detector) containing the same pixels obtained by scanning Fig. 7.39c, which is called staring imaging in infrared.

1. **Parallel scanning camera mode** [5]

Parallel scanning is to form dozens to hundreds of detectors into a row, as shown in Fig. 7.39a. Each pixel of the detection scans one line, and n-pixels of the detection can realize wide-frame mn scanning through m times of rows scanning.

The SNR of the single pixel detector at parallel scanning

$$U_s = PRG$$

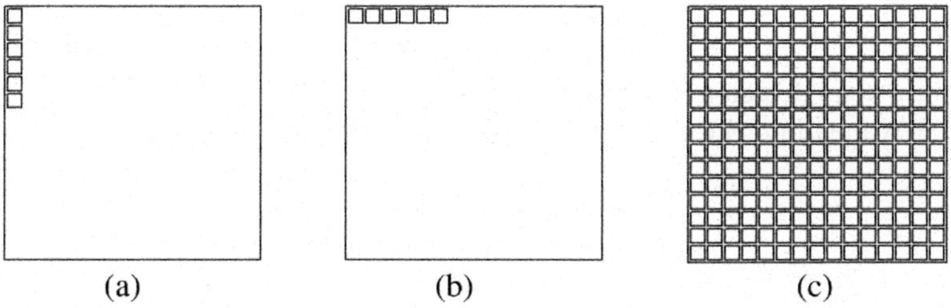

Fig. 7.39 Camera modes

where

U_s output voltage of single pixel detector;
P the radiated power incident to the single pixel detector;
R responsivity of unit detector;
G gain of the preamplifier.

$$U_n = \frac{RG\sqrt{A_d \Delta f}}{D^*}$$

where

U_n root mean square noise voltage output by single pixel detector;
D^* normalized detectivity of single pixel detector;
A_d pixel area of the detector;
Δf the equivalent noise bandwidth of the system.

$$\Delta f = \frac{\pi}{4t_1}$$

where

t_1 single pixel detector dwell time.

$$\frac{U_s}{U_n} = \frac{PD*}{\sqrt{A_d \Delta f}}$$

The SNR of single pixel detector at n-pixels parallel scanning

$$t_d = nt_1$$

where

t_1 single pixel detector dwell time;
t_d dwell time of detectors at n-pixels parallel scanning.

$$\Delta f_d = \frac{\Delta f}{n}$$

Then, the SNR of single pixel detector at n-pixels parallel scanning

$$\frac{U_s}{U_n} = \sqrt{n}\frac{PD^*}{\sqrt{A_d \Delta f}}$$

Thus, the SNR of n-pixels parallel scanning detector is $\sqrt{n}$ times higher than that of single pixel scanning detector.

2. **Serial scanning camera mode** [5]

String multiple pixels in a row, as shown in Fig. 7.40, 1 for the time delay line, 2 for the preamplifier, 3 for the single pixel detector, 4 for the delay direction and 5 for the scanning direction.

The detector scans the target in the direction of the scan; after delayed by the delay line, the detector signals of each unit are overlapped synchronously, and the overlapped signals are sent to the main amplifier. The scene scans each detector in turn, and the row sweep speed and frame sweep speed are the same as the unit detector. The signal-to-noise ratio (SNR) of each detector is improved by delay stacking.

Suppose the responsivity R_i, normalized detectivity D_i*, gain G_i of the preamplifier can be different for each pixel, and the radiated power P incident to each pixel detector is the same.

$$U_{si} = PR_i G_i$$

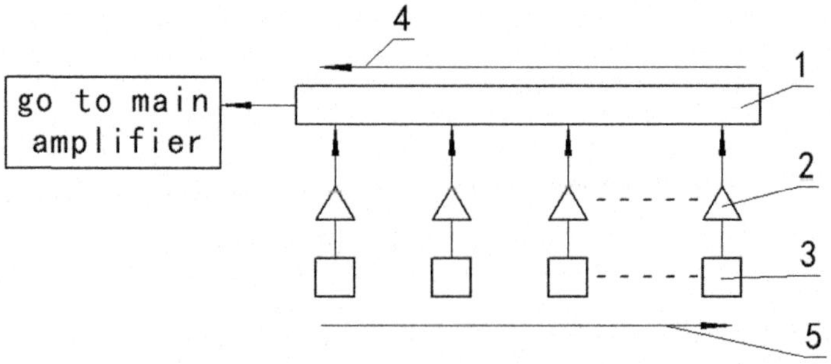

Fig. 7.40 Serial scanning mode

where

U_{si} output voltage of single pixel detector.

$$U_{ni} = \frac{R_i G_i \sqrt{A_d \Delta f}}{D_i^*} = \frac{R_i G_i}{D_i}$$

where

A_d pixel area of single pixel detector;
Δf equivalent noise bandwidth of the system;
D_i detection rate of single pixel detector;
Δf the bandwidth of the electronic system.

SNR of single pixel detector

$$\frac{U_{si}}{U_{ni}} = P D_i$$

After signal stacking

$$U_s = P \sum_{i=1}^{n} R_i G_i$$

U_s output signal voltage of detector group after signal stacking.

$$U_n = \sqrt{\sum_{i=1}^{n} \left(\frac{R_i G_i}{D_i} \right)^2}$$

U_n root mean square noise voltage of detector group after signal stacking.

SNR of detector group after signal stacking

$$\frac{U_s}{U_n} = P \frac{\sum_{i=1}^{n} R_i G_i}{\sqrt{\sum_{i=1}^{n} \left(\frac{R_i G_i}{D_i} \right)^2}}$$

Calculate the condition of maximum SNR of detector group.

If R_i and D_i of n-pixels are fixed, G_i can be adjusted, find the relation between SNR and G_i.

Relative gain $g_i = G_i/G_1$, $g_1 = 1$, then the above equation can be rewritten as

$$\frac{U_s}{U_n} = P\frac{\sum_{i=1}^{n} R_i g_i}{\sqrt{\sum_{i=1}^{n}\left(\frac{R_i g_i}{D_i}\right)^2}} \tag{7.7}$$

Choose g_i appropriately to maximize U_s/U_n, even if it is

$$\frac{\partial}{\partial g_k}\left(\frac{U_s}{U_n}\right) = 0$$

$$\frac{\partial}{\partial g_k}\left(\frac{U_s}{U_n}\right) = \frac{P R_k}{\left(\sum_{i=1}^{n}\left(\frac{R_i g_i}{D_i}\right)^2\right)^{1/2}} - \frac{P R_k^2 g_k}{D_k^2}\cdot\frac{\sum_{i=1}^{n} R_i g_i}{\left(\sum_{i=1}^{n}\left(\frac{R_i g_i}{D_i}\right)^2\right)^{3/2}} = 0$$

$$\sum_{i=1}^{n}\left(\frac{R_i g_i}{D_i}\right)^2 = \frac{R_k g_k}{D_k^2}\sum_{i=1}^{n} R_i g_i$$

Calculate g_k

$$\left(\frac{R_k g_k}{D_k}\right)^2 + \sum_{i\neq k}\left(\frac{R_i g_i}{D_i}\right)^2 = \frac{R_k g_k}{D_k^2}\left[R_k g_k + \sum_{i\neq k} R_i g_i\right]$$

Arranged

$$g_k = \frac{D_k^2 \sum_{i\neq k}\left(\frac{R_i g_i}{D_i}\right)^2}{R_k \sum_{i\neq k} R_i g_i}$$

It is known that n equations have to be solved, for example, $n = 2$

$$g_2 = \left(\frac{D_2}{D_1}\right)^2 \frac{R_1}{R_2}$$

$$g_1 \equiv 1$$

Substitute this group of solutions into the SNR in Formula 7.7

$$\frac{U_s}{U_n} = P\frac{\sum_{i=1}^{n} D_i^2}{\sqrt{\sum_{i=1}^{n} D_i^2}} = P\sqrt{\sum_{i=1}^{n} D_i^2}$$

It shows that the maximum SNR of the best gain of the series sweep detector group is the square root of the sum of squares of SNR of single pixel detector. If the detection rate D_i of each pixel detector is equal, it means that the series scanning system has normalized detectivity, i.e.,

$$D_0^* = \sqrt{\sum_{i=1}^{n} D_i^{*2}} \tag{7.8}$$

By the mean value theorem

$$D_0^* = \sqrt{n} D_{\text{rms}}^*$$

$$D_{\text{rms}}^* = \sqrt{\frac{\sum_{i=1}^{n} D_i^{*2}}{n}}$$

Formula 7.8 shows that, if the normalized detectivity D_i^* of each pixel detector in the detector group of n-pixels series scanning is equal, the SNR of the system in the detector group of n-pixels series sweep is $\sqrt{n}$ times higher than that of the single pixel detector (Table 7.6).

3. Staring imaging mode

Staring imaging has been made possible by the rapid development of area array imaging devices, infrared area devices in the area array middle waveinfrared detector have reached 640 × 512 pixels, as shown in Fig. 7.41a, and thermal infrared 640 × 480 pixels, as shown in Fig. 7.41b. In the visible and near infrared spectrum, it has reached a scale of 9 K × 9 K, as shown in Fig. 7.42.

So staring imaging has become one of the main imaging methods of infrared camera, and it is more and more important.

Table 7.6 Comparison of two camera methods

	Parallel scanning	Series scanning
Scanning mode	Convergent beam	Must be parallel beam
Scanner moving	Parts speed is low, easy to achieve	High-speed moving parts, short motor life, high line number and high frame frequency difficult to achieve
Electronic circuit band	Narrow band	Wide band
Refrigerating effect	Array long, poor effect	Array short, good effect
Signal and noise	Theoretically improved $\sqrt{n}$ times	Theoretically improved $\sqrt{n}$ times
1/f noise zone	Bandwidth reduction is dependent on the $1/f$ noise region	The low-frequency end should be higher to avoid the $1/f$ noise area
Detection element	Uniformity requirements high	No uniformity requirements

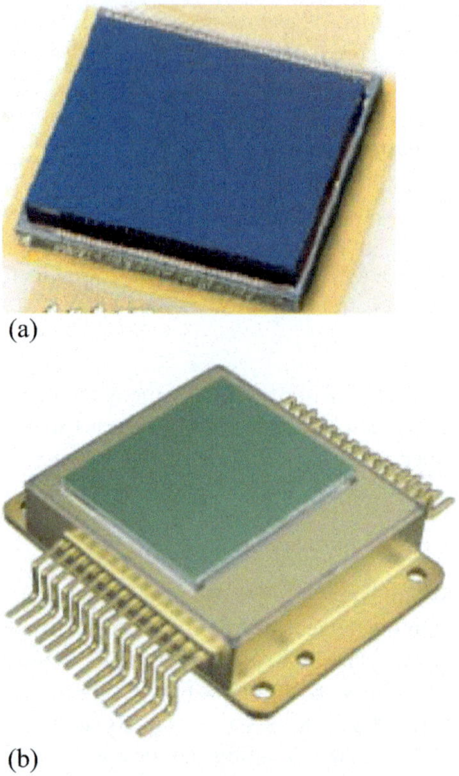

Fig. 7.41 Infrared area array detector

Fig. 7.42 Visible and near infrared area array detector

In theory, the SNR of staring imaging is higher than that of scanning imaging because it is $\sqrt{n}$ times that of single pixel detector scanning, where n is the number of pixels of area array detector. The parameter design and calculation of staring imaging infrared camera are more similar to that of visible light imaging camera.

For details, please refer to Sect. 7.5.8.2 quantization evaluation of infrared camera performance.

Staring imaging is the same as visible light imaging in principle, but it uses infrared spectrum, the target brightness is lower than visible light, so more attention needs to be paid to noise suppression. One measure is to cool the detector. With refrigeration, the technical difficulty and cost of infrared camera are very high, so the non-refrigerating infrared imaging technology has been promoted and developed.

The development of infrared technology from the 1960s of the single pixel or line array detector light machine scanning infrared imaging, known as the first generation of infrared imaging technology. In the 1980s, it was developed into area array infrared detector imaging without opto-mechanical scanning, which is called the second generation of infrared imaging technology. However, these two generations of infrared imaging technologies required refrigeration and worked at liquid nitrogen temperature of 77 K, which was expensive. In the mid-1990s, the United States invented a solid infrared focal plane device working at room temperature, which overcame the problem of refrigeration and greatly reduced the cost, known as the third generation of infrared imaging technology or non-refrigeration infrared imaging technology. In 2011, the U.S. installed a non-refrigerating infrared sensor as carrying equipment on the Aquarius satellite.

The normalized detectivity D^* value of the non-refrigerating infrared detector is nearly 2 orders of magnitude lower than that of the refrigerating infrared detector with liquid nitrogen refrigerating to 77 K, which is the limitation of the non-refrigerating infrared imaging technology.

7.5.7 Refrigerating Methods

The refrigerating of infrared detector includes Dewar refrigerating, phase change material refrigerating, semiconductor refrigerating, Stirling refrigerating machine refrigerating and space radiator refrigerating. Due to the need to consume liquid nitrogen and other refrigerants, Dewar bottle refrigerating is mainly used in space research, as a space duty working equipment whose life is limited, so the latter four methods or their combination are adopted.

1. Phase change material refrigerating

Phase change material refrigerating is more accurately known as phase change material refrigerating storage, which is to store the cold amount of other refrigerators for future use. Materials used for detector refrigerating should have appropriate phase change temperature, large latent heat of phase change, high thermal conductivity, small volume change of phase change, good compatibility with containers and other characteristics, as shown in Table 7.7.

The thermal conductivity of low-temperature phase change materials is also an important parameter of phase change materials used in detector refrigerating. The thermal conductivity of some phase change materials increases from 0.24 to 4–7 W/

Table 7.7 Thermodynamic parameters of low-temperature phase change materials

Material	Melting point/K /K	Specific heat/kJ kg^{-1}
Hydrogen	13.8	57.8
Nitrogen	63.2	25.8
Propane	85.3	79.9
Butane	134.9	80.2
Pentane	143.4	116.3
Ethanol	156	108
Methanol	175	99
Ammonia	195.4	351.7
Zinc chloride	211	116.8
Carbon dioxide	216.5	180.9
Calcium chloride	218	164.9
ferric chloride	218	155.5

m K after graphite is added, and the thermal conductivity increases by one order of magnitude. In recent years, nanoporous materials are also a research direction to achieve high thermal conductivity of low-temperature phase change materials.

2. Semiconductor refrigerating

n-type semiconductor and p-type semiconductor are connected to form a galvanic couple, as shown in Fig. 7.43, to form a closed circuit. Figure 7.43a is the schematic diagram of semiconductor refrigerating. Under the action of external electric field, electrons and holes in one joint produce separation motion, absorb energy and become cold, and at the other joint produce recombination, release energy and become hot. Its refrigerating capacity depends on the properties of semiconductor materials and the current in the loop. Bismuth telluride and its solid melt alloy are better semiconductor materials. The single-stage semiconductor refrigerator can obtain a temperature difference of about 60 °C, and the multistage series is shown in Fig. 7.43b, that is, the hot junction of one thermocouple forms a good thermal contact with the cold junction of the next thermocouple. The low-temperature of the third-stage series can reach 190 K, and the low-temperature of the six-stage and eight-stage refrigerator can reach 170 and 145 K, and the effect of increasing the series is not obvious. It can only be used for the refrigerating of infrared detectors such as lead selenide and lead sulfide whose refrigerating temperature is not too low.

The advantages of semiconductor refrigerator are simple structure, small size, light weight, high reliability, long life, no mechanical vibration and impact noise. The disadvantage is that the refrigerating temperature is not low enough to meet the

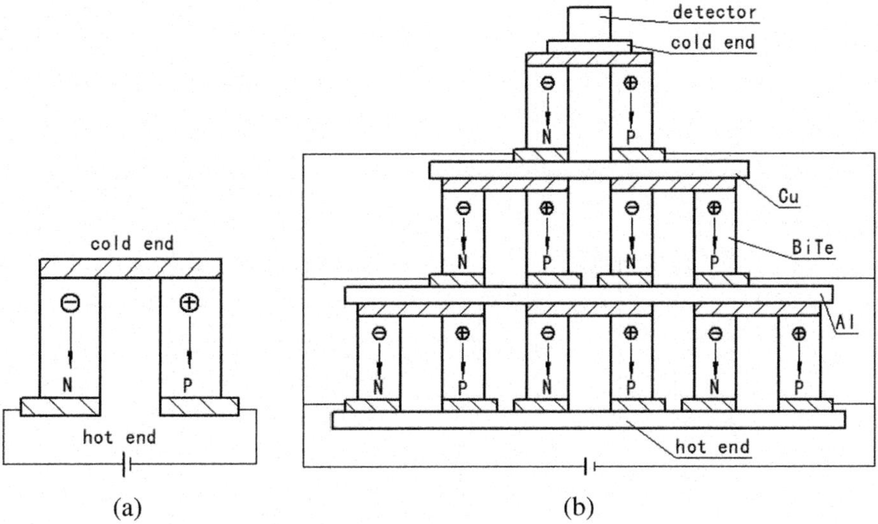

Fig. 7.43 Schematic diagram of semiconductor refrigeration

refrigerating needs of HgCdTe, InSb and other high-performance infrared detectors, and the efficiency is also low, generally about 10%.

3. Stirling refrigerating

In 1816, Stirling proposed a closed thermodynamic cycle consisting of two isothermal processes and two constant volume heat recovery processes, called Stirling cycle, also known as constant volume heat recovery cycle.

Fig. 7.44 shows an ideal Stirling cycle schematic. Figure 7.44a is the schematic diagram of the structure and the axial distribution of temperature.

The refrigeration machine is composed of regenerator *R*, cooler *A*, cooling heat exchanger *C* and two cylinders and two pistons. The expansion piston is on the left side, and the compression piston is on the right side. Two cylinders and the piston form two working cavities: cold chamber (expansion chamber) V_c and room temperature (compression) chamber V_g, connected by the regenerator *R*, the two pistons are driven by a coaxial crank with a certain initial phase angle for intermittent motion. It is assumed that a temperature gradient has been formed in the regenerator under stable working conditions. The temperature of the cold chamber is T_c, and the temperature of the room temperature chamber is T_g. Figure 7.44b is a schematic diagram of piston motion. Starting from state 1 in the figure, both the compression piston and the expansion piston are at the right dead center. There is a certain amount of gas in the cylinder, the pressure is P_1, and the volume is V_1. Pressure *P*, volume *V* and temperature *T* in the cycle process are shown in Fig. 7.45.

The cycle process is as follows:

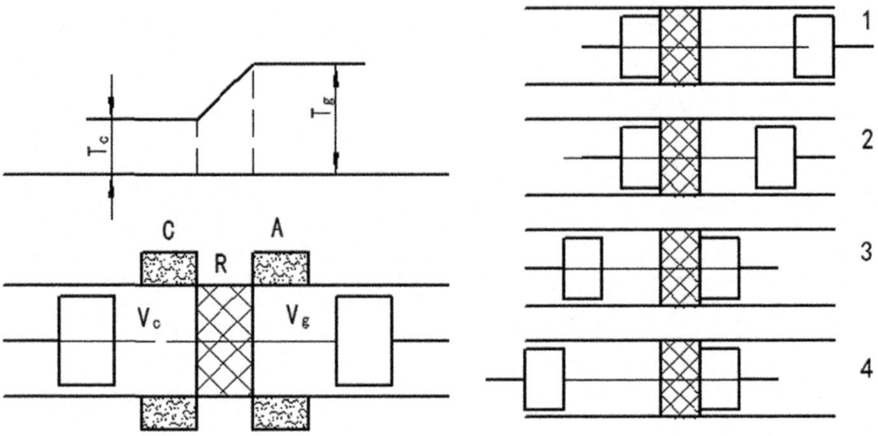

Fig. 7.44 Stirling refrigerator cycle working process

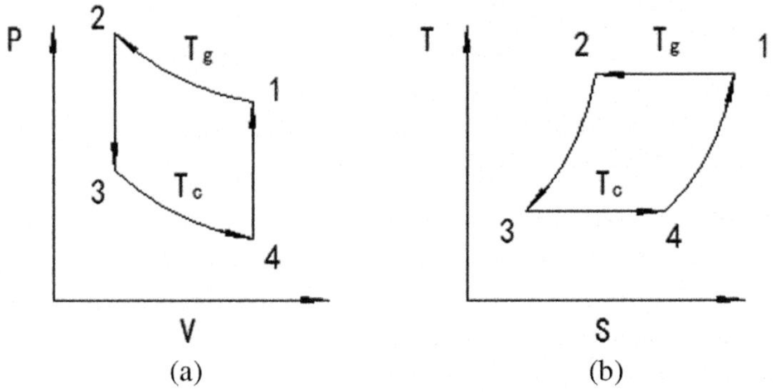

Fig. 7.45 Pressure volume diagram and temperature entropy diagram of Stirling refrigeration cycle

Isothermal compression process 1–2: The compression piston moves to the left while the expansion piston does not. The gas is isothermal compressed, the compression heat is transferred to the cooling medium (water or air) through cooler *A*, the temperature remains constant T_g, the pressure increases to P_2, and the volume decreases to V_2.

Constant volume heat release process 2–3: Both pistons move to the left at the same time, and the volume of the gas remains constant until the compression piston reaches the left dead center. As the gas passes through the regenerator *R*, heat is transferred to the packing, so that the temperature decreases from T_g to $T_{c,}$ and the pressure decreases from P_2 to P_3.

Isothermal expansion process 3–4: The compression piston stops at the left dead center, and the expansion piston continues to move to the left until the left dead center. The gas with the temperature of T_c expands isothermal and absorbs a certain amount of heat Q_c from the low-temperature heat source (cooling object) through the cooling heat exchanger C. Volume increases to V_4 and pressure decreases to P_4.

Constant volume endothermic process 4–1: Both pistons move to the right at the same time until the right dead center, and the gas volume remains unchanged, $V_1 = V_4$, returning to the starting position. When the gas at temperature T_c flows through R, it absorbs heat from the regenerator R packing, and the temperature rises to T_1, while the pressure increases to P_1. The heat absorbed by the gas is equal to the heat released by the gas in the process 4 to 1.

In the cycle, process 3–4 is the cooling process, and process 1–2 is the exothermic process.

Since the heat transfer in regenerative process 2–3 and 4–1 belongs to internal heat transfer and has nothing to do with the energy consumption of the whole cycle, the work consumed in the cycle is equal to the difference between the compression work and the expansion work.

Stirling refrigeration machine is a small cryogenic refrigeration machine in the most in-depth research, the most widely used, the most mature development, the most variant, characterized by compact structure, wide operating temperature range, fast starting, high efficiency, easy to operate.

The development and changes of Stirling refrigerator mainly include the following aspects:

(1) The cooling temperature ranges from general cold to deep cold, with the lowest temperature reaching 3 K;
(2) The cooling capacity develops to micro (milliwatt level) and large (46.8 kW) simultaneously;
(3) The development of multicylinder refrigeration machine;
(4) From single stage to multistage, 5 stage refrigeration machines have appeared;
(5) From single action to multiaction refrigerating machine;
(6) Development from integral to divided type;
(7) Developed a variety of driving methods. Such as crank-connecting rod mechanism, shaking plate drive, swash plate drive, diamond drive, hydraulic drive, electromagnetic drive, pneumatic and so on;
(8) Various forms, such as double piston, push piston, parallel arrangement, angular arrangement, coaxial arrangement and so on.

Table 7.8 shows the performance parameters of a Stirling refrigeration machine with internal opposite compressor and expander separated [7].

4. Radiator refrigerating

When the spacecraft is working in space, the outer space is in a high vacuum and deep low temperature, and the cold space temperature is only 4 K. The infrared detector can be refrigerated by radiation heat transfer. Radiation refrigerator is a passive refrigerator that requires almost no energy. It is composed of cold plate, radiator, edge, insulation multilayer, solar shield, etc. In order to obtain different cooling temperatures, a plurality of radiation refrigerators can be connected in series

Table 7.8 Refrigeration machine performance parameters

Item	Parameters
Cooling capacity/temperature	4.8 W/95 K, 14.5 W/150 K
The weight of the	8.5-kg
Power consumption	(124 × 2)W
Working frequency	60 Hz
The input voltage	24 V_{rms}
The microvibration	≤ 3.5 N_{rms}
Life	≥ 45,000 h
Compressor dimension	Φ94 mm × 260 mm
Expander dimensions	Φ55 mm × 120 mm

to form two or three stage refrigerators. Figure 7.46 is a radiation refrigerator on the European ESA satellite that cools the infrared detector to 95 K.

The advantage of the radiation refrigerator is that it requires no energy, has no moving parts and therefore no shock and vibration and has a long life. The disadvantage is that the radiation refrigerator should always face the cold space, which increases the limitation on the attitude of the spacecraft, and does not allow sunlight and infrared radiation of the earth to enter the radiator of the refrigerator.

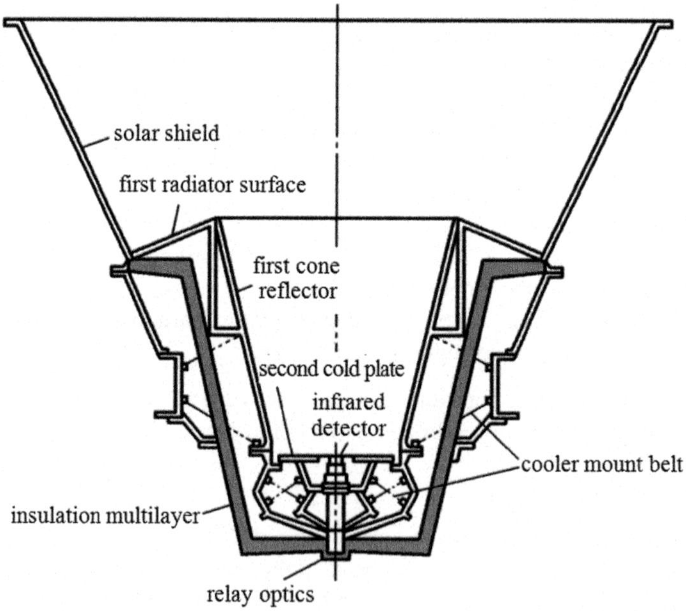

Fig. 7.46 Radiation refrigerator

Which kind of refrigerating method should be adopted according to the required performance of the infrared detector. The main parameters are the operating temperature and refrigerating capacity required by the infrared detector, as well as the starting working time, power consumption, life span, size and so on. With the improvement of spatial resolution of infrared detector, the requirement of shock vibration of refrigerator will be further improved.

7.5.8 *Comprehensive Evaluation of Infrared Detector Performance*

7.5.8.1 Noise Equivalent Temperature Difference (NETD)

The noise equivalent temperature difference is defined and tested using a standard test graph in Fig. 7.47, which takes a square target with uniform temperature T_T K and angular size $\alpha \times \beta$ in a uniform temperature T_B K background. Both target and background are assumed to be blackbody. The infrared system under test detects this graph. When the ratio of the output signal voltage peak value of the system to the RMS value of the noise voltage is 1, the temperature difference between the blackbody target and the blackbody background on the test graph is the noise equivalent temperature difference of the system under test.

During actual measurement, in order to obtain good results, the angular size of the blackbody target made is several times larger than the instantaneous field of view of the system α and β, and the temperature difference between the target and the background is ten or dozens of times larger than the required NETD, so as to ensure

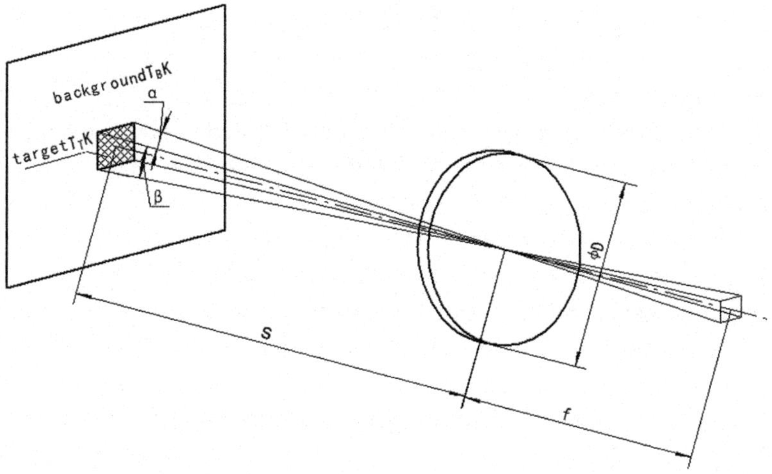

Fig. 7.47 NETD test graph

that the peak signal voltage U_s output by the electronic system is much larger than the root mean square noise voltage U_n value. In this way, the following formula is used to calculate NETD

$$\text{NETD} = \frac{\Delta T}{U_s/U_n}$$

The derivation of NETD formula is as follows:

The assumption of derivation [5];

1. Target and background are the blackbody;
2. Atmospheric transmission loss is ignored;
3. The responsivity of the whole detector is uniform and consistent;
4. The D^* of the detector is only a parameter of the detector itself and has nothing to do with other parameters of the system;
5. The electronic circuit does not bring additional noise.

Setting the target as a blackbody, and its spectral radiation is $M_\lambda(T)$, and the unit is W cm^{-2} μm^{-1}. Then, the spectral radiance is $L_\lambda = M_\lambda(T)/\pi$, and the unit is W cm^{-2} $\mu\text{m}^{-1}\text{Sr}^{-1}$.

Setting the target as a blackbody, and its spectral radiant existence is $M_\lambda(T)$, and the unit is W cm^{-2} μm^{-1}. Then, the spectral radiance is $L_\lambda = M_\lambda(T)/\pi$, and the unit is W cm^{-2} $\mu\text{m}^{-1}\text{Sr}^{-1}$.

Suppose the distance between the target and the system is s, the clear aperture φD, and the effective receiving area of the system is $A_0 = \pi D^2/4$, then the solid angle that the system extends to the target is A_0/s^2. Therefore, the radiant flux within the unit spectral width of the unit area of the target received at the aperture of the optical system is

$$\varphi_\lambda = \frac{L_\lambda A_0}{s^2} = \frac{M_\lambda(T)\, A_0}{\pi s^2} \quad \text{unit/cm}^{-2}\mu\text{m}^{-1}$$

The instantaneous FOV plane angles of the system are α and β, the target area corresponding to $\alpha \times \beta$ is $\alpha\beta s^2$, and the spectral radiation power radiated to the effective receiving surface of the system within a unit spectral width is

$$P_\lambda = \varphi_\lambda \alpha\beta s^2 = \frac{1}{\pi} M_\lambda(T) A_0 \alpha\beta \quad \text{unit/W}\mu\text{m}^{-1}$$

The transmittance ratio of the optical system $\tau_o(\lambda)$, then the spectral radiation power on the detector is

$$P_\lambda = \frac{1}{\pi} M_\lambda(T) A_0 \alpha\beta \tau_o(\lambda) \quad \text{unit/W}\mu\text{m}^{-1}$$

One is interested in the change of the radiant power reaching the system detector when the target temperature is different from the background temperature or between adjacent parts of the target, and the rate of change is

$$\frac{\partial P_\lambda}{\partial T} = \frac{1}{\pi} A_0 \alpha \beta \tau_o(\lambda) \frac{\partial M_\lambda}{\partial T} \quad \text{unit/W}\mu\text{m}^{-1}\text{K}^{-1}$$

Assuming that the responsivity of the system detector is $R(\lambda)$, the incident radiation is transformed into the output signal voltage by the detector, and the change rate of the signal voltage with the target temperature is

$$\frac{\partial U_s(\lambda)}{\partial T} = \frac{1}{\pi} A_0 \alpha \beta \tau_o(\lambda) R(\lambda) \frac{\partial M_\lambda}{\partial T} \quad \text{unit/V}\mu\text{m}^{-1}\text{K}^{-1}$$

$$R(\lambda) = \frac{U_n D^*(\lambda)}{\sqrt{ab\Delta f}} \quad \text{unit/V/W}$$

where

- $D^*(\lambda)$ spectral normalized detection rate of the detector, unit /W^{-1} cm $\text{Hz}^{1/2}$;
- U_n root mean square value of noise voltage output of the electronic system under test, unit /V;
- a, b pixel size of detector, unit /cm;
- Δf equivalent noise bandwidth of the electronic system under test, unit /Hz.

$$\frac{\partial U_s(\lambda)}{\partial T} = \frac{1}{\pi} A_0 \alpha \beta \tau_o(\lambda) \frac{U_n D^*(\lambda)}{\sqrt{ab\Delta f}} \frac{\partial M_\lambda}{\partial T}$$

Integrating over all wavelengths to get the total rate of voltage change

$$\frac{\partial U_s}{\partial T} = \frac{A_0 \alpha \beta U_n}{\pi \sqrt{ab\Delta f}} \int_0^\infty D^*(\lambda) \tau_o(\lambda) \frac{\partial M_\lambda}{\partial T} \mathrm{d}\lambda$$

And using the small-signal approximation method, replace $\Delta U_s/\Delta T$ with $\partial U_s/\partial T$, so:

$$\frac{\Delta U_s}{\Delta T} = \frac{A_0 \alpha \beta U_n}{\pi \sqrt{ab\Delta f}} \int_0^\infty D^*(\lambda) \tau_o(\lambda) \frac{\partial M_\lambda}{\partial T} \mathrm{d}\lambda$$

Arrange

$$\frac{\Delta U_s}{U_n} = \frac{A_0 \alpha \beta \Delta T}{\pi \sqrt{ab\Delta f}} \int_0^\infty D^*(\lambda) \tau_o(\lambda) \frac{\partial M_\lambda}{\partial T} \mathrm{d}\lambda$$

Let $\Delta U_s/U_n = 1$, ΔT can be gotten, that is, NETD

$$\mathrm{NETD} = \frac{\pi\sqrt{ab\Delta f}}{A_0\alpha\beta\int_0^\infty D^*(\lambda)\tau_o(\lambda)\frac{\partial M_\lambda}{\partial T}\mathrm{d}\lambda} \quad \mathrm{unit/K}$$

This is the basic equation of the infrared imaging system NETD.

The differential $\partial M_\lambda/\partial T$ in the equation is the derivative of the spectral emission $M_\lambda(T_\mathrm{B})$ with respect to temperature at the relative background temperature T_B.

The spectral emission of the target radiation is

$$M_\lambda = \frac{c_1}{\lambda^5(\mathrm{e}^{c_2/\lambda T} - 1)}$$

Differentiate, so:

$$\frac{\partial M_\lambda}{\partial T} = M_\lambda\frac{c_2\mathrm{e}^{c_2/\lambda T}}{\lambda T^2(\mathrm{e}^{c_2/\lambda T} - 1)}$$

where c_1 and c_2 are the first and second radiation constants, respectively.

Substitute into the basic equation of NETD, so:

$$\mathrm{NETD} = \frac{\pi\sqrt{ab\Delta f}}{\frac{c_2A_0\alpha\beta}{T^2}\int_0^\infty D^*(\lambda)\tau_o(\lambda)M_\lambda\frac{\mathrm{e}^{c_2/\lambda T}}{\lambda(\mathrm{e}^{c_2/\lambda T}-1)}\mathrm{d}\lambda} \tag{7.9}$$

Infrared imaging usually has $\mathrm{e}^{c2/\lambda T} \gg 1$, so

$$\frac{\partial M_\lambda}{\partial T} \approx \frac{c_2}{\lambda T^2}M_\lambda$$

For photodetectors, the ideal spectral normalized detection rate is

$$\begin{aligned} D^*(\lambda) &= \tfrac{\lambda}{\lambda_p}D^*(\lambda_p) & \lambda \le \lambda_p \\ D^*(\lambda) &= 0 & \lambda > \lambda_p \end{aligned}$$

If the system is limited to work within a certain spectral range, for most optical materials, $\tau_\mathrm{o}(\lambda)$ is considered to be a constant τ_o in the selected wavelength range. For other wavelengths $\tau_\mathrm{o}(\lambda) = 0$, so that τ_o is mentioned outside the integral sign, the equation can be simplified without affecting the accuracy.

The $D^*(\lambda)$ in the integral sign is taken as the relative value $D^*(\lambda)/D^*(\lambda_p)$ according to $D^*(\lambda_p)$ of the peak wavelength λ_p, so the integral part of the basic equation of NETD is

$$\int_{\lambda_1}^{\lambda_2} D^*(\lambda)\tau_o(\lambda)M_\lambda\frac{\mathrm{e}^{c_2/\lambda T}}{\lambda(\mathrm{e}^{c_2/\lambda T} - 1)}\mathrm{d}\lambda = \tau_o\frac{D^*(\lambda_p)}{\lambda_p}\int_{\lambda_1}^{\lambda_2} M_\lambda\mathrm{d}\lambda$$

Then,

$$\text{NETD} = \frac{\pi \lambda_p T^2 \sqrt{ab\Delta f}}{c_2 A_0 \alpha \beta \tau_o D^*(\lambda_p) \int_{\lambda_1}^{\lambda_2} M_\lambda \mathrm{d}\lambda} \tag{7.10}$$

When the background temperature T_B, the working wavelength λ_1 and the detector cut-off wavelength λ_p are known, the integral value of the above formula can be obtained by referring to the blackbody radiation data table. Finally get

$$\text{NETD} = \frac{\pi \lambda_p T_B^2 \sqrt{ab\Delta f}}{c_2 A_0 \alpha \beta \tau_o D^*(\lambda_p) \int_{\lambda_1}^{\lambda_p} M_\lambda (T_B) \mathrm{d}\lambda} \quad \text{unit/K}$$

Sometimes, it is desirable to know what effect a single parameter has on NETD, and for this reason, the NETD equation is changed to take the form of the basic parameter representation of the system.

Set the optical system clear aperture φD_o, focal length f, field of view $A \times B$, frame rate F', and the total horizontal and vertical scanning efficiency η, then

$$A_O = \frac{\pi}{4} D_o^2$$

Detector area $ab = \alpha \beta f^2$
N-pixels sweep $t_d = \frac{n\alpha\beta\eta}{ABF'}$.
System equivalent noise bandwidth $\Delta f = \frac{\pi}{4t_d} = \frac{\pi ABF' f^2}{4nab\eta}$.
Again so:

$$\text{NETD} = \frac{2\lambda_p T_B^2 f \sqrt{\pi ABF'/n\eta}}{D_o^2 \alpha \beta \tau_o D^*(\lambda_p) c_2 \int_{\lambda_1}^{\lambda_2} M_\lambda (T_B) \mathrm{d}\lambda}$$

Discuss:

The three quantities NETD, ab and F' in the previous formula represent the three main parameters of the infrared system performance when $D_o, f, t_o, D^*(\lambda_p)$ and the radiation characteristics of the scene are all determined. They are contradictory, such as increasing the temperature resolution NETD will sacrifice the spatial resolution ab or reduce the frame rate F'.

Under the premise that the spatial resolution of the system is expressed in angle quantities, the NETD is independent of the distance from the system to the target (ignoring the influence of atmospheric transmittance).

There are the following approximations in the NETD derivation process:

1. The target and the background are regarded as blackbodies, which can be satisfied in the laboratory, but in practical applications, they are not blackbodies, but gray

body, and the emissivity $\varepsilon \neq 1$. Different objects have different ε, reflectivity and reflect different environmental radiation. There are temperature differences, emissivity differences and reflectivity differences between gray body targets. Even two gray bodies with the same temperature may have different radiations, the above derivation does not consider the impact of these factors on NETD;
2. The influence of atmospheric attenuation is not considered;
3. The NETD formula only reflects the characteristics of the optical system, detectors and a small parts of circuits and does not consider the noise source or filtering effect from the measurement point to the display;
4. The temperature difference of each part of the scene causes the signal change. This signal change often has a transition process. This transition process will be confused with the signal-to-noise ratio of the stable signal in the measurement of the signal-to-noise ratio, which affects the accuracy of the reading.
5. The influence of the observer's eyes is not considered.

It can be seen that NETD cannot fully reflect the quality of the system and has limitations. However, its physical meaning is clear, and it has practical significance for understanding and mastering the characteristics of infrared systems.

In addition to NETD, there are two basic concepts in the performance evaluation of infrared systems: minimum resolvable temperature difference (MRTD) and minimum detectable temperature difference (MDTD).

Minimum Resolvable Temperature Difference (MRTD). For some infrared systems to observe the scene directly, it is necessary to establish the relationship between the scene-instrument-eye to evaluate the performance of the system as a whole. Since NETD only reflects the sensor noise and does not relate the temperature resolution to the spatial resolution, the concept of minimum resolvable temperature difference needs to be further established.

The standard test graph used to define and test the minimum resolvable temperature difference is shown in Fig. 7.48a, and it has four strips with an aspect ratio of 7:1.

The MRTD test device as shown in Fig. 7.48b consists of three parts. The first component is the blackbody radiation source 1, the temperature is continuously adjustable from 0 to 100 °C, the adjustment accuracy is better than 0.02 °C (20–25 °C laboratory environment), and the emissivity of the radiation source is > 0.95. The second component is the target generator 2, i.e., the standard test pattern and the ambient temperature slice 3. Ambient temperature slice is made of high mass density material with the same emissivity as the radiation source. The ambient temperature slice can be filled with test patterns of different sizes, these patterns are corroded from a thin metal plate, which is coated with paint and has the same emissivity as the ambient temperature slice and the radiation source. The patterned metal plate has good thermal contact with the ambient temperature slice. In this way, the temperature of the patterned metal plate is the ambient temperature, and the temperature of the corroded four-rod is the temperature of the blackbody radiation source. Adjusting the temperature of the blackbody radiation source can cause the difference temperature between the target and the background. The third component is the collimator 4

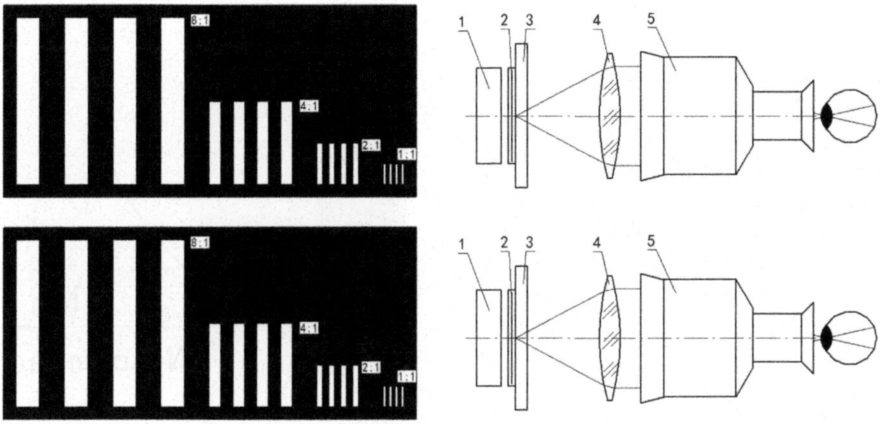

(a) (b)

Fig. 7.48 Infrared system MRTD test

whose exit pupil is larger than or equal to the entrance pupil of the system 5 under test.

The spacing or density of the black and white stripes in the standard test chart represents the high or low visible frequency of the target, as shown in Fig. 7.49. The period of the black and white stripes is T_x, unit is mm. The distance from them to point o is s, and the unit is m. Then, the opening angle of T_x to o is $\varphi_x = T_x/s$, unit is mrad. The spatial frequency of the test chart is $N_T = 1/\varphi_{\mathrm{x}} = s/T_x$, unit is c/mrad.

The measurement steps are as follows: Place the pattern with the lowest spatial frequency in front of the ambient temperature slice, start with zero temperature difference between the pattern and the radiation source, tune the system so that the noise is clearly visible on the display, and gradually increase the temperature

Fig. 7.49 Spatial frequency

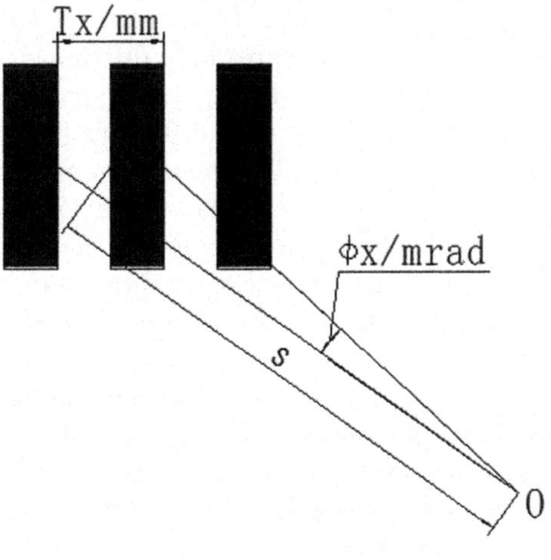

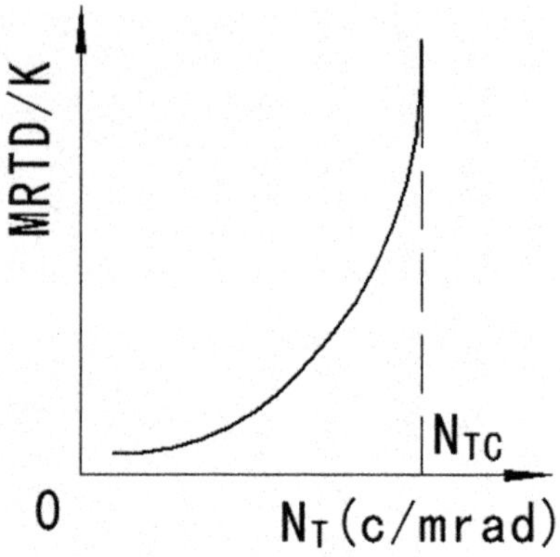

Fig. 7.50 MRTD curve

difference until the observer can indeed identify these patterns from the display, the temperature difference ΔT at this time is the minimum distinguishable temperature difference at the spatial frequency.

Place the four-bar pattern with a slightly higher spatial frequency in front of the ambient temperature slice, increase the temperature difference between the pattern and the radiation source until the pattern is clear, and note the spatial frequency and temperature difference.

In this way, the spatial frequency is continuously increased, and the temperature difference is continuously increased for measurement until the four-rod pattern cannot be distinguished even if the temperature difference is increased. At this time, the spatial frequency of the pattern has reached a limit value N_{TC}. Since beyond this spatial frequency, the experimenter cannot observe the target pattern, MRTD $\rightarrow \infty$. A typical curve of MRTD is shown in Fig. 7.50, and the abscissa is the spatial frequency N_{T} of the pattern. During observation, the brightness of the display and the distance between the observer and the display can be adjusted to the optimum value arbitrarily.

Minimum Detectable Temperature Difference (MDTD). Some infrared systems need to know the temperature difference ΔT between an individual target (unlike MRTD, which is used a target with spatial frequency) and the background in order to detect it. This temperature difference just enough to detect an individual target from the background is called the minimum detectable temperature difference (MDTD).

When measuring MDTD, a single square target is used, and its size is much smaller than that corresponding to the instantaneous field of view of the system, so it can be regarded as a point source target. The minimum detectable temperature difference represents the system's ability to detect non-decomposable point source targets.

The difference between MDTD and MRTD is that four-rod-shaped targets are used in MRTD measurement, and the spatial resolution of the four-rod pattern is constantly increasing. When the limit of spatial resolution is reached, MRTD $\rightarrow \infty$, and there is an asymptote. When measuring MDTD, a square target is used, and it's just one piece. There is no problem that the four-bar pattern becomes denser and denser so that no matter how high the temperature difference is, it cannot be distinguished when the spatial resolution reaches the limit. As long as the temperature is high enough, it can always be detected regardless of the size of the target. Therefore, MDTD, although is a function of the size of a target, there is not an asymptote. The calculation of these two concepts involves the specific design of the electronic system of the infrared system and the more specific performance parameters of the detector, see [5, 6] for details. For the calculation of the performance of MRTD, the method of "quantization evaluation of infrared camera performance" can also be borrowed below, and the calculation of MDTD can refer to the later section of "the working distance of infrared system".

7.5.8.2 Quantized Evaluation of Infrared Camera Performance

The quantization evaluation of infrared camera performance will describe the influence of various factors of infrared system on detection results from the perspective of photoelectron quantization conversion. From this perspective, the description of infrared detection system performance is more direct and clear than NETD, MRTD and MDTD.

Compared with visible light cameras, space infrared cameras are used to image or detect weak targets. Based on the comparison between target and background, this kind of detection can be divided into two categories, one is the detection with the earth as the background, and the other is the infrared radiation detection with the deep space as the background. The temperature of the earth–atmosphere system should be considered in the range of 220–300 K with the earth as the background. If infrared longwave detection is adopted, the influence of earth–atmosphere system on detection should be fully considered. When the temperature of the detected point target is low, the target signal is easily submerged in the earth background noise, because the earth's background temperature is comparable to the target temperature, so that the space infrared camera cannot detect it.

For this reason, the infrared detection of objects on the earth's surface should use the near infrared or mid-infrared spectrum where the scene reflects the solar radiation as much as possible.

As the background temperature of deep cold space is about $T_b = 4$ K, infrared detection based on it can use longwave detection, mid-infrared and near infrared detection, which mainly depends on the radiation characteristics of the detected target. Since the background temperature of deep cold space is very low, the background noise can be ignored under this condition. The noise of space infrared camera mainly comes from the camera itself, and its detection capability is mainly limited by the noise generated by the camera itself.

With the rapid development of optoelectronic technology and craft level, staring-type area detectors are more and more applied in the space field. How to use the parameters of the detector and the optical system to more accurately evaluate the performance of the detection system is one of the main problems in target detection.

From a signal processing point of view, the main factor affecting the probability of detection is the signal-to-noise ratio (SNR). The signal-to-noise ratio is defined as the ratio of the output voltage of the target pixel to the system noise voltage.

For infrared detection systems, the usual sensitivity calculation method is based on the specific detection rate of the detector. The signal-to-noise ratio of the infrared detector is the ratio of the target signal voltage to the noise voltage. The noise voltage takes V_n in $D^*(\lambda)$, which is the noise voltage inside the detector. The noise is relatively fixed within a certain temperature range and is the noise exhibited by the detector when measured in the laboratory. Generally, the measured value in the laboratory is the detector response under a certain temperature blackbody background (usually 500 K), which is only suitable for targets with a temperature close to 500 K.

In practical applications, the temperature of the detected targets varies greatly. If the target temperature is as high as several thousand degrees, the atmospheric background temperature can also be as low as about 250 K. Therefore, it is necessary to multiply the temperature correction coefficient or adopt other methods during calculation. The D^* method is used to calculate the SNR, which implicitly assumes that the noise level is constant under any circumstances. This calculation model is simple and highly maneuverable and is suitable for rough estimation of detector performance, but it cannot be used to calculate the noise of each part of the system, so that corresponding technical measures can be taken to improve the detection performance of the system.

It is one of the ways to analyze the detection ability of satellite-borne infrared detector system from the perspective of photoelectron number. Let A_0 be the entrance pupil area of the optical system, τ_{int} be the integral time, h be Planck constant, c_0 be the speed of light, $\eta(\lambda)$ be the quantum efficiency, and $\tau_o(\lambda)$ be the transmittance of the camera system. When detecting the point target, s is the distance, $I(\lambda)$ is the radiation intensity of the point target, and the number of electrons generated on the area array detector is

$$N = \frac{A_o \tau_{\text{int}}}{s^2 h c_o} \int I(\lambda)\eta(\lambda)\tau_o(\lambda)\lambda \mathrm{d}\lambda$$

The noise of satellite-borne infrared staring detection system mainly includes photon noise, fixed figure noise, readout noise and dark current noise. Quantization noise and $1/f$ noise are ignored here.

Photon noise is mainly formed by the random generation of photogenerated electrons caused by the random incidence of photons. The number of incident photons satisfies the Bose–Einstein statistical distribution, which approximates the Poisson distribution when a large number of photons are incident, and the mean and variance are equal, so there is

$$\overline{n}_{\text{ph}} = \sqrt{N_{\text{ph}}}$$

$\overline{n_{\text{ph}}}$ is the root mean square value of photon noise, and N_{ph} is the number of incident photons.

In the focal plane array, due to the difference in offset or response of each detector pixel, each pixel of the detector will produce different output noise for the same input, which is called fixed pattern noise (FPN). Response non-uniformity is defined as

$$U = \frac{\sigma_R}{R}$$

where

U is the non-uniformity of pixel response, such as 0.1%;
σ_R is the mean square error of the response rate of each pixel of an infrared detector;
R is the average responsivity of each pixel.

Since there is still residual response non-uniformity after non-uniformity correction, which is represented as fixed pattern noise. The fixed pattern noise of each pixel is described by the number of electrons as

$$\overline{n}_{\text{FPN}} = U N_{\text{ph}}$$

Readout noise describes all the electronic noise in the whole process from focal plane electrons to amplified current to *A–D* conversion. The detailed calculation is more complicated, and it can also be obtained by laboratory measurement. The formula is quoted here.

$$\overline{n}_{\text{read}}^2 = \overline{n}_{\text{floor}}^2 + v\overline{n}_{\text{rms}}^2$$

$\overline{n}_{\text{floor}}$ is the bottom charges number, which is generally given by the detector, such as 5 e^-/pixel when the temperature is 80 K;
$\overline{n}_{\text{msr}}$ is the readout noise when the pixel rate is equal to 1 MHz, such as 50 e^-/pixel;

$v = \frac{N_1 N_2}{\tau_{\text{int}}} \times 10^{-6}$ is the channel pixel rate (MHz), and N_1 and N_2 are the number of horizontal and vertical pixels on the area detector, respectively.

Dark current noise is a quantity closely related to detector temperature, i.e., 30 e^-/pixel for a detector of 40 μm × 40 μm pixels. When there is no signal input, a certain temperature detector generates a dark current offset due to the irregular thermal motion of electrons, forming the so-called dark current noise. The theoretical calculation formula is

$$\overline{n}_{\text{dark}}^2 = \frac{2kA_d}{q^2} \cdot \frac{\tau_{\text{int}} T}{R_0 A_d}$$

where

T is focal plane temperature;
R_0A_d is the product of the detector impedance and the photosensitive area (Ω cm^2);
K is the Boltzmann constant;

q is the amount of electron charges.

Therefore, the total noise of the system can be obtained as

$$\overline{n} = \sqrt{\overline{n}_{\text{ph}}^2 + \overline{n}_{\text{FPN}}^2 + \overline{n}_{\text{read}}^2 + \overline{n}_{\text{dark}}^2}$$

The contribution of different noise sources to the overall detection performance can be calculated by calculating the signal-to-noise ratio using the photoelectron number method such as the influence of temperature of optical system on photon noise and the influence of non-uniformity of detector response on fixed graph noise. In addition, it can accurately reflect the noise size in different detection scenes. However, the model of this method involves many detector parameters, so it is not suitable to use this method when the corresponding parameter values are not known.

7.5.8.3 The Working Distance of the Infrared System

The concept of MDTD is to detect the minimum temperature difference ΔT between a point source target and the background. This means that the solid angle exerted by the target to the system is smaller than the instantaneous field of view of the system. In this way, the system can detect how far away the target is from the system. The farther away the system is, the less energy it can receive from the target and the larger the temperature difference is required.

Suppose that the noise of the system is limited only by the noise of the detector, and the point source target is a blackbody.

The irradiance of the infrared radiation emitted by the target reaching the incident aperture of the system is E_λ, then

$$E_\lambda = \frac{I_\lambda \tau_a(\lambda)}{s^2} \tag{7.11}$$

where

I_λ spectral radiation intensity of the target;
$\tau_a(\lambda)$ atmospheric spectral transmittance ratio on the propagation path;
s the distance from the target to the detector;

Spectral radiant power incident on the detector P_λ

$$P_\lambda = E_\lambda A_o \tau_o(\lambda) \tag{7.12}$$

where

A_o entrancc pupil area of the optical system;
$\tau_o(\lambda)$ spectral transmission ratio of optical system.

The output voltage of the detector is $U_{s\lambda}$

$$U_{s\lambda} = R(\lambda)P_\lambda \tag{7.13}$$

where

$R(\lambda)$ spectral responsiveness of the detector.

The signal voltage U_s in a specific spectral interval λ_1–λ_2

$$U_s = \int_{\lambda_1}^{\lambda_2} R(\lambda)P_\lambda d\lambda \tag{7.14}$$

Substitute 7.11 and 7.12 into 7.14, so:

$$U_s = \frac{A_o}{s^2}\int_{\lambda_1}^{\lambda_2} I_\lambda \tau_a(\lambda)\tau_o(\lambda)R(\lambda)\mathrm{d}\lambda$$

The spectral response $R(\lambda)$ of the detector in the above formula can be expressed by the normalized detection rate $D^*(\lambda)$

$$R(\lambda) = \frac{U_n D^*(\lambda)}{\sqrt{A_d \Delta f}}$$

Substitute in the above equation

$$\frac{U_s}{U_n} = \frac{A_o}{s^2\sqrt{A_d \Delta f}}\int_{\lambda_1}^{\lambda_2} I_\lambda \tau_a(\lambda)\tau_o(\lambda)D^*(\lambda)\mathrm{d}\lambda \tag{7.15}$$

The average normalized detection rate D^* is given by infrared detectors in general. The transmittance $\tau_o(\lambda)$ of the optical system in the spectral interval λ_1–λ_2 is expressed by its average τ_0. Equation 7.15 is expressed in terms of detection distance as

$$s = \sqrt{\frac{A_o \tau_o D^* \int_{\lambda_1}^{\lambda_2} \tau_a(\lambda) I_\lambda \mathrm{d}\lambda}{\frac{U_s}{U_n}\sqrt{A_d \Delta f}}} \tag{7.16}$$

Denote $\tau_a(\lambda)$ by the average value in its spectral range λ_1–λ_2, namely τ_a, and let

$$I_{\lambda 1-\lambda 2} = \int_{\lambda_1}^{\lambda_2} I_\lambda \mathrm{d}\lambda$$

Arrange 7.16 as

$$s = \sqrt{\frac{A_o \tau_a \tau_o D^* I_{\lambda 1-\lambda 2}}{\frac{U_s}{U_n}\sqrt{A_d \Delta f}}} \tag{7.17}$$

If the instantaneous field of view (solid angle) ω is set, focal length f, and aperture of the optical system D, then

$$A_d = \omega f^2$$

$$A_o = \frac{\pi}{4} D^2$$

$$\frac{A_o}{\sqrt{A_d}} = \frac{\pi}{4\sqrt{\omega}} \frac{D}{f} D = \frac{\pi D}{4F^{\#}\sqrt{\omega}}$$

$$s^2 = \frac{\pi \tau_a \tau_o D D^* I_{\lambda 1-\lambda 2}}{4F^{\#}\frac{U_s}{U_n}\sqrt{\omega \Delta f}} = \frac{\pi}{4} \quad \tau_a I_{\lambda 1-\lambda 2} \quad \frac{\tau_o D}{F^{\#}} \quad D^* \quad \frac{1}{\sqrt{\omega}} \quad \frac{1}{\frac{U_s}{U_n}\sqrt{\Delta f}}$$

coefficients target and atmosphere optical system detector resolution electronics

The above equation clearly shows that the best way to improve the detection ability of infrared system is to increase the normalized detection rate D^* of the detector, followed by small $F^{\#}$ system or larger aperture of the optical system. Decreasing bandwidth and resolution can improve the detection range slowly.

7.5.9 *Interpretation of Infrared Images*

The interpretation of the target through the image is related to the minimum resolvable temperature difference (MRTD) of the imaging system, and MRTD is related to the actual temperature difference between the target and the background and the atmospheric transmittance.

Generally, the ability to distinguish the target is divided into three levels: discovery, classification and clear recognition. Discovery is to distinguish the target from the background on the screen; classification and recognition are to determine the category of the target from the image, such as people, vehicles and ships; clear recognition is to be able to determine which type of target the determined category belongs to. Such as a certain type of tank aircraft and so on.

Table 7.9 Target interpretation probability and resolution (n_0)

Probability	Discovery	Recognition (conservative/optimal)	Clear recognition
1.0	3	12/9	24
0.95	2	8/6	16
0.8	1.5	6/4.5	12
0.5	1.0	4/3	8
0.3	0.75	3/2.25	6
0.1	0.5	2/1.5	4
0.02	0.2	1/0.75	2
0	0	0/0	0

Table 7.10 Target interpretation and resolution (n_0)

Category	Detection	Classification	Recognition	Identification
Truck	0.9	1.25	4.5	6.0
Tank	0.75	1.20	3.5	6.0
Vehicles without caterpillar	1.00	1.50	4.0	5.0
Jeep	1.20	1.50	4.5	5.5
Command vehicle	1.20	1.50	4.3	5.5
Individual	1.50	1.80	3.8	8.0
105 howitze	1.00	1.50	4.8	6.0

In the design, it is often based on the number of cycles that the target can be distinguished for each level of judgment of the target obtained from the experiment, and then according to the conditions of use, design and calculate the parameters of the system.

Let the target size be H and the distance be s, then the opening angle of the target to the system is H/s. If the limiting spatial frequency of the system is N_c (the number of cycles per unit angle), the number of resolvable cycles to the target is $n_0 = (H/s)\ N_c$. The following are two sets of data for target interpretation [5], as shown in Tables 7.9 and 7.10.

7.5.10 Typical Application of Infrared Cameras

The early warning satellites are equipped with infrared detectors and television cameras. When a ballistic missile is launched from the ground or underwater, the infrared detector with high sensitivity can detect the infrared radiation of the engine tail flame during the active phase of the missile flight and give an alarm. It is an important means of defense in modern war. With the development of infrared focal plane technology, the missile early warning satellite system has gradually shifted from strategic application to tactical application. Modern early warning satellite systems with infrared cameras are basically composed of two parts: One is a network of several geostationary orbit satellites to detect missile launches around the world; the second is a constellation of low-orbit satellites used to track missiles in mid-flight, while the latter uses cold space as a background to find and track targets. Such a satellite system would be used to detect and track not only strategic and tactical missiles but also cruise missiles and strategic aircraft. For early warning, the infrared early warning satellite located in the geostationary orbit can realize the detection and early warning of strategic missiles, tactical ballistic missiles and aircraft in afterburner work.

In addition to Russia and the United States, which have 24-h uninterrupted global missile early warning monitoring capabilities, the French National Aerospace Corporation believes that Europe has the ability to develop such a system, and two stationary satellites can cover Europe and serve as a missile early warning system, which can predict the trajectory of incoming missiles and the impact zone of possible attack.

The United States, which is the leader in this aspect, has already passed through the Midas polar-orbiting satellite and the three-generation DSP early warning satellite, as shown in Table 7.11. The space-based infrared system SBIR, which uses scanning and staring dual-mode imaging, has also been put into operation. Judging from the working status of missile early warning satellites, more than 1000 missile launches (including long-range and submarine-launched missiles) have been observed from the Soviet Union, the United States, France and China since the DSP satellite network was established to the end of the twentieth century. In the Gulf War, the two DSP satellites of the United States detected the "Scud" missile within 30 s after it was launched and immediately transmitted to the launch position of the "Patriot" missile. The phased array radar immediately began to search, capture and track the "Scud" missile. It enables the "Patriot missile" to intercept and destroy the "Scud" within 5 min of its flight. The infrared telescope on the DSP plays a key role.

7.6 Imaging Spectrometers

Imaging spectrum technology is the frontier of international remote sensing technology development in the 1990s, and it is one of the focuses of contemporary remote sensing. Its special requirement is to perform spectral analysis on each imaging position, so spectral resolution and spectral range, as well as strict radiometric response calibration of the receiver, are important technical indicators. Its spectral range is generally visible to infrared spectrum, such as $\lambda = 0.4$–$2.5\ \mu m$, and thermal infrared (8–14 μm) spectral region, the spectrum is subdivided into tens or even hundreds of spectral segments, and the same ground object is imaging to obtain a large amount of narrow-spectrum remote sensing image information.

Depending on the spectral resolution, imaging spectrometers can be classified into three types: multispectral, hyperspectral and ultra-spectral. The spectral resolution of the multispectral type is about 1/10 of the corresponding wavelength, that is, $\Delta\lambda$

Table 7.11 Approximate parameter table of typical foreign infrared early warning cameras

	Midas	647 (DSP) satellite		
		First generation (1–7)	Second generation (8–13)	Third generation (14–22)
Start and end time (years·month)	1960.2–1963.7	1970.11–1977.2	1978.6–1987.11	1989.6–
Satellite height (cm)	914.4	654	670	990
Diameter (cm)	152.4	278	278	420
In-orbit weight (kilogram)	3000	1130–1150 907 (net weight) 450 (payload)	1149.8	2381.4 180 (payload)
Attitude control	Gravity gradient stabilization	Spin stabilized	Three axis stabilized	Three axis stabilized
Nominal Orbit height (km)	3600	36,000	36,000	36,000
Nominal period (hour)	2.5	24	24	24
Networking (number of pieces)	8–12	3	3 (two more for backup)	3 (two more for backup)
Working life (years)	Hours and days	1–2	3–5	7–9
Detector	(Single pixels) lead sulfide	2000 pixels lead sulfide device (2D structure close to linear array)	PbS4000 (2000)	6000 pixels PbS detector is equipped with a set of mercury cadmium telluride two-color detectors at the end
Working temperature (K)	Normal temperature	195		77
Work spectrum (microns)	2.3	2.7	2.735–2.89	2.7–2.9, 4.3–4.4
Spectral bandwidth (microns)		0.07–0.1	0.155	
Ground resolution		About 3 km	1.5 km	1 km

$= \lambda/10$, and generally, there are dozens of spectral bands; the spectral resolution of the hyperspectral type is about 1/100 of the corresponding wavelength, that is, $\Delta\lambda = \lambda/100$, there are generally hundreds of spectral bands; the spectral resolution of the ultra-spectral type is about 1/1000 of the corresponding wavelength, that is, $\Delta\lambda = \lambda/1000$, and generally, there are nearly a thousand spectral bands.

The multispectral type covers a wide spectrum range, and the spectral segments are selected at the places that can best reflect the radiation characteristics of the target. The number of spectral segments is usually about 10–20, which is suitable for zone classification and land use assessment. The hyperspectral type has narrow spectrum coverage and a resolution of 50–300 spectrum segments, which is mainly used in agriculture, forest, mineral, land, river basin survey and coastal area analysis. The ultra-high spectral type has the narrowest spectral range and the highest spectral resolution, generally more than 300 spectral segments, which can be used for the study of particles and gas components.

A large number of experiments have shown that most substances on the earth's surface have diagnostic spectral features in the spectral ranges of 0.4–2.5 μm and 8–14 μm in the atmospheric window. These features can identify most substances on the earth's surface, as shown in Table 7.12. Using spectral features can improve the detection and recognition ability to the target. With the wide application of earth observation optical remote sensing technology in national defense and national economy and the continuous deepening of research, users have higher and higher requirements for the ability to identify substances, and multispectral remote sensing technology has become more and more difficult to meet the needs of users. With the growing demand, hyperspectral remote sensing technology provides the possibility to meet the needs of users.

Since the imaging spectrometer is designed with spectral resolution as the goal, the focal length of the camera is not long, such as the high-resolution imaging spectrometer $f = 600$ mm. The imaging spectrometer is formed by connecting the imaging objective lens and the spectrometer. In addition to the spectral resolution perpendicular to the slit direction, the imaging spectrometer also requires the imaging resolution along the slit direction, which is different from ordinary spectrometers. The higher the spectral resolution, the narrower each spectral band and the weaker the energy. In order to obtain a suitable exposure, it is generally necessary to have a larger relative aperture of the imaging objective.

Another key point of imaging spectroscopy technology is spectroscopic technology. Traditional imaging spectrometers use prisms and gratings to split light. Interferometric hyperspectral imaging techniques of the Fourier transform type have now been developed.

7.6.1 Dispersing Prism Splitting

For dispersive prism splitting, Fig. 7.51 shows the typical application of dispersive prism in imaging spectrometer. The incident slit is located on the front focal plane

Table 7.12 Spectral and substance identification

Spectral/ μm	Substance	Spectral/ μm	Substance
0.40–0.42	Strong absorption of plant chlorophyll/water color/metal ions	0.78–0.80 0.80–0.82	Divalent copper, serpentine absorption
0.42–0.44	Plant chlorophyll a, strong absorption of carotenoids, absorption of ferrous minerals	0.82–0.84 0.84–0.86	Vegetation high reflection/red edge top
0.44–0.46	Strong absorption of plant chlorophyll/metal ions/aerosol	0.86–0.88	Mineral absorption of ferric iron
0.46–0.48	Carotenoids are strongly absorbed	0.88–0.90	Vegetation reflection peaks, identifying ice and snow particle size, ferric absorption
0.48–0.50	Strong absorption of carotenoid, absorption of ferric iron	0.90–0.92 0.92–0.94 0.94–0.96	Divalent and trivalent iron mineral absorption
0.50–0.52 0.52–0.54	Strong reflection area of plant chlorophyll / strong absorption of phycoerythrin in phycobilin		
0.54–0.56	Mineral absorption of ferrous iron and trivalent copper, strong reflection area of plant chlorophyll	0.96–0.97 0.97–0.98 0.98–1.00 1.00–1.02 1.02–1.04 1.04–1.06 1.07–1.08 1.08–1.10 1.10–1.11	Ferric iron mineral absorption/ atmospheric water vapor content/ water strength absorption valley point Identify ice and snow particle size/ vegetation type Vegetation reflectance peaks to identify ice and snow particle size Identify ice and snow particle size/ Fe ion uptake
0.56–0.58 0.58–0.60 0.60–0.62 0.62–0.64	Strong absorption of phycoerythrin in phycobilin		
0.64–0.66	Strong absorption of phycoerythrin in phycobilin, strong absorption of plant chlorophyll		
0.66–0.68	Strong absorption of plant chlorophyll, most vegetation reflectivity valleys	1.11–1.12 1.12–1.14 1.14–1.16 1.16–1.18 1.18–1.19 1.19–1.20 1.20–1.22 1.22–1.24 1.24–1.25 1.25–1.26 1.26–1.28	Identify the size of ice and snow particles, the strong absorption valley point of ice Identify ice and snow particle size/ vegetation type/rock mineral water absorption Identify snow and ice particle size vegetation types/atmospheric water vapor content/pyrophyllite absorption
0.68–0.70	Vegetation red edge/copper trivalent, yangqi stone absorption		
0.70–0.72 0.72–0.74 0.74–0.75	Vegetation red edge information/ weak absorption of water		
0.75–0.76	Strong absorption valley point of water and oxygen, plant red edge information		
0.76–0.78	Plant red edge information/ pyroxene		

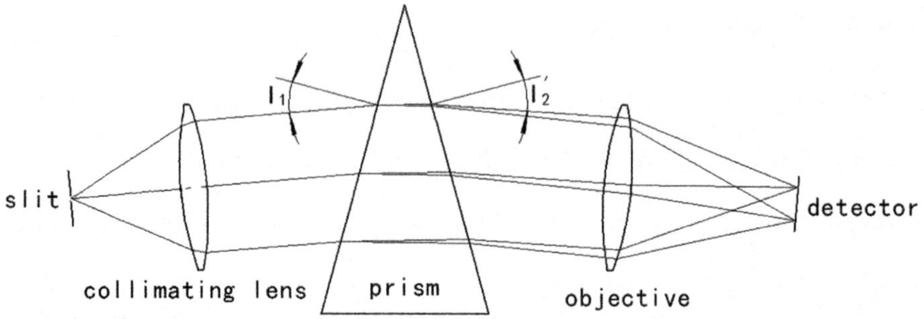

Fig. 7.51 Prism splitting

of the collimation system. After the incident light is collimated by the collimating system, the slit is imaged on the focal plane detector by the imaging system through the prism according to the wavelength.

Analysis of the dispersion principle of dispersive prism splitting, set the incident angle of the incident ray I_1, its refraction angle I_1', the incident angle of the prism I_2, its refraction angle I_2', the refractive index of the prism to the wavelength of light n and the top angle of the prism α.

According to the law of refraction

$$\sin I_1 = n \sin I_1'$$

$$n \sin I_2 = \sin I_2'$$

$$\sin \frac{I_1 - I_2'}{2} = \frac{n \sin \frac{\alpha}{2} \cos \frac{I_1' + I_2}{2}}{\cos \frac{I_1 + I_2'}{2}}$$

$$= \frac{n \sin \frac{\alpha}{2} \cos \frac{\arcsin \frac{\sin I_1}{n} + \arcsin \frac{\sin I_2'}{n}}{2}}{\cos \frac{I_1 + I_2'}{2}}$$

According to this set of equations, the position and direction of the incident light from the prism can be obtained. The refractive index n of the prism for different wavelengths is different, and the convergence position is different on the area array detector. Therefore, the spectral image of the slit is generated on the detector, that is, the segmented spectral imaging is realized.

7.6.2 Diffraction Splitting

Diffraction grating splitting. The classic application method of diffraction grating is the same as that of dispersive prism, which are all located in the collimated light path. As shown in Fig. 7.52, the incident slit is located on the front focal plane of the collimating system. After the incident light is collimated by the collimating system and split by the grating, the slit is imaged by the imaging system on the area array detector in the focal plane according to the wavelength.

Another usage of the diffraction grating is to place it in a diverging beam, as shown in Fig. 7.53, the light incident from the slit is directly incident on the diffraction grating without going through the collimation system, and the target slit is obtained after diffraction by the grating. The imaging system images the slit on the area array detector in the focal plane by wavelength. This imaging technique was used in the conceptual design of the tactical remote sensor for the OrbView-4 satellite.

At present, the mature aerospace dispersive imaging spectrometers in the world are based on diffraction gratings. JPL whisk broom imaging spectrometer AVIRS, with 224 spectrum and 30° field of view, was successfully flighted in 1987. CASI in Canada and AISA in Finland also have carried out research work. In August 1997, the world's first spaceborne imaging spectrometer (HIS and LEISA split with grating and wedge filters, respectively) was launched by NASA-TRW and Lewis. HIS was developed by TRW Company, 0.4–2.5 μm, 384 channels. LEISA was developed by the Lewis Company and failed to launch. The hyperion high-resolution imaging spectrometer, launched in November 2000 by the United States, is equipped with a 220-channel grating spectrum of 0.4–2.5 μm, a spatial resolution of 30 m, coverage of 7.5 km and a push-sweep operation.

In terms of military applications, the United States plans to carry a high-performance imaging spectrometer Coastal Ocean Imaging Spectrometer (COIS) on the Naval Earth Mapping Observer (NEMO) satellite for environmental monitoring. It is mainly used for coastal sea water monitoring, agricultural remote sensing,

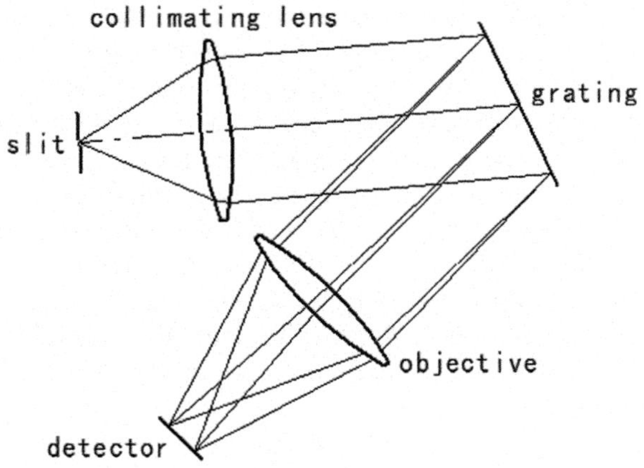

Fig. 7.52 Diffraction grating splitting

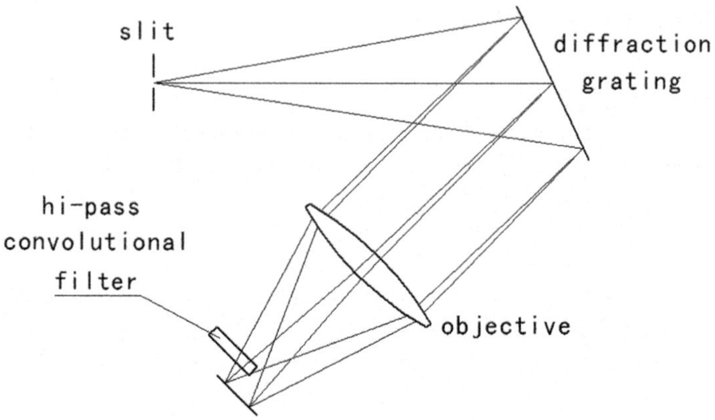

Fig. 7.53 Divergent light path diffraction grating splitting

resource survey and disaster assessment. The ground resolution is 30 m (compensated) 160 m (non-compensated), the coverage is 30 km, the spectrum is 0.4–2.5 μm, 210 channels, the spectral resolution is 10 nm, and the panchromatic camera has a resolution of 5 m.

7.6.3 Binary Optical Splitting

Using the splitting technique of binary optical elements in Fig. 7.54, Enise Lgons of the Roma Laboratory of the American Optical Quantum Center proposed an imaging spectrometer using binary optical elements. The binary optical element is both an imaging element and a dispersive element. The area array CCD detector is used to scan the imaging range of the required spectral band along the optical axis direction, and each position corresponds to the corresponding wavelength imaging area.

According to the dispersion properties of the binary optical element, the binary optical element converges the incident light like an ordinary lens, but it is based on the principle of diffraction rather than the principle of refraction. The effective focal length of chromatic aberration due to diffraction is inversely proportional to the wavelength $f(\lambda) = f_0\lambda_0/\lambda$, where f_0 is the focal length of the design wavelength λ_0. Unlike prisms or grating elements, which disperse along the direction perpendicular to the optical axis, binary optical elements disperse along the axis. The spectral resolution of imaging spectrometers using binary optics is determined by the size of the detector. The imaging spectrometer has a compact structure and high diffraction efficiency. Pacific Advanced Technology has developed a number of imaging spectrometers based on image multispectral sensing (IMSS) technology of binary optical elements.

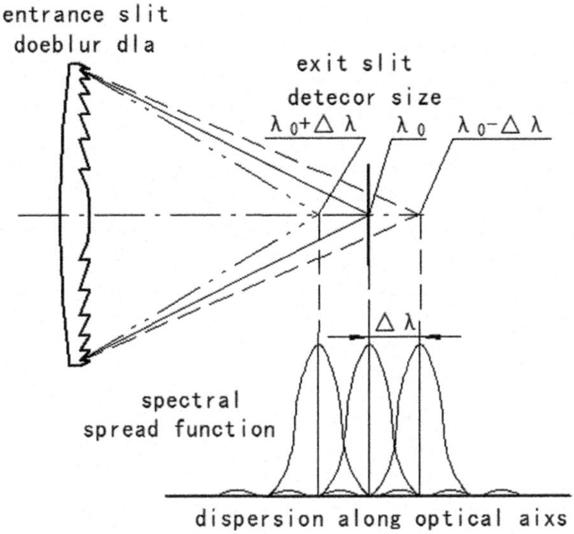

Fig. 7.54 Binary optical splitting

7.6.4 *Interference Splitting*

The spectral resolution and spatial resolution of the dispersive imaging spectrometer are restricted by the slit, so that the light energy passing through is limited, and the signal-to-noise ratio of the obtained spectrum is not very ideal. Therefore, high-throughput and multichannel interferometric imaging spectrometers are gradually developed. Due to its huge potential, many countries have carried out research work in interferometric imaging spectroscopy.

7.6.4.1 Time-Modulated Interference Splitting

Michelson interferometry. Research on interferometric imaging spectroscopy began in 1980s, when most of the imaging spectrometer schemes based on Michelson interferometers were used. As shown in Fig. 7.55, the slit is located on the focal plane of the collimating mirror, the collimated beam is split into two beams by the beam splitter, one beam is returned by the stationary mirror, and the other beam is returned by the moving mirror, the two returning beams are superimposed and coherent on one side of the lens and then imaged on the detector by the lens. As the position of the moving mirror is different, the optical path difference of the two beams of light on the detector is different to form images of different wavelengths, that is, a spectrum map is obtained. Since the interferogram of the imaging spectrometer based on the Michelson interferometry is collected over time, it is also a time-modulated imaging spectrometer.

The main advantages of this imaging spectrometer are high sensitivity and high spectral resolution. The main difficulty is that it is sensitive to disturbance and

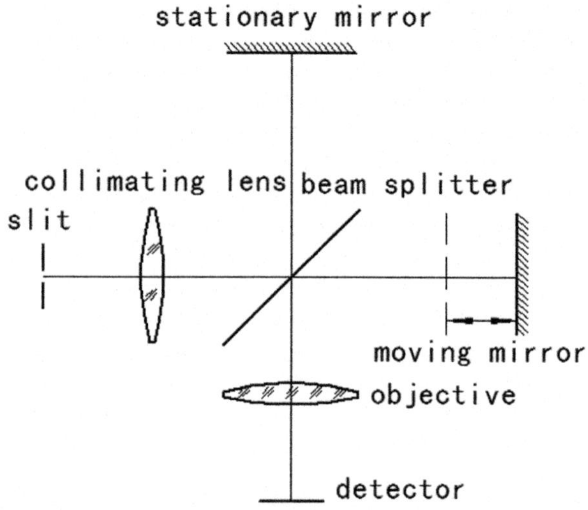

Fig. 7.55 Michelson interferometry

require a moving mirror scanning system with high stability. In 1991 and 1995, the French Space and Strategic Systems Division and the American Livermor Laboratory developed the Michelson interferometric time-modulated imaging spectrometer prototype.

7.6.4.2 Transverse Shear Interference Splitting

Sagnac interferometry. In order to avoid the difficult problem of dynamic scanning mirror stability in time-modulated interferometric imaging spectrometer, the Stationary Imaging Fourier Transform Spectrometer becomes a new research focus. Imaging spectrometers based on transverse shear interferometry are one of them. At present, Sagnac transverse shear interferometry is the most widely used transverse shear interferometry technique in imaging spectrometer design.

The so-called Sagnac beam splitter is to translate a reflection surface in the Sagnac interferometer for a certain distance, so that the two original coincident outgoing beams are separated in parallel, as shown in Fig. 7.56. When the object is at infinity, that is, a parallel beam of light enters the Sagnac beam splitter and is laterally sheared into two coherent beams that are parallel to each other. The distance between the two parallel beams is called the lateral shearing amount. In a space modulated interferometric imaging spectrometer, the object (slit) is located at finite distance (front focal plane of the FTL), and the Sagnac beam splitter divides the slit into two virtual slits (virtual images) along its width. The two virtual images are located on the front focal plane of the Fourier mirror. After passing through the Fourier lens, they become two parallel beams with a certain angle. When the two beams of the same wavelength in the parallel light of the same light source meet, the light an interference phenomenon will occur. Lights of different wavelengths have different interference positions on

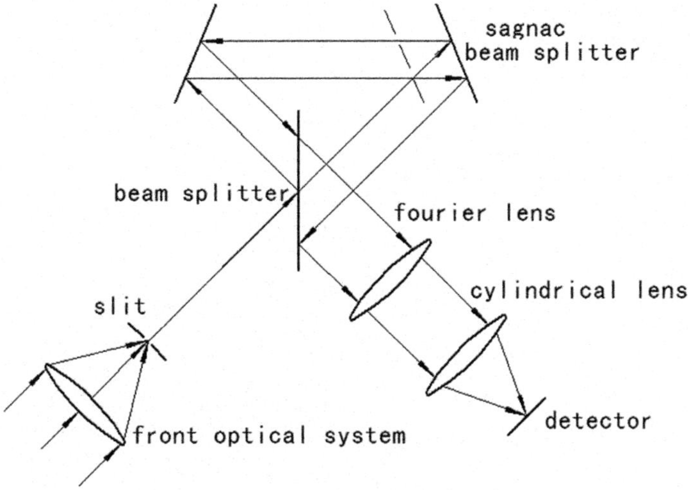

Fig. 7.56 Transverse shear interference splitting

the vertical plane of the two beams of light, forming an interference spectrum, which is imaged on the photosensitive surface of the detector by a cylindrical mirror to obtain a spectral image of the target.

The transverse shear optical path analysis of the Sagnac beam splitter is shown in Fig. 7.57. Firstly, the optical path difference of the two beams of Sagnac beams splitter transverse shearing transformation is analyzed by taking a special case of incident at an incident angle of 45° with the beam splitter. After the light is incident at point A, it is divided into beams, one way to $ACBE$, and the other way to $AB'C'Q$. As $B'D \| CC'$, $LD \perp AB'$, the isosceles right triangle $\triangle A'B'C' \cong$ isosceles right triangle $\triangle aDC$, that is, coincides with the shadow. The first path of light subtracts the side length of the shadow part from A to E and notices that $AA' = OE$, $A'Q = QO$, then the remaining length of the first path of light is $b + b + BD$. The second path of light subtracts the side length of the shadow part from A to Q and notices that $AA' = OE$, $A'Q = QO$, then the remaining length of the second light is $a + a$. If the two optical paths are equal, only needs $a = b + B'D/2$.

Because CC' and BB' are static and moving mirrors symmetrically arranged relative to the beam splitter.

Therefor $2\angle 1 = \angle BB'D$.

And because AC and CB are the incident light and reflected light of the mirror CC'.

Therefore, $\angle HCA = \angle BCC' = \angle BDB' = 45° + \angle 1 = 45° + \angle LB'D$.

Therefore, $\angle 1 = \angle LB'D$.

Connect B', p cross DB to F.

Right angle $\triangle DFB' \cong$ Right angle $\triangle BFB'$.

$\angle BB'D/2 = \angle FB'D = \angle 1 = \angle LB'D$.

Therefore, right angle $\triangle LB'D \cong$ Right angle $\triangle FB'D$.

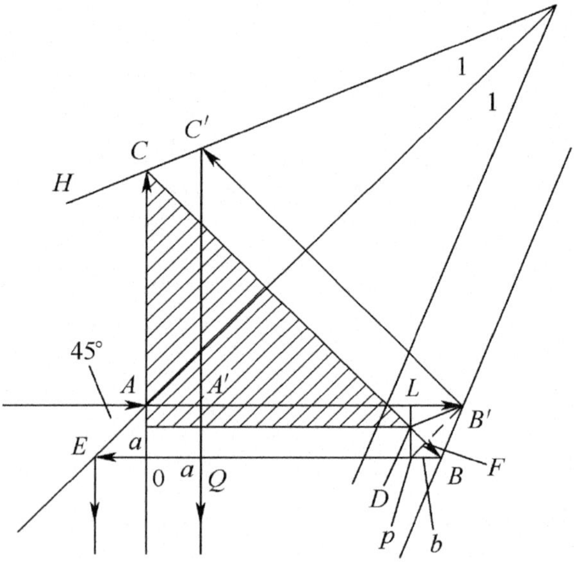

Fig. 7.57 Transverse shear optical path of the Sagnac beam splitter

Therefore, $LD = BD/2$

$$\text{Because } a = LD + b$$
$$= BD/2 + b$$

Therefore, $b + b + BD = a + a$.

That is to say, the light enters from point A, exits from points E and Q and is transversely sheared by $2a$, and the optical path difference is 0.

It can be proved that when the incident angle is other general angles, as long as the symmetrical configuration of the beam splitter and the reflector as shown in Fig. 7.56 remains unchanged, the iso-optical path characteristics of the two shear beams remain unchanged.

The Sagnac-type space modulated interferometric imaging spectrometer is mainly composed of a front optical system, a slit, a Sagnac-type beam splitter, a Fourier lens FTL, a cylindrical mirror CL, an area array detector and a signal processing system. The front optical system images the target on the front focal plane of the Fourier lens, where a slit is placed perpendicular to the paper surface. After being split by the Sagnac beam splitter, two slit virtual images are formed, which are both located on the front focal plane of the Fourier lens and parallel to the original slit, as shown in Fig. 7.58, EF is the iso-optical path plane of the beam S_1, PQ is the iso-optical path plane of the beam S_2, O is the center point of the two intersecting beams and is also the origin of the ξ, η coordinate system. At point O, the optical paths of the two beams are equal. From point O to the positive direction of the ξ axis, the optical path of the beam S_1 to the ξ axis (plane) gradually decreases, while the optical path of the beam S_2 to the ξ axis (plane) gradually increases. The optical path difference of the

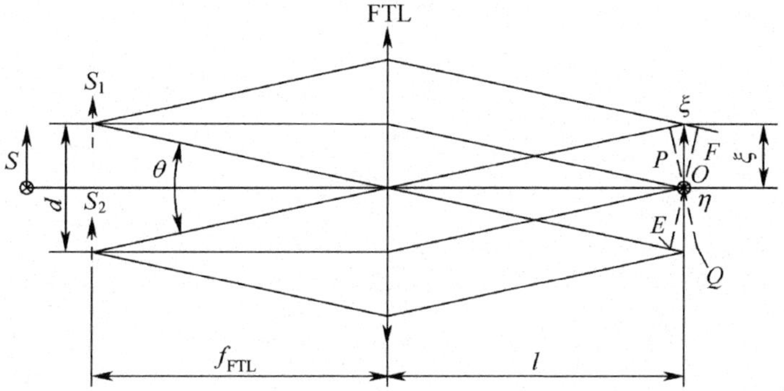

Fig. 7.58 Transverse shear interference principle

two beams reaching the ξ axis (image plane) gradually increases along the positive direction of the ξ axis.

After passing through the Fourier lens, the virtual images of the two slits become two parallel light beams with a tiny angle, in which the light of the same wavelength will interfere after meeting. In the previous analysis, it has been known that along the positive direction of the ξ axis, the optical path difference of the two beams S_1 and S_2 on the image plane gradually increases. After passing through the cylindrical mirror, the beam converges in the width direction of the slit and is imaged on the detector to form an interference spectrum.

Compared with the time-modulated interferometric imaging spectrometer, the space modulated interferometric imaging spectrometer has the following basic performance parameters:

1. Spectral resolution

$$\delta\nu = \frac{f_{FTL}}{\varepsilon dN} \quad \delta\lambda = \lambda^2\delta\nu$$

2. Maximum measurable wavenumber

$$\nu_{max} = \frac{f_{FTL}}{2\varepsilon d} = \frac{N}{2}\delta\nu$$

3. Maximum resolution

$$R = \frac{\nu_{max}}{\delta\nu} = \frac{N}{2}$$

where

N the number of sampling points of the interferogram;
ε the sampling interval of the interferogram, that is, the spacing of the detector pixels;

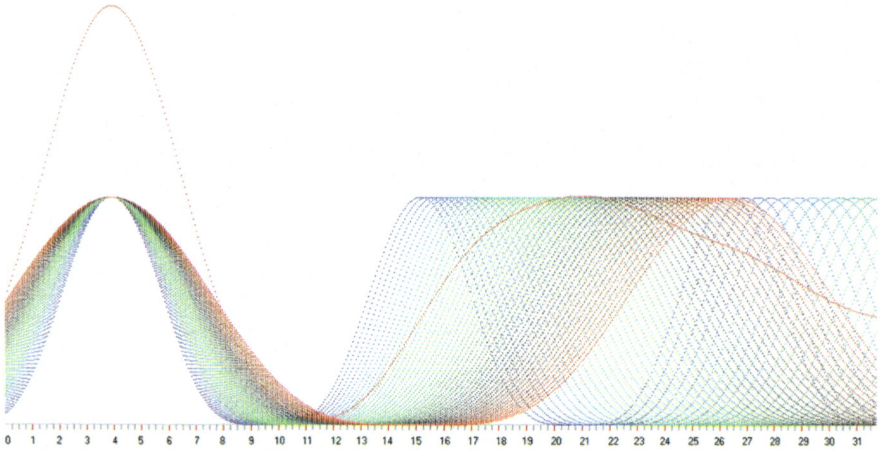

Fig. 7.59 Simulation of transverse shear interferences spectra

d the distance between the two slit virtual images shown in the figure.

According to the principle of transverse shearing interferometric spectrometer as described above, in the detector configuration shown in Fig. 7.58, the geometric center of the photosensitive surface of the detector is located at the coordinate origin O. Considering the symmetry of the interferogram, a detector with 512 rows of pixels is used in the simulation. The configuration is shown in Fig. 7.59. The origin of the coordinates is located at the 4th scale point in the figure. The abscissa in the figure has a total of 32 large grids, each large grid is divided into 4 small grids, and each small grid represents 4 pixels, and a total of $32 \times 4 \times 4 = 512$ rows of pixels. The computer simulation results of transverse shear interference imaging are shown in Fig. 7.59. The abscissa corresponds to the ξ axis, and the peak height at the fourth scale point represents the energy at this coordinate point, which is the panchromatic spectral point.

Since the spectral characteristics of the target are not set, the peak heights of the spectral curves of each color in the figure are equal. The highest curve corresponding to the 4th scale, line type …, represents the sum of the various spectral energies at each pixel. Figure 7.60 shows the imaging spectrum of the target static interference imaging spectrometer under natural illumination, and the situation near the panchromatic spectral point is very similar to the computer simulation.

In 1993, with the support of the US Office of Naval Research (ONR), Hawaii University and Florida Institute of Technology jointly developed a space modulated interferometric imaging spectrometer spatial modulation imaging Fourier transform spectrometer (SMIFTS) using the Sagnac transverse shearer as a splitter element. The spectral range is 1–5 μm, and the spectral resolution is 100–1000 cm^{-1}. In 1995, with the support of the US Air Force, Kestrel Corporation cooperated with Florida Institute of Technology and other units to improve the previously developed SMIFTS and developed an airborne Fourier transform imaging spectrometer FTVHSI, whose

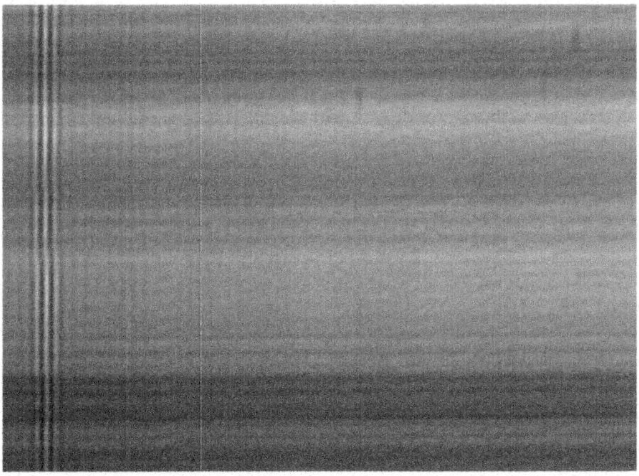

Fig. 7.60 Spectrum of static interference imaging spectrometer under natural illumination

spectral range was 0.44–1.1 μm, number of spectral bands was 256, field of view was 15°, and instantaneous field of view was 0.8 mrad.

On July 19, 2000, the interferometric imaging spectrometer FTHSI, based on the Sagnac transverse shear interferometer spectroscopy, carried on the powerful satellite Mightysat 2.1, which was successfully launched from Vandenberg Air Force Base in California, USA, performed very well in the subsequent detection, which fully proved that the superior performance of space modulated interferometric imaging spectrometers.

7.6.4.3 Crystal Shear Interference Splitting

The principle of steady-state polarization interference imaging spectrometer (SPIS) based on Savart deflection light transverse shear interferometer is shown in Fig. 7.61. It is mainly composed of collimating mirror L_1, polarizer P_1, Savart deflection light transverse shearer, analyzer P_2, imaging mirror L_2, area array detector (placed on the focal plane) and data image processing system.

Savart optical splitter is composed of two uniaxial positive crystals (or negative crystals) with a thickness of t. The optical axis of the front piece (left board) is in the paper plane (*XZ* plane) and forms an angle 45° with the positive *X* and *Z* axes; the optical axis of the second piece (right board) is in the horizontal plane (*YZ* plane) and forms an angle 45° with the positive *Y* and *Z* axes. The polarization directions of P_1 and P_2 are all at an angle of 45° with the positive *X* and *Y* axes (the maximum shear amount can be obtained).

The target *S* is located on the focal plane of the collimating mirror L_1. After passing through P_1, a beam of natural light emitted from *S* becomes a linearly polarized beam vibrating in the polarization direction of P_1, which is incident on the left plate of the Savart optical splitter and enters the birefringent crystal. After that, it is divided into

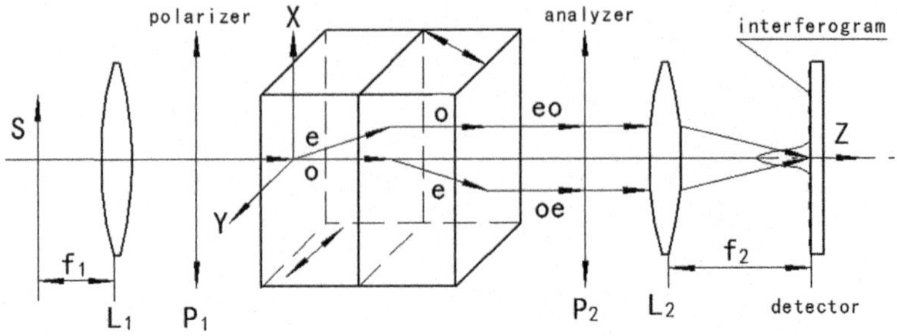

Fig. 7.61 Crystal shear splitting

ordinary light (o light) and extraordinary light (e light). The *o* light propagates along the original direction, and the e light is deflected.

After entering the second crystal plate after passing through the interface of the two plates, the original *o* light becomes the e light deflection, and after being deflected by the rear surface of the Savart optical splitter, it exits along the direction parallel to the incident light (oe light); the original e light becomes *o* light and is deflected and then exits in the direction parallel to the incident light (eo light). That is, the outgoing light is two beams of linearly polarized light with a certain distance (transverse shear amount) parallel to the propagation direction of the original incident light with the vibration directions perpendicular to each other. After passing through P_2, it becomes two beams of linearly polarized light with the same vibration direction (along the polarization direction of P_2), which meet on the plane of the area array detector through the imaging mirror and satisfy the coherence conditions (same frequency, same vibration direction and fixed phase difference) to form an interferogram as a target image. After the interferogram is analyzed and processed (Fourier transform), the spectral information of the incident light source can be obtained.

In 1992, Osaka University in Japan developed a multichannel Fourier transform infrared spectrometer based on Savart optical splitter, with a wavelength range of 2–5 μm and a spectral resolution of 27.6 cm^{-1} (11–44 nm).

7.6.4.4 Rotating Moving Mirror Interference Spectrometer

The time-modulated interferometric imaging spectrometer is formed on the basis of the Michelson interferometer and uses the moving mirror to scan to generate the interference spectrum. The linear moving mirror scanning system has high requirements on the precision of the driving system of the moving mirror, and the real-time performance is not good. But when the translation of the moving mirror is converted into rotation, such as the high-speed rotating mirror spectrometer proposed by the famous scholar Griffiths, it provides a new way for the development of spectrometers with high spectral resolution.

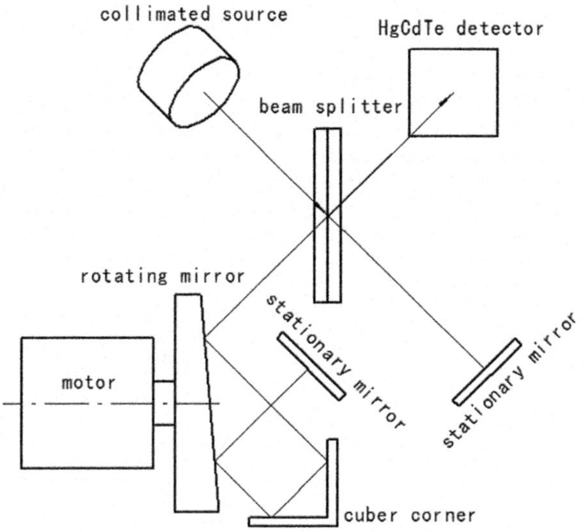

Fig. 7.62 Time-modulated interferometric splitting

Figure 7.62 shows the ultra-rapid-scanning Fourier transform infrared spectrometry proposed by Griffiths.

The system adopts a relatively stable rotating mirror system and uses a cuber corner structure, so that the wavefront of the incident light after multiple reflections by the high-speed rotating mirror, corner reflector and flat reflector can be returned to the beam splitter according to the incident light path, maintaining good coherence. Griffiths was also awarded the 2003 American Bomem Michelson Outstanding Contribution Award for Analytical Chemistry and Applied Spectroscopy for proposing this novel imaging spectrometer scheme. In this structure, when the rotation frequency of the motor is 500 revolutions, 1000 scans per second can be achieved. The scanning speed of this scheme can reach at least 2000 scans per second. The optical path difference can reach 0.25 cm. This structure is suitable for the measurement of fast spectral transformation process.

For imaging spectrometer, the key lies in the choice of spectrometer scheme, from the traditional dispersion type to the later interference type. Generally speaking, the spectral resolution of the dispersive type is lower, and the spectral resolution of the interference type is higher. The most notable is the transverse shear static interferometric imaging spectrometer, which has high reliability and is suitable for aerospace remote sensing because there are no moving parts.

References

1. Zhang F, Sun X. Engineering optics. 2nd ed. Tianjin: Tianjin University Press; 1990.
2. Yimo Z, editor. Applied optics. Beijing: Publishing House of Electronics Industry; 2010.

3. Tao J, Wang S. Analysis on the large-scale space three-linear array camera precision sensitive factors. Opt Tech. 2006.
4. http://sjfw.sasmac.cn/index/wxcp.jsp.
5. Boruo C. Infrared system. Beijing: Ordnance Industry Press; 1995.
6. Li J, Li Y, Jin W. Low light level and infrared imaging technology. Beijing: Beijing Institute of Technology Press; 1995.
7. Huo Y, Yan C, Zhang A, Zhang X. Research on a new type of double temperature zone large cooling capacity Stirling refrigeration machine. In: Proceedings of the 19th China Symposium on Remote Sensing (2014).

Chapter 8
Platforms of Optical Remote Sensing

Space optical remote sensing platforms have been serialized, and by the end of twentieth century, about 6000 satellites have been launched. There are space platforms such as Voyagers 1 and 2, which have already flown out of the solar system, and space platforms for deep space exploration. Space-based optical telescopes such as Hubble have carried out long-term high-precision and high-resolution astronomical observations and achieved unprecedented observation results. There are also some space platforms include manned space station, space laboratory, recoverable satellite and space shuttle. In addition, there are large satellites with comprehensive applications and small satellite clusters with specific thematic goals. Satellites of different altitudes and applications constitute multiangle and multiperiod earth observation.

8.1 Main Types of Remote Sensing Satellites

Remote sensing satellite has two major fields, military and civil. Civil satellite is easier to be serialized and typified, and its development direction and purpose are clear and stable. The development of military remote sensing satellite has a leapfrog, and the serialization and type are not obvious. A certain performance requirement is always prominent and clear, which has great traction on technical performance of remote sensing satellite. Both of them have relatively high functional similarity in terms of use, such as imaging, mapping and spectral imaging. Therefore, optical remote sensing satellites tend to be shared by military and civilian, as civilian in peacetime and military in wartime. China's remote sensing satellite platform has developed into series, and the technical performance has approached international level. The following is an introduction to optical remote sensing satellite from perspective of its application.

J. Tao, *Space Optical Remote Sensing*, Advances in Optics and Optoelectronics,
https://doi.org/10.1007/978-981-99-3318-1_8

8.1.1 Imaging Remote Sensing Satellites

In terms of imaging satellites, Gaofen-2, which was successfully launched by a Long March-4B rocket from Taiyuan Satellite Launch Center at 11:15 on August 19, 2014, entered the preset orbit smoothly. The scanning width of the satellite is 45 km, and the focal length of camera on satellite is 7 m. The design life of the satellite is five years, as well as the target life is eight years, and 98% of the satellite was domestically produced. It consists of more than 10 subsystems and more than 200 devices.

The development of Gaofen-2 satellite has achieved submeter spatial resolution, acquired multispectral optical remote sensing data, solved the difficulties in design of long focal length, light camera and satellite system, achieved the high-precision attitude maneuver and high-precision image positioning with high stability and improved long life and high reliability of low orbit remote sensing satellites. It can cooperate with the Gaofen-1 satellite in orbit to provide service support for dynamic monitoring of land use, monitoring and evaluation of urban and rural planning, transportation network planning, forest resource survey, desertification monitoring and other industries and regional applications in the Capital Economic Circle.

Figure 8.1 is an in-orbit image of Gaofen-2 satellite with a high spatial resolution imaging camera and a multispectral camera. The measurement and control system has star sensor, sun sensor, omnidirectional beacon antenna and separation switch; propulsion system has thruster, engine and propellant tank; and data transmission system has transmitting and receiving antenna, etc.

Figure 8.2 is a fusion image of Beijing with panchromatic 0.8 m and multispectral 3.2 m resolution taken by Gaofen-2 satellite on September 27, 2014, with a scale of 1:4000.

The civil (commercial) satellite with the highest resolution in the world is GeoEye-1 satellite of DigitalGlobe Corporation of United States, as shown in Fig. 8.3. Its spatial resolution is 0.41 m, and its positioning accuracy can reach within 5 m.

Fig. 8.1 Gaofen-2 satellite

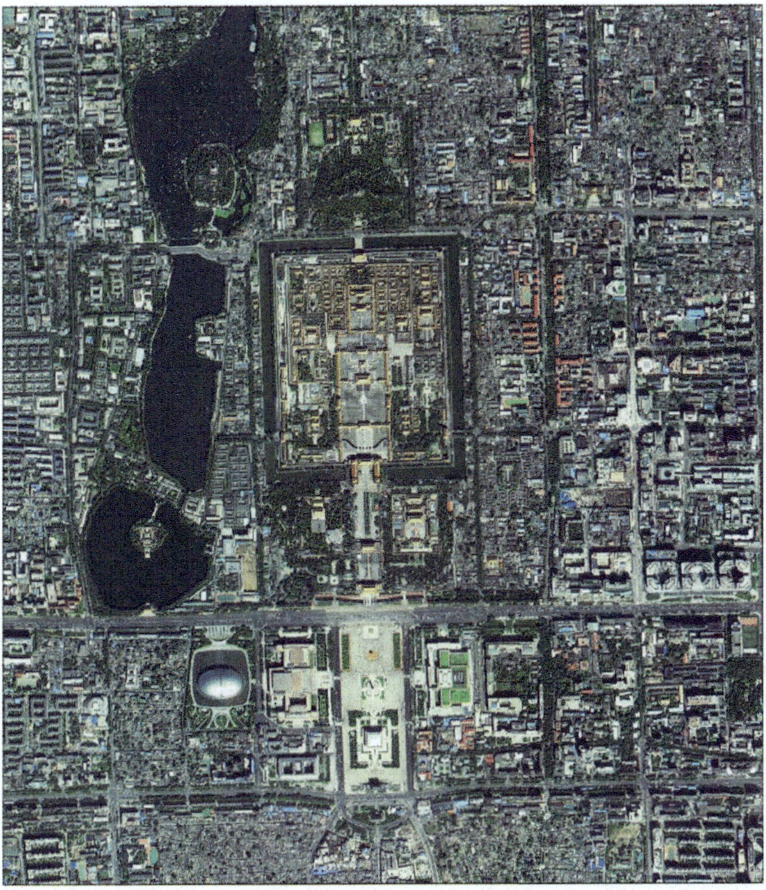

Fig. 8.2 Fusion satellite image with spatial resolution of 0.8 m and multispectral 3.2 m

Fig. 8.3 GeoEye-1 satellite

8.1.2 Optical Stereo Mapping Satellites

Tianhui-1 01 and 02 satellites were successfully launched and put into operation on August 24, 2010, and May 6, 2012, respectively. Tianhui-1 03 was launched on October 26, 2015, making China become one of the few countries in the world to master complete sets of satellite mapping technology. Figure 8.4a and b are stereo mapping satellite and stereo mapping camera, respectively.

As China's first generation of transport stereoscopic mapping satellites, Tianhui-1 is mainly used for scientific research, land and resources census, mapping and other scientific experiments. It adopts CAST 2000 satellite platform developed by DFH Satellite Co., Ltd. The payload includes integrated three-linear CCD camera, 2 m resolution panchromatic camera and 10 m resolution multispectral camera. The camera is equipped with three star sensors and two measuring GPS receivers (cold backup for each other), and the payload accounts for about 50% of dry weight of the satellite that weighs about 1000 kg. And it can obtain 3D geographic information, the establishment of global photography measuring control network, the implementation of 1:50,000 scale map and also can get blue, green, red and near infrared spectrum of 4 bands of multispectral imaging, as well as quantitative inversion of physical properties of ground objects. By multispectral image and panchromatic image, fusion processing produces colorful projective products. It can also obtain 2 m resolution panchromatic ground object image [1]. Tianhui-1 realized transformation of mapping satellite from recoverable film type to CCD transmission type in China. Image data were processed by ground system, and the relative accuracy plane of 12 m and elevation of 6 m (both 1σ) were close to American SRTM under the condition of no ground control point.

Among the three types of optical remote sensing equipment in Tianhui-1, two main optical remote sensing technologies were adopted, namely total reflection off-axis three-mirror optical system technology and three-linear array stereo mapping technology. The former was used in the imaging camera and multispectrum camera,

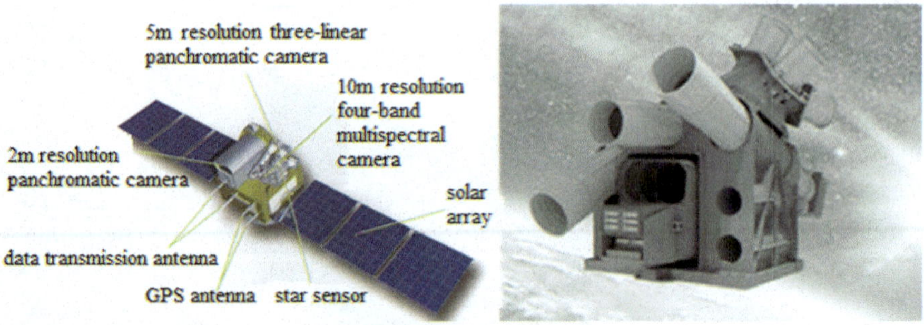

(a) Stereo mapping satellite

(b) Stereo mapping camera on Tianhui-1

Fig. 8.4 High-resolution stereo mapping satellite

while the latter was used in the three-linear array panchromatic stereo camera, which was the core payload of Tianhui-1 satellite. Main parameters are shown in Table 8.1.

1. **Off-axis three-mirror reflection optical system technology**

Both the high resolution and the multispectral cameras used off-axis three-mirror anastigmat (Cook-TMA) all reflection optical system. The mapping camera, one of the core payloads on Tianhui-1 satellite, had ground coverage of 60 km. As a high-resolution camera and a multispectral camera on the same platform, it had the same coverage as the mapping camera, which was conducive to satellite data collection and data utilization. Thus, three cameras were required to have the same field of view, and the mapping camera had a field of view of 7°. According to the introduction of optical

Table 8.1 Performance data of Tianhui-1 satellites

Satellites	Tianhui-1 01	Tianhui-1 02
Launch time	August 24, 2010	May 6, 2012
Altitude (km)	500	500
Orbital inclination (°)	97.3	97.3
Orbital eccentricity	0	0
Camera type	Panchromatic camera Multispectral camera Three-linear array panchromatic stereo camera	Panchromatic camera Multispectral camera Three-linear array panchromatic stereo camera
Sub-satellite pixel resolution	Panchromatic camera, 2 m Multispectral camera,10 m Three-linear array panchromatic stereo camera, 5 m	Panchromatic camera, 2 m Multispectral camera, 10 m Three-linear array panchromatic stereo camera, 5 m
Side view (°)	0	± 10
Width (km)	60	60
Panchromatic/stereo: spectral band	0.51–0.69 μm	0.51–0.69 μm
Multispectral spectrum (μm)	Blue: 0.43–0.52 Green: 0.52–0.61 Red: 0.61–0.69 Near infrared: 0.76–0.90	Blue: 0.43–0.52 Green: 0.52–0.61 Red: 0.61–0.69 Near infrared: 0.76–0.90
Regression period/day	58	58
Photographic coverage	Between 80° north and south latitude	Between 80° north and south latitude
Local time at descending intersection	13:30	13:30
Business programming ability	Have	Have
Imaging capability (km^2/day)	1.5 million	1.5 million

selection of remote sensor, there are two kinds of optical systems that can achieve 7° field of view: off-axis three-mirror reflection system and refraction system. High-resolution cameras require a long focal length, so the three-mirror reflection system can shorten the tube length. Multispectral camera requires a wide spectrum, so it is advisable to choose an off-axis three-mirror system with total reflection ability.

The high-resolution camera achieved requirements of high pixel resolution and wide coverage by using off-axis three-mirror anastigmat reflection optical system. The multispectral camera achieved requirements of wide field of view and wide spectrum and achieved requirements of blue, green, red and near infrared four-spectrum imaging on a single camera.

2. **Three-linear array stereo mapping technology**

The CCD transmission three-linear array stereo mapping cameras were used in Tianhui-1 satellite. The mapping cameras were composed of three cameras with forward-looking, face-looking and rear-looking, all of which adopt refraction optical system. The cameras adopt three-linear array photography to realize three-dimensional mapping. The two ends of the linear array of face-looking camera were supplemented by four small area arrays for image correction symmetrically, forming the LMCCD mapping system, which was conducive to solving the problems of distorted three-dimensional model of track and poor elevation accuracy in the dynamic photogrammetry of the three-linear array mapping camera. And meanwhile, it was beneficial to improve the capability of high-precision photogrammetry without ground control points of the real-time data transmission optical photogrammetry satellite.

1. Main technical indicators of stereo mapping camera

   (1) Types of mapping: three-linear array stereo mapping;
   (2) Optical system type: refraction system;
   (3) Field of view: 7°;
   (4) Spectrum: 0.51–0.69 μm;
   (5) MTF: 0.23 @ 77 1p/mm;
   (6) Distortion: 0.003%;
   (7) Base height ratio: 0.6;
   (8) Pixel resolution: 5 m;
   (9) Orbital altitude: 500 km.

The three-linear array stereo mapping camera can take three-linear array stereo images of the same orbit with a resolution of 5 m, and the ratio of the height and the baseline of look forward and look back was about 0.6, which was conducive to stereoscopic observation and automatic image matching.

It can be seen from the chapter of main optical remote sensor types that the three-linear array stereo mapping camera is a metering device in essence. In order to achieve high precision of mapping, some technical methods were also adopted in the camera design.

2. Targeted design technology of stereo mapping camera

    (1) Optical system: image space telecentric optical path;
    (2) Focal plane of formal view camera: LMCCD configuration;
    (3) Area array CCD: using TDI driving technology;
    (4) Structural design: set up a mapping optical platform with high strength, high stiffness and high stability.

Image space telecentric optical path was beneficial to reduce the defocus sensitivity of mapping accuracy. Five independent CCD devices, one linear array and four area arrays, namely LMCCD mapping system, were designed and integrated on the focal plane of face-looking camera to improve the elevation accuracy of target positioning. The geometric position relationship between CCD devices was strict and can be stable after experiencing the corresponding dynamics and thermal environment. In order to improve the image signal-to-noise ratio (SNR) of dark ground objects, TDI imaging driving technology was used in the area array CCD of stereo mapping camera.

Mapping optical platform selected casting titanium alloy ZTC4 material, with precision solvent, high-pressure centrifugal casting titanium alloy material to achieve molding. The installation datum and testing datum parts were machined with manual grinding to ensure the high accuracy of geometric angle requirements of mapping tasks. It was beneficial for the mapping optical platform to integrate three mapping cameras, three star sensors and one multispectral camera into one, ensuring the high precision and high stability of relationship between star sensitivity and camera and the geometric angle between cameras. Three medium-precision star sensors were equipped and provided long-term, highly reliable and high-precision three-axis pointing for mapping tasks.

3. Calibration technology of stereo mapping camera

    (1) Focal length of parallel optical tube: 7.5 m;
    (2) Accuracy of two-dimensional rotary table: 0.5";
    (3) Calibration accuracy of main point position: 0.2 pixels;
    (4) Calibration accuracy of main distance: 20 μm;
    (5) Calibration accuracy of intersection angle: 2";
    (6) Distortion: geometric calibration.

High-precision geometric calibration of three-linear array stereo mapping camera made the calibration accuracy of orientation elements in the forward-looking, face-looking and rear-looking cameras better than 0.2 pixels.

4. Main technical indicators of stereo mapping satellites related to optical loads

(1) GPS time synchronization accuracy: better than 0.1 ms;
(2) Data rate: 1.62 Gbit/s.

In order to meet the requirements of mapping tasks for high-precision external orientation elements, the satellite was equipped with a metrology GPS receiver, which combined geometry and orbital dynamics positioning methods to achieve high-precision real-time orbit determination. At the same time, the original GPS measurement data can be transmitted to ground user through the data transmission channel to realize ground secondary orbit determination and further improve the accuracy of orbit determination.

The Tianhui-1 satellite was equipped with a high-precision payload time system. High-precision hardware second pulse signal was provided through GPS receiver, and whole second time corresponding to the above second pulse of GPS was broadcast through the bus. Each information source related to the mapping task took the GPS second pulse signal as the timing reference to generate their own high-precision time scale and ultimately ensure that the time scale and GPS time synchronization accuracy of each related information was better than 0.1 ms.

In terms of data transmission, the code rate was high, the data source was numerous, the data source rate was different from each other, the difference was great, and the total original data rate was 1.62 Gbit/s. Mapping transmission channel data processor (AOS1) sets six virtual channel, as well as three channel image data of three-linear mapping camera with 4:1 image data compression, one channel image data of multispectral camera with 4.5:1 compression, one channel image data of four small array without compression, one channel data of GPS original measurement/whole satellite telemetry without compression and carried on reconnection and transmission of the above data and non-image data. The high-resolution camera data transmission channel data processor (AOS2) had eight virtual channels to realize the reconnection and transmission of 6:1 compressed image data of high-resolution camera. According to the importance of different cameras in mapping tasks, different data compression rates were adopted.

5. Main technical indicators of ground system of stereo mapping satellite

(1) Standard landscape image: 60 $\times$ 60 km;
(2) Elevation resolution: 25 m;
(3) Measuring scale: 1:50 thousand;
(4) Revised measurement scale: 1:25,000.

The ground application system is an important part of system engineering of Tianhui-1 satellite, which realizes data reception, operation and control management, production and application services and is the main part of exerting satellite application efficiency. Its main function is to complete satellite photography mission planning and satellite payload operation management; receiving satellite downlink data; preprocessing satellite downlink data to generate satellite image products; complete data storage, management and distribution; detecting main parameters of satellite photography system and calibrating the image characteristics; conducting emergency surveying and mapping support processing and generating emergency surveying and mapping support products; complete balancing and positioning of satellite images; producing 1:50,000 scale digital topographic maps, digital elevation models and orthophotograph maps; revising 1:25,000 scale digital topographic map. Tianhui-1 satellite–ground application system has the capability of large-scale production, producing a variety of satellite image products and geographic information products. The archived data are enhanced by a certain image preprocessing algorithm. The accuracy of 1:50,000 digital elevation model (DEM), digital line graphic (DLG) and regional network balancing can meet the requirements of 1:50,000 mapping in China. Theoretically, the highest precision interval DEM of 25.0 m can be obtained from three-linear array image data. DEM obtained from satellite image data has good expression of geomorphic details and high precision, which is conducive to automatic collection of 1:50,000 DEM data.

6. Technical coordination of satellite system

The satellite adopted a high-precision unified time system, which provided a unified time better than 0.1 ms for each data source of the satellite's orbit measurement, attitude measurement and optical camera through the surveying GPS receiver, which provided a unified high-precision time scale for obtaining high-precision correction information of the ground data processing system, outer azimuth element and inner azimuth element. From the point of view of satellite–earth large-scale system, technical coordination was carried out, which provided favorable conditions for obtaining high-precision mapping results.

The optical system of mapping camera adopted image space telecentric optical path, the focal plane of face-looking camera adopted LMCCD configuration, the area array CCD adopted TDI driving technology and the unified mapping optical platform was used as the structural benchmark, which provided conditions for reducing the sensitivity of mapping camera precision to platform environment, time and attitude.

This was a practical application of the overall design idea of space remote sensing system.

7. Application of stereo mapping satellite

It provides a large number of remote sensing image data at many levels for 120 users of land mapping, transportation, urban planning, resource census and disaster monitoring. They are also widely used in land and resources survey, agricultural yield estimation, forestry survey, mapping, marine environment monitoring, urban land use and other fields. In response to major natural disasters, emergency response mechanisms were activated for Luobei flood in Heilongjiang Province and Dingxi earthquake in Gansu Province, photographic plans were made, and data were received

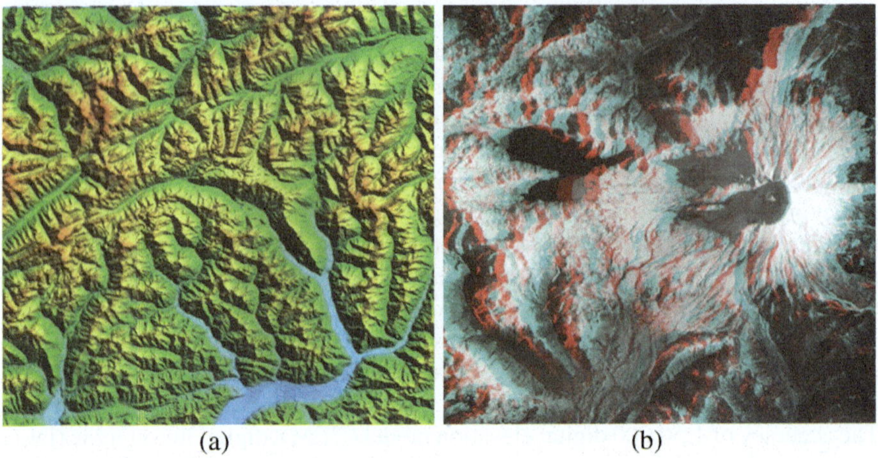

Fig. 8.5 Mapping image of Tianhui-1 satellite

overnight, as well as satellite image data were provided to the Ministry of Water Resources, the Ministry of Land and Resources, the National Emergency Response Office and other relevant ministries and commissions in the second half of 2013.

Figure 8.5 is the stereo image of Tianhui-1 satellite. Figure 8.5a is the digital elevation model of the Alps, showing three-dimensional sense of stereo mapping image. Figure 8.5b is a pair of red-blue stereoscopic images of Mount Rainier in United States. Stereoscopic vision can be produced by using the left-eye red filter and right-eye blue filter of both eyes.

8.1.3 Wide Coverage Optical Remote Sensing Satellites

There are two types of wide coverage satellites. One is high orbit satellites such as infrared early warning satellites, which cover about 40 percent of earth's surface area. The other is low orbit satellite with wide field of view, its coverage width in more than hundreds to thousands of kilometers. Such satellites generally have a specific purpose and are usually used by military or military/civilian sharing.

1. **Infrared early warning satellite**

Earth observation by using geostationary orbit has three advantages over other orbits. In terms of synchronous observation over a large area, geostationary orbit is a high orbit, and a single satellite can cover more than 40% of earth's surface. In terms of timeliness, the repeated observation period of the same region is the shortest in geostationary orbit, which is limited only by integration time, data acquisition speed and data downlink capacity. In other orbits, the repeated observation period varies from

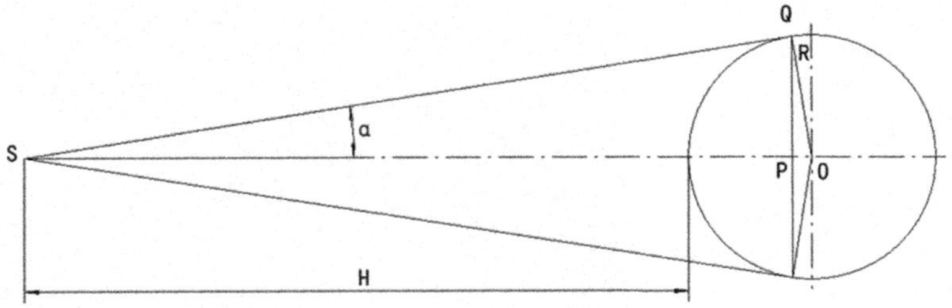

Fig. 8.6 Field angle covering the earth disk

a few days to dozens of days, depending on orbital altitude. Economically, geostationary orbit is far from earth, satellites do not have the problem of losing altitude due to atmospheric resistance, and the satellites have a long life span, allowing three satellites to achieve global coverage except for the poles. As well as remote sensing satellites can also be used for communication. Therefore, the problems related to earth observation in geostationary orbit have been paid much attention. In addition, staring imaging can be used in geostationary orbit, which avoids the problems of contrast transfer function reduction because mismatching of the sweep imaging between charge transfer and target image movement and improves the signal-to-noise ratio by extending integration time. Therefore, it is significant to study earth observation in geostationary orbit.

As large, high-precision, aspheric optical mirror manufacturing implementation, high transmittance, wide spectral band, wide field of total reflection type optical system has also developed, combined with small pixel large-area array CCD manufacturing technology development, makes it possible to use optical remote sensor in geostationary orbit of staring imaging.

The field of view of optical remote sensor in geostationary orbit staring imaging demands.

Take the earth as a sphere, radius $R_e = 6378.145$ km and geostationary orbit radius 42,164.26 km. Then, $2\alpha = 17.40^{\circ}$ is easily obtained in Fig. 8.6.

This is the $17.5 \times 17.5^{\circ}$ field angle that optical remote sensor must have to realize the coverage range of a single satellite as shown in Fig. 8.6. This is already a large field of view for a total reflective optical system.

(1) Infrared early warning scanning imaging

In order to reduce the development difficulty of wide-field infrared optical warning camera, infrared early warning satellites generally use scanning imaging method, such as the typical American DSP infrared early warning satellite scanning imaging global warning with a satellite spin. Figure 8.7 is the DSP infrared early warning satellite, the infrared camera visual axis and satellite spin axis offset 7.5°, as well as the camera's field of view is 3 °.

(a)

(b)

Fig. 8.7 DSP infrared warning satellite

(2) Scanning and staring dual imaging for infrared early warning

American SBIRS infrared early warning satellite, following DSP satellite, adopts the early warning system of scanning and staring dual-infrared camera imaging, as shown in Fig. 8.8. The scanning camera uses a large linear field of view to scan the earth disk and confirms and tracks the suspected target obtained by the scanning camera by the area array staring camera with a higher signal-to-noise ratio, which can greatly reduce the false alarm rate and improve the prediction accuracy of target trajectory, as well as the coverage area is much larger than DSP.

SBIRS system consists of four high elliptical orbit (HEO) satellites and five geostationary orbit (GEO) satellites. SBIRS GEO satellite is based on Lockheed's A2100 satellite platform with a 12-year design life. The platform is three-axis stabilized, has a power supply of about 2800 watts and weighs about 4500 kg. And the infrared sensor used as the payload weighs about 450 kg. Since 1996, when the U.S. Department of Defense approved the Space-Based Infrared High Orbit System program, the progress of SBIRS has been constantly delayed. Originally, the first satellite of SBIRS GEO was launched in 2004, but the contract was adjusted in 2002,

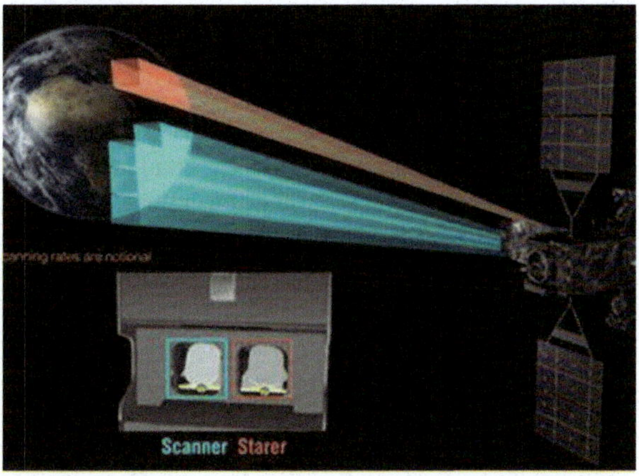

Fig. 8.8 SBIRS infrared warning satellite

the launch was postponed to 2006, then the launch was postponed again to 2007, and finally the launch was postponed to 2011. As a result, there have been major overruns, and budget requirements have doubled from $2.1 billion for the 1996 contract to $7.5 billion. High elliptical orbit (HEO) satellite, with the transfer of the system (SBIRS-Low) from the US Air Force to the Ballistic Missile Defense Agency in 2001, the system was renamed Space Tracking and Surveillance System (STSS). Now known as SBIRS system generally refers to the original SBIRS-HIGH, and high orbit satellites do not communicate with each other, but can communicate with low orbit to achieve relay tracking. STSS satellites are located in sun-synchronous orbits in three different planes. These low orbit satellites are equipped with wide-field scanning detectors and narrow-field gaze multispectral detectors. Wide-field scanning probe can capture ballistic missile plume below horizon, as soon as possible to complete tracking task passed by the high orbit satellite. The narrow-field multispectral detector has medium and long wavelength and visible light detection capabilities and can lock onto target and track the entire midcourse and re-entry trajectory. Using a very sensitive multispectral detector, STSS can detect cold targets such as warhead in the mother compartment after the booster has burned out, track warhead separation in clutter and noise and have ability to distinguish warhead, warhead mother compartment, light and heavy optical radar decoys. The accurate positioning of the ballistic missile warhead by STSS system is based on premise of simultaneous detection and tracking by four STSS satellites, with high positioning accuracy. The combined detection of SBIRS and STSS for long-range and intercontinental missile can detect and track ballistic missile in boost phase, rise period, middle and re-entry segment, through precise positioning for interceptors to provide coordinates. And the interceptor missile can be launched before the incoming missile enters the detection range of land-based and sea-based radar to achieve multilayer interception and improve the success rate of interception.

(3) Staring imaging of infrared warning

Because the SBIRS project has always had problems, the U.S. Department of Defense began implementing a parallel program in 2006, called the alternative infrared satellite system (AIRSS), which applied staring imaging technology to monitor the earth's disk using wide-field infrared cameras instead of relying on scanning imaging.

Both SBIRS-high scanning camera and AIRSS staring camera require a large field of view of nearly 10° or 10° × 10°.

2. **Low orbit wide coverage optical remote sensing satellite**

On March 22, 2010, it was reported that "Remote-sensing 9 surveillance satellite" was sent from Jiuquan Satellite Launch Center on March 5. It will mainly be used in scientific experiments and other civilian fields, possibly for optical detection of earth [1].

The satellite will be "mainly used for scientific experiments, land and resources surveys, crop yield estimation and disaster prevention and mitigation" and can also be used by the military and civilians.

Remote-sensing 9 consists of three satellites, one main satellite and two small satellites. The main satellite has an orbital altitude of 1085 × 1099 km, with two small satellites in its vicinity and three satellites with an orbital inclination of 63.4°. This multisatellite layout and orbit is similar to the approach reportedly taken by the U.S. Navy's Ocean Surveillance Satellite System (NOSS).

Remote-sensing 9 is believed to carry millimeter-wave radar to help three satellites stay in formation in low-earth orbit, infrared sensors to detect ships and antennas to receive electronic transmissions [2].

The size of an aircraft carrier formation is usually 200 × 200 km–400 × 400 km, so the coverage width of low orbit wide coverage optical remote sensing satellite is more than 100 km. Figure 8.9 is the US aircraft carrier formation.

It can be seen that optical remote sensing satellites with some special functional requirements have the characteristics of wide coverage, and their field of view is generally more then 10°, belonging to the wide coverage optical remote sensing satellites.

8.2 LEO Optical Remote Sensing Platform

The altitude of low orbit is below 2000 km, the altitude of middle orbit is between 2000 to 20,000 km, and the altitude of earth synchronous orbit is 36,000 km.

Optical remote sensing platform is the equipment that carries and provides in-orbit working environment for the payload. The advantage of setting a platform is

Fig. 8.9 US aircraft carrier formation

relatively independent of the payload that lies in the standardization, generalization and serialization of the platform, and the advantage of standardization, generalization and serialization is that it can improve the reliability. Reliability of space products can be treated as the life of space products. Secondly, standardization, generalization and serialization of platform can reduce cost of satellite launch. Standardization and generalization can make different platforms share components, thus greatly reducing cost of design and test verification. Serialization can save launch costs while meeting payload and engineering requirements.

Optical remote sensing platform should carry the payload and provide energy, thermal control environment, attitude pointing service, orbit keeping service, telemetry and remote control service, communication and data transmission service for the payload. Remote sensing platform is required to have structure system, energy system, thermal control system, attitude and orbit control system, telemetry and remote control system, propulsion system, communication and data transmission system and data management system for satellite normal operation.

Take the China Brazil Earth Resources Satellite (CBERS) platform as an example. The platform consists of eight subsystems, including structure, thermal control, power, attitude and orbit control, propulsion, telemetry and remote control, on-board data management and data transmission. The data transmission subsystem is sometimes referred to as a payload system, mainly because it is closely related to the payload, and depending on the payload, its specifications vary, unlike the other seven subsystems, which have relatively fixed technical specifications.

8.2.1 Specifications of CBERS Platform

1. **Structure subsystem**

Structure subsystem is the main support part of the satellite. The CBERS-1 platform adopted box structure, with the external outline as shown in Fig. 8.10. The part from middle plate to service cabin was platform structure part. Above the platform was the payload bay. The structure of the service cabin and the payload bay together form the main structure of the whole satellite. All kinds of instruments and equipment were installed on the main structure of satellite according to requirements of mass characteristics and field of view of instrument, forming the basic configuration of the whole satellite. Structure subsystem shall bear the dynamic conditions and various statics conditions of the whole satellite and meet the mass (weight) requirements.

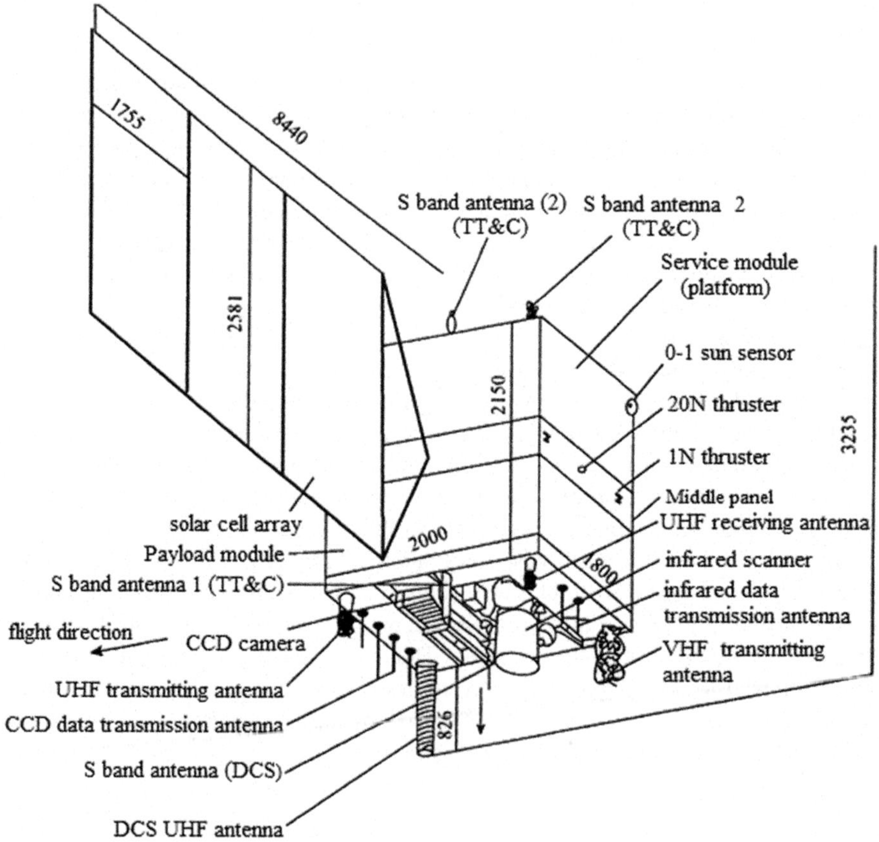

Fig. 8.10 Outline diagram of CBERS-1

2. **Thermal control subsystem**

Thermal control subsystem is an important system to ensure the normal in-orbit operation temperature of satellite platform and payload. The orbit of CBERS-1 (ZY-1) satellite and the characteristics of remote sensor made the distribution of external heat flux on satellite surface very uneven and periodically drastic changes, while the optical remote sensor had a very strict requirement for small temperature change range. Hence, the satellite adopted passive thermal control methods mainly consisting of heat pipe and OSR and active thermal control hardware mainly consisting of heater with temperature sensor. Sufficient ground thermal balance test and thermal vacuum test proved the correctness of thermal control design.

3. **Power subsystem**

The power subsystem of CBERS consisted of solar array and nickel-cadmium battery for joint power supply. Solar array was divided into two categories, charging array and power array, and there were two charging arrays for two sets of batteries, respectively. The battery in each group is made up of two battery block series, and each battery block consisted of 18 monomer battery series. Each battery had a capacity of 30 AH and a maximum discharge depth of 20%. The power supply array was divided into six independent solar cell arrays connected to six shunt regulators. Both battery power supply and solar array power supply stabilized the supply voltage at 18 $\pm$ 0.6 V through shunt regulator switch.

4. **Attitude orbit control (AOCS) subsystem**

Remote sensing satellites have strict requirements on pointing accuracy and stability of satellite attitude. CBERS satellite adopted three-axis stability control and earth orientation. Control precision was

Pointing precision $|\varphi, \theta, \psi| \leq 0.3°$ (3σ);
Measurement precision $|\varphi, \theta, \psi| \leq 0.15°$ (3σ);
Stability $|\dot{\varphi}, \dot{\theta}, \dot{\psi}| \leq 0.001°/s$ (3σ);
Jitter $\leq 0.0001°$ (σ);

The attitude control subsystem adopted four offset momentum wheels to form zero-momentum control system of whole satellite, and magnetic torquer was used to unload the momentum wheels. When the amount of attitude adjustment was large, 16 (8×2) 1 N single-component hydrazine thrusters were used for attitude adjustment. The attitude measurement was completed by two conical scanning infrared horizon detectors, two groups of six orthogonal mounted liquid floating integrating gyros and three kinds of sun sensors (two digital, three analog, two 0/1). GPS-assisted orbit determination was added on 02B satellite to further improve the orbit determination accuracy [3].

CBERS remote sensing satellite adopted solar synchronous regression freezing orbit, which had strict requirements on orbit accuracy.

Semi-major axis $|\Delta a| \leq 35$ m;
Eccentricity $|\Delta e| \leq 0.00007$;

Orbital inclination $|\Delta i| \leq 0.007^{\circ}$;
Perigee angle $|\Delta \omega| \leq 3.5^{\circ}$;
Orbital altitude $H = 778$ km.

Orbit maintenance was accomplished by two 20 N single-component hydrazine thrusters.

5. **Propulsion subsystem**

The propulsion subsystem of CBERS remote sensing satellite is composed of storage tank, tube valve, 1 N single-component hydrazine thruster, 20 N single-component hydrazine thruster, pipeline and so on, which provides power for attitude and orbit control.

6. **Telemetry tracking and control subsystem**

It is the general name of telemetry subsystem and remote control subsystem. Telemetry subsystem monitors working state of instruments on board, such as working temperature, power supply state, selection of master and backup, gain size, satellite attitude pointing, attitude rate, orbital coordinate, on-board time, fault state and fault location information.

Remote control subsystem is the command channel that ground controls the satellite in orbit. The early remote control is mainly switch command, such as power supply on and off, signal on and off. However, with the increasingly functions of satellite, more and more data need to be uploaded to the satellite, such as system reconstruction, on-board computer software maintenance, control program upload and working parameters adjustment.

The hardware is mainly ground omnidirectional telemetry antenna, ground omnidirectional remote control antenna, back ground omnidirectional telemetry antenna, back ground omnidirectional remote control antenna, modulator, demodulator, etc.

The satellite adopts unified S-band and VHF/UHF, both of which are capable of tracking and measuring the orbit, telemetry and remote control. VHF also acts as data transmission for space environment monitoring (SEM, an aurora particle detector, in-satellite particle detector and CMOS radiation effect detector of CBERS-1 for monitoring the orbital space environment).

7. **On-board data management subsystem**

Data management subsystem is composed of central terminal unit (CTU), remote terminal unit (RTU), serial data bus (SDB), telemetry control unit (TCU) and command decoding unit (CDU). The system adopts dual-unit cold backup. Data management subsystem has the ability of data acquisition, instruction sending, time correction, information exchange and information storage of subsystem. In order to prevent interference of high-energy particles in space, software and hardware protection, error correction circuit and periodic data refresh are adopted.

8. **Data transmission subsystem**

CBERS-1 remote sensing satellite has three sets of x-band data transmission system, CCD-1 and CCD-2 have a code rate of 53 M bit/s and adopt QPSK carrier mode, and each uses one set of data transmission system. Two sets adopt four 32 W TWT, two primary and two standby.

The other is used to transmit data and beacon signals for infrared scanners and wide-field imagers. The infrared scanner modulated the carrier in BPSK mode with a code rate of 6.127 Mbit/s, and the wide-field imager modulated the carrier in QPSK mode with a code rate of 1.1 Mbit/s. This set of data transmission system uses two 4 W solid-state amplifiers, one primary and one standby.

8.2.2 On-Orbit Performances of CBERS Platform

From the second day to the seventh day after the satellite was placed in orbit, each payload was opened successively, and orbit adjustment began on the 15th day. Solar synchronous regression freezing orbit was realized with high precision through seven orbit changes. After that, orbit maintenance was carried out once about 2 months, so that east–west deviation between sub-satellite point trajectory and ground trajectory of standard regression orbit was less than 10 km. On the 24th to 28th day, open the protective cover of radiator to heat and decontaminate the radiator. The satellite completed 85 in-orbit tests, achieving or exceeding design targets.

- Attitude orbit control (AOCS) subsystem

Independent fault diagnosis and system reconstruction performance of the subsystem have been tested during in-orbit period, and a high performance has been achieved.

Attitude angle error: $(\varphi, \theta, \psi) \leq 0.1°$;
Stability: $(\dot{\varphi}, \dot{\theta}, \dot{\psi}) \approx 5 \times 10^{-4} \sim 1 \times 10^{-3}$ (°/s).

After several adjustments, the error value of satellite orbit and the designed 778 km solar synchronous regression frozen orbit is

Semi-major axis $|\Delta a| \leq 10$ m;
Eccentricity $|\Delta e| \approx 0.00001$;
Perigee angle $|\Delta\omega| \approx 1°$.

- Propulsion subsystem

The satellite was launched with 80 kg of hydrazine propellant, with 60 kg remaining after initial adjustment and subsequent orbital maintenance within 8 months. Orbit maintenance was carried out about every 2 months, consuming 0.1 kg of fuel hydrazine each time.

- Power subsystem

The telemetry indicators of power subsystem were as follows:

Bus voltage	28 ± 0.6 V;

Solar charging array output power	732 W;
Solar supplying array output power	649 W;
Discharge regulator output current (eclipse area)	15.5 A;
Battery average discharge depth	10.7%.
Battery temperature	3–6 °C.

- On-board data management subsystem

The subsystem works well in orbit, and measures against space high-energy particles are effective. When time interval between centralized and uniform calibration is more than 15 days, time error is less than 5 ms. Switching and transmission of various telemetry and remote control information and its information format are normal, and transmission of image auxiliary data to optical remote sensing loads is normal. Especially, when orbital computer is disturbed by high-energy particles in space, it can upload important data from on-board data manager immediately, so that attitude control system can keep normal.

- Data transmission subsystem and other subsystem

It works normally.

- Optical remote sensing payload subsystem

Except that CCD camera has stripes in the image which need to be processed and corrected due to CCD splice mode and CCD device and B_8 spectrum segment of infrared scanner needs to be interpolated as the second and fifth pixel response is close to zero after nearly 1800 laps of satellite flight, the rest of them all meet or exceed the design objective.

8.3 Performances of GEO Optical Remote Sensing Platform

GEO optical remote sensing platform and LEO optical remote sensing platform are the same in terms of efficiency and also bear and provide in-orbit working environment for the payload. But there is also a need for platform serialization. GEO optical remote sensing platform is also composed of structural system, power system, thermal control system, attitude and orbit control system, telemetry and remote control system, propulsion system, communication and data transmission system and satellite integrated electronic system.

8.3.1 Characteristics of Optical Remote Sensing on GEO/LEO

As the resolution of space optical remote sensing has been greatly improved, the resolution of commercial images of low orbit optical remote sensing has been better than 1 m. For example, Pleiades 1 satellite of EADS-Astrium has reached a resolution of 0.5 m on orbit of 694 km [4]. And the resolution of new generation of commercial satellites, such as GeoEye-1, is 0.41 m on orbit of 684 km [5]. Low orbit optical remote sensing satellite mainly adopts sun-synchronous orbit with altitude between 200 and 1000 km, and general civil optical remote sensing satellite orbit is about 700 km. Low orbit optical remote sensing satellites using solar synchronous orbit can achieve global observation, but the revisiting period can be long, generally between a few days to dozens of days.

Optical remote sensing in geosynchronous orbit can continuously observe about 1/3 of the earth surface at sub-satellite point, but the resolution of pixel is greatly reduced because geosynchronous orbit is as high as 35,786 km. For this reason, the current optical remote sensing in geosynchronous orbit is mainly used for meteorology and early warning, and the resolution of space pixel is at the kilometer level [6, 7]. Such spatial pixel resolution has great limitations in land observation, ocean observation and disaster observation. However, according to the development level of spatial resolution of current low orbit optical remote sensing, such as GeoEye-1, when applied to high orbit optical remote sensing, simple calculation shows that its sub-satellite pixel resolution can reach 21.5 m. Such spatial resolution has high application value in many areas, especially in geosynchronous orbit, optical remote sensing can realize continuous observation for a specific range and get a certain space resolution, and then it complements the low temporal resolution and high spatial resolution of low orbit global observations, so many countries have carried out the research work in this aspect [8].

In 2007, India released its "11th Five-Year" space development plan, which covers the period from fiscal year 2007 to 2012. Twelve new projects were planned in the field of remote sensing, including the Geostationary Orbit High-Resolution Imaging Satellite (GEO-HR IMAGER). GEO-HR IMAGER weighs 1200 kg and is equipped with "Advanced Wide Field Imaging Sensor" (AWFIS), covering visible, near infrared and shortwave infrared spectrum bands, with the highest resolution of 60 m and a width of more than 700 km. The satellite can cover all of India with a resolution of 60 m every half hour. Satellites are mainly used for crop monitoring, resource surveys and disaster monitoring.

In 2000, Thales Alenia Space Corporation in Europe carried out research of high-resolution optical earth remote sensing in geostationary orbit. In about 2005, it put forward the preliminary scheme of 20 m resolution optical earth imaging satellite in geostationary orbit. The satellite is based on SpaceBus4000 platform and has been adapted to it. The total launch weight of the satellite is about 2900 kg, the camera is about 600 kg, and the ground pixel resolution of sub-satellite point is 20 m. The satellite is supposed to be launched from Kourou using the Soyuz rocket of Russia.

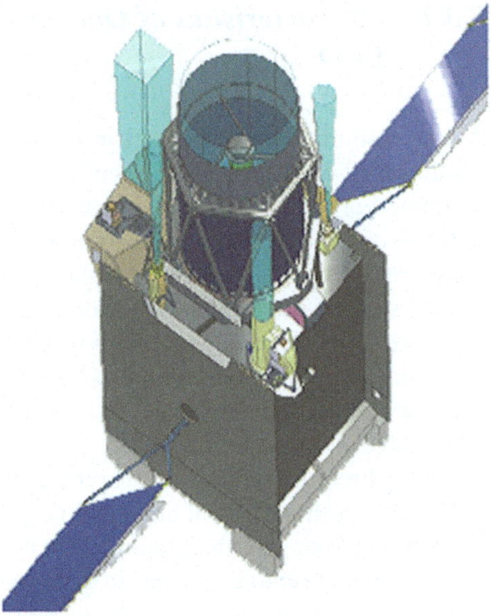

Fig. 8.11 Astrium's 20 m resolution satellite in geostationary orbit

Astrium has also put forward the idea of geostationary orbit 20 m resolution optical imaging satellite on the basis of Eurostar series communication satellite platform, as shown in Fig. 8.11. According to report, the satellite has a camera aperture of 1.5 m and will be launched directly into orbit, greatly reducing the weight of the satellite, which is expected to weigh no more than 2000 kg.

The European Research Organization, led by EADS-Astrium, has done geostationary orbit earth observation concept and satellite performance research for Africa, with sub-satellite point 25 m resolution.

The National Reconnaissance Office of United States held a meeting in the summer of 2004 to invite manufacturers to discuss space-deployable telescope concepts. Set out to develop an optical telescope that would be folded at launch but deployable once in space. It is expected that an imaging satellite equipped with such a telescope will be put into the geostationary orbit to achieve staring observation in the next 20 years, with resolution of 1 m, aperture of 30 m and an estimated focal length of about 250 m.

It can be seen that it is feasible to achieve optical remote sensing better than 20 m spatial resolution in geostationary orbit at the current international level, but it also has its own particularity.

(1) High orbit requires rockets with large carrying capacity, which will lead to difference in carrying dynamic environment between geostationary optical remote sensing satellites and low orbit optical remote sensing satellites.
(2) In-orbit environment of geostationary optical remote sensing satellites is different from that of low orbit optical remote sensing satellites.

(3) Geostationary optical remote sensing satellites and low orbit optical remote sensing satellites are different in-orbit imaging methods.
(4) Above differences will lead to different requirements for optical remote sensing in geostationary orbit.

1. **Transport environment**

The launch of geostationary orbit optical remote sensing satellite requires heavy rocket. Take CZ-3C rocket as an example, equipped with 4000 F fairing, inner diameter envelope size φ3650 mm, satellite mass no more than 3900 kg, transverse first-order frequency no less than 12 Hz and longitudinal first-order frequency no less than 35 Hz. The design limit load requirements for satellites are shown in Table 8.2.

The above parameters of CZ-3B rocket are the same as CZ-3C rocket, except that the satellite mass is not more than 5100 kg.

The launch of low orbit optical remote sensing satellite adopts rocket with low thrust. Take CZ-2C rocket as an example, it is equipped with 3350F (split type) fairing, inner diameter envelope size is φ3000 mm, satellite mass is not more than 3900 kg, transverse first-order frequency is not less than 12 Hz, and longitudinal first-order frequency is not less than 35 Hz. The design limit load requirements for satellites are shown in Table 8.3.

Table 8.2 Ultimate load requirements of CZ-3C rocket for satellite design unit: g

Load/g	Load type	Transonic and maximum dynamic pressure state	Before boosters separated	After separation of two stages
Vertical overload	Static	+ 2.2	+ 5.3	+ 1.0
	Dynamic	+ 0.8/-08	+ 0.8/−3.6	+ 2.7/−3.6
	Combination	+ 3.0	+ 6.1	+ 3.7/−2.6
Lateral overload		1.5	1.0	1.0

Note "+" means pressure, and "−" means pull

Table 8.3 Ultimate load requirements of CZ-2C rocket for satellite design unit: g

Load /g	Load type	Transonic and maximum dynamic pressure state	Primary engine shutdown	After separation of two stages	Secondary main engine shutdown
Vertical overload	Static	+ 2.2	+ 4.6	+ 0.8	+ 6.7
	Dynamic	+ 0.4/−04	+ 1.0/ −1.0	+ 3.0/−3.0	+ 0.5/−0.5
	Combination	+ 2.6	+ 5.6	+ 3.8/−2.2	+ 7.2
Lateral overload		1.0	0.6	0.8	0.4

Note "+" means pressure, and "−" means pull

It can be seen that the lateral load of high orbit satellite is 1.5 g, and that of low orbit satellite is 1.0 g. The longitudinal loads of high orbit satellites are + 6.1 g/ −3.6 g and + 7.2 g/−3.0 g for low orbit satellites.

2. **The in-orbit environment of geostationary optical remote sensing satellites is different from that of low orbit optical remote sensing satellites**

GEO orbit is high, without influence of atmospheric ionosphere, and is more likely to be affected by high-energy particles than low orbit. It is less affected by earth J_2 term, but more affected by ellipsoid J_{22} term in equatorial plane. Low orbit has influence of atmospheric ionosphere, is less likely to be affected by high-energy particles than high orbit and is more affected by earth J_2 term.

More importantly, the thermal cycle of GEO orbit satellites is 24 h long, while the thermal cycle of low orbit satellites is shorter than about 1.5 h. The thermal distribution gradient of high orbit satellites is generally more difficult to control than that of low orbit satellites, and the thermal deformation of structure is also larger than that of low orbit satellites.

The solar array angle variation of high orbit satellite is smaller than that of low orbit satellite, and the variation period is long. The period of high orbit is 24 h, and the period of low orbit is about 1.5 h. This makes energy changes of high orbit satellites smoother than that of low orbit satellites.

3. **Imaging mode of high-resolution optical remote sensing in geostationary orbit**

As the geostationary satellite is relative stationary to earth, so frame imaging is the preferred method for satellite imaging. This requires that the optical remote sensor of geostationary orbit satellite must be a plane field of view. However, when there is orbit relative motion between low orbit optical remote sensing satellite and the earth, sweep imaging scheme is generally adopted in low orbit. So the field of view of low orbit optical remote sensing satellite is approximately one dimension, because the other field dimension is very small. The volume and weight of geostationary optical remote sensors will increase correspondingly due to the requirement of two-dimensional field of view. As the field of view for camera is difficult to be very large, in order to image a large range, frame imaging of the observed range must be carried out [9], as shown in Fig. 8.12.

4. **Geostationary orbit optical remote sensor**

According to the above analysis, in order to achieve high spatial resolution optical imaging in geostationary orbit due to the long distance of ground target, the camera needs a long focal length, a small field of view and a frame imaging mode. Long focal length and high spatial resolution mean large aperture, so high orbit optical remote sensor is large in volume and weight, which will reach 1.5 m in diameter and nearly 1 ton in weight. Such a huge optical remote sensor has long development cycle, high cost and great difficulty. When high orbit optical remote sensor carries out dynamic experiment of transport environment, the experiment period is long, the cost is high, and the risk is big.

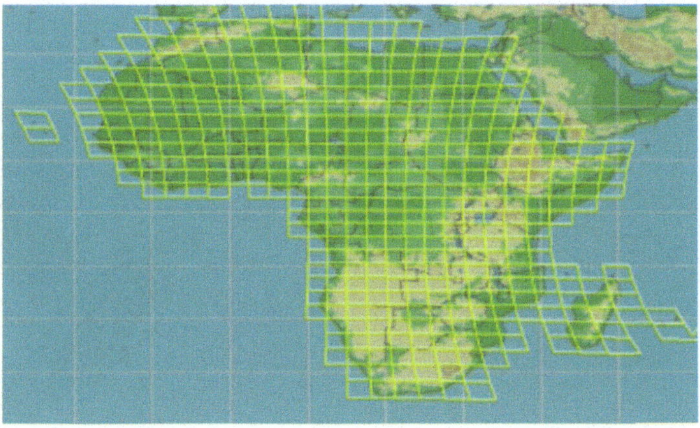

Fig. 8.12 Frame imaging mode of high-resolution optical remote sensing in geostationary orbit

8.3.2 Specifications of Optical Remote Sensing Platform on GEO

High orbit optical remote sensing platform adopts frame imaging due to its high resolution and high orbit, which requires the platform to be flexible and stable in attitude. Figure 8.13 shows the structure of platform for high orbit communication satellite. This platform can be equipped with optical remote sensing with high space resolution through adaptive adjustment to realize the comprehensive application of high orbit optical remote sensing and communication. Experimental satellites have been launched into orbit. General configuration of special high orbit remote sensing platform is shown as Astrium platform in Fig. 8.11 and "Fengyun4" remote sensing platform in Fig. 8.14. High orbit platforms generally have long life, such as 10 years.

Specifications of each subsystem of high orbit remote sensing platform are generally as follows:

1. **Structure subsystem**

Structure subsystem is responsible for overload and gravity during satellite launch and integration.

(1) Transverse first-order frequency (solar array undeployed state): $\geq$ 12 Hz;
(2) Longitudinal first-order frequency (solar array undeployed state): $\geq$ 35 Hz;
(3) Dynamic load requirements refer to Table 8.2;
(4) Outer profile of satellite meets rocket envelope conditions such as 4000F (or 4200F) fairing, and the inner diameter envelope size is φ3650 mm;
(5) Payload: no more than 1200 kg;
(6) Weight limit of satellite for launch mass: no more than 5500 kg.

2. **Power subsystem**

Provide energy for satellites to work in orbit. It is mainly composed of primary power supply subsystem and overall circuit subsystem. The primary power supply

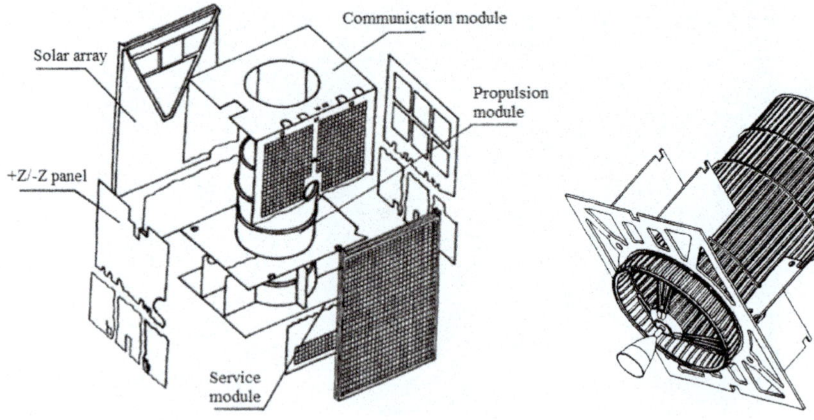

(a) (b)

Fig. 8.13 Main structure of high orbit communication satellite platform

Fig. 8.14 High orbit "Fengyun4" remote sensing platform

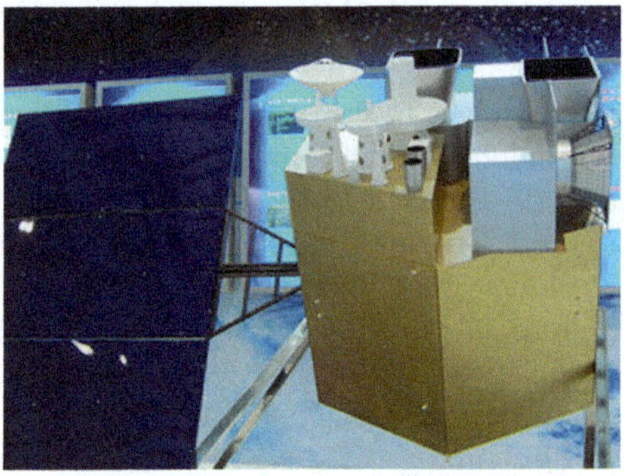

subsystem is composed of solar cell array and battery group and adopts a fully regulated, single bus topology form. Distribution appliances are set up in each cabin to uniformly manage the power supply and distribution of each subsystem equipment nearby, and the secondary power supply is combined with centralized and decentralized transformation. Main technical indicators are as follows:

(1) Bus type: fully adjusted bus;
(2) Bus voltage: 100.3 $\pm$ 0.29 V;
(3) Solar cell array: gallium arsenide cells, solar cell array output power 3000–6000 W (EOL);
(4) Battery: 65 Ah nickel-hydrogen battery, maximum discharge depth of 80% (EOL).

3. **Thermal control subsystem**

Thermal control subsystem is a temperature control system to ensure instruments and equipment on board are kept within allowable temperature range in all working modes from launch, transfer orbit and geosynchronous orbit until the end of their life.

Thermal control of optical remote sensing satellite should be simple and reliable on the premise of meeting requirements of optical remote sensing instrument. Thermal control of the whole satellite mainly adopts passive thermal control technology, supplemented by necessary active thermal control (electric heater and thermistor). The general design principles are as follows:

(1) Passive thermal control and active thermal control are combined, with passive thermal control as the main and active thermal control as the auxiliary.
(2) Subdivision design method is adopted to facilitate design modification of each compartment.
(3) As light as possible.
(4) Power consumption of the heater should be as small as possible.
(5) Improve capability of satellite temperature self-controlling and reduce ground operation.
(6) Take south/north side of satellite as heat dissipation surface.
(7) Surface of satellite is fully covered with multilayer heat insulation components (MLI) except for heat dissipation surface.
(8) Isothermal design is adopted inside to strengthen radiation and heat conduction.
(9) Reduce radiation and conduction thermal coupling between outer parts and satellite body.

4. **Attitude and orbit control subsystem**

The basic function of attitude and orbit control subsystem is to complete attitude control and orbit control of satellite in each mission stage from satellite and rocket separation to orbit operation and the end of its life. Attitude and orbit control subsystem consists of sensors, controllers and actuators.

The main technical indicators of attitude and orbit control subsystem are as follows. In order to realize large-area earth observation, it is necessary to constantly move observation field of camera. Two-dimensional movement of camera observation field is realized by rolling and pitching of satellite platform.

Moment gyro or reaction wheel is used as actuator and meets the demand of fast maneuver control for torque. In order to improve ability of rapid response based on payload work mode, real-time planning paths of attitude maneuver and method and improving system bandwidth are adopted to improve response speed.

(1) Attitude maneuver and stability include the following processes: acceleration, uniform speed, deceleration and stability.
(2) Attitude pointing accuracy: $\pm$ 0.03° for all three axes;
(3) Attitude determination accuracy: pitch, yaw, roll $\pm$ 3";
(4) Attitude stability: all three axes are better than 5×10^{-4}°/s(3σ);
(5) Orbit keeping accuracy: $\pm$ 0.05° in the north–south and east–west directions;

(6) Long-term bias capacity: pitching/rolling ± 8.5°;
(7) Fast maneuvering ability: number of images /min.

5. **Telemetry and telecontrol subsystem**

The ground telemetry and tracking and command (TT&C) station mainly completes telemetry, tracking, remote control and management of satellite in stages of take-off and launch, orbit transfer and attitude adjustment, as well as part of TT&C and management in the operation stage and provides various complete parameters and post-processing data files. The satellite TT&C subsystem generally adopts international C-band unified carrier TT&C scheme, as well as S and Ka bands. Uplink remote control signal and ranging signal are frequency modulated to the same carrier in time division mode, and downlink telemetry signal and forwarding ranging sound are phase modulated to the same downlink carrier together. Wide beam coverage is provided by TT&C antenna before and after the fixed point to ensure the normal operation of up and down omnidirectional TT&C channels. TT&C subsystem consists of three parts: tracking subsystem, telemetry subsystem and remote control subsystem. The main technical indicators are as follows:

(1) Ranging accuracy: on-board equipment is generally better than 30 m.
(2) Bit error rate: uplink Pe $\leq 1 \times 10^{-5}$, downlink Pe $\leq 1 \times 10^{-5.}$
(3) Instruction miss probability: under the condition of big loop instruction verification, instruction miss probability is less than $1 \times 10^{-6.}$
(4) Crosstalking instruction probability: $\leq 1 \times 10^{-9.}$

6. **Propulsion subsystem**

Propulsion subsystem accomplishes apogee maneuver, fixed point capture in synchronous orbit, position holding, attitude control and adjustment at various stages of satellite.

Propulsion subsystem of high orbit satellite generally adopts a two-element unified propulsion scheme. The impulse required by apogee maneuver is provided by apogee engine, which is installed on $-Z$ plane of satellite and its nominal thrust direction is along $+ Z$ axis and passes through the centroid of satellite at the time of apogee ignition. Multiple 10 N thrusters provide power needed for attitude and orbit control. Two tanks with column sections provide fuel needed for propulsion. Usually, the system consists of the following:

(1) 490 N engine: 1;
(2) 10 N thruster: multiple sets;
(3) Oxidizer storage tank: 1;
(4) Combustion agent storage tank: 1;
(5) High-pressure helium bottle: 50 × 3 L.

7. **Data transmission subsystem**

Spacecraft data transmission should consider atmospheric absorption, ionospheric interference, cloud and rain attenuation and other factors. In order to obtain higher data transmission speed, high orbit data transmission generally adopts Ka band.

Since high orbit optical remote sensing satellite adopts frame imaging mode and data transmission antenna and optical imaging remote sensor are on the same platform, the attitude step brought by frame imaging will inevitably affect the link connection between data transmission system and ground station, so the data transmission antenna of high orbit optical remote sensing platform also needs to be able to track the ground station. In order to slow down wear of transmission mechanism of antenna and prolong service life of satellite, the field of view of antenna can be appropriately expanded, and starting frequency of transmission mechanism of antenna can be reduced by matching the field of view of antenna with stride of attitude maneuver required by optical imaging, so as to prolong service life of transmission mechanism of antenna.

1. Main functions of data transmission subsystem

    (1) Compress and encode image data output by CCD optical remote sensor;
    (2) Unified high-speed data duplication and CCSDS AOS coding for CCD optical remote sensor data;
    (3) Encrypt or harass the transmitted data stream;
    (4) Modulate and amplify signal power for the data stream of data processor or solid-state memory replay, so as transmit to ground receiving station;
    (5) Data processing equipment and transmission channel equipment have the function of primary and backup switchover.

2. Data transmission subsystem generally has three basic working modes

(1) Real-time transmission mode

When camera is turned on for imaging, the image data of CCD camera is directly transferred to ground data receiving station through data transmission subsystem. In this mode, data processing part, data transmission channel part and data transmission antenna work.

(2) Real-time transmission mode

Image data of the camera is compressed (optionally not compressed) and stored in memory.

(3) Real-time transmission mode

Data processor provides playback clock to solid-state memory, and the playback data is transmitted to ground data receiving station through down link.

3. Main technical indicators of data transmission subsystem of geostationary optical platform

    (1) Transmission frequency: Ka band;
    (2) Number of channels: single channel transmission;

(3) Data transmission rate: 200–300 Mbps;
(4) Modulation mode: QPSK;
(5) Encryption: security level;
(6) Scrambling: efficiency.

4. Data transmission subsystem usually consists of following equipment:

Data transmission subsystem mainly includes data processing part, data transmission channel part, data transmission control part and data transmission antenna.

(1) Data processing

- Compression encoder;
- Data processor;
- Solid-state memory.

(2) Data transmission channel

- Modulation transmitter;
- Ka up-converter;
- Ka filter;
- Ka band traveling wave tube (TWTA);
- Waveguide switch;
- Duplexer.

(3) Data transmission antenna

- Reflective surface;
- Feedsource;
- Rotational joint;
- Two-dimensional rotating mechanism;
- Antenna controller.

8. **Composition of integrated electronic subsystem**

Integrated electronic system is a new system of organization form for high orbit satellite business management, which is responsible for exchange of information and command transmission between control subsystem, TT&C subsystem, payload subsystem, power subsystem, data transmission subsystem and thermal control subsystem, as well as between each subsystem and integrated electronic system. It is a kind of high-efficiency information management system on satellite. It consists of three parts: central management unit (CMU), platform integrated service unit (ISU), payload integrated service unit, CSB internal bus, 1553B bus and so on.

1. Composition of integrated electronic subsystem

(1) On-board computer;
(2) Emergency task processing module;
(3) Monitoring and reconstruction module;

(4) Power module;
(5) Interface management module;
(6) Internal bus.

2. Main modules of platform integrated service unit

(1) Bus interface management module;
(2) Analog quantity acquisition and instruction output module;
(3) 100 V heater power distribution module;
(4) 7 V heater power distribution module;
(5) Hybrid power distribution module;
(6) Explosive management module;
(7) Antenna deployment control module;
(8) CSB internal bus.

3. Main modules of platform integrated service unit

(1) Bus interface management module;
(2) Matrix instruction and matrix telemetry module;
(3) Analog quantity acquisition and instruction output module;
(4) Payload distribution module;
(5) 100 V heater power distribution module;
(6) CSB internal bus.

8.4 Development of Optical Remote Sensing Platforms

The development of optical remote sensing platform has obvious characteristics of diversity. In addition to the above-mentioned differences of high and low orbits, it is also reflected in diversified development characteristics of large-scale, microminiaturization and dedicated platforms.

1. **Platform enlargement**

Enlargement trend of high orbit satellite platforms is obvious, and the launching weight of satellites composed of large platforms is 10 tons. Foreign platforms of this kind have been proved to be mature platforms by flight, such as FS1300 and BSS702 of the United States, and SpaceBus3000, SpaceBus4000 of France, which are advanced large communication satellite platforms.

China's high orbit large optical remote sensing platform has also entered the finalized development stage, including DFH-5 (Dongfanghong-5) large communication and remote sensing integrated platform of China Academy of Space Technology under the China Aerospace Science and Technology Corporation. Figures 8.15a and b are the model of DFH-5 platform exhibited at the 8th Zhuhai aerospace exhibition on November 15, 2010. It is equipped with two-dimensional deployable solar array, deployable thermal radiator, energy system of lithium ion battery and integrated propulsion system of new plasma electric thruster and chemical thruster [10].

(a)

(b)

Fig. 8.15 DFH-5 large communication remote sensing integrated platform

The main technical specifications of DFH-5 platform are shown in Table 8.4.

The SAST9000 platform of the Eighth Academy of China Aerospace Science and Technology Corporation has completed mechanical tests, aiming at important fields such as high orbit optical remote sensing and second-generation electronic reconnaissance and building a high orbit remote sensing satellite platform with generalization, strong expansibility and large bearing capacity. The platform can be built into the future high-resolution geostationary orbit large optical satellite, geostationary orbit large electronic reconnaissance satellite and future super-synthetic aperture radar satellite, belonging to a large photoelectric integrated remote sensing platform. Figure 8.16 shows the SAST9000 platform in vibration mechanics test [11].

The adoption of electric propulsion technology is a major aspect in the development of high orbit large satellite platform. Electric propulsion technology can

Table 8.4 Main technical specifications of DFH-5 platform

Item	Index
Launch weight	6500–9000 kg
Platform weight	1800–2750 kg
Satellite power	10,000–30,000 W
Bearing capacity	1200–2200 kg
Power consumption of payload	4000–22,000 W
Data transmission	360 Mbit/s
Life	12–15 years (based on configuration)

Fig. 8.16 SAST9000 platform in dynamic test

significantly increase specific impulse of fuel and improve the working life of satellite in same weight of fuel, which is particularly important for large satellites. The application of electric propulsion technology in China is still in initial stage, and the sputtering and contamination of electric propulsion plume on satellite surface [12] remain to be studied. It is necessary to study modeling method of electric propulsion for the satellite as a whole [13] because of the large size of the satellite and the complex exterior profile formed by instruments and equipment on the outer surface, hence to predict the influence of electric propulsion plume on satellite surface characteristics, thermal performance, electrical performance, etc. [14]. Additionally, the structural thermal stability of satellite platform is crucial for high-precision optical remote sensing [15].

2. **Microminiaturization of platform**

   (1) The driving force for development of micro-minisatellites

Huge expense of space launch and high cost of the spacecraft have been restricting the development of space science and technology. It has been calculated that launching a 1000 kg satellite costs at least $100 million, which many countries cannot afford. After the end of Cold War, space science and technology has gradually shifted from military application to economic construction. As a result, commercial operation makes space science and technology to reduce costs and improve benefits. In this background, on the basis of rapid development of science and technology, microsatellite has been rapidly developed.

Demand of space market stimulated development of microsatellite technology. In the early twenty-first century, the cost of satellite manufacturing increased, but the technological advancement decreased with the increase of manufacturing cycle. Therefore, modern small satellites with lightweight, good performance, short development period, low cost, especially nanosatellites bring new opportunities to development of space technology. In fact, modern small satellites have been widely used in communication, remote sensing, electronics, reconnaissance and other fields. It attracts widespread attention from aerospace, military, industrial and general research institutions, becoming one of the important directions of current space technology development and has good military, economic and social benefits.

More and more countries attach importance to development of microsatellite technology, especially its launch, which does not need large rockets and large launching sites, can be launched by small rockets anytime and anywhere. Therefore, many countries have included it in the national research plan, as a growth point of technology and economic development in the twenty-first century.

(2) Subdivision of micro-minisatellites

Satellites weighing less than 1000 kg are collectively known as micro-minisatellites. It can be further subdivided into "smallsat", weight 1000–100 kg; "microsat", weight 100–10 kg; "nanosat", weight 10–1 kg; "picosat", weight 1–0.1 kg; and "femtosat" weight less than 0.1 kg. Micro, nano, pico and femto are the international system of units for decimal multiples of 10^{-6}, 10^{-9}, 10^{-12} and 10^{-15}, respectively, which are used to classify micro-minisatellites by weight and do not have the actual meaning of its value [16].

(3) The practice of microsatellites

Microsatellite is a new generation of satellites with definite uses in space science and technology. As early as May 1990, the United States used scout rockets to launch two "multichannel communications satellites" weighing 6.8 kg each, which were prototypes of microsatellites. America's Lewis and Clark small satellites use GPS positioning, fiber optic data buses, nickel-hydrogen batteries in common containers and advanced processors and memory.

Subsequently, more than 20 countries and regions in the world have carried out research and development of microsatellites, and there are more than 300 microsatellites in space operation. Microsatellite is characterized by new technology content, short development period (about one year), low development cost (tens of millions of RMB) and can be further networked to form distributed constellation to form "virtual large satellite". The advantage of microsatellite is lightweight, small size, coupled with batch production, low production cost; it can be launched with a small rocket or as a allocate payload for a large rocket with low launch cost. It can be launched from a fighter aircraft (such as the F-22 or F-15) or even from a balloon or by ground (water) artillery, meeting the need for rapid response.

Internationally, the United States has launched a variety of small satellites weighing less than a few hundred kilograms and less than 10 kg of experimental nanosatellite and picosatellite; Britain and Sweden also launched nanosatellite in 2000; France, India, Argentina, Chile, Brazil, South Korea, Thailand, Pakistan and other countries already have small satellites of their own. In addition, Indonesia, Malaysia, Philippines and China's Taiwan region are cooperating with space powers to develop small satellites or microsatellites.

NASA experts point out that the "orbital revolution" in space in twenty-first century mainly refers to low-cost launch vehicles and development of microsatellites less than 100 kg.

(4) Technological traction of microsatellite development

Development of microelectronic technology. Lightweight materials and appearance of high power solar array cells have created conditions for research and development of microsatellites. The key to miniaturization of satellite is the miniaturization of various instruments and equipment. Since the 1980s, researchers of United States, Russia, Germany, France and other countries have successfully developed microgyroscope and microaccelerometer such as miniature inertial instrument and inertial measurement unit. Multifunction structure has been developed with bearing equipment, protection function, function of transmission power and information and thermal control. The wireless connection is realized. The successful development of microelectromechanical system and opto-electromechanical system has greatly improved integration of satellite. Researchers also use integrated design techniques, using the design ideas and manufacturing processes of large-scale integrated circuits, to integrate sensors, microprocessors, inertial measurement instruments and electronic, optical and other devices like electronic circuits. The success of these processes and technologies has ensured miniaturization of satellite and facilitated satellites with small size, lightweight, low energy consumption and high reliability. In particular, the development of nanotechnology and application of scanning tunneling microscope processing technology have made it possible for people to fabricate various microelectromechanical devices at the atomic and molecular levels according to their own will and assemble them into nanosatellites smaller than microsatellites.

Micro-minisatellites are mainly used for communication, remote sensing to earth, interplanetary exploration, scientific research and technological experiments, and their development is still driven by demand and technology. Wider application requires revolutionary breakthroughs and innovations in key technologies. These new technologies mainly include electric propulsion technology, multifunctional structure, MEMS, integrated design, advanced memory and computer software technology, orbit control technology, etc. As these technologies are constantly conquered, microsatellite will become a large category of spacecraft, and as a supplement of large spacecraft, it will be widely used in military and national economy departments.

The successful development and wide use of microcomputers have led to a revolution in information technology. The success and use of micro-minisatellites have also led to a technological revolution in space. It makes satellite users from a single country, into a country, departments, units and even individual person with diversified pattern. Its development unit will also expand from a few national space departments to many universities. Micro-minisatellites technology provides technical support for three-dimensional information superhighway with integration of space and earth networks and presents a new pattern of sustainable development in the fields of communication, space, environment and resources in the twenty-first century. In communication, remote sensing and other aspects, micro-minisatellites will play an important role. In military field, such as reconnaissance, communications, command, decision-making, logistics and weapons equipment, micro-minisatellites will play an important role in adapting to the needs of modern warfare.

(5) Development of micro-minisatellites in China

In China, as early as 1995, Chinese Academy of Sciences proposed to independently develop China's first low orbit data communication satellite and its communication system weighing less than 100 kg according to the country's future development needs of satellite–ground communication technology. In 1996, Institute of microsystems of Chinese Academy of Sciences submitted a report on development of low orbit data communication small satellite and its communication system. At the end of 1997, Chinese Academy of Sciences officially approved a special support for project "storage and forwarding communication small satellite and its application system", preparing to develop a two-way data communication small satellite "Chuangxin 1". The research task is mainly undertaken by Shanghai Institute of Microsystems and Shanghai Institute of Technical Physics. On October 21, 2003, the small communication satellite "Chuangxin 1", a major knowledge innovation project of Chinese Academy of Sciences, was successfully launched into orbit, realizing non-real-time low orbit two-way data communication worldwide. The communication payload of satellite adopts spread spectrum communication technology to improve anti-interference and enhance confidentiality. The satellite is a hexahedron structure, using body mounted solar cells, with gravity gradient plus magnetic torquer active attitude control and supplemented by micromomentum wheel attitude control scheme, had total weight of more than 80 kg and average power consumption of 30 watts. It is the first microsatellite of less than 100 kg independently developed by China and also the first generation of low orbit data communication microsatellites in China. It has played an important role in research and development of microsatellites in China.

In September 2008, the companion satellite of Shenzhou-7 manned spacecraft flew into space. It is China's first space companion microsatellite developed on the basis of inheriting the mature technology of "Chuangxin 1" microsatellite developed by Chinese Academy of Sciences. Consequently, "Chuangxin 1(02)" was successfully launched in November of the same year [16].

China successfully launched remote sensing satellite 14 into space on a Long March-4B rocket from Taiyuan Satellite Launch Center at 15:06 on May 10, 2012. At the same time, Tiantuo-1 was successfully launched as a allocate microsatellite, which marked an important breakthrough in the field of microsatellites in China.

There are also many universities and companies involved in developing microsatellites in China, such as: Tsinghua university and China aerospace electrical group jointly developed the "Tsinghua 1" microsatellite; China satellite, Shenzhen Aerospace Science and Technology Innovation Research Institute and Harbin Industrial University National University Science Park Ltd. jointly set up Aerospace Dongfanghong Hit Satellite Company Ltd. developed "Shiyan 1" and "Shiyan 3" satellite; Zhejiang University developed "Pixing 1A"; and Nanjing University of Aeronautics and Astronautics launched "Tianxun 1" microsatellite. Figure 8.17a is the microsatellite of "Tsinghua 1" and "Tiangxun 1".

Microsatellites will develop in two main directions. One is development of lightweight single satellites, and this kind of microsatellites is already carrying out earth observation missions and providing images at military resolution. The US Air Force's future global positioning system (GPS) satellites will weigh less than 100 kg each. The other is to form microsatellites into constellations and fly in formation instead of expensive single large satellites, such as space-based radar (SBR) constellations, long baseline signals intelligence (SIGINT) constellations and communications satellites linked to small ground terminals.

Since 2012, many domestic units have participated in microsatellite productization technology innovation and made progress. China's space authorities have made it clear that procurement services will replace previous procurement satellites in microsatellite area.

3. **Specialized platform**

In addition to development of platform to two extremes of large and small size, specific purpose platform will also be a development direction of space remote sensing. A concrete example is Hubble Space Telescope, as shown in Fig. 8.18. The platform body, antenna, solar array and optical remote sensor have integrated into one, without clear boundary between platform and payload. The design of this specialized platform enables performance of entire satellite to be fully utilized and achieved, which is undoubtedly a best example and achieving world-renowned scientific observations.

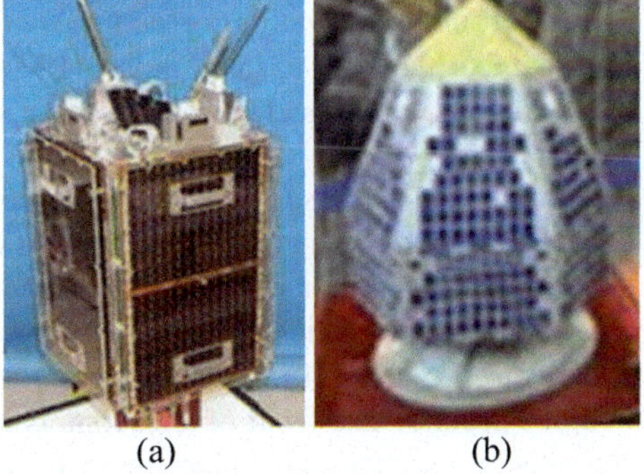

(a) (b)

Fig. 8.17 Tsinghua 1 and Tianxun 1 microsatellites

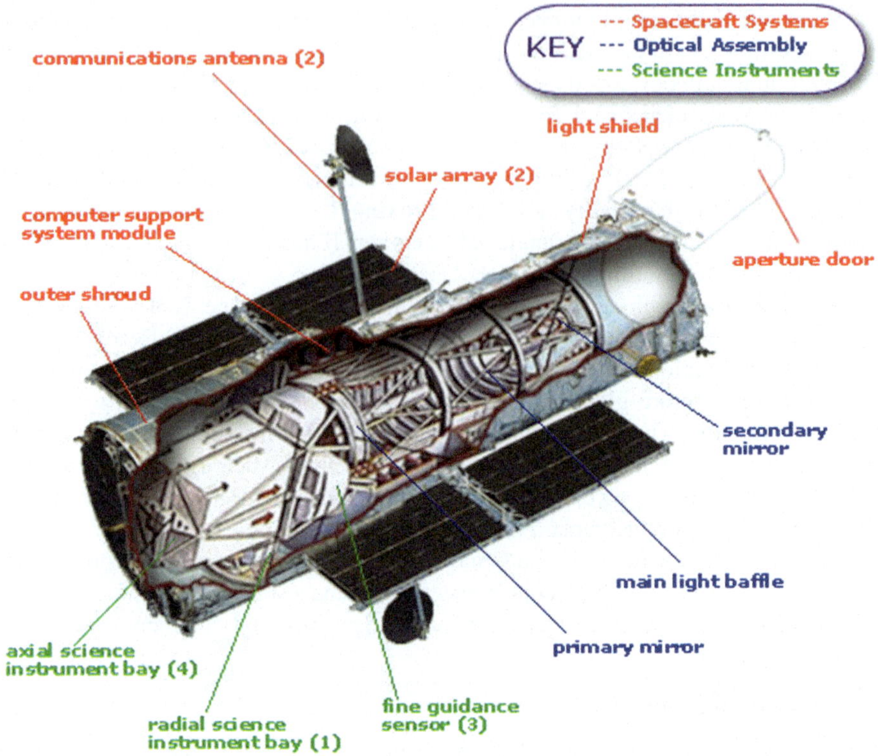

Fig. 8.18 Hubble space telescope

Hubble Space Telescope, named after astronomer Hubble, is a telescope in orbit around earth. Its position above earth's atmosphere gives it advantages that the imaging was not influenced by the atmospheric turbulence disturbance, and it has excellent visibility without background light from atmospheric scattering, as well as it can observe ultraviolet light absorbed by the ozone layer. Launched in 1990, it has become the most important instrument in the history of astronomy. It has filled in the gaps of ground observation, helped astronomers solve many fundamental problems and enabled mankind to have more understanding of astrophysics. Hubble's ultra-deep field of view is the deepest (and most sensitive) optical image astronomers have ever obtained.

Hubble Space Telescope is the largest telescope ever put into orbit. It is 12.8 m in length, 4.27 m in diameter and 11 tons in weight. It is composed of three parts: the first part is optical part, the second part is scientific instrument, and the third part is auxiliary system, including two solar arrays with 11.8 m in length and 2.3 m in width, which can provide 2.4 kW power, and two parabolic antennas for communication with the ground. The front of telescope is optical section, and the back is an annular chamber in which a set of scientific instruments are placed on the focal plane behind the telescope's primary mirror. The solar array and antenna protrude from the middle part of telescope.

Optical part is the heart of the instrument. It uses a Cassegrain reflector system, consisting of two hyperboloid mirrors, a primary mirror with a diameter of 2.4 m and a secondary mirror with a diameter of 0.3 m installed about 4.5 m in front of the primary mirror. The light projected onto the primary mirror is first reflected to the secondary mirror and then from the secondary mirror to the central hole of the primary mirror, which passes through the central hole to the focal plane of optical system, forming high-quality images for precision processing by various scientific instruments. The resulting data is sent back to the ground via the relay satellite system.

In addition to the optical part, the other major part of the telescope is eight scientific instruments mounted on the focal plane, which are the wide field of view planetary camera, the dark celestial camera, the dark celestial spectrograph, the high-resolution spectrograph, the high-speed photometer and three precision-guided remote sensors.

References

1. http://baike.baidu.com/link?url (2014).
2. http://news.xinhuanet.com/mil/2010-03/24/content_13235799.htm. (2015).
3. Qingjun Z, Shijun M, Achievements and progress of china-brazil earth resource satellite, Spacecraft Eng. 2009:18(4):1-8.
4. http://www.geoconnexion.com/geo_news_article/EADS-Astrium.
5. http://www.mapmart.com/products/satelliteimagery/geoeye1.aspx.
6. Krimchansky A, Machi D, Cauffman SA, Davis MA. Next generation geostationary operational environmental satellite (GOES-R Series): a space segment overview. SPIE 2004;5570:155–164.
7. Schmit TJ, Gurka J, Menzel WP, et al. Possible scanning scenarios of the GOES-R HES(Hyperspectral Environmental Suite). SPIE 2005; 5806:642–652.
8. Xia S. Research on precise orientation technology of geostationary satellite image. Remote Sens Inform. 2003;3:26–30.
9. Qiang G, Xiaohu Z, Image shift compensation modeling and analysis for two-dimensional scanning on geostationary orbit. Acta Optica Sinica. 2007;27(10).
10. http://www.cast.cn/castcn/show.asp?articleid=46392, 2015;1779–1787.
11. http://www.guancha.cn/Project/2015_01_26_307570.shtml.
12. Xiaoxiong L, Zheng W, Jiasheng T. Simulation analysis of deposition effect of molybdenum atoms from ion thruster plume on satellite OSR thermal properties. Spacecraft Eng. 2016;25(3):52–26.
13. Xiaoxiong L, Zheng W, Jiasheng T. A method for numerical simulation of ion thruster's plume based on unstructured mesh. Spacecraft Environ Eng. 2016;33(5):484–9.
14. Xiaoxiong L, Jiasheng* T, Zheng W. Influence of plume deposit of ion thruster on thermal control propertiesof satellite. J Rocket Propulsion. 2017;43(2):9–17.
15. Yue C, Jiasheng T, Xiaoxiong L, Jiandong Z, Thermal deformation analysis of high-precision large GEO platform structure. In: Proceedings of the 19th China symposium on remote sensing; 2014. 375–381.
16. http://baike.baidu.com/view/4448202.htm.

Chapter 9
System Overall Parameters Design

9.1 Overview

Using space optical remote sensing cameras in the orbits can image ground's targets that are hundreds of kilometers away. The resolution of the image is affected by many factors, such as carrier platform's control technology, orbit design, sun, earth, atmosphere, target's characteristics and many vibration sources of the satellite. The influence rule of various factors on image resolution is discussed a lot. The influence of the sampling discretization of the analog quantity of the CCD detector on the image resolution, the improvement of the oversample imaging on the image resolution, the impact of the space optical remote sensing camera on the image quality when the aerospace optical remote sensing camera is disturbed by vibration when working in orbit, these are all research on a certain aspect of the problem. There are also researches on calculation methods for large systems aimed at improving the image quality of space optical remote sensing cameras.

It is able to design and manufacture with nanometer-level precision a space optical remote sensing camera with μard angular resolution. The speed stability of the carrier platform is as high as 10^{-4}°/s. The accuracy is so high that the imaging of the optical remote sensing camera will be adversely affected easily, whether it is the motion of the spacecraft relative to the earth or its vibration. Therefore, not only the static image quality of the space optical remote sensing camera, but also the dynamic factors related to the working state must be considered when considering the in-orbit imaging resolution of space optical remote sensing cameras. The impact on the image quality of the space optical remote sensing camera is all closely related to image shift, whether it is motion or vibration. Their impact on image quality is also closely related to exposure time. It is beneficial to improve the imaging quality of space optical remote sensing camera, as long as the factors that are conducive to shorten the exposure time, from the perspective of space optical remote sensing camera image shift.

J. Tao, *Space Optical Remote Sensing*, Advances in Optics and Optoelectronics,
https://doi.org/10.1007/978-981-99-3318-1_9

They are mainly related to the following three aspects:

1. The types of camera optical systems: reflection, catadioptric and refraction, which are closely related to the transmittance of the camera.
2. The relative aperture and blocking ratio of the camera are directly related to the image plane illuminance of the camera.
3. The choice of the time delay integration CCD (TDICCD) stages will directly determine the length of the exposure time. The sampling and discretization of the CCD detector imaging will also have a direct impact on the resolution of the camera.

What it has been seen above is the side with many factors and complicated problems in optical remote sensing of the earth. In terms of high accuracy, the accuracy of earth optical remote sensing is also very high. The space optical remote sensor can reach 10^{10} in size span, which is a typical high-precision mechanical design and manufacturing problem. It can be said that the overall research of the space optical remote sensor is a high-precision and complicated topic.

9.2 Remote Sensing Method

The main light source of space optical remote sensing is the sun, and electromagnetic waves radiate by the target itself, such as infrared rays. The electromagnetic spectrum used in remote sensing is extremely broad, from gamma rays with a wavelength of only 0.001 nm to microwaves with a wavelength of 1 m, which spanning 12 orders of magnitude. For different electromagnetic waves, the sensors that receive the electromagnetic waves are also different. The working principles of different sensors and the technical fields involved are also very different. For example, some nuclear physics instruments, such as Geiger counters, scintillation counters and nuclear latex, are used as the sensors for the gamma ray and X-ray band with wavelengths of 0.001–10 nm. The optical instruments, such as cameras and spectrometers, are used as sensors for the near-ultraviolet–visible-infrared band with a wavelength of 0.3–14 μm. The microwave instruments, such as microwave radiometers, microwave scatterometers and synthetic aperture radars, are used as sensors for the microwave band with a wavelength of 1 mm–1 m.

Cameras, spectrometers, etc. are passive sensors that detect by receiving the target's own radiation or the reflection of solar radiation. The active sensors, which have transmitters that emit electromagnetic waves to the target, and then receive the electromagnetic waves reflected from the target, such as laser radar and synthetic aperture radar. The classification is shown in Fig. 9.1.

From the perspective of the development history of space remote sensing, the frame imaging method is the first application, as shown in Fig. 9.2. At that time, the space remote sensing cameras were all film cameras. The image of the film-type remote sensing camera on the satellite cannot be transmitted to the ground on time and only can be received together with the capsule returned back to ground. Special

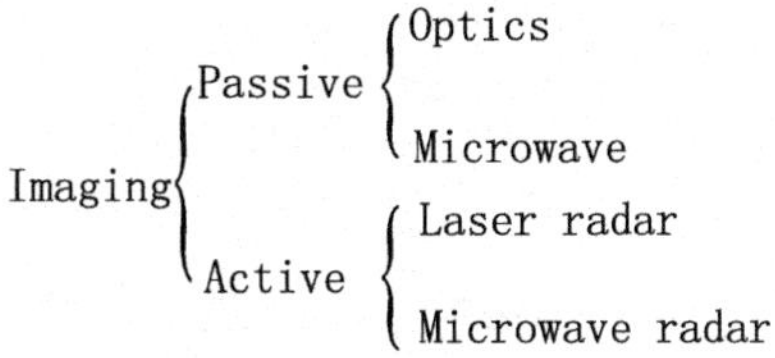

Fig. 9.1 Imaging categories and methods

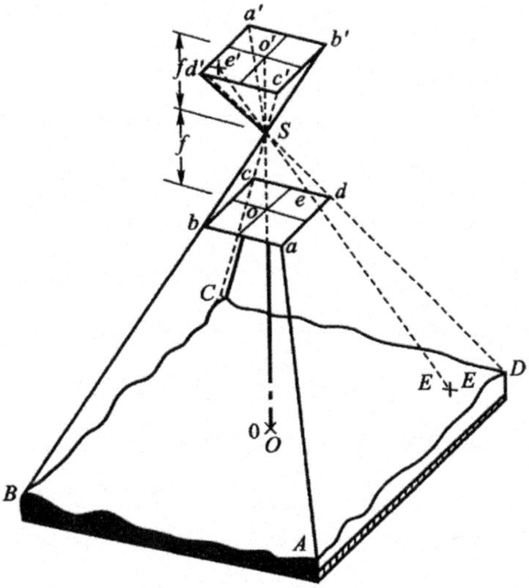

Fig. 9.2 Frame imaging principle diagram

space return and reception procedures are required, which are complex, difficult and risky from technical perspective. Therefore, the film-type remote sensing camera not only be unable to receive information on time, but also has a high risk on technology. Therefore, the real-time transmission cameras for photoelectric scanning imaging have been developed.

With the development of photoelectric technology, the two-dimensional photoelectric scanning imaging methods have been adopted to achieve real-time transmission and extend the life of optical remote sensors and satellites. Figure 9.3 is a multispectral scanner (MSS) on the US Landsat. Figure 9.3a is its scanning optical channel and detector configuration, and Fig. 9.3b is the scanning coverage of each channel. Channels 4, 5, 6 and 7 have 6 photomultiplier tube arranged perpendicular to the scanning direction to achieve parallel scanning. Channel 8 has 2 HgCdTe infrared devices that arranged in a diagonal manner to achieve 2-element parallel scanning, and the infrared devices are cooled to 100 K. The scanning frequency of the swing mirror is 13.65 Hz, and the detector of each channel uses the same filter. The detectors of channels 4, 5 and 6 are composed of 18 photomultiplier tubes, and channel 7 is composed of 6 silicon photodiodes. The spectrum of each channel is as follows:

Channal 7, 0.8~1.04μm;

Channal 8, 10.4~12.6μm;

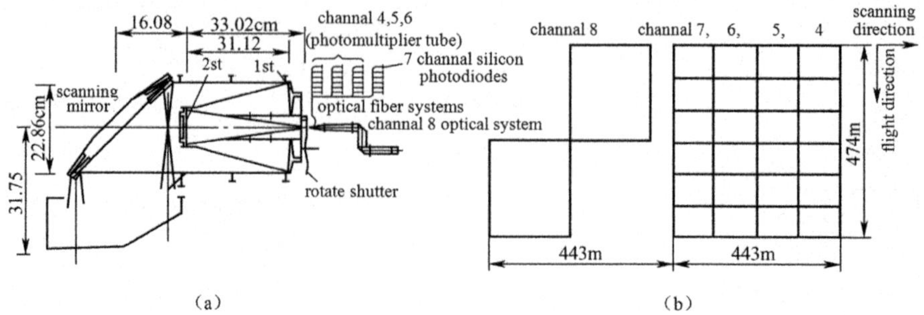

Fig. 9.3 Two-dimensional scanning imaging principle diagram

Channel 4, 0.5–0.6 μm;
Channel 5, 0.6–0.7 μm;
Channel 6, 0.7–0.8 μm;
Channel 7, 0.8–1.04 μm;
Channel 8, 10.4–12.6 μm;

The spectral arrangement of each channel is shown in Fig. 9.4, Figure 9.4a is the spectral band pass of the four channels of visible light and near infrared, and Fig. 9.4b is the spectral band pass of the infrared channel.

The structure of the two-dimensional scanning optical-mechanical system is complex, and the reliability and accuracy are difficult to maintain. With the development of CCD, space optical cameras mainly use one-dimensional scanning imaging. See Fig. 9.5, Figure 9.5a is an image of satellite push-broom imaging in orbit, and Fig. 9.5b is the working principle of push-broom. The push-broom imaging of the satellite in orbit is a one-dimensional scan that realizes the push-broom by the orbital

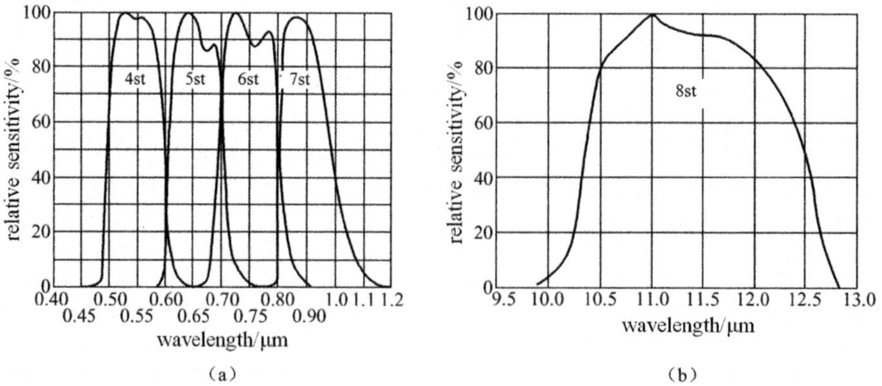

Fig. 9.4 MSS spectral characteristics

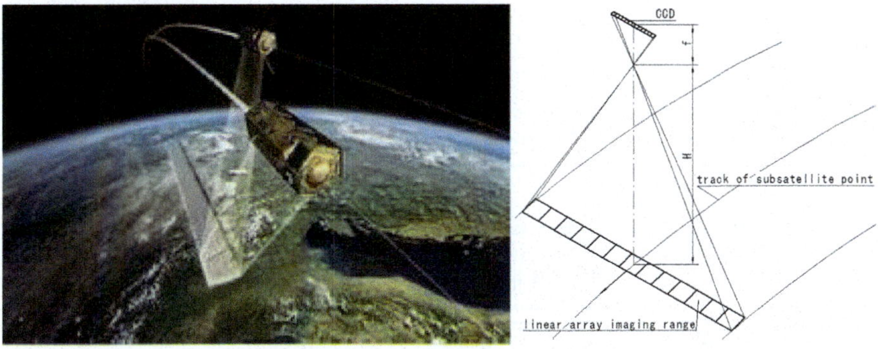

(a) (b)

Fig. 9.5 Push-broom imaging principle

motion of the satellite and does not require a scanning of structure system, which simplifies the mechanism and improves the reliability. However, for this push-broom imaging method to achieve global seamless imaging, the camera needs a larger field of view, otherwise a long return period is required.

Take CBERS-1 as an example, the orbit height is 778 km, the orbital period is 100.36 min, and the orbit regression period is 26 days. The CCD camera has a focal length of 520 mm, a field of view angle of 8.4°, the lens $F^{\#} = 4$ and 5 imaging bands. The wavelength in the blue spectrum is 0.45–0.52 μm, in the green spectrum is 0.52–0.59 μm, in the red spectrum is 0.63–0.69 μm, the near infrared spectrum is 0.77–0.89 μm, and panchromatic band is 0.51–0.73 μm.

The orbit setting of the CBERS-1 satellite requires a camera field of view angle of $2\omega = 8°$ to achieve global coverage. The field of view of the CCD camera is 8.4°, and each side overlaps by 0.2°.

It can be seen that although the regression period has reached 26 days, it still needs a larger camera field of view of 8.4°. The CCD camera of the CBERS-1 satellite uses a refraction lens, as shown in Fig. 9.6. In order to shorten the revisit period, the camera has a swing mirror with a swing angle of ± 32°, so that the camera can revisit and image a certain ground scene every 3 days.

With the development of space optical remote sensing, optical imaging remote sensing has developed from low orbit to high orbit, and the imaging method has returned to the frame imaging method again. However, the image reception has developed from a film to an area array photoelectric detector.

9.3 Energy of Optical Remote Sensing Systems

The sun is the main light source of space optical remote sensing. The solar light passes through the atmosphere of earth to the ground. After being reflected by the ground, it passes through the earth's atmosphere again and travels to the entrance of

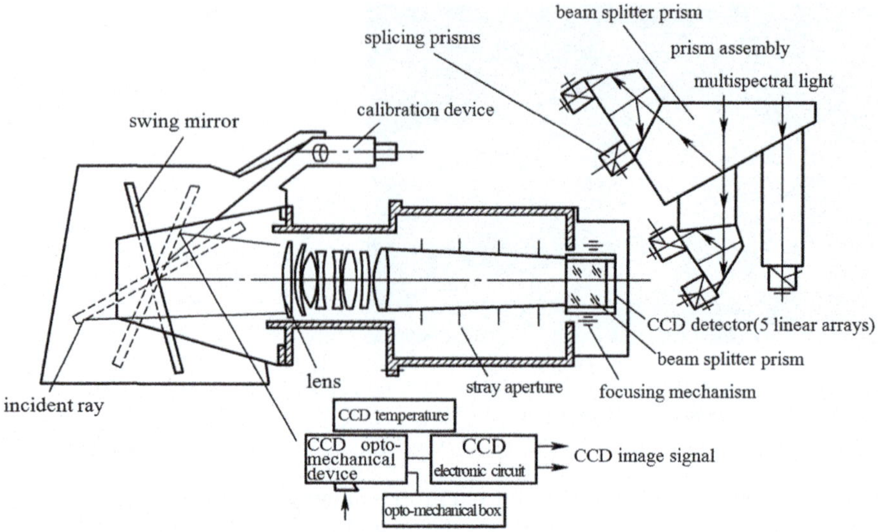

Fig. 9.6 CCD camera of CBERS-1 satellite

the optical remote sensor and then reaches the photodetector through the lens. After the photoelectric conversion of the photodetector, the optical signal is converted into electric and digital signal. This process is shown in Fig. 9.7. In this process, the optical signal undergoes three main attenuations, one downward radiation through the atmosphere, one upward radiation through the atmosphere and the other one passes through the camera lens involving a total of three optical transmittances and one ground object scattering.

The orbits of space optical remote sensing satellites range from 200 to 2000 km in low orbits and 35,786 km in high orbits. One point on the ground object whose

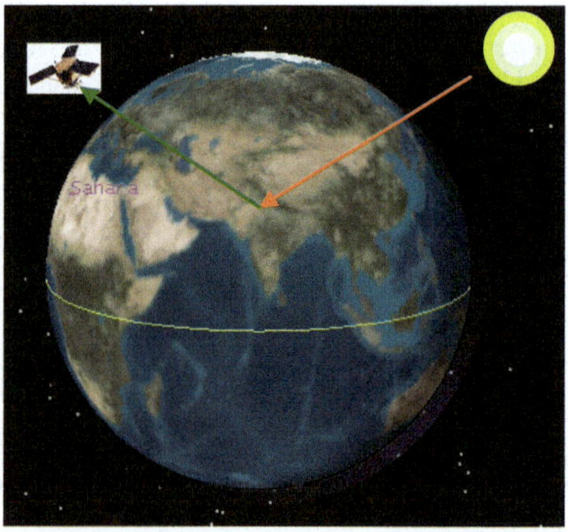

Fig. 9.7 Process of the energy propagation

diffusing the light can reach the camera detector is very small. Take the CCD camera of the CBERS-1 satellite as an example, the orbit is 778 km, the focal length is 520 mm, $F^{\#} = 4$ and the diameter $D = 0.13$ m. Then the solid angle of the camera to an object point on the ground is

$$\Omega = \frac{0.13^2\,\pi}{4 \times 778^2 \times 10^6}$$
$$= 2.1929 \times 10^{-14}\,Sr$$

If an object point is regarded as a diffuse point and its diffuse light is directed into a space of 2π steradian, then the energy of a certain object point on the ground obtained by such a camera is only about $1/10^{15}$. The received energy is so weak, that the energy calculation of the space optical remote sensing system is an important aspect of the calculation of the parameters of the space optical remote sensing system.

9.3.1 Target Radiance

The target irradiance is related to the season, latitude, local time, weather and atmospheric type. The target radiance is related to the target's irradiance, surface reflectivity, observation altitude angle, observation azimuth angle, atmospheric scattering, etc.

Figure 9.8 shows the radiation situation when the visibility is 23 km, and the sun zenith angle is 30°, under vertical observation. From the near infrared side of the figure, the bottom curve is the radiance curve of atmospheric backscatter, the second curve from bottom to top is the radiance curve of the ground target, the third curve from bottom to top is the radiance curve of the earth–atmosphere system, and the top curve is the transmittance curve of the entire atmosphere.

Figure 9.9 shows the changes of the various curves after reducing the spectral resolution under the conditions of Fig. 9.8. The changes of the radiation and absorption peaks are obvious. For example, the absorption peak near 0.43 μm has a transmittance of 0.55 in a graph with higher spectral resolution and a transmittance of 0.62 in a graph with lower spectral resolution. The change is obvious. The radiance of the radiation peak near 0.45 μm is 0.0194 W cm^{-2} Sr^{-1} μm^{-1} in the graph with higher spectral resolution, while its radiance is 0.0221 W cm ^{2} Sr ^{1} μm ^{1} in the graph with lower spectral resolution. The difference between them approaches to 14%, which also shows the change is obvious.

Figure 9.10 shows the radiation situation when the visibility is 5 km, and the sun zenith angle is 30°, under vertical observation. From the near infrared side of the figure, the bottom curve is the radiance curve of the ground target, the second curve from bottom to top is the radiance curve of atmospheric backscattering, the third curve from bottom to top is the radiance curve of the earth–atmosphere system, and the top curve is the transmittance curve of the entire atmosphere.

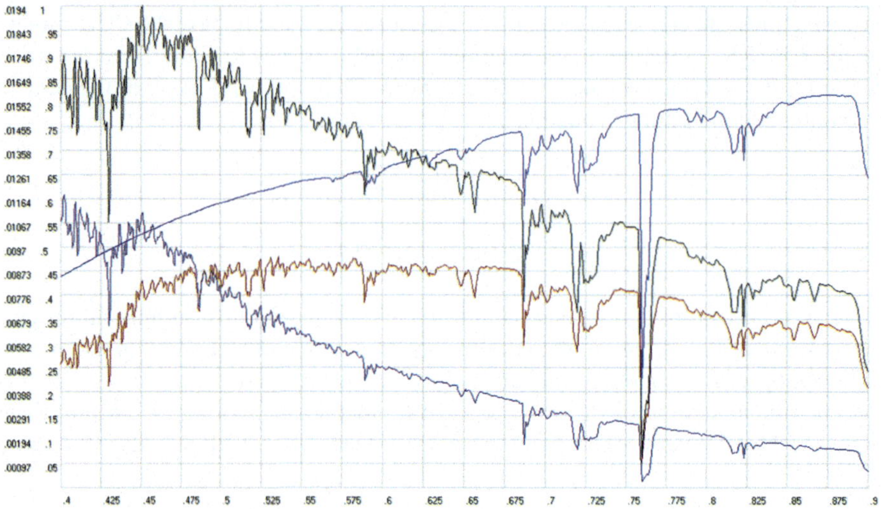

Fig. 9.8 Radiation situation during vertical observation at a visibility of 23 km

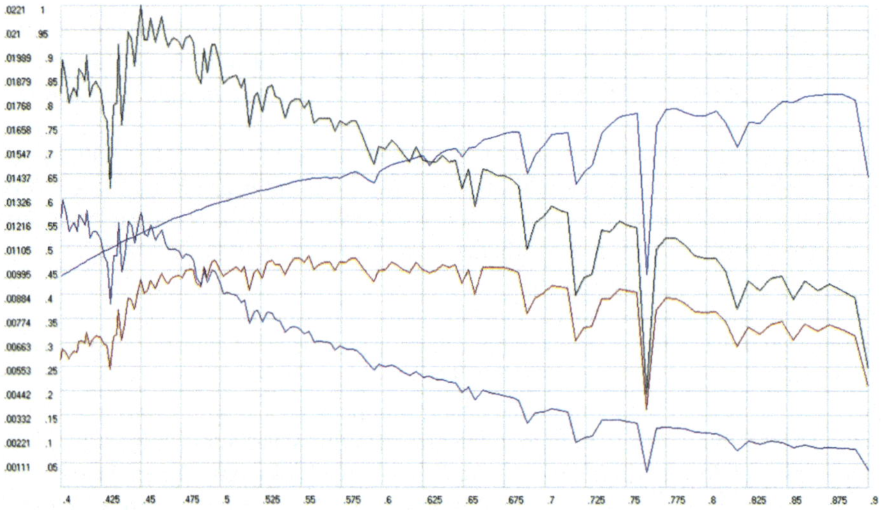

Fig. 9.9 Radiation situation during vertical observation on the low spectral resolution at a visibility of 23 km

It can be seen from the above figures that the target radiance of space optical remote sensing is not only related to the directional reflection factor ρ'', which includes the incident light's altitude angle, incident azimuth angle, reflected altitude angle and reflection azimuth angle, but also related with target reflectivity, weather condition, Sun irradiation (local time), atmospheric type (such as rural aerosol type, urban aerosol type), geographic latitude and many other objective factors. It is also related

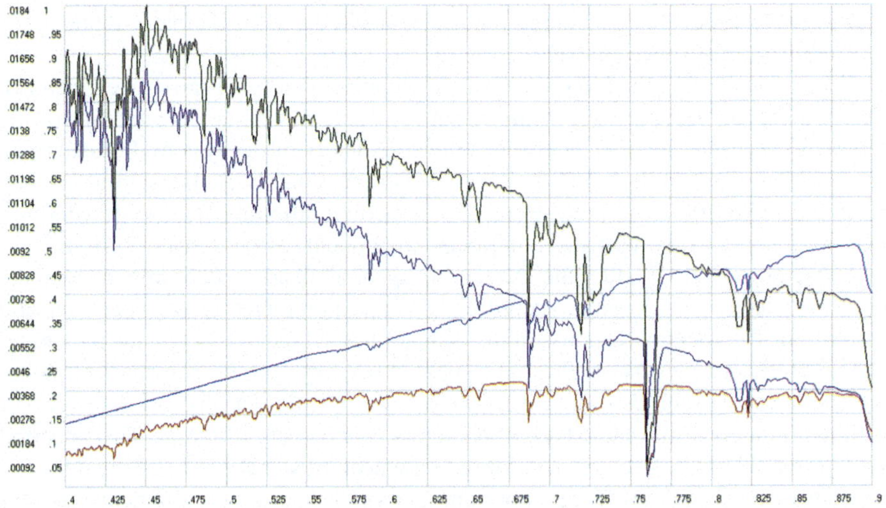

Fig. 9.10 Radiation situation during vertical observation at a visibility of 5 km

to our subjective factors, such as the spectral resolution selected when describing radiance and transmittance.

The calculation of target radiance is complicated based on the above analysis and description. Therefore, the calculation of target radiance in space optical remote sensing is generally solved by atmospheric radiation calculation software such as LOWTRAN or a look-up table such as the "American Standard Atmospheric Manual". Through the look-up table supplemented by methods such as interpolation calculation, the specific problems related to optical remote sensing, such as atmospheric radiation transmission and target radiance calculation, are solved.

9.3.2 Remote Sensing Spectrum Bands

The spectrum used in optical remote sensing is determined firstly by user requirements. Some users have clear requirements and can be executed according to the requirements. Some users have only a general description then need to be demonstrated. In the process of demonstration and refinement, relevant spectra theory is needed, and it should be also referred to the existing experiences. The following will share some spectral detection characteristics and the spectral applications status of satellites that have worked in orbit, as well as the spectral requirements of some special applications for reference. Table 9.1 is correspondence table of spectrum bands and measured elements, Table 9.2 is the setting and application of domestic satellites in orbit, Table 9.3 is the setting and application of foreign satellites in orbit, all the above three table's main reference [1], and Table 9.4 is thematic applications of satellite optical spectrum bands.

While considering the requirements, the choice of detector is also a key factor in determining the detection spectrum. It not only determines the detectability of the required spectrum, but also determines the complexity of the system. For example, the electronic part of the CCD detector needs more complicated design more than CMOS detector needs, and mid-infrared and thermal infrared detectors may require cooling system, which will also complicate the optical remote sensing system.

Detector selection: after the detection, spectral band is determined, and the detector needs to be selected according to the requirements of the spectrum and resolution. The spectral response of the detector has a direct impact on the parameter design of the optical system. From the spectral response curves of various photodetectors, it can be seen that the spectral responses of different types of detectors, or different models of the same type of detectors, can be very different. The selection of detector should be decided according to the need of signal power of the detector. The signal power is one of the important factors of determining the parameters of the remote sensing system.

9.3.3 Optical System Parameters

Generally, since the choice of spectrum is more determined by the user's needs, the detector is a purchased part, which is more determined by the manufacturer's product situation. The both above design freedom are limited. After that, the selection and parameter design of optical system have more choice. The parameters of the optical system are determined by integrating the needs of spatial resolution, MTF and energy. The spatial resolution involves the focal length f, the pixel size a, the orbit height H and the ground pixel resolution r_g, and the relationship is relatively simple. MTF is determined by $F^{\#}$ and wavelength λ and has a clearer mathematical expression. However, the determination of energy is systematic and directly related to the type of optical system, $F^{\#}$, obscuration ratio, transmittance and detector response.

If the form of the optical system chooses the transmission type, it is generally related to the $F^{\#}$ number and transmittance. The $F^{\#}$ can be selected by design. The transmittance is not only related to the thickness of the lens and the quality of the coating, but also related to the transmittance of the selected optical material. The transmission spectrum of some optical materials is shown in Table 9.5.

For the catadioptric hybrid optical system, it is not only related to the $F^{\#}$ and the transmittance of the optical material, but also to the reflectivity and blocking ratio.

For a reflective optical system, it is related to the $F^{\#}$ and the reflectivity of the optical mirror, and for an obstructed optical system, it is also related to the obscuration ratio.

Table 9.1 Spectrum detection characteristics table

No.	Band	Spectrum/ μm	Detection element	Application
	UV	0.16–0.4	Ozone vertical distribution Total (0.3–0.36 μm, FY-3)	Meteorological
	Panchromatic	0.52–0.90	Spatial high-resolution grayscale image	Geology
	Blue-green	0.45–0.52	The penetration depth of clean water is 10–20 m, interpret shallow water terrain and offshore cement sand; Distinguish soil vegetation; 0.503–0.52 μm reflectance is sensitive to chlorophyll concentration	Hydrology Water quality Soil Plant
	Green	0.52–0.60	Identify vegetation, located near the maximum peak of chlorophyll reflection and detect the reflectance of healthy green vegetation; Located on the longwave side of the minimum attenuation of the water, it can detect the turbidity of the water and the water depth of 10–20 m; 0.52–0.564 μm reflectance is more sensitive to changes in chlorophyll mass concentration	Plant Water quality
	Red	0.63–0.69	0.672–0.676 μm chlorophyll absorption area; Geological research and sediment content research in water Good observation of road, bare soil, vegetation species	Water quality, Geology Plant, soil
	Near infrared	0.7–0.8	Soil water content; Different boundary of land and water, the reflectance of water and ground is obviously different, and the lake water absorbs almost all the infrared radiation 0.703–0.709 μm chlorophyll a reflection area, the most sensitive detection; 0.705–0.745 μm (red edge Worldview-2), assists in analyzing plant growth, directly reflecting plant health information, estimating biomass and distinguishing health from plant diseases and insect pests	Plant Microorganism Soil Water quality Geology
	Near infrared	0.8–1.1	0.8–0.89 μm can penetrate the atmosphere well; Geological research Measure the biomass and monitor the growth of crops; the vegetation performance is particularly bright; Different boundary of land and water, the reflectivity of water and background is obviously different, and the water absorbs almost all infrared radiation	Plant Microorganism Geology

(continued)

Table 9.1 (continued)

No.	Band	Spectrum/ μm	Detection element	Application
	Near infrared	1.55–1.75	Have a good ability to penetrate the atmosphere, clouds and fog; Can use to distinguish roads, bare soil and water; Good contrast between different vegetations; The humidity of soil and plants can also be analyzed from the gray scale of this spectrum, making it easier to interpret soil types and monitor the growth stage of vegetation and crops	Plant Hydrology Water quality Soil Geology
	Near infrared	2.08–2.35	Useful for distinguishing rocks and minerals; Also used to identify vegetation cover and moist soil	Geology, soil plant
	Mid-infrared	3–6	4.45–4.57 μm monitor N_2O (FY-3); Monitoring forest fires, volcanic eruptions and straw burning	Meteorological
	Thermal infrared	6–15	Monitor thermal radiation targets; 6–14 μm detect water temperature, water emissivity ≈ 1 (approximately black body) and use measured brightness temperature to represent water temperature; 13.35–14.95 μm CO_2 monitoring (FY-3)	Water quality Air quality
	L-Band C-Band	1.5–3 mm 38–75 mm	Radar. Used to monitor changes in the speed and structure of glaciers, ocean surface and vertical structure distribution of changes in the earth, etc.	Geology
	P-Band	0.3–1 m	Radar. Global forest height and biomass changes, carbon storage estimates	Geology
	True color compositing	No.3 + 4 + 5	Close to the naked eye; Used to distinguish the boundaries of mining areas, vegetation, water (contaminated and clean), highways and large mining areas (such as Tonglushan open-pit mining); Used for coastal area research, observe underwater since visible light can penetrate the water surface	Hydrology Geology
	Near infrared synthesis	No.4 + 5 + 6	Vegetation has high reflectivity in the near infrared, and chlorophyll has high reflectivity in this spectrum. The vegetation in the image shows different shades of red, and different types of plants have different red tones; Water absorbs almost all near infrared radiation, and the water surface is almost black	Plant
	Near infrared synthesis	No.5 + 6 + 9	The reflection of the infrared spectrum mainly depends on the water content of the surface of the object. This type of image can be used for vegetation protection and land research	Plant Soil

Table 9.2 The Setting and application of domestic satellites in orbit

Type	Satellite	Load	Application
Resource satellite	Resource No.1 01, 02 satellites	CCD camera (blue, green, red, near infrared and panchromatic) Wide field of view camera (red light, near infrared, width 890 km, cycle 5 days) Infrared scanner (4 bands)	The observation range is large and can quickly observe information on the earth's surface. It is used to prospect land usage, measure arable land, estimate forest stock, crop growth, yield rate, pasture stock, underground resources detection, disaster monitoring, coastal beaches usage, aquaculture, environmental pollution, etc.
	Resource No.1 02B satellite	CCD camera (19.5 m, width 226 km, 4 bands) HR camera (2.36 m, width 45 km)	With orbital maneuverability, the camera has a side swing angle of ± 32° and can observe the interest target for 3 days consecutively. It is used for land and resources prospecting, environmental monitoring and protection, urban planning, crop yield estimation, disaster prevention and mitigation, etc.
	Resource No.3	Three-line scan camera (panchromatic 0.50–0.80 μm, normal view 2.1 m, front and rear view 3.5 m) Multispectrum camera (5.8 m, 4 bands) 0.45–0.52 μm, 0.52–0.59 μm 0.63–0.69 μm, 0.77–0.89 μm	The first civil three-dimensional surveying and mapping satellite, integrated surveying and resource inquiry functions can produce 4D products and various thematic products such as slope maps, aspect maps and other three-dimensional products, providing services for land resources, agriculture, forestry and other fields
Environmental disaster mitigation	HJ-1A	30 m, width 700 km; 4 bands 0.43–0.52 μm, 0.52–0.60 μm 0.63–0.69 μm, 0.76–0.9 μm Hyperspectral camera 100 m, width 50 km 0.45–0.95 μm divided into 110–128 spectrum	Special satellite for environmental monitoring, 30 m is a medium–high-resolution satellite (meteorological 2 km), revisited in 48 h, take care of both high resolution and wide coverage. The hyperspectral imager can analyze the characteristic spectra of different ground objects and can carry out ecological environment, water and gas environment monitoring

(continued)

Table 9.2 (continued)

Type	Satellite	Load	Application
	HJ-1B	CCD camera (same as HJ-1A) Infrared camera 150 m, width 700 km, 4 bands 0.75–1.10 μm, 1.55–1.75 μm 3.50–3.90 μm, 10.5–12.5 μm	Infrared camera, mid-infrared can detect the earth surface temperature of 500 K, suitable for the detection of disasters such as forest fires and volcanic eruptions
	HJ-1C	SAR S-band strip mode 4–6 m, width 40–50 km S-band scanning mode 40–60 m, width 400–500 km	
Ocean	HY-1B	Ocean water color scanner (1100 m, 10 bands) 0.402–0.422 μm, 0.433–0.453 μm 0.480–0.5 μm, 0.51–0.53 μm 0.555–0.575 μm, 0.660–0.680 μm 0.74–0.76 μm, 0.845–0.885 μm 10.3–11.4 μm, 11.4–12.5 μm Coastal zone imager 250 m, width 512 km, 4 bands 0.433–0.453 μm, 0.555–0.575 μm 0.655–0.675 μm, 0.675–0.695 μm	It can detect chlorophyll, suspended sediment, soluble organic matter and ocean surface temperature, pollutants, shallow ocean depth, underwater topography, etc. and can also monitor the dynamic changes of the coastal zone; Observe the Bohai Sea, the Yellow Sea, the East China Sea, the South China Sea and the coastal zone in real time; Observation data of other regions is stored on the satellite and can be replayed and downloaded
	HY-2	Radar altimeter, microwave scatterometer, scanning microwave radiometer and calibrated microwave radiometer as well as DORIS, dual-frequency GPS and laser rangefinder	China's first marine dynamic environment monitoring satellite, which monitors the marine dynamic environment and provides information for marine disaster prevention and mitigation, early warning of disastrous marine conditions, marine scientific research, environmental forecasting and global climate change research

(continued)

Table 9.2 (continued)

Type	Satellite	Load	Application
Meteorological	FY-2C/D/E FY-3A/B	Scanning radiometer (0.45–12.5 μm, divided into 4 visible spectrum bands, 1 near infrared band, 2 shortwave infrared bands, 3 medium and longwave infrared bands) Infrared spectrometer (0.69–15 μm, divided into visible light, near infrared, shortwave infrared, medium wave infrared, longwave infrared) Medium-resolution imaging spectrometer (0.40–12.5 μm, divided into 10 visible spectrum bands, 7 near infrared spectrum bands, 2 shortwave infrared bands and 1 thermal infrared band) There are also microwave imagers, microwave thermometers/hygrometers, total ozone detectors, ozone vertical detectors, earth/solar radiation monitors, space environment monitors, etc.	FY-3 can monitor small and medium-scale convective cloud clusters and fine features of the land surface, improve the quantitative inversion accuracy of cloud characteristics, aerosols, land surface characteristics, vegetation, ocean water color and low-level water vapor and realize multispectral continuity of the atmosphere, land and ocean comprehensive observation. The 10-channel scanning radiometer can detect vegetation, sediment, land surface temperature, sea surface temperature, total water vapor, ice and snow, cloud images, etc.; the 20-channel medium-resolution imaging spectrometer can detect ocean water color, aerosols, total water vapor, vegetation, surface land characteristics, surface temperature, cloud characteristics, 26-channel infrared spectrometer can detect O_3 content, CO_2 concentration, aerosol, atmospheric temperature and humidity, cloud parameters, ice and snow, precipitation, etc.

9.4 Selection of System Design Objectives

The goal of the system design is to select one core problem of the space optical remote sensing systems, after comprehensively considering factors such as the range and number of spectral bands, the performance of the detector, the type of orbit, the working mode of the platform and the number of spectral channels of the optical remote sensor. Then the design of each link of the space optical remote sensing system will be carried out with this problem as the core.

Table 9.3 Spectral band settings and applications of foreign satellites in orbit

Type	Belonging	Satellite	Load	Features
Land satellite	USA NASA	Landsat-5	Multispectral scanner TM, 7 bands Blue 0.45–0.50 μm Green 0.52–0.60 μm Red 0.63–0.69 μm Near infrared 0.76–0.94 μm Shortwave infrared 1.55–1.75 μm Shortwave infrared 2.08–2.35 μm Thermal infrared 10.4–12.5 μm	To investigate underground mineral deposits, marine resources and groundwater resources, monitor the reasonable use of agriculture, forestry, animal husbandry and water resources, predict crop yields, study natural plant growth and forecast natural disasters and environmental pollution; TM spectrum 3rd monitors the degree of water pollution, and the 3rd, 4th and 5th comprehensive use distinguishes vegetation, bare land, grassland, mining area, etc.
			Optical-mechanical scanner MSS, width 185 km, 4 bands	
		Landsat-7	Enhanced subject imager ETM + , width 185 km Blue 0.45–0.515 μm, 30 m Green 0.525–0.605 μm, 30 m Red 0.63–0.69 μm, 30 m Near infrared 0.75–0.90 μm, 30 m Shortwave infrared 1.55–1.75 μm, 30 m Thermal infrared 10.4–12.5 μm, 60 m Shortwave infrared 2.09–2.35 μm, 30 m Panchromatic 0.52–0.90 μm, 15 m	Wide coverage, short period, wide spectrum, moderate resolution, suitable for the resource and environmental surveys on a scale of ≤ 1:200,000

(continued)

Table 9.3 (continued)

Type	Belonging	Satellite	Load	Features
EOS Earth observation satellite EOS	USA NASA	Terra observe land	Moderate Resolution Imaging Spectrometer (MODIS), 0.25–1 μm divided into 36 spectral bands	MODIS has a high time resolution of 4 to 8 times coverage per day at the same place, with a medium resolution of 100–250 m;
		Aqua observe the water cycle	MODIS (cooperate with Terra on time update frequency, plus transiting the data at night, at least twice the day and twice at night to update data)	Monitor the terrestrial atmosphere and ocean, such as land and ocean temperature, land surface coverage, clouds, aerosols, water vapor and fire conditions Monitor land use, vegetation index, natural disasters, ecological environment, crop yield estimation, urban heat island
Ground observation system	USA	WorldView-2	Blue 0.45–0.51 μm Green 0.51–0.58 μm Red 0.63–0.69 μm Near infrared 0.77–0.895 μm Coastal zone 0.40–0.45 μm Yellow 0.585–0.625 μm Red edge 0.705–0.745 μm Near infrared 0.860–1.04 μm (2 band)	Coastal spectrum supports plant identification and analysis, as well as coastal sounding research based on aquatic plant chlorophyll and water seepage rate. Due to the influence of atmospheric scattering, atmospheric correction technology is required; The yellow band assists in correcting the chromaticity to conform to visual habits; Red edge spectrum can assist in analyzing plant growth and indicates plant health information; The two NIR spectra, which are not affected by the atmosphere, were used for plant analysis and biological quantitative study

(continued)

Table 9.3 (continued)

Type	Belonging	Satellite	Load	Features
		WorldView-3	Track 617 km, revisit period is 1 day Panchromatic 0.31 m Multispectrum 1.24 m Shortwave infrared 3.7 m Panchromatic 0.45–0.80 μm Coastal blue 0.40–0.45 μm Blue 0.45–0.51 μm Green 0.51–0.58 μm Yellow 0.585–0.625 μm Red 0.63–0.69 μm Red edge 0.705–0.745 μm Near infrared 1 0.77–0.895 μm Near infrared 2 0.86–1.04 μm Shortwave1 1.195–1.225 μm Shortwave2 1.55–1.590 μm Hortwave3 1.64–1.68 μm Shortwave4 1.71–1.75 μm Shortwave5 2.145– 2.185 μm Shortwave6 2.185–2.225 μm Shortwave7 2.235–2.285 μm Shortwave8 2.285–2.365 μm	The near infrared spectrum can detect land, coast, cloud boundaries and temperature abrupt changes; Inherited the detection features of WorldView-2; Coastal blue is more sensitive to water depth than blue light, which can improve the measurement of port water depth; The shortwave infrared spectrum extends the spectrum coverage and increases the number of spectrum bands to facilitate material identification; Shortwave infrared is benefit to imaging through haze, smoke, aerosols and other air particles
		IKONOS-2	Panchromatic (1 m, 0.45–0.90 μm) Multispectral (4 m, 4 bands)	High spatial resolution for 1:10,000–1:50,000 survey and monitoring

(continued)

Table 9.3 (continued)

Type	Belonging	Satellite	Load	Features
		QuikBird	Panchromatic (0.61 m, 0.45–0.90 μm) Multispectral (4 m, 4 bands)	High spatial resolution images are used to distinguish crop species, monitor agricultural irrigation conditions and monitor post-flood disasters, which is benefit to the development of refined agriculture
Earth observing system	CNES French space research center CNES	SPOT-1,2	2 visible light remote sensors HRV (panchromatic10 m; multispectrum 20 m; width 117 km)	Stereo imaging is possible, a single satellite can be revisited in 1–3 days
		SPOT-4	HRV adds a shortwave infrared band of 1.58–1.75 μm and changes the original 0.61–0.68 μm red light to 0.49–0.73 μm, panchromatic 10 m, 20 m multispectrum; Wide field of view vegetation detector VGT 1.15 km, width 2250 km, 5 bands Blue 0.43–0.47 μm Green 0.5–0.59 μm Red 0.61–0.68 μm Near infrared 0.79–0.89 μm Shortwave infrared 1.58–1.75 μm	Shortwave infrared has high atmospheric transmittance, more definition topography and high contrast. Soil and plant moisture can be analyzed from the image gray level, which is easy to interpret the soil types and monitor the growth of vegetation and crops; VGT is used for global and regional continuous monitoring of natural vegetation and crops, which is meaningful for large-scale environmental changes, meteorology, ocean and other applied research
		SPOT-5	2pcs high-resolution geometric imaging equipment HRG 1pcs high-resolution stereo camera HRS 1pcs VGT	Real-time stereo imaging, subpixel processing to improve resolution, up to 2.5 m
		SPOT-6	Panchromatic 0.455–0.745 μm, 1.5 m Multiband 6 m, 4 bands	Using Reference3D, automatic orthophotographs with a positioning accuracy of 10 m (CE90)

(continued)

Table 9.3 (continued)

Type	Belonging	Satellite	Load	Features
Earth observation satellite	Japan	ALOS	Stereo mapping camera PRISM 2.5 m, 0.52–0.77 μm; Visible and near infrared radiometer AVNIR-2 10 m, width 70 km, 4 bands L-band synthetic aperture radar PALSAR 3 modes: high resolution, scanning, polarization	PRISM three-dimensional surveying and mapping; AVNIR-2 is used for land and coastal area observations to provide land coverage and usage classification maps for environmental monitoring; PALSAR obtains wider coverage than ordinary SAR; Purpose: surveying and mapping, regional environmental observation, disaster monitoring, resource survey, etc.
Earth observation satellite	Canadian space agency and MDA	Radarsat-2	SAR, C-band, full polarization, 11 beam patterns, coverage width 50–500 km, resolution 3–100 m, regression period 24 days	Disaster prevention: flood, geological disaster and oil spill monitoring; Agriculture: crop classification, crop growth monitoring and yield rate estimation; Forestry: forest classification, forestry resource assessment and monitoring; Hydrology: soil moisture monitoring, marshland identification; Ocean: sea ice-type identification, iceberg monitoring; Geology: Lithological structure
Indian resources satellite	Indian space agency	IRS-1A,B	Line scan image scanner LISS	High resolution, panchromatic 5.8 m, suitable for fusion with TM multispectral data for 1:25,000 to 1:50,000 scale resource and environmental surveys
		IRS-1C,D	LISS Wide field of view scanner WiFs Panchromatic camera PAN, 5.8 m, coverage 70 km, 0.5–0.75 μm, stereo imaging	

Table 9.4 Special application table of optical spectrum

Application	Observation elements	GSD (m)	Spectral band (μm)	Observation elements	GSD (m)	Spectral band (μm)
Disaster mitigation and prevention	Vegetation coverage	30–100	0.6–1.3	Flood impact range and intensity	10–50	0.4–1.3
	Land usage	30–100	0.4–1.3	Scope and intensity of fire impact	30–50	0.65 and 0.86 3–5, 8–14
	Soil moisture	30–100	0.4–1.3 3–5 8–12	Scope and intensity of drought impacts	50–100	0.72–1.30 1.30–3.00 10.780–1.280
	Surface temperature	30–100	3–5 8–12	Scope and intensity of effects of pests and diseases	100	0.4–2.5 8–14 At least 50 bands
	Submerged range	100–200	0.4–1.3	Scope and intensity of earthquake	10–30	0.4–1.3
	Snow covered	50–100	0.72–1.30 1.30–3.00 10.780–12. 80	Scope and intensity of snow disaster	100	0.4–1.3 3–15
	Fire area	100–500	0.65 and 0.86 3–5, 8–14			
Land resources	CH_4	50–100	Visible light, near infrared, shortwave infrared, thermal infrared	Fracture, alteration zone	50	Visible light, near infrared
	Oil film	100	Visible light, near infrared, shortwave infrared, thermal infrared, microwave	Land type change	20–50	Visible light, near infrared
	Temperature	50	Thermal infrared	Mountain change	20–50	Visible light, near infrared, microwave

(continued)

Table 9.4 (continued)

Application	Observation elements	GSD (m)	Spectral band (μm)	Observation elements	GSD (m)	Spectral band (μm)
	Vegetation	50	Visible light, near infrared	Soil vegetation changes	50	Visible light, near infrared
Atmospheric environment monitoring	Remote sensing monitoring of particulate matter	30–100	0.45–0.52 0.53–0.60 0.63–0.69 2.10–2.15 0.89–0.92 0.93–0.94 0.91–0.96	Remote sensing monitoring of straw burning	10–100	0.45–0.52 0.53–0.60 0.63–0.69 0.76–0.90 2.10–2.15 3.50–3.90 10.78–11.28 11.77–12.27
	Ozone	100	9.580–9.880	Remote sensing monitoring of sand and dust storms	30–100	0.45–0.52 0.53–0.60 0.63–0.69 0.76–0.90 3.50–3.90 10.5–12.0
	Fog pollution monitoring	100	0.45–0.52 0.53–0.60 0.63–0.69 0.89–0.92 0.93–0.94 0.91–0.96 1.55–1.75 2.10–2.15			

(continued)

Table 9.4 (continued)

Application	Observation elements	GSD (m)	Spectral band (μm)	Observation elements	GSD (m)	Spectral band (μm)
Water environment monitoring	Transparency	30	0.480–0.500	Water temperature	30	10.78–11.28 11.770–12.270
	Chlorophyll a	30	0.555–0.575 0.640–0.660 0.670–0.690	Red tide (water bloom)	10	0.459–0.479 0.545–0.565 0.620–0.670 0.841–0.876
	Petro	10–30	Mainly microwave, ultraviolet, visible, infrared,	Oil spill	20–30	microwave
	Suspended sediment	20	0.670–0.690 0.690–0.710 0.720–0.740			
Ecological environment monitoring	Vegetation index (NDVI)	10–30	Visible light, near infrared	Desertification intensity	10–30	Visible light, near infrared
	Leaf area index (LAI)	10–30	Visible light, near infrared	Landscape type	10–30	Visible light, near infrared
	Evapotranspiration (ET)	10–30	Visible light, near infrared, thermal infrared	Surface exposure rate	10–30	Visible light, near infrared
	Net primary productivity (NPP)	10–30	Visible light, near infrared, thermal infrared	Land degradation index	10–30	Visible light, near infrared
	Surface albedo (Albedo)	10–30	Visible light, near infrared, thermal infrared	Degree of salinization	10–30	Visible light, near infrared
	Ecological classification	10–30	Visible light, near infrared	Spatial distribution of solid waste	10–30	Visible light, near infrared
	Land surface temperature (LST)	60–1000	Thermal infrared(10–14)	Urban corridor distribution	10–30	Visible light, near infrared

(continued)

Table 9.4 (continued)

Application	Observation elements	GSD (m)	Spectral band (μm)	Observation elements	GSD (m)	Spectral band (μm)
	Vegetation cover status and changes	10–30	Visible light, near infrared	Urban heat island	60–300	Thermal infrared(10–14)
	Biomass	10–30	Visible light, near infrared	Eco-environmental quality status of nature reserves	10–30	Visible light, near infrared
	Bioabundance index	10–30	Visible light, near infrared	Eco-environmental quality status of important ecological function areas	10–30	Visible light, near infrared
	Landscape pattern index	10–30	Visible light, near infrared	Carbon flux	10–30	Visible light, near infrared, thermal infrared
	Valley density distribution	10–30	Visible light, near infrared	Soil moisture	30	Visible light, thermal infrared Microwave

Table 9.5 Transmittance of optical materials with a thickness of 10 mm [2]

wavelength λ/nm	Spectral transmittance $\tau_{i\lambda}$															
	QK1	QK3	Kl	K2	K3	K5	K6	K7	K8	K9	K10	K11	BaK1	BaK2	BaK3	BaK4
700	0. 997	0.996	0. 997	0. 996	0. 995	0. 996	0. 993	0. 993	0.996	0. 999	0. 996	0. 996	0. 997	0. 998	0. 995	0. 990
650	0. 997	0. 993	0. 995	0. 994	0. 995	0. 996	0.992	0. 994	0. 997	0. 998	0.996	0. 996	0. 997	0. 999	0. 995	0. 988
600	0. 998	0.993	0. 995	0. 994	0. 998	0. 996	0. 991	0. 994	0. 998	0. 998	0. 996	0. 996	0. 998	0. 999	0. 995	0. 992
550	0. 996	0. 992	0. 993	0. 996	0. 998	0. 994	0. 990	0. 994	0. 999	0.998	0. 995	0. 996	0. 998	0. 999	0. 997	0.993
500	0. 994	0.992	0. 994	0.993	0. 997	0.992	0. 992	0. 992	0. 999	0. 998	0. 994	0. 995	0. 996	0. 997	0.992	0. 988
480	0. 992	0.992	0. 994	0. 992	0. 995	0. 992	0. 991	0. 992	0. 999	0. 998	0. 994	0. 994	0. 996	0. 996	0. 993	0. 986
460	0. 989	0. 992	0. 993	0. 991	0. 992	0. 991	0. 990	0. 992	0. 999	0. 996	0. 992	0. 992	0. 994	0. 995	0. 991	0. 984
440	0. 984	0. 990	0. 992	0. 990	0. 991	0. 986	0. 985	0. 991	0. 996	0. 996	0. 991	0. 989	0. 992	0. 990	0.991	0. 980
420	0. 983	0. 990	0. 992	0. 988	0. 989	0. 985	0. 985	0. 991	0.994	0. 996	0. 990	0. 989	0.991	0. 990	0. 990	0.978
400	0. 988	0. 990	0. 992	0. 994	0. 993	0. 981	0. 985	0.990	0.996	0. 994	0. 967	0. 988	0. 989	0. 988	0. 990	0. 973
390	0. 984	0. 990	0. 986	0. 987	0.991	0. 978	0. 983	0. 985	0. 994	0. 992	0. 965	0. 985	0. 985	0.978	0. 975	0. 955
380	0. 964	0. 968	0. 974	0. 956	0. 973	0. 963	0. 969	0. 971	0. 988	0. 988	0. 951	0. 972	0.972	0. 958	0. 950	0. 923
370	0. 960	0. 970	0. 975	0. 971	0.973	0. 955	0. 968	0. 965	0.982	0. 985	0. 905	0. 972	0.970	0. 960	0. 938	0.910
360	0.952	0. 958	0.971	0. 964	0. 966	0. 941	0. 962	0.951	0. 982	0. 980	0.812	0. 964	0. 955	0.935	0. 885	0. 835
350	0.922	0. 935	0. 955	0. 925	0. 938	0. 906	0. 842	0.918	0. 970	0.969	0. 630	0. 935	0. 922	0. 885	0. 800	0. 722
340	0. 852	0.892	0. 922	0. 859	0. 899	0. 832	0.896	0. 852	0. 920	0. 942	0. 360	0. 882	0. 856	0.778	0. 662	0. 522
330	0. 740	0. 883	0. 855	0. 742	0.810	0. 680	0. 800	0.731	0. 810	0. 890	0. 110	0. 791	0. 745	0. 600	0. 452	0. 300
320	0. 532	0.713	0.672	0. 551	0. 650	0.470	0. 610	0. 502	0.570	0.772	0. 009	0. 590	0.552	0. 342	0. 235	0. 100
310	0. 252	0. 500	0. 355	0. 262	0. 400	0. 300	0. 340	0. 240	0. 275	0. 590	0. 004	0.310	0. 302	0.112	0. 062	0.015
300 290 280	0. 052	0. 220 0. 050	0. 060 0. 005	0. 04	0.150 0. 02	0. 120 0. 02	0. 085 0. 01	0. 028 0. 005 0. 005	0. 040 0. 002	0. 300 0. 080 0. 005	0. 004	0. 070 0. 003	0. 085 0. 009	0.012	0. 006	

For example, when pursuing a short revisit period with minimum operating costs, the main consideration is to use a single satellite to meet the demand. If the form of constellations is considered, it will inevitably lead to increased costs. Therefore, it is necessary to comprehensively select the regression coefficient $Q = I \pm C/N$, the camera field of view 2ω, the attitude adjustment ability of the platform around the rolling axis x_b and the stability of the entire 3-axis attitude, and the reasonable selection of the regression coefficient and the camera field of view 2ω is especially important, and it determines the ability of the satellite to revisit the target in a short time and the realization of seamless global coverage (except the poles).

Taking the $Q = 14 \pm 3/7$ orbits which has been discussed as an example, they can also be more favorable to achieve the global 3-day revisit, which can be seen from the figure below, both regression coefficients can be in 1d, 2d, 3d orbits by swinging a single track (except for individual tracks, see the track arrangement diagram in the regression track section for details). Track coverage width γ, and the target at any location in the world that you want to observe can be observed within 3 days.

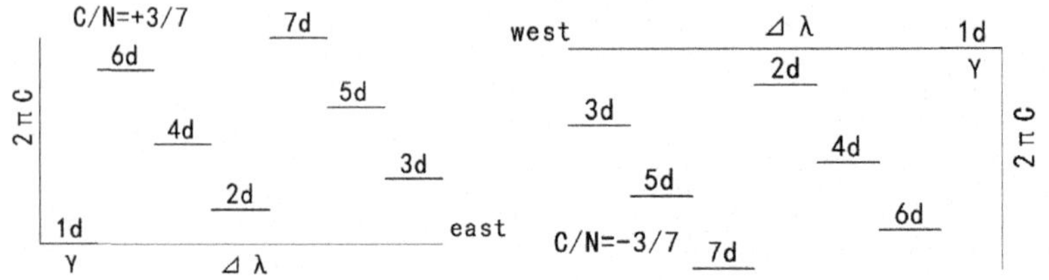

However, the requirements for the optical camera are quite different. When $Q = 14 + 3/7$, the orbit height is 743.4 km and the coverage width is 396.78 km. The required camera field of view is

$$\begin{aligned} 2\omega &= 2 \arctan \frac{396.78}{2 \times 743.4} \\ &= 29.88^{\circ} \end{aligned}$$

When $Q = 14\text{–}3/7$, the orbit height is 1040.8 km, the coverage width is 421.84 km, and the required camera field of view is

$$\begin{aligned} 2\omega &= 2 \arctan \frac{421.84}{2 \times 1040.8} \\ &= 22.91^{\circ} \end{aligned}$$

It can be seen that the requirements for the camera field of view of the orbit when $Q = 14 + 3/7$ are much greater than that of the orbit when $Q = 14\text{–}3/7$, which is nearly 1/3 larger.

However, from the focal length of the camera, when achieving the same spatial resolution on the ground, the focal length of the camera in the orbit when $Q = 14-3/7$ is 1.4 times that when $Q = 14 + 3/7$.

And the focal plane scale of the camera in the orbit when $Q = 14-3/7$ is also slightly larger than the focal plane scale of the camera when $Q = 14 + 3/7$.

In the operation of the satellite, the roll angle required when $Q = 14 + 3/7$ is nearly 1/3 larger than that when $Q = 14-3/7$, which will reduce the imaging resolution of the camera by about 13%.

It can be seen that the choice of the design target has a great influence on the parameter design of the whole system and also has a great influence on the effect of practical application.

Take the MODIS optical remote sensor of the US earth observation satellite as an example, and let us take a look at the selection of optical remote sensor types. MODIS has 36 spectral bands, as shown in Table 9.6. There are three types of spatial resolution, 250, 500 and 1000 m depending on the spectral band. The scanning length perpendicular to the orbital directions is 2330 km, and the scanning width along the orbital direction is 10 km.

Judging from the number of 36 spectral bands, it has the characteristics of a multispectral camera, and from the perspective of the specific spectral distribution, except for the serial numbers 5, 6 and 7 in the spectral bands No. 1 to 19, the spectral range is 0.405–0.965 μm, and all visible light detectors can be used to receive a total of 16 spectral bands.

The spectral bands 20–26, plus the band 5, 6 and 7, that is 10 spectral bands the spectral range of 1.23–4.549 μm, which can be received by mid-infrared detectors such as InSb detectors.

The spectral bands 27–36 have a total of 10 spectral bands, with a spectral range of 6.535–14.385 μm, which can be received by thermal infrared detectors such as HgCdTe detectors.

MODIS satellite orbit height $H = 705$ km can be obtained

$$\text{Orbital period, } T_n = 98.8\,\text{min};$$
$$\text{Ground speed, } V = 6.76\,\text{km/s}.$$

The along-track scanning coverage width of MODIS is 10 km, the scanning efficiency is calculated as 40%, and the scanning length is 2330 km.

Each scan time, $T_d = 1.48$ s;

Scanning speed, $V_s = 3937.6\,\text{km/s}$.

The spatial resolution of MODIS is 250, 500 and 1000 m, and the corresponding exposure times are 0.063, 0.127 and 0.254 ms, respectively.

MODIS has a total of 36 spectral bands, including 16 spectral bands for visible light channels. In terms of the number of channels, there are two types of optical remote sensors to choose, namely multispectral bands imaging cameras and imaging spectrometers. If you choose the imaging spectrometer class, take spectral numbers 2 and 16 as examples, and the spectral bands are 0.841–0.876 μm and 0.862–0.877 μm,

Table 9.6 Performance parameters of MODIS

Basic use	No.	Band[a]	Spectral radiance[b]	SNR requirement[c]	Resolution
Land/cloud/ aerosol boundary	1	620–670	21.8	128	250 m
	2	841–876	24.7	201	
Land/cloud/ aerosol properties	3	459–479	35.3	243	500 m
	4	545–565	29.0	228	
	5	1230–1250	5.4	74	
	6	1628–1652	7.3	275	
	7	2105–2155	1.0	110	
Ocean color/ phytoplankton/ biogeochemistry	8	405–420	44.9	880	1000 m
	9	438–448	41.9	838	
	10	483–493	32.1	802	
	11	526–536	27.9	754	
	12	546–556	21.0	750	
	13	662–672	9.5	910	
	14	673–683	8.7	1087	
	15	743–753	10.2	586	
	16	862–877	6.2	516	
Atmospheric water vapor	17	890–920	10.0	167	
	18	931–941	3.6	57	
	19	915–965	15.0	250	
Surface/cloud temperature	20	3.660–3.840	0.45 (300 K)	0.05	1000 m
	21	3.929–3.989	2.38 (335 K)	2.00	
	22	3.929–3.989	0.67 (300 K)	0.07	
	23	4.020–4.080	0.79 (300 K)	0.07	
Atmospheric temperature	24	4.433–4.498	0.17 (250 K)	0.25	
	25	4.482–4.549	0.59 (275 K)	0.25	
Cirrus cloud water vapor	26	1.360–1.390	6.00	150 (SNR)	
	27	6.535–6.895	1.16 (240 K)	0.25	
	28	7.175–7.475	2.18 (250 K)	0.25	
Cloud features	29	8.400–8.700	9.58 (300 K)	0.05	
Ozone	30	9.580–9.880	3.69 (250 K)	0.25	
Surface/cloud temperature	31	10.780–11.280	9.55 (300 K)	0.05	
	32	11.770–12.270	8.94 (300 K)	0.05	
Cloud top height	33	13.185–13.485	4.52 (260 K)	0.25	
	34	13.485–13.785	3.76 (250 K)	0.25	
	35	13.785–14.085	3.11 (240 K)	0.25	
	36	14.085–14.385	2.08 (220 K)	0.35	

[a]No. 1–19 units nm; No. 20–36 units μm
[b]Spectral radiance units (W/m^2–μm-Sr)
[c]SNR signal-to-noise ratio
[d]NETD noise equivalent temperature difference
Note The achieved performance is 30–40% better than the required performance

respectively. If the existing spectral requirements are still to be well achieved, and the spectral resolution in the 0.862–0.877 μm spectral range should be achieved 0.015 μm when the imaging spectrometer solution is used:

$$0.877\,\mu\text{m} - 0.862\,\mu\text{m} \approx 0.015\,\mu\text{m}$$

If the transverse shearing interferometric imaging spectrometer is specifically adopted, the spectral resolution of spectral bands 1, 2, 3, 4, etc., will be greatly improved, which will lead to a great reduction of exposure and a reduction of signal-to-noise ratio. Especially for spectral bands 1 and 2, the spatial resolution is high, the exposure time is only 0.063 ms, and the problem of insufficient exposure will be more prominent.

From the spectral settings and resolution differences in Table 9.6, it can be inferred that MODIS adopts the multispectral bands imaging method of separate channel settings.

It can be seen from the selection analysis of the type of optical remote sensor that takes MODIS as an example; although MODIS has 36 spectral bands, there are already many spectral bands, but from the comprehensive analysis results of various factors, separate channel settings imaging is still used. The method is better than the imaging spectrometer method. And engineering development should also consider factors such as inheritance.

From the selection of system design goals in the above two examples, it can be seen that the selection of goals needs to consider multiple factors. After the system design goals are reasonably selected, it is a more detailed system parameter optimization problem.

9.5 Main Parameters Comprehensive Analysis and Performance Evaluation of the System

Since space systems are usually non-maintainable, high reliability must be guaranteed, and redundant design is also a common method. Larger-scale satellite design in engineering usually relies on inheritance and adopts the methods of experience, analogy and analysis. The design focuses on how to ensure the progress and realize the normal operation on orbit and the design life. For this reason, the contractor who undertakes the task of subsystem or component development will spare no expense to ensure the reliability of the product and the planned node. The coordination of the satellite system, the comprehensive balance and application of various designable variables in the process of realizing the design goal are insufficient.

1. The Purpose of DSP Early Warning Satellite

DSP early warning satellites are reconnaissance satellites launched to monitor and detect enemy ballistic missiles. They usually run in geostationary orbits or large elliptical orbits with a period of 12 h. Generally, several satellites form an early warning network with a large coverage area. The early warning satellite is equipped with infrared detectors and TV cameras. When encountering ground or underwater ballistic missiles, it can detect the infrared radiation of the engine tail flame of the active section of the missile and issue an alarm. Both Gulf Wars occurred at night, which shows the importance of infrared early warning.

2. Selection of System Design Goals

This is an optical detection system with clear requirements. The design goal is to detect high-temperature infrared radiation targets. There are two requirements for early warning problems: false alarm rate and missed alarm rate. False alarm rate is the probability of alarm without target. The missed alarm rate is the probability that there is a target but no alarm. Both of which are directly related to the signal-to-noise ratio (SNR) in weak signal detection, so the signal-to-noise ratio is selected as the target of system design. The geostationary orbit and the apogee of the 12-h large elliptical orbit are relatively far away from the detection target, and the signal is very weak. If the SNR is too large, it is difficult to achieve technically. If the SNR is too small, it is prone to miss alarms and false alarms. Recommend SNR = 5–8.

3. Selection of Early Warning Satellite Orbits

Following the example of DSP satellites adopting the geostationary orbit as the orbit of the early warning satellite, only three satellites can achieve the coverage of most of the earth's surface except for the poles, as shown in Fig. 9.11.

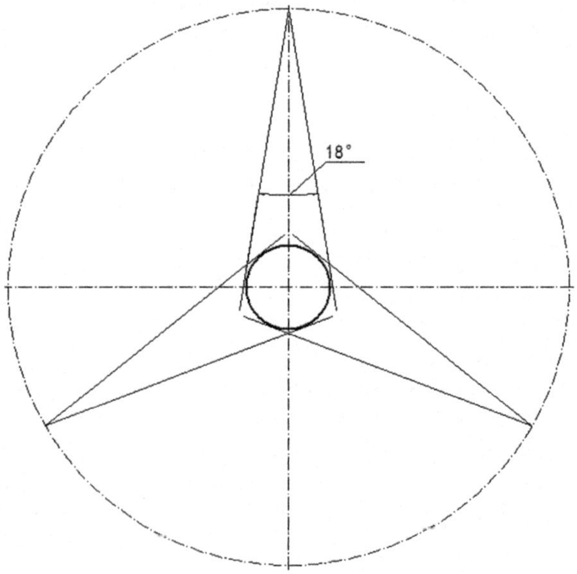

Fig. 9.11 Geostationary orbit three-satellite network

Compared with the four-star network, this three-star network has the problem of reducing the coverage latitude at the intersection of the satellite observation area, but it is more economical.

4. Target Characteristics of Early Warning Satellite Observation Objects

Strategic missiles are missiles used to strike strategic targets. Offensive strategic missiles have a range of more than 8000 km and carry nuclear warheads. They mainly target important strategic targets such as political and economic centers, military and industrial bases, nuclear weapons arsenals and transportation hubs. According to the missile flight conditions and the possibility of finding, tracking and destroying it, the trajectory can be divided into four main stages: the active stage, the warhead separation stage, the orbital stage and the final stage. Destroying the missile in the active segment is most effective because it disables all warheads and false targets in the ICBM's nose compartment. Otherwise, during the missile re-entry into the dense atmosphere, a single target can be decomposed into complex ballistic targets composed of hundreds of objects, causing great difficulties for selecting and destroying them. The last stage is used to make the warhead finally aim at the predetermined target, and the height is from the missile re-entry into the dense atmosphere until it reaches the target.

Tactical missile

Missiles are used to directly support the battlefield and strike targets within the tactical depth of the campaign. The range is less than 1000 km. These include ground-to-ground, ground-to-air, ship-to-ground, shore-to-ship, ship-to-ship, air-to-ship, ground-to-air, air-to-air, anti-radar missiles, anti-tank missiles, etc. However, only certain these tactical missiles could be detected effective by early infrared warning satellites in geostationary orbits.

The spectral characteristics of the missile plume

Generally, the strong infrared radiation region of the gas plume is located in the shortwave infrared region, and its precise wavelength depends on the actual composition of the gas. Typical peak wavelengths are at 2.7 μm (SWIR) and 4.3 μm (MWIR), with radiant energy up to megawatts/steradian. Although there is a thrust system after the boost stage, the radiation is only a few hundred watts/steradian. From the middle section to the re-entry section is near room temperature (20 °C or 293 K) in the surface of the missile. The peaking at 10 μm (LWIR), only on the order of a few watts/steradian, the contrast to the earth background is extremely poor, and these targets are only detectable against the cold space background (4 K).

The radiation of missiles at low altitude is mainly from thermal radiation sources such as the nozzle of the engine, the jet flow (plume) and the aerodynamic heating of the aircraft surface. Although the temperature of the missile plume is very high (thousands of degrees), it is relatively short in the dense atmosphere at low altitudes, and its high-temperature region is within 100–200 m. At this time, it is indeed too far away for the infrared camera in geostationary orbit (35,786 km), and its image in the camera can only be a very small point. At low altitudes, the temperature of the rocket engine drops rapidly due to the continuous mixing of the surrounding atmosphere with the jet plume of the rocket engine. At the altitude of 30–40 km, due to the thin atmosphere, the mixing effect is weakened, and the length of the tail flame will be greatly increased.

Strategic aircraft.

There are four main sources of infrared radiation for strategic aircraft, namely the hot metal cavity of the engine combustion chamber, the hot exhaust gas, the radiation from the surface of the casing and the reflection to the environment. The cruising altitude is about 10 km, and the speed is 800–900 km per hour. The engine nozzle temperature is generally 850–900 K in the non-afterburning state, and 1800–2000 K in the afterburning state. The temperature in the afterburning state is close to the nozzle temperature of the missile engine. Since the atmospheric background radiation has a strong absorption band of water vapor near 2.7 μm, the background spectral radiation formed by the ground background radiation and the sky scattered sunlight radiation near 3–4 μm is also near a minimum value, while the aircraft jet engine produces similar to missile engines, and the radiation spectrum also has peaks around 2.7 and 4.3 μm wavelengths.

Through the above analysis, the following conclusions can be drawn. The use of infrared cameras for missile and strategic aircraft early warning should select the detection spectrum at 2.7 and 4.3 μm.

If the thermal radiation of the aircraft skin and the identification of the true and false targets of the missile are considered, the wavelength range of 8–24 μm is also an important spectral region, but the radiance of the background light of the geo-atmospheric system in this spectral region is close to the normal temperature black body, so it is very difficult to detect and identify targets in this spectrum by using a geostationary orbit infrared camera due to the poor contrast.

The radiation power of ballistic missiles is 4000–5000 kW, and the temperature is 2000–3000 °C. According to the spectroscopic measurement of the rocket flow of Rosenburg burning kerosene liquid oxygen cited by Little Hudson, the radiation spectrum of the undisturbed cone in the rocket jet is continuous, and it is almost consistent with the black body radiation of 2000 K, temperature $T = 2000$ K, and take full radiation power as $P = 4500$ kw. Also set the radiation source area S, then the radiation emittance of the whole spectrum is $M = P/S$, and the emittance between the wavelengths of 4.3–4.4 μm is:

$$M_{4.3-4.4} = \frac{P}{S}[F(\lambda_2 T) - F(\lambda_1 T)]$$

As the point target, the radiated power between 4.3 and 4.4 μm is:

$$\begin{aligned} P_{4.3-4.4} &= M_{4.3-4.4} \cdot S \\ &= P[F(\lambda_2 T) - F(\lambda_1 T)] \end{aligned}$$

The engine tail flame can only radiate to the 2π space in the injection direction. Considering that there is a certain amount of backscattering, the radiation range will be greater than 2π, but less than 4π, and the average is 3π.

The radiation intensity between 4.3 and 4.4 μm when the point target is:

$$I_{4.3-4.4} = \frac{P_{4.3-4.4}}{3\pi}$$
$$= \frac{P}{3\pi}[F(\lambda_2 T) - F(\lambda_1 T)]$$
$$F(\lambda_2 T) - F(\lambda_1 T) = F(4.4 \times 2000) - F(4.3 \times 2000)$$
$$= F(8800) - F(8600)$$
$$= 0.88413 - 0.87786$$
$$= 0.00627$$
$$I_{4.3-4.4} = \frac{4500000 \times 0.00627}{3 \times \pi}$$
$$= 2993.7\,\mathrm{W/Sr}$$

When used as a point target, the radiation intensity between 2.7 and 2.9 μm is:

$$I_{2.7-2.9} = \frac{P_{2.7-2.9}}{3 \times \pi}$$
$$= \frac{P}{3\pi}[F(\lambda_2 T) - F(\lambda_1 T)]$$
$$F(\lambda_2 T) - F(\lambda_1 T) = F(2.9 \times 2000) - F(2.7 \times 2000)$$
$$= F(5800) - F(5400)$$
$$= 0.72013 - 0.68033$$
$$= 0.0398$$
$$I_{2.7-2.9} = \frac{4500000 \times 0.0398}{3 \times \pi}$$
$$= 19003.1\, W/Sr$$

It can be seen that the infrared radiation intensity at 2.7 μm is much greater than the radiation intensity at 4.3 μm, so when the 2.7 μm and 4.3 μm dual spectrum bands are used, the camera aperture will depend on the calculated value at 4.3 μm spectrum band.

5. Determination of Background Radiance

The selection of the infrared camera's working spectrum must consider both the spectral characteristics of the target and the spectral characteristics of the background. For infrared cameras operating in geostationary orbits, the background is the infrared radiation of the earth and its atmospheric system.

After calculation by atmospheric software, the radiance of the earth–atmosphere system in the wavelength range $\lambda = 2.7$–2.9 μm is 1.873×10^{-7} W cm^{-2} Sr^{-1}. While the radiation spectrum of solid-state rocket engines and the radiation spectrum of aircraft jet engines has their peaks near 2.9 μm. Therefore, 2.7–2.9 μm is selected as the working spectrum of the infrared camera.

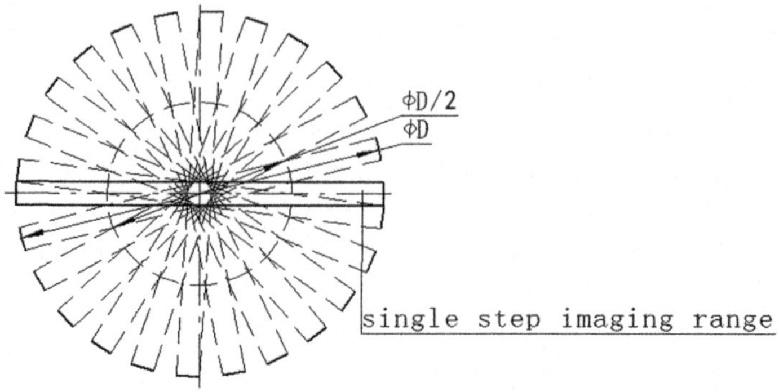

Fig. 9.12 Step staring imaging

After calculation by atmospheric software, the radiance of the earth–atmosphere system within the wavelength ranges $\lambda = 4.3–4.4\ \mu m$ is 4.495×10^{-7} W cm^{-2} Sr^{-1}. The working spectrum of the infrared camera is determined to be 4.3–4.4 μm.

6. Infrared Early Warning Camera Solution

The DSP adopts the scanning imaging scheme, and the SBIRS system in the United States adopts a combination of linear scanning and area staring, while the AIRSS system after SBIRS adopts a complete staring scheme. The complete staring scheme is nearly ideal, but the scale of the detector is too large to be solved, so a compromise can be made to the step-gaze scheme, as shown in Fig. 9.12.

When a network of three satellites is used, the maximum field of view of the area observed by a single satellite is

Geostationary orbit radius 42,164.17 km;
The earth's equatorial radius is 6378.145 km;
Maximum field of view $2\alpha_m = 17.40°$.

From this point of view, in order to achieve complete coverage of the entire observation area of the earth in the geostationary orbit, the design of the wide coverage camera optical system and the choice of detectors are the two main issues. The off-axis three-mirror optical system can achieve the field of view 18°, and the detector needs to be spliced.

The advantage of this scheme is that it can give full play to the advantages of the rapid development of the area array CCD and the strong detection ability of staring imaging, and staring imaging conforms to the development direction of infrared early warning CCD cameras.

7. Detector Selection

It can be seen from the introduction of photodetectors that PbS and HgCdTe detectors in infrared detectors should be the first choice detectors. The detection rate of PbS is higher than that of HgCdTe detectors near the wavelength of 2.7 μm, and the working temperature of PbS detectors is high, but due to PbS has coarse grains, and its detector pixel size is larger than that of HgCdTe detectors, and its number of pixels is less than that of HgCdTe detectors. The development of HgCdTe detectors is rapid, and it is moving toward longwave and bispectral large area arrays and high operating temperature. Taking the French area array CCD as an example, the pixel size $L_{\text{row}} = 30\ \mu\text{m}\ L_{\text{col}} = 30\ \mu\text{m}$, the number of pixel rows $N_{\text{row}} = 256$, the number of columns $N_{\text{col}} = 320$, the detectivity of the HgCdTe detector is 4.3–4.4 μm and has reached (3.5–10.2) × 10^{11} cm $\text{Hz}^{1/2}\ \text{W}^{-1}$, the operating temperature can reach 130 K, and the frame rate is 400 frames per second. And the US satellite infrared early warning camera is also developing in this direction. Due to the rapid development of HgCdTe large area arrays, if the infrared wide-field camera can realize stepping staring imaging, then the use of large area array HgCdTe detectors will be beneficial to improve the detection frequency and spatial resolution of satellites. When the focal plane of the camera adopts a combination of detectors with two wavelengths of 2.7 and 4.3 μm, it will also help to identify real and fake targets.

8. Detector Cooling Mode Selection

The HgCdTe detector can work at 130 K and adopt secondary radiation cooling, up to 77 K, and thermoelectric cooling can reach a maximum temperature difference of 130–140 K. When the HgCdTe detector uses a three- or four-stage thermoelectric cooling component, it can be used for thermal imaging and radiometers. When using radiant cooling, a single stage can achieve a cooling temperature of 140 K. However, the radiant cooling cooler has a larger diameter. Therefore, a combination of radiation cooling and electric cooling can be considered. Stirling cycle refrigeration can also be used, but with longevity, reliability and vibration issues.

9. Comprehensive Analysis of the Main Parameters of the Infrared System

Referring to the angular velocity of the US DSP spinning satellite of 5.7 rpm, the satellite spin period of the stepping staring infrared detection can be taken as 20 s.

Take the sub-satellite point ground resolution $r_g = 0.8$ km, focal length F, focal plane length L_f, field of view $2\alpha_m = 17.401°$, $H_{\text{geo}} = 35{,}786$ km, then

The focal length

$$\begin{aligned} f &= \frac{L_{\text{row}} \cdot H_{\text{geo}}}{rg} \\ &= \frac{30 \times 35786}{0.8} \\ &= 1.342\ \text{m} \end{aligned}$$

The focal plane length

$$\begin{aligned} Lf &= 2 \cdot \tan \frac{\omega}{2} \cdot f \\ &= 2 \times \tan \frac{17.401}{2} \times 1.342 \\ &= 410.73\ \text{mm} \end{aligned}$$

The number of area array CCD

$$
\begin{aligned}
N_{ccd} &= \frac{L_f}{N_{\text{col}} \cdot L_{\text{row}}} \\
&= \frac{410.73}{320 \times 30} \\
&= 42.78
\end{aligned}
$$

Considering the radian of the earth, in order to achieve a spatial resolution of not less than 0.8 km in a certain coverage area and also for the symmetry of the camera structure.

$$
N_{\text{ccd}} = 44 \,\text{pices.}
$$

The focal plane length

$$
L_f = 44 \times 30 = 422.4 \,\text{mm.}
$$

The focal length

$$
F = 1.36 \,\text{m.}
$$

The field of view

$$
\begin{aligned}
2\alpha &= 17.6544^\circ \\
r_a &= 6371.23 \,\text{km;} \\
r_s &= 42,164.17 \,\text{km;}
\end{aligned}
$$

Since the earth is spherical, the resolution of ground pixels varies from the sub-satellite point to the edge of the field of view. When the focal length is 1.36 m, the range of pixel resolution of 0.8 km is:

The average radius of the earth $r_a = 6371.23$ km;

Geostationary orbit radius $r_s = 42{,}164.17$ km;

The object distance corresponding to 0.8 km resolution is

$$
\begin{aligned}
L_{0.8} &= \frac{0.8\,F}{30 \times 10^{-3}} \\
&= \frac{0.8 \times 1.36 \times 10^3}{30 \times 10^{-3}} \\
&= 36266.67 \,\text{km}
\end{aligned}
$$

The corresponding satellite's opening angle to the earth is as shown in Fig. 9.13

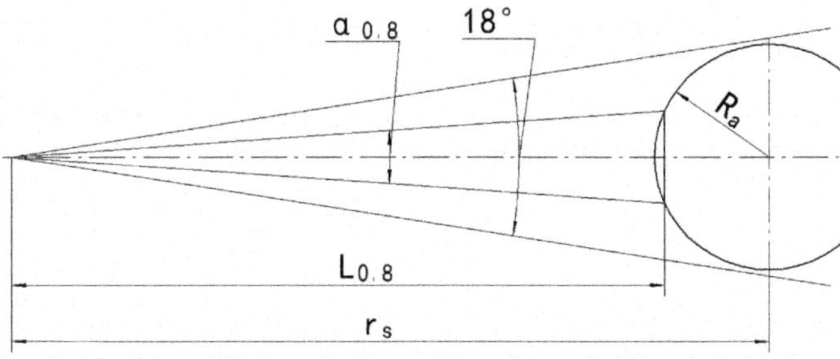

Fig. 9.13 0.8 km spatial resolution coverage

$$\tan\frac{\alpha_{0.8}}{2} = \frac{\sqrt{r_a^2 - (r_s - L_{0.8})^2}}{L_{0.8}} = \frac{\sqrt{6371.23^2 - (42164.17 - 36266.67)^2}}{36266.67}$$

$$\alpha_{0.8} = 7.6063^\circ$$

$$\frac{\alpha_{0.8}}{2\alpha} = \frac{7.6063}{17.401} = 43.7\%$$

The area accounting for about 43.7% of the imaging range can achieve a ground pixel resolution of 0.8 km.

The jet of the rocket engine is 106 km long and 53 km wide at an altitude of 36,280 km. At low altitudes, the temperature of the jet plume of the rocket engine drops rapidly due to the continuous mixing of the surrounding atmosphere with the jet plume of the rocket engine. Considering the state of the plume at high and low altitudes, taking the projection of the plume on the ground as 0.8 × 0.5 km, the area of the background within a pixel range is 0.8 × 0.3 km. Calculated according to such an area, its area is

$$0.8 \times 0.3 = 0.24 \times 10^{10}\,\text{cm}^2$$

The radiation energy of the earth–atmosphere system in the area covered by it in the wavelength range $\lambda = 2.7$–$2.9\ \mu\text{m}$ is

$$1.873 \times 10^{-7} \times 24 \times 10^8 = 499.52\,\text{W/Sr}$$

The radiation energy in the range of wavelength $\lambda = 4.3$–$4.4\ \mu\text{m}$ is

$$4.495 \times 10^{-7} \times 24 \times 10^8 = 1078.8\,\text{W/Sr}$$

In the range of wavelength $\lambda = 2.7$–$2.9\ \mu\text{m}$, the increment to the incident energy of the camera formed by the radiant energy of the target is:

$$19{,}003.1 - 499.52 = 18{,}503.58\,\text{W/Sr}$$

In the range of wavelength $\lambda = 4.3$–$4.4\ \mu\text{m}$, the increment to the incident energy of the camera formed by the radiation energy of the target is:

$$2993.7 - 1078.8 = 1914.9\,\text{W/Sr}$$

Of the 256 lines of pixels in the CCD, 36 lines are used for time delay integration, 5 lines of pixels on each side are used for field overlap, and 210 lines are used for imaging.

The highest imaging frequency of the selected CCD is 400 Hz, and the camera spin period is 20 s/r, then the opening angle of the 36 pixels to the center of the scanning circle is

$$\begin{aligned}\alpha p &= 2 \cdot \text{arctg}\frac{30 \times 36}{1000 \times L_f} \\ &= 0.292989^\circ\end{aligned}$$

Exposure integration time

$$\tau_j = 20 \times 0.292989/360 = 16.27717\,\text{ms}.$$

The central angle corresponding to 210 pixels

$$\begin{aligned}\alpha f &= 2 \cdot \text{arctg}\frac{30 \times 210}{1000 \times L_f} \\ &= 1.708980^\circ\end{aligned}$$

Dwell time corresponding to 210 pixels.

$$t_f = 20 \times 1.708980/360 = 94.9434\,\text{ms}.$$

It can be used to perform image motion compensation setup time.

$$94.9434 - 16.27717 = 78.67\,\text{ms}.$$

Dwell time corresponding to each pixel

$$\begin{aligned}\alpha p &= 2 \cdot \text{arctg}\frac{30}{1000 \times L_f} \\ &= 0.0081386^\circ \\ \text{T}_p &= 20 \times 0.0081386/360 \\ &= 0.4521447\,\text{ms}.\end{aligned}$$

In order to realize the CCD delay integration, the imaging charge of the area array CCD must be moved down one line within the dwell time of one pixel, and the moving time of one line of the selected CCD is:

$$T_C = 1/(400 \times 256) = 9.715 \times 10^{-6} S << T_p, \text{ can be implemented.}$$

The determination of the camera aperture, because the infrared early warning camera works in a geostationary orbit, is 35,786 km away from the ground, the infrared radiation source has become a point radiation source for the camera, and its working method is for the purpose of discovery. Under the premise of the detector noise limit, the working distance of the camera is:

$$S = \sqrt{\frac{\int_{\lambda_1}^{\lambda_2} I_\lambda \tau_a(\lambda) d\lambda \cdot \tau_0 A_0 D^*}{(A_d \Delta f)^{1/2} (U_s/U_n)}}$$

where A_0 is the clear aperture area of the optical system, and if the clear aperture is D, then

$$A_0 = \frac{\pi}{4} D^2$$

Substitute into the above formula to get:

$$D = S \sqrt{\frac{(A_d \Delta f)^{1/2} (U_s/U_n)}{\int_{\lambda_1}^{\lambda_2} I_\lambda \tau_a(\lambda) d\lambda \cdot \tau_0 D^*} \cdot \frac{4}{\pi}}$$

λ_1 initial wavelength;
λ_2 cutoff wavelength;
I_λ spectral radiation intensity of point target;
$\tau_a(\lambda)$ atmospheric transmittance on the propagation distance from the target to the infrared system;
D the clear aperture of the optical system;
Δf equivalent noise bandwidth;
Ad detector pixel area;
U_s/U_n the signal-to-noise ratio of the system;
τ_o transmittance of optical system;
D^* the normalized detection rate of the detector;
where

$$\Delta f = \frac{\pi}{2} \cdot \frac{1}{2\tau_d}$$

τ_d pixel dwell time;
λ_1 4.3 μm;

λ_2 4.4 μm;

$\tau_a(\lambda)$, when $\lambda = 4.3$–4.4 μm, it is the strong absorption band of CO_2, the height of the missile or its plume should be detected above the dense atmosphere, that is, 11 km. The altitude is above, and the infrared radiation of $\lambda = 2.7$ μm is just the absorption band of water, so it can only be detected after the missile has flown out of the cloud layer (about 10 km in height), and the strategic aircraft is generally in the stratosphere. It is active (altitude 20–50 km), so take the initial height of 30 km, and after Lowtran7 calculation, the average atmospheric transmittance in the range of $\lambda = 4.3$–4.4 μm should be $\tau_a(\lambda) = 0.8$.

In the range of 3.7–4.8 μm, the $D^* = 10.2 \times 10^{11}$ cm Hz$^{1/2}$ W^{-1} of the area array HgCdTe detector.

$$U_s/U_n > 5{-}8, \quad \text{take } 8.$$

τ_o, the optical system adopts a three-mirror system, one surface reflection, and the reflection ratio is 0.98. The transmittance of the interference filter and Si transmission window is 0.8, then

$$\tau_0 = 0.98^4 \times 0.8 = 0.73789$$
$$A_d = 0.003^2$$

From the design of the aforementioned scheme, the CCD pixel dwell time $\tau_d =$ 16.27717 ms, then

$$\begin{aligned}\Delta f &= \frac{\pi}{2} \cdot \frac{1}{2 \cdot \tau_d} \\ &= \frac{\pi}{2} \cdot \frac{1}{2 \times 16.27717} \\ &= 48.2515\,\text{Hz}\end{aligned}$$

As mentioned above.

$$I_{4.3\sim4.4} = 1914.9\,\text{W/Sr}$$

Clear aperture of optical system

$$\begin{aligned}D &= S\sqrt{\frac{(A_d\Delta f)^{1/2}(U_s/U_n)}{\int_{\lambda_1}^{\lambda_2} I_\lambda \tau_a(\lambda)d\lambda \cdot \tau_0 D^*} \cdot \frac{4}{\pi}} \\ &= 41664.76\sqrt{\frac{(0.003^2 \times 48.2515)^{1/2} \times 8}{1914.9 \times 0.8 \times 0.73789 \times 10.2 \times 10^{11}} \cdot \frac{4}{\pi}} \\ &= 565.3\end{aligned}$$

Since the background luminous flux will also generate noise; instead of simply deducting the background radiation intensity, the actual aperture is larger than the calculated result.

Take $D = 600$ mm

The radiation intensity at 2.7 μm is much larger than that at 4.3 μm, so the camera aperture at the wavelength of 2.7 μm does not need to be calculated.

10. The Data Rate of the Camera

The infrared detector of the camera takes 94.9 ms to take an infrared image, each slice has 210 rows and 320 columns, and the gray level is 10. There are 44 slices in total, and the data rate is

$$
\begin{aligned}
N_{\text{bit}} &= \frac{210 \times L_{\text{col}} \times 10 \times 44}{t_f} \\
&= \frac{210 \times 320 \times 10 \times 44}{94.9} \\
&= 311.564\,\text{Mbits/s}
\end{aligned}
$$

Such data rate satellite data transmission system can be achieved.

11. Performance Evaluation

In the comprehensive analysis of the system parameters mentioned above, taking strategic missiles, strategic aircraft and some tactical missiles as early warning targets and taking the US infrared early warning satellite DSP as a reference, a general comprehensive analysis of the parameters of the geostationary infrared early warning system is carried out, and the satellite is obtained. A set of preliminary parameters for orbit are satellite networking, spin platform, infrared detector, infrared camera, payload data rate, detection object, detection background, detection spectrum, detection initial height, spatial resolution, signal-to-noise ratio, etc. This set of parameters is only a preliminary determination of the composition of the system and the most basic technical indicators of each component and provides a demonstration object for the next feasibility analysis.

After selecting the existing set of parameters, with this preliminary framework, the specific evaluation of the detection performance can be carried out for the specific target such as strategic missiles, strategic aircraft and tactical missiles that may be warned by the system. For example, the caliber of the infrared camera is determined by the radiation characteristics of the target and background in the 4.3–4.4 μm spectrum, and the detection altitude is determined by the cruising altitude of the strategic aircraft, while the parameters of the 2.7–2.9 μm spectrum are not reflected in the determination of the caliber of the infrared camera, so the performance of this spectrum needs to be evaluated for specific strategic missiles, strategic aircraft and the tactical missiles that may be early warning targets. This will be a process of further calculation of more detailed, more complex and more specific components of the system and more specific parameters of each target. This evaluation process is similar to the above-mentioned comprehensive analysis process. Due to the limitation of space and some infrared technical parameters, it will not be repeated here. It is only used as a reference example of comprehensive analysis of parameters of a space optical remote sensing system.

12. Discussion

This is an example of a complex satellite remote sensing system analysis. It can only reflect some basic ideas of comprehensive analysis of satellite remote sensing systems. Its limitations are obvious and discussed below.

(1) Limitations of design target analysis

In the comprehensive analysis of the main parameters of the above-mentioned geostationary infrared early warning satellite, the signal-to-noise ratio of the system is taken as the analysis object, and the influence of the signal-to-noise ratio is not only the noise of the detector and the infrared camera electronic, but also the influence of stray light and other elements. A more difficult problem is to quantify the background noise. Due to the geostationary orbit infrared warning, the distance is long, the target is a non-cooperative target, and the detected signal is very weak and easily overwhelmed by the background noise. The processing here is only to subtract the background radiance from the target radiance. It is considered that the 4.3–4.4 μm spectrum is the strong absorption spectrum of CO_2, and CH_4 has an absorption peak at 3.3, and 2.7–2.9 μm spectral band is the strong absorption band of H_2O, so the atmosphere is basically opaque to these two bands, so it can be regarded as a uniform background.

However, the radiation intensity of this uniform background is still very large. Taking the 4.3–4.4 μm spectrum as an example, the radiance of the target is 2993.7 W/Sr, the radiance of the background is 1078.8 W/Sr, and the background radiation is about 40% of the target radiation. The noise caused by this 40% energy is not taken into account in the comprehensive analysis target (SNR) of this paper. It is difficult to calculate this effect theoretically. It can be solved by experimental measurement.

(2) Decomposition of technical indicators

The rough parameter comprehensive analysis of the geostationary orbit infrared early warning system has obtained a set of the most basic preliminary technical indicators, through which the main technical indicators of each subsystem should be formed and the indicator demonstration and scheme design of each subsystem should be carried out. For example, the technical indicators of the optical system given by the comprehensive analysis of the existing satellite system parameters are:

Field of view: $2\omega = 18^{\circ}$;
Focal length: $f = 1.36$ m;
Diameter: $D = 0.6$ m
Knowing that $F^{\#} = f/D = 1.36/0.6 = 2.27$.

This is a reflective optical system with a small $F^{\#}$ (that is, a large relative aperture) and a large field of view. Its design and implementation is very difficult. If it needs to be adjusted, for example, if the $F^{\#}$ is to be increased, this must be fed back to the large size of the infrared warning. In the comprehensive analysis of the system, a new round of comprehensive balance of large system parameters is carried out.

(3) Scheme feasibility of each component

In addition to the factors of technical indicators, there are also factors of engineering technology. In the comprehensive analysis of the large system of the infrared early warning satellite, the spin platform selected by the DSP satellite is referred to, and the spin speed of the platform is determined. The heat conduction of the heat pipe in the thermal control has an impact. If the thermal control heat pipe cannot work due to the centrifugation generated by the spin of the platform and the thermal control system cannot find other solutions, this problem will also be fed back to the synthesis analysis of the large infrared warning system. In the analysis, a new round of comprehensive balancing of large system parameters is carried out.

(4) Motion compensation scheme

Although the concept of stepping staring imaging conforms to the development direction of large-scale area array infrared detectors, how to realize the image movement compensation during time delay integration is still a link to be concreted, because of the image movement in the form of rotation, compensation is still relatively difficult. It can be mechanical or opto-mechanical, but the difficulties are different and need to be studied in depth. It may also lead to a new round of comprehensive adjustment of large system parameters and balance of difficulties.

(5) Iteration from part to whole

After each subsystem completes the design of technical indicators and the feasibility demonstration of the scheme, along with the adjustment of the indicators of each subsystem, the comprehensive analysis of the large-scale system parameters of infrared early warning needs to carry out new adjustments and comprehensive balances to alleviate some local special difficulty.

It can be seen from the above that the overall design of the space optical remote sensing system is a complex and delicate system engineering, which not only involves many disciplines, but also involves complex engineering and technical issues. What is shown here is only the analysis of the parameters of the infrared early warning satellite system. Confirmation already reflects this complex and delicate feature.

References

1. Yuan H, Dan C, Jian W, Huoping D, Satellite application, 2014;(08):56–64.
2. Shixian L, Lin L. Optical design manual, Beijing: Beijing Institute of Technology Press; 1996. p. 32.

Chapter 10
Resolution of CCD Sampling Imaging

The magnitude of the MTF of the space optical remote sensing camera is first determined by the optical system used. The optical system of the space optical remote sensing camera mainly adopts a reflective system. There for the basic factors, including the uniformity and transmittance of the optical material, that restrict the quality of the optical system is the secondary factors. The occlusion problem of the optical system has become one of the main factors affecting its MTF. Here, the quantitative relationship between various occlusion ratios and the MTF of the optical modulations degree transfer function of the optical system will be obtained. Besides, some characteristics of the resolving power of sampling optical remote sensors will be analyzed [1].

10.1 Basic Theory of Optical Modulation Transfer Function

From the basic theory of optical transfer function(MTF), an imaging optical system with an image quality close to the diffraction limit can be regarded as a spatially invariant system within the scale range of the point spread function $P(x, y)$ [2]. When using coherent light imaging, the optical transfer function of the optical system is the Fourier transform of the amplitude point spread function $u(x, y)$, namely

$$H\left(v_x, v_y\right) = \mathscr{F}[u(x, y)] \tag{10.1}$$

The amplitude point spread function is the Fourier transform of the pupil function $g(\xi, \eta)$, namely

$$u(x, y) = \mathscr{F}[g(\xi, \eta)] \tag{10.2}$$

Then

J. Tao, *Space Optical Remote Sensing*, Advances in Optics and Optoelectronics,
https://doi.org/10.1007/978-981-99-3318-1_10

$$H(v_x, v_y) = \mathscr{F}\{\mathscr{F}[g(\xi, \eta)]\} \tag{10.3}$$

From the properties of the Fourier transform, it is known

$$H(v_x, v_y) = g(-\xi, -\eta)$$

When the coordinate system of $H(\nu_x, \nu_y)$ is taken opposite to the coordinate axis of the coordinate system of $g(-\xi, -\eta)$, then there is

$$H(v_x, v_y) = g(\xi, \eta) \tag{10.4}$$

That is, the optical transfer function of the coherent optical system is equal to the pupil function of the optical system,

where

- ξ the abscissa of the pupil function,
- η the ordinate of the pupil function,
- x the abscissa of the amplitude point spread function,
- y the ordinate of the amplitude point spread function,
- ν_x the abscissa of the optical transfer function,
- ν_y the ordinate of the optical transfer function.

The coordinate relationship of pupil function $g(\xi, \eta)$, amplitude point spread function $u(x, y)$ and optical transfer function $H(\nu_x, \nu_y)$ is shown in Fig. 10.1.

The imaging of the space optical remote sensing camera is incoherent imaging. For incoherent imaging, the optical system transmits light intensity information, and the light intensity point spread function $p(x, y)$ is

$$p(x, y) = |u(x, y)|^2$$

For incoherent imaging, the optical transfer function of the optical system is the Fourier transform of the light intensity point spread function [2], namely

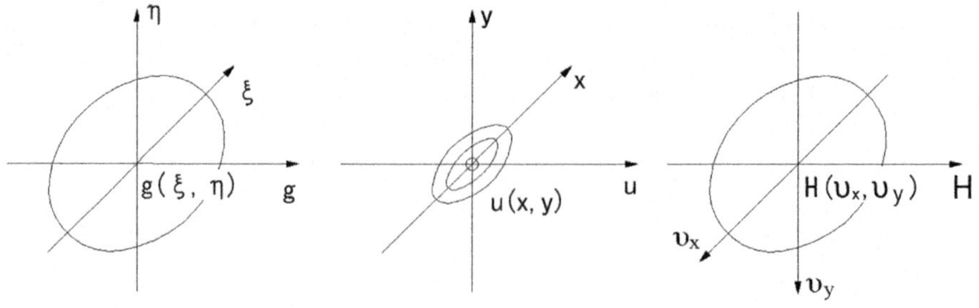

Fig. 10.1 Pupil function, the amplitude point spread function and the coordinate relationship of the optical transfer function

$$K\left(\upsilon_x, \upsilon_y\right) = \mathscr{F}\left[u(x, y) \cdot u^*(x, y)\right], \quad {}^*: \text{conjugate} \tag{10.5}$$

In information optics [2], the Fourier transform of the function u(x, y)

$$\mathscr{F}[f(x)] = F(u), \text{ that is } f(x) \Leftrightarrow F(u)$$

Then
$f(x) \cdot f^*(x) = f^*(x) \cdot f(x) \Leftrightarrow$ F(u) ☆F(u), ☆: Correlation.
Then the formula 10.5 is

$$K\left(\upsilon_x, \upsilon_y\right) = \mathscr{F}\left[u(x, y) \cdot u^*(x, y)\right] = \mathscr{F}[u(x, y)]\mathscr{F}[u(x, y)] \tag{10.6}$$

Substitute 10.2 into 10.6

$$\begin{aligned} K\left(\nu_x, \nu_y\right) &= \mathscr{F}\{\mathscr{F}[g(\xi, \eta)]\}\mathscr{F}\{\mathscr{F}[g(\xi, \eta)]\} \\ &= \mathrm{g}(-\xi, -\eta) \star \mathrm{g}(-\xi, -\eta) \end{aligned}$$

Similar to Fig. 10.1, taking the opposite directions of the $K(\nu_x, \nu_y)$, $g(\xi, \eta)$ coordinate axes, so:

$$K\left(\upsilon_x, \upsilon_y\right) = g(\xi, \eta) g(\xi, \eta) \tag{10.7}$$

This formula shows that the optical transfer function $K(\upsilon_x, \upsilon_y)$ of the incoherent optical system is the autocorrelation of the pupil function $g(\xi, \eta)$.

In the optical system of coherent imaging, the system transmits the vibration information of light, and the image formed on the image surface is the superposition of light vibrations of various points on the object surface. In the optical system of incoherent imaging, the system transmits the intensity information of light, and the image formed on the image surface is the superposition of the light intensity of each point on the object surface. However, the optical system is a linear system in coherent light imaging and incoherent light imaging and is a linear invariant system in the isohalo region. Therefore, in incoherent imaging, the optical transfer function of the optical system is the autocorrelation of its pupil function. After normalization, this is

$$\begin{aligned} D(\nu_x, \nu_y) &= D\left(\frac{\xi'}{\lambda R}, \frac{\eta'}{\lambda R}\right) \\ &= \frac{\int_{-\infty}^{\infty}\int_{-\infty}^{\infty} g(\xi + \xi', \eta + \eta') g * (\xi, \eta) \mathrm{d}\xi \mathrm{d}\eta}{\int_{-\infty}^{\infty}\int_{-\infty}^{\infty} |g(\xi, \eta)|^2 \mathrm{d}\xi \mathrm{d}\eta} \end{aligned}$$

where

ξ The abscissa of the pupil function;
η The ordinate of the pupil function;

ξ' The distance that the pupil is staggered in the abscissa direction;
η' The distance that the pupil is staggered in the ordinate direction.

$D(\nu)$ represents a complex number, namely the optical transfer function $D(\nu) = T(\nu)e^{-i\theta(\nu)}$, its mode is the optical modulation transfer function (MTF), its argument angle is the phase of the optical transfer function, that is the phase transfer function.

In

$$g(\xi, \eta) = \begin{cases} B(\xi, \eta)e^{-ikw(\xi,\eta)} & \text{in the pupil} \\ 0 & \text{out of the pupil} \end{cases}$$

$B(\xi, \eta)$ amplitude transmittance within the pupil,
$W(\xi, \eta)$ wavefront aberration within the pupil.

What is to be analyzed here is the influence of various obscuration ratios of the space optical remote sensing camera on its optical modulation transfer function, so let

$$B(\xi, \eta) = 1$$
$$W(\xi, \eta) = 0$$

where

$$g(\xi, \eta) = \begin{cases} 1 \text{ in the pupil} \\ 0 \text{ out of the pupil} \end{cases}$$

So the optical transfer function formula becomes

$$\begin{aligned} D(\nu_x, \nu_y) &= D(\frac{\xi'}{\lambda R}, \frac{\eta'}{\lambda R}) \\ &= \frac{\int_{-\infty}^{\infty}\int_{-\infty}^{\infty} g(\xi + \xi', \eta + \eta')g * (\xi, \eta)\mathrm{d}\xi\mathrm{d}\eta}{\int_{-\infty}^{\infty}\int_{-\infty}^{\infty} |g(\xi, \eta)|^2\mathrm{d}\xi\mathrm{d}\eta} \\ &= \frac{\int_{-\infty}^{\infty}\int_{-\infty}^{\infty} g(\xi + \xi', \eta + \eta')g * (\xi, \eta)\mathrm{d}\xi\mathrm{d}\eta}{A} \\ &= \frac{\mathrm{S}}{A} \end{aligned} \tag{10.8}$$

$$S = \int_{-\infty}^{\infty}\int_{-\infty}^{\infty} g(\xi + \xi', \eta + \eta')g * (\xi, \eta)\mathrm{d}\xi\mathrm{d}\eta$$

That is the pupil $g(\xi, \eta)$ the overlapping area after the pupils are shifted by the abscissa ξ ordinate η'.

$A = \int_{-\infty}^{\infty}\int_{-\infty}^{\infty} 1\, d\xi d\eta$, is the pupil area.

Make

$$l = \sqrt{\xi'^2 + \eta'^2}$$

$$\nu_x = \frac{\xi'}{\lambda R}$$
$$\nu_y = \frac{\eta'}{\lambda R}$$

where λ the wavelength of light in the working spectrum of the optical system; R the distance from the exit pupil of the optical system to the image plane, for space optical remote sensing, is equivalent to the object distance being infinite, at this time there is $R = f$, which is the focal length of the camera.

$$\begin{aligned}\nu &= \sqrt{\nu_x^2 + \nu_y^2} \\ &= \sqrt{\left(\frac{\xi'}{\lambda R}\right)^2 + \left(\frac{\eta'}{\lambda R}\right)^2} \\ &= \frac{l}{\lambda f}\end{aligned}$$

l is the distance that the pupil is staggered. The numerator S of the optical transfer function formula $D(\nu_x, \nu_y)$ is the overlapping area under this distance of the pupil staggering, and the denominator A of $D(\nu_x, \nu_y)$ is the pupil area. Their ratio S/A is the MTF of the optical system at this spatial frequency ν. For a space optical camera, the maximum value of the distance l that the pupil is staggered is generally the aperture D of the camera, and D/f is the relative aperture.

10.2 Example of MTF Calculation Under Various Obscuration Ratios

The following is a design example of a space camera, camera parameters:

Focal length: $f = 6.5$ m;

The optical aperture: $D = 650$ mm;

Spectral range: $\lambda = 0.50 \sim 0.9$ μm.

When the camera has a central obscuration, its pupil is shown in Fig. 10.2. In the figure, D_2 is the diameter of the light-transmitting area, and D_1 is the diameter of the obscuration area. This is the pupil diagram of a typical space camera, and it is of practical significance to study its influence on the MTF.

Let the obscuration ratio be K

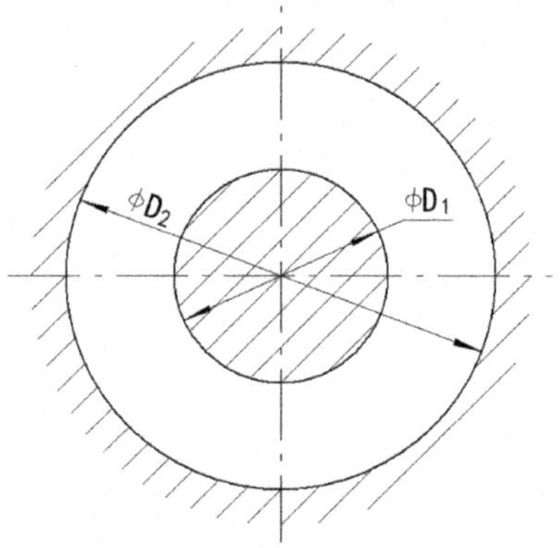

Fig. 10.2 Obstructed pupil

$$K = \frac{D_1}{D_2}$$

When K is constant, stagger the pupil shown in Fig. 10.2 by a distance l, as shown in Fig. 10.3. The shaded area in Fig. 10.3 is the value of the numerator of formula 10.8. The shaded part shown in Figs. 10.3, 10.4, 10.5, 10.6, 10.7 and 10.8 is the light-passing area of the optical system, which is different from Fig. 10.2. In Fig. 10.2, the non-shaded area is the light-passing area, which is also the value of the denominator of Eq. 10.8, then the value $m = S/A$ of the modulation transfer function is obtained.

As long as the two overlapping pupils shown in Fig. 10.2 are gradually staggered by a distance of l, as shown in Fig. 10.3, the MTF corresponding to different frequencies ν can be obtained. In the process of l gradually increasing from 0 to the maximum value D_2, the overlapping part of the pupil, that is the shadow part, is also changing, and the shadow part is the solution threshold. Figures 10.2, 10.3, 10.4, 10.5, 10.6, 10.7, 10.8 and 10.9 shows the typical links of this process. From Figs. 10.2, 10.3, 10.4, 10.5, 10.6, 10.7, 10.8 and 10.9, it can be seen that this is a complex change process, which reflects many stages of change, and each stage has its own graphic features.

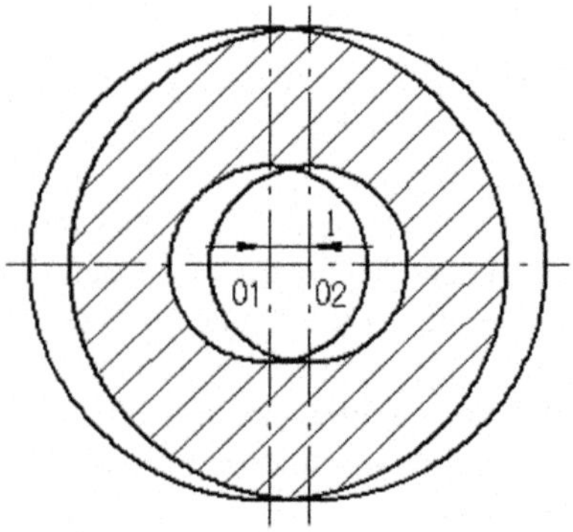

Fig. 10.3 Obstructed pupil

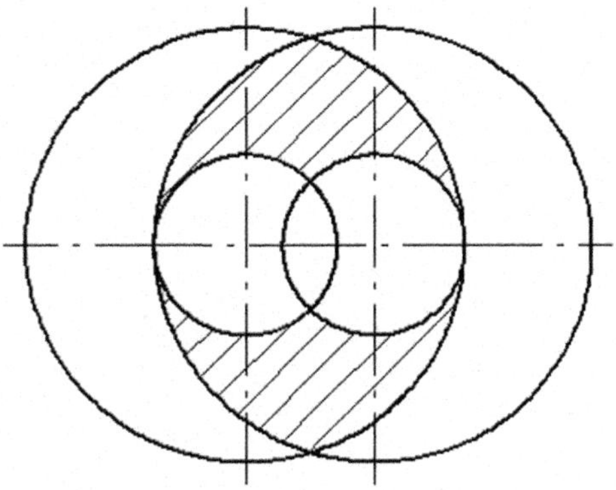

Fig. 10.4 Obstructed pupil

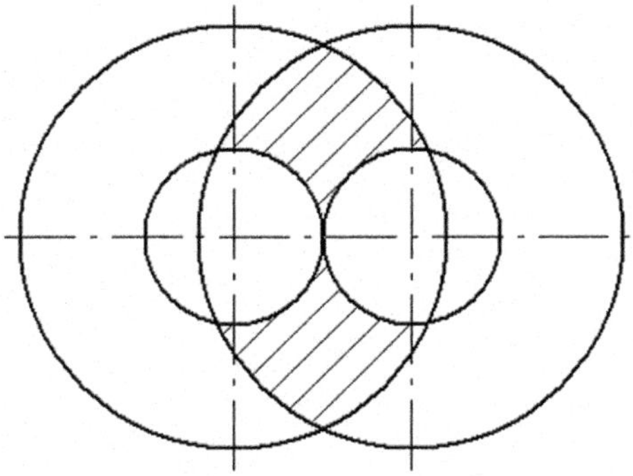

Fig. 10.5 Obstructed pupil

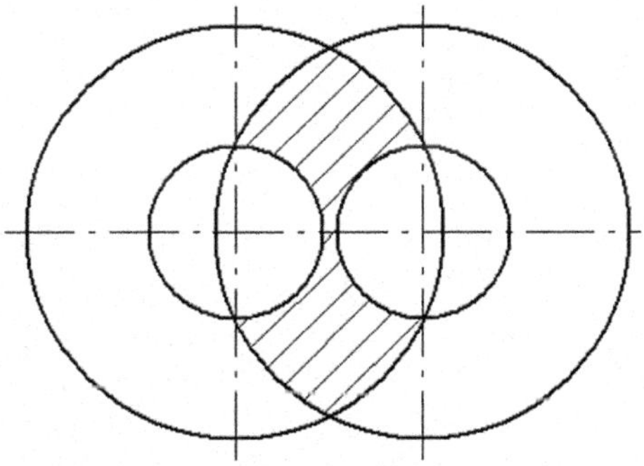

Fig. 10.6 Obstructed pupil

The division of the above stages is only a preliminary analysis. When solving the corresponding MTF within the solution threshold of each stage, the function needs to be solved segmentally for different l variation ranges.

The above has analyzed the process of solving the MTF under the condition of a certain obscuration ratio K. When the obscuration ratio K changes, the change process of the overlapping pupils should be analyzed in detail corresponding to each K value and the change stages of each pupil should be re-divided. The whole

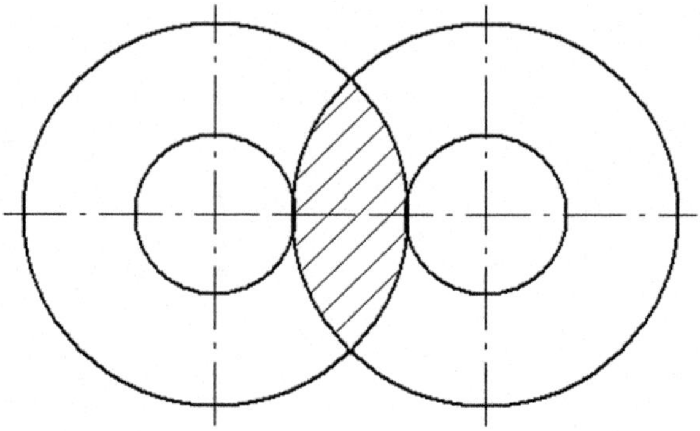

Fig. 10.7 Obstructed pupil

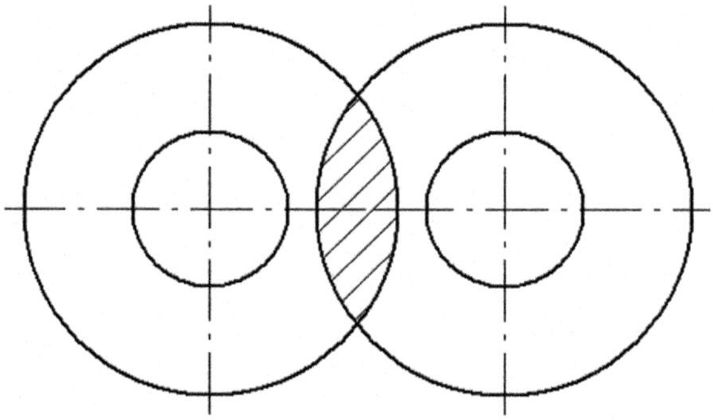

Fig. 10.8 Obstructed pupil

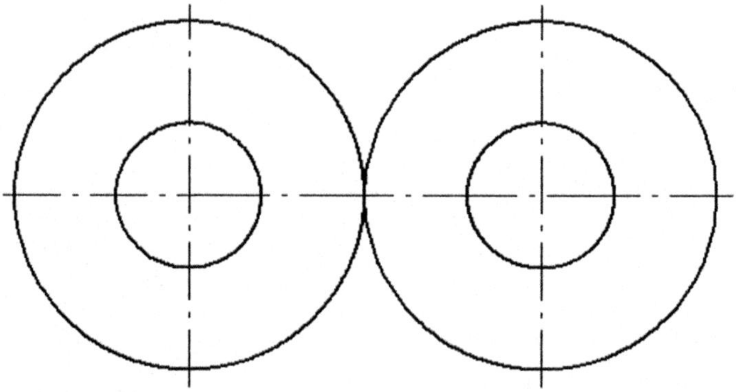

Fig. 10.9 Limit of pupil movement

analysis and calculation process is complex and must be solved by means of computer programming.

For the optical system of the incoherent space optical remote sensing camera, the cutoff frequency of the optical modulation transfer function is

$$\nu_{\mathrm{lim}} = \frac{D}{\lambda f} \tag{10.9}$$

It can be seen from the cutoff frequency formula of the optical modulation transfer function that different wavelengths of light have different cutoff frequencies even for the same optical system, which is inconvenient for describing and analyzing the optical modulation transfer function. Therefore, the spatial frequency is normalized, that is, take

$$\begin{aligned} \gamma &= \frac{\nu}{\nu_{\mathrm{lim}}} \\ &= \frac{\frac{l}{\lambda f}}{\frac{D}{\lambda f}} \\ &= \frac{l}{D} \end{aligned} \tag{10.10}$$

It can be seen from the formula 10.10 that through the normalization of the coordinates, the normalized abscissa has nothing to do with the wavelength.

Take the normalized spatial frequency γ as the abscissa, and take the optical modulation degree transfer function value m as the ordinate, to describe the modulation degree transfer function. Under the condition of

$$\begin{aligned} B(\xi, \eta) &= 1 \\ W(\xi, \eta) &= 0 \end{aligned}$$

The optical modulation transfer function curve of each wavelength λ has exactly the same shape, which is very beneficial to analyze the influence of various obscuration ratios on the optical modulation transfer function.

The computer solves the curve of the MTF under various obscuration ratios K as shown in Fig. 10.10.

10.3 The Relationship of the Pixel Size of the CCD Detector, Obscuration Ratios and the MTF

From the optical modulation transfer function curve, it can be seen that the overall trend of the optical modulation transfer function decreases with the increase of the obscuration ratio. The curve has an inflection point near the normalized frequency

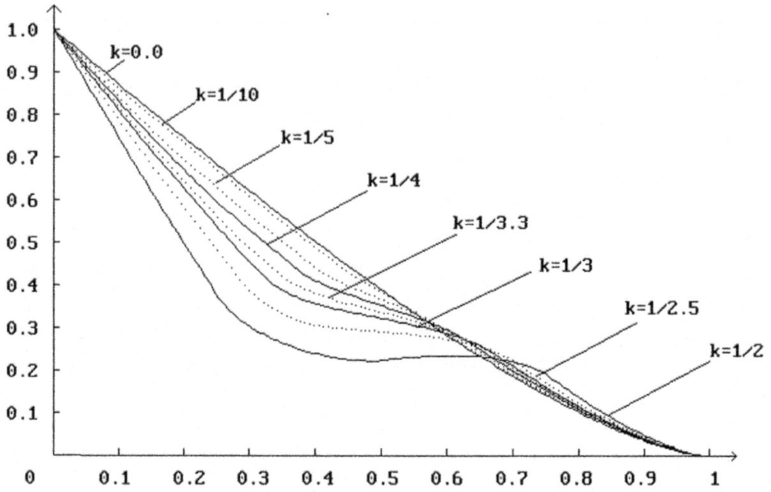

Fig. 10.10 Normalized curve of optical modulation degree transfer function under various obscuration ratios K

$\gamma = 0.6$. When γ is less than 0.6, the optical modulation transfer function has a relatively large decrease. When k is greater than 1/2, the value of the MTF on the side where $\gamma > 0.6$ is greater than the value of the MTF in $\gamma = 0.6$ slightly.

They also hardly contribute to the imaging quality of the space optical remote sensing camera in this part where the MTF increases. There are two reasons for this.

First, the Nyquist frequency of space optical remote sensing cameras is generally difficult to reach $\gamma = 0.6$. For example, in the example here, take its central wavelength $\lambda = 0.7\ \mu\text{m}$, then

$$\begin{aligned} \nu_{\text{lim}} &= \frac{D}{\lambda f} \\ &= \frac{650}{0.7 \times 6500} \\ &= 143 \end{aligned}$$

On the $\gamma = 0.6$

$$\begin{aligned} \nu &= \nu_{\text{lim}} \times \gamma \\ &= 143 \times 0.6 \\ &= 86 \end{aligned}$$

Even if the pixel size of its CCD adopts $a = 7\ \mu\text{m}$, its Nyquist frequency is $\nu = 71.4$ lp/mm, which cannot reach 86. Therefore, the optical modulation transfer function above the inflection point of $\gamma = 0.6$ cannot be used by space cameras, and the pixel size of the CCD detectors of the space optical remote sensing cameras currently in use is larger than 7 μm, that is to say its Nyquist frequency cannot reach above 86 lp/mm.

Second, since CCD detector imaging is a sampling discretization process, when its Nyquist frequency cannot reach the inflection point of $\gamma = 0.6$, the inflection MTF of above the $\gamma = 0.6$ higher, the more unfavorable the imaging quality of the space optical remote sensing camera.

From this, it can be seen that the optical modulation transfer function of the space optical remote sensing camera in the case of obscuration is not only related to the obscuration ratio, but also closely related to the pixel size of the CCD detector used. In order to compare the v the MTF under various obscuration ratios and the pixel size of various CCD detectors, regard the Nyquist frequency corresponds to the size of various CCD detectors as the cutoff frequency to calculate the figure of area under the curve of MTF in Fig. 10.10 with various K, use it as a judge standard to analyze and compare K the utilization of MTF under various K of the space optical remote sensing camera.

The pixel size of the CCD detector commonly used in space optical remote sensing cameras and the pixel size of the CCD detector in this example are listed as follows:

Using the Nyquist frequency in Table 10.1 to integrate the area under the MTF curve under various obscuration ratios in Fig. 10.10 by computer programming, the values obtained in the integration frequency range $[0, \nu_{\text{lim}}]$ are listed in Table 10.2.

The spatial frequency 143 in Table 10.2 is the cutoff frequency corresponding to the optical system when the center wavelength $\lambda = 0.7$ μm in this example. The value in the table is the MTF integral values for 8 kinds of obscuration ratios from $K = 0$ to $K = 1/2$. The spatial frequency 71 is the Nyquist frequency of the optical system when the pixel size of the CCD is $a \times a = 7$ μm × 7 μm. The value in

Table 10.1 Nyquist frequency corresponding to the pixel size of the CCD detector

Pixel size (μm)	7	8.75	10	13
Nyqist frequency	71.43	57.14	50	38.46

Table 10.2 MTF integral table for various obscuration ratios

K	$\nu_{\max}$				
	143	71	57	50	38
0	61.131	49.708	43.470	39.756	32.436
1/10	60.364	48.854	42.403	39.194	32.042
1/5	58.240	46.576	40.974	37.689	31.004
1/4	56.735	45.015	39.693	36.651	30.442
1/3.3	54.959	43.224	38.227	35.449	29.772
1/3	53.631	41.919	37.169	34.573	29.285
1/2.5	50.644	39.093	34.923	32.707	28.166
1/2	45.36	34.498	31.300	29.540	25.968

this column in the table indicates that when the working wavelength of the optical system is $\lambda = 0.7$ μm, the integral frequency range is [0, 71]. The meaning of the other three columns is the same.

Analyze the situation with a certain blocking ratio, when $K = 0$, there is no obscuration:

When the pixel size of the CCD detector is 13 μm, the ratio of the integral value of the MTF in the interval [0, 38] to the integral value in the [0, 143] interval is

$$\frac{32.436}{61.131} = 0.53$$

When the pixel size of the CCD detector is 10 μm, the ratio of the integral value of the MTF in the interval [0, 50] to the integral value in the interval [0, 143] is

$$\frac{39.756}{61.131} = 0.65$$

When the pixel size of the CCD detector is 7 μm, the ratio of the integral value of the MTF in the interval [0, 71] to the integral value in the interval [0, 143] is

$$\frac{49.708}{61.131} = 0.81$$

When $K = 1/3.3$, which is equivalent to 1/10 of the area obscuration:

When the pixel size of the CCD detector is 13 μm, the integral ratio is

$$\frac{29.772}{54.959} = 0.54$$

When the pixel size of the CCD detector is 10 μm, the integral ratio is

$$\frac{35.449}{54.959} = 0.65$$

When the pixel size of the CCD detector is 7 μm, the integral ratio is

$$\frac{43.224}{54.959} = 0.79$$

When $K = 1/2$, which is equivalent to 1/4 of the area obscuration:

When the pixel size of the CCD detector is 13 μm, the integral ratio is

$$\frac{25.968}{45.36} = 0.57$$

When the pixel size of the CCD detector is 10 μm, the integral ratio is

$$\frac{29.54}{45.36} = 0.65$$

When the pixel size of the CCD detector is 7 μm, the integral ratio is

$$\frac{34.498}{45.36} = 0.76$$

It can be seen from the comparison of the above three sets of figures that when the obscuration ratio is constant, with the reduction of the pixel size of the CCD detector, the utilization rate of the MTF of the space optical remote sensing camera is greatly improved. This shows that from the perspective of the MTF utilization, the pixel size of the CCD detector should be as small as possible to improve the information detection capability of the space optical remote sensing camera, which can also improve the resolution of the space optical remote sensing camera. Of course, the problem of MTF value decline due to the reduction of CCD pixel size should also be considered in the engineering practice.

Comparison of the MTF at different blocking ratios. Next, the influence of various obscuration ratios on the MTF of the space optical remote sensing camera will be analyzed. For this reason, divide the integral value of the MTF of all obscuration ratios in the previous table by the integral value of the MTF of $K = 0$ in the corresponding column.

For example, in the last column of Table 10.2, when the pixel size of the CCD detector is 13 μm, the corresponding cutoff frequency $\nu_{\text{lim}} = 38$. When $K = 0$, the integral value of the corresponding MTF is 32.436, and when $K = 1/2$, the corresponding integral value of the MTF is 25.968. The corresponding ratios in Table 10.3 are

$$\frac{25.968}{32.436} = 0.800592$$

The meaning is the same when K is equal to other values.

Table 10.3 Comparison table of integral value of MTF for the same cutoff frequency

K	$\nu_{\max}$				
	143	71	57	50	38
0	1	1	1	1	1
1/10	0.987453	0.98282	0.975454	0.985864	0.987853
1/5	0.952708	0.936992	0.942581	0.948008	0.955852
1/4	0.928089	0.905589	0.913112	0.921899	0.938525
1/3.3	0.899036	0.869558	0.879388	0.891664	0.917869
1/3	0.877313	0.843305	0.855049	0.86963	0.902855
1/2.5	0.82845	0.786453	0.803382	0.822693	0.868356
1/2	0.742013	0.694013	0.720037	0.743032	0.800592

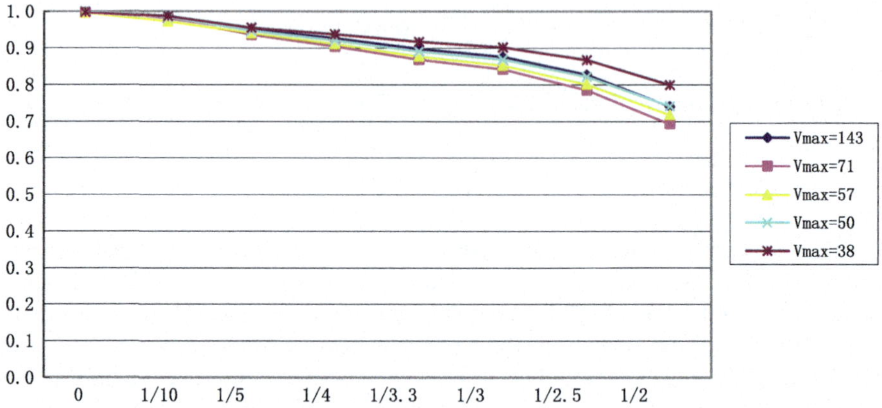

Fig. 10.11 Ratio curve of the integral value of the MTF under different pixel sizes and obscuration ratios

When $K = 0$, its ratio is

$$\frac{32.436}{32.436} = 1$$

Similarly, when the pixel size of the CCD detector is other values, the ratio can be obtained in this way, as listed in Table 10.3.

For further analysis, take the obscuration ratio K in table 10.3 as the abscissa, and take the ratio of the MTF as the ordinate. The ratio curve of the integral value of the MTF about the different pixel size of CCD detectors and the different obscuration ratio K can be drawled, as Fig. 10.11.

It can be seen from Fig. 10.11 that the utilization rate of the MTF decreases by 30%, when the obscuration ratio increases from $K = 0$ to $K = 1/2$, at the smallest pixel size of the CCD detectors being analyzed.

It can be seen from Table 10.3 that regardless of the pixel size of the CCD detector, the utilization rate of the MTF decreases with the increase of the obscuration ratio K. And with the reduction of the pixel size of the CCD detector, the utilization rate of the MTF is also reduced, and the trend of this reduction is obvious. When the pixel size of the CCD detector decreases from 13 μm, at 7 μm, the utilization ratio of its MTF decreases

$$0.800592 - 0.694013 = 0.106579$$

It shows that the smaller the pixel size of the CCD detector, the more sensitive to the obscuration ratio of the optical system. Therefore, the higher the resolution of the space optical remote sensing camera, the more sensitive the obscuration ratio of the optical system is.

The combined effect of occlusion ratio and pixel size on MTF. The effects of different obscuration ratios and different pixel sizes of CCD detectors on the different MTF have been analyzed previously. In order to comprehensively analyze the influence of these two factors on the MTF, the integral value of the MTF is divided by the integral value of MTF with an obscuration ratio of $K = 0$ and a cutoff frequency of $\nu_{\text{lim}} = 143$ in Table 10.2, the quotients are filled in Table 10.4. For the obscuration ratio from $K = 0$ to $K = 1/2$, the utilization rate of the MTF decreases by a maximum of 30% (see Table 10.3). When the pixel size of the CCD detector is reduced from 13 to 7 μm, the use ratio of the MTF most increases by 0.81–0.53 = 28% (see Table 10.4). It can be seen that both of them have great influence on the MTF utilization.

It can be seen from Table 10.4 that the utilization rate of the MTF is relatively low in the current technical level when there are dual effects of the obscuration ratio and the pixel size of the CCD detector. From $K = 1/2$ and $\nu_{\text{lim}} = 38$ in Table 10.4, it can be seen that the use rate of its MTF is only 42.4%. Among them, obscuration affects 1 – 0.742 = 26%, and the effect of CCD detector pixel size is 1 – 0.53 = 47%. The current space optical remote sensing camera has a CCD detector pixel size of less than 13 μm, even if 7 μm is taken, the influence of pixel size of CCD detector is 1 – 0.81 = 19%, and both effects are obvious.

Space optical remote sensing cameras generally use sun-synchronous orbits, and its light source is the sun. The sunlight passes through the atmosphere and shoots toward the target at different altitudes and azimuths, then the sunlight is reflected by the target, and passes through the atmosphere again, and through the lens reaches the receiver for imaging. In this information transfer process, the first is that sunlight passes through the atmosphere to reach the target at different altitude and azimuth angles, and this process is affected by the absorption, scattering, turbulence and transmittance of the atmosphere. The reflection by the target is also affected by the reflectivity of the target. The spectral reflection characteristics of deserts, snow-capped mountains, wetlands, vegetation, and oceans have a huge impact on the information transfer. The sunlight reflected by the target carries the information to be

Table 10.4 Comparison table of integral value of MTF for different obscuration ratios and pixel sizes of CCD detectors

K	$\nu_{\max}$				
	143	71	57	50	38
0	1	0.813139	0.711096	0.650341	0.530598
1/10	0.987453	0.813139	0.693642	0.641148	0.524153
1/5	0.952708	0.761905	0.670265	0.616528	0.507173
1/4	0.928089	0.736369	0.64931	0.599549	0.49798
1/3.3	0.899036	0.707072	0.625329	0.579886	0.48702
1/3	0.877313	0.685724	0.608022	0.565556	0.479053
1/2.5	0.82845	0.639496	0.571281	0.535031	0.460748
1/2	0.742013	0.564329	0.512015	0.483225	0.424793

obtained. When this part of the reflected light passes through the atmosphere again, it will be affected by absorption, scattering, turbulence and transmittance again, which has a great influence on the information propagation of the target. When light passes through the lens, it will be affected by the absorption, reflection, transmission, stray light, focal length and aperture of the lens, and finally reaches the image plane and is received by the receiver. Of course, the response efficiency and noise of the receiver are also involved here. There are problems of the stability and measurement and control accuracy of the carrier platform. This is only the influence of external factors for the space optical remote sensing cameras, and the lens manufacturing of space optical remote sensing cameras is an important factor in determining the MTF. It involves many links such as optical design, optical processing, mechanical design, mechanical processing, mechanical assembly, optical adjustment, etc., and the camera is also affected by vibration and thermal deformation when the camera is in orbit. These series of factors will affect the MTF of the space optical remote sensing camera. However, only the obscuration of the space optical remote sensing camera and the pixel size of the CCD detector can lose about 50% of the use rate of the MTF. It can be seen that they are two important problems of the space optical remote sensing camera, that's why it makes sense to study them.

10.4 Computer Simulation Analysis of Detector Static Image Resolution

The previous section introduced the influence of the obscuration ratio and the size of the CCD pixel on the utilization rate of the MTF. Here, the resolution of the CCD sampling and discrete imaging to the target with a certain size of the pixel is further analyzed. The photoelectric imaging method of CCD detectors has been widely used, especially in the field of space reconnaissance, so that the resolution of this sampling discretization imaging has been paid attention to. A prerequisite for obtaining the MTF in the theory of information optics is the imaging optical system is regarded as a spatially invariant system (see Sect. 10.1 Fundamental theory of optical transfer function) within a small area of the finite size of the image plane, which is no longer strictly true, that is, the imaging of the CCD detector does not have the property of spatial invariance in the isohalo region. In order to describe and analyze the image quality and resolution of the photoelectric imaging system with the CCD detector as the receiver in practice, here the computer simulation of the sampling discretization imaging of CCD is carried out. Considering that the imaging of the CCD detector is digital imaging, it is also beneficial to use the simulation method to study, which is not only a supplement in theory, but also has practical significance in engineering practice.

When evaluating the resolution performance of a CCD camera, the pixel resolution of the camera is generally used, and the mean MTF at Nyquist frequency under various phases between the black and white stripes of the resolution identification rate plate and the CCD column pixels, namely the MTF of the square wave signal, the same as below, which is used as the an image quality evaluation index for evaluating CCD cameras, because it is beneficial to theoretical derivation and usße in CCD imaging optical systems. But it cannot be distinguished under under special conditions (such as Fig. 10.12c).

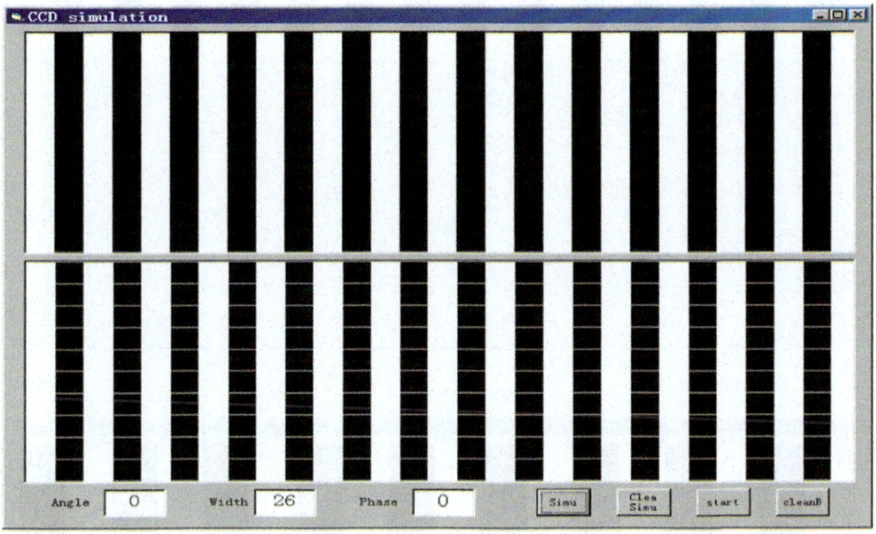

(a)

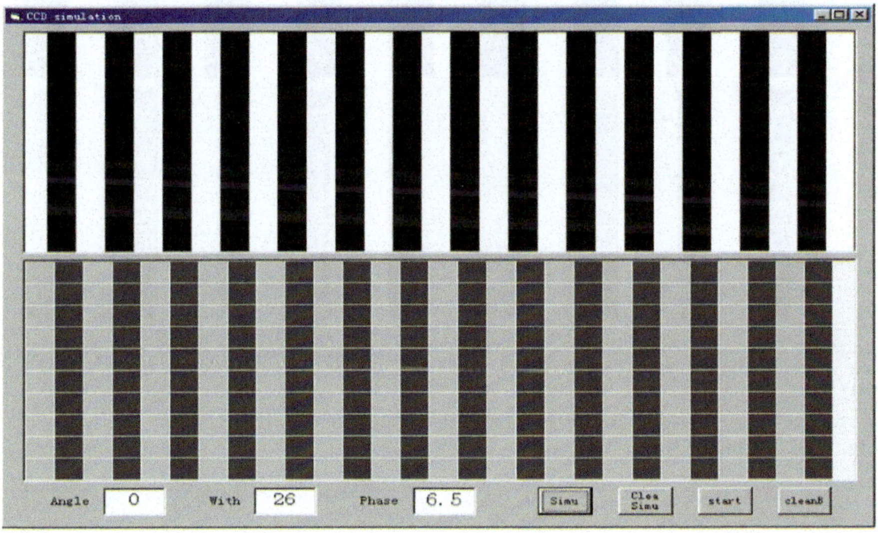

(b)

Fig. 10.12 Simulation picture of static imaging of CCD detector

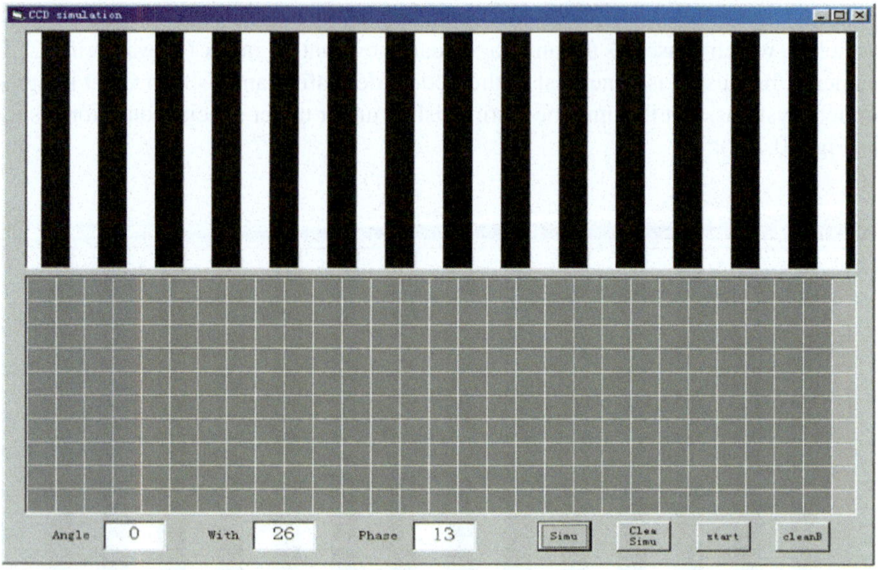

(c)

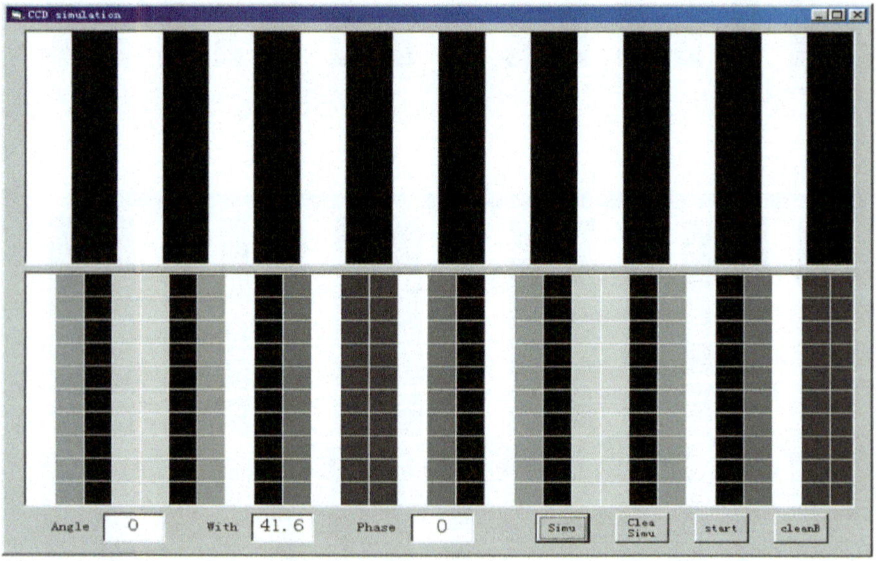

(d)

Fig. 10.12 (continued)

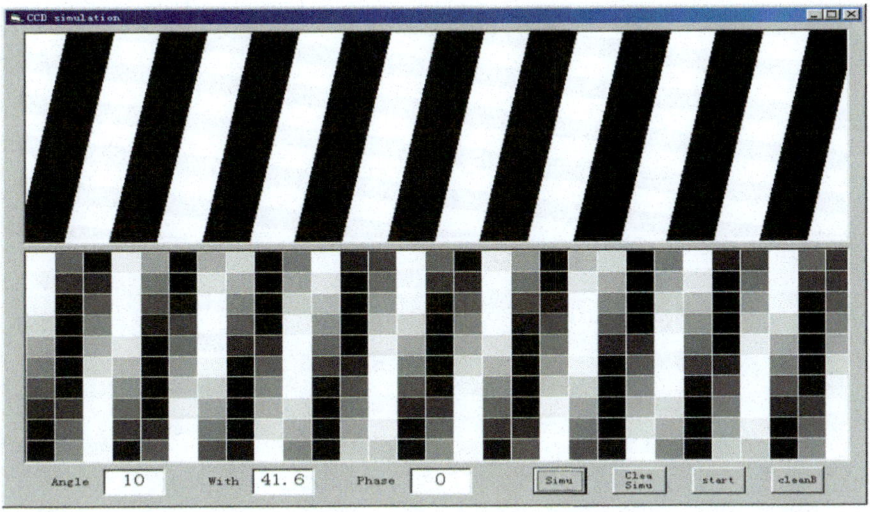

(e)

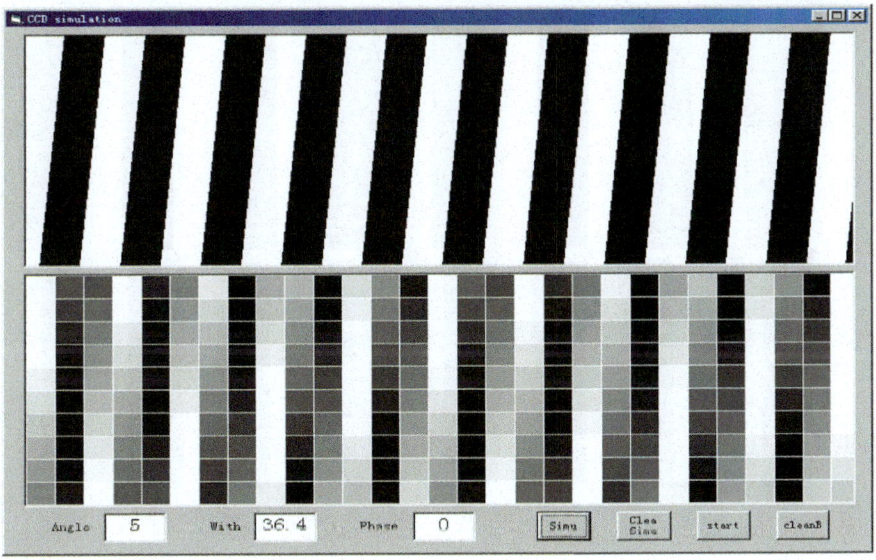

(f)

Fig. 10.12 (continued)

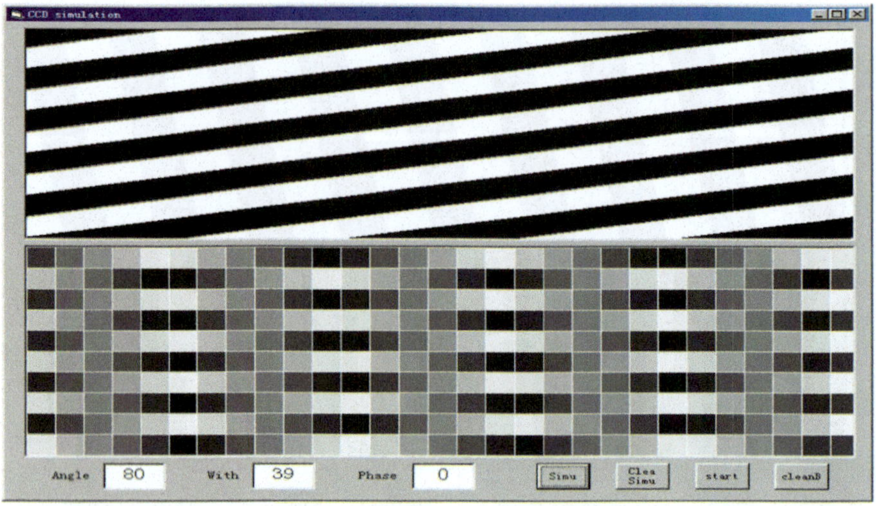

(g)

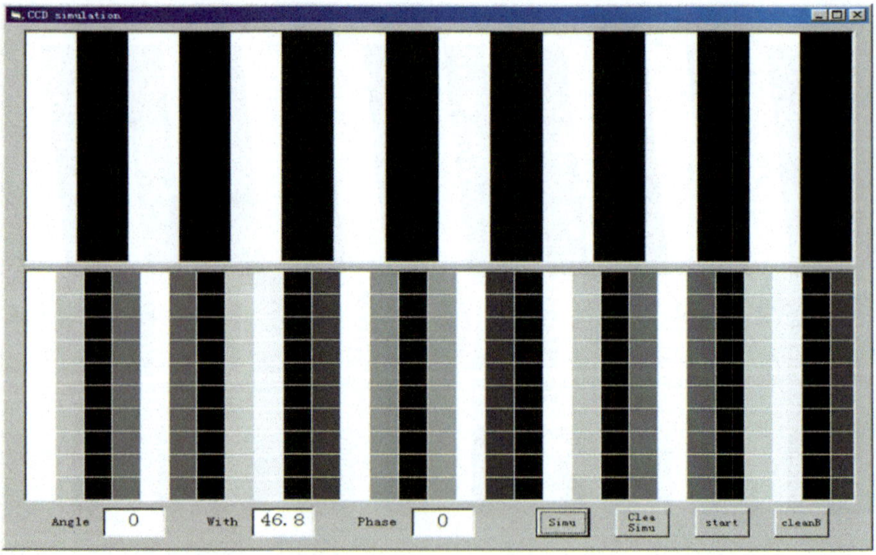

(h)

Fig. 10.12 (continued)

(i)

(j)

Fig. 10.12 (continued)

(k)

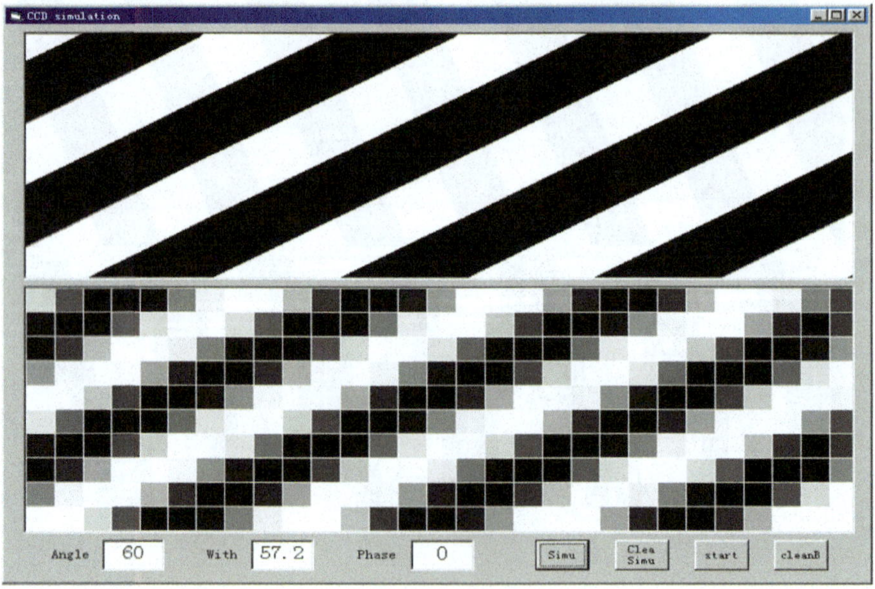

(l)

Fig. 10.12 (continued)

Since the MTF calculation of CCD imaging using the convolution theory [3], there is no parameter similar to the phase P that represents the relative positional relationship between the strips and the CCD pixels, so only a mean MTF can be obtained by using this method. Here, the computer simulation method is used to study the image resolution of CCD imaging, which can resolve the target under the condition of spatial shift change and in most cases. The meaning of the most cases refers to the extent to which the method of computer simulation can be achieved. Because the relative position relationship between the object image and the CCD pixels has three parameters, two translations and one rotation, they are all continuous quantities, and the simulation can only obtain a limited number of discrete quantities, although the simulation results are continuously analyzed and simulated continuously, but not exhaustive in all cases.

The resolution of CCD image is considered to be pixel resolution multiplied by 2 in the literature [4]. This is because the CCD detector describes the target with the limited geometric size of the CCD pixel when imaging, and actually displays the continuous target image in a discontinuous form, which will reduce the clarity of the target image. At the same time, the CCD detector also displays the grayscale of the target image in a discontinuous form when imaging, which will also reduce the clarity of the target image. This reduction in image clarity will inevitably reduce the resolution of the image. There is also sampling in the electronics theory to study the imaging process of the CCD detector, but the sampling in the electronics, the sampling time $\Delta t \ll$ sampling period T, and the sampling of the CCD detector is $\Delta t = T$. It is necessary to discuss the relationship between CCD image resolution and pixel resolution of CCD imaging, which is one aspect of the problem.

From another aspect of the problem, the image formed by the CCD detector must ultimately be interpreted by the human eye, and the human subjective factor also plays a role that cannot be ignored. This problem was published in 1964 [5], and the literature [6] was also published for the purpose of object recognition with discretized sampled images formed by CCD detectors. However, most of these literatures are aimed at specific target recognition, such as how many lines are needed to distinguish a car, how many lines are needed to distinguish a tank; or how much the ratio of the graphic area to the imaging area is required to distinguish a specific triangle or hexagon. However, specific things are inexhaustible.

Here, the resolution of the CCD camera is evaluated from another perspective, that is, for identify the characteristics of the target image, how much the resolution scale of the CCD camera is necessary. This avoids the exhaustion of various specific things and has universal significance and strong practicability.

In view of the fact that the optical modulation transfer function is mostly used in the theory of space optical remote sensing cameras to analyze and discuss issues, the concept of MTF is still used here to analyze the effect of the sampling discretization of the CCD detector on the imaging contrast and image resolution of the space optical remote sensing camera. The MTF of the CCD detector sampling discretization imaging discussed here is obtained directly from the simulation.

From the definition of the MTF

$$\mathrm{MTF} = \frac{M_g(\nu)}{M_o(\nu)}$$

$M_g(\nu)$ Modulation of the image.
$M_o(\nu)$ Modulation of objects.

In the simulation of CCD detector imaging, as long as the modulation degree $M_o(\nu)$ of the subject and the modulation degree $M_g(\nu)$ of the image after the simulated imaging are known, the sampling discretization of the CCD detector reduces the degree of the MTF of the space remote sensing camera, which is also obtained.

The imaging process of a space optical remote sensing camera with a CCD detector as a receiver is an imaging process under dynamic conditions. There are many factors affecting this process. For the convenience of analyzing the problem, the static imaging situation of the CCD detector is simulated first.

Definition of image resolution of CCD detectors. The image resolution of the CCD detector mentioned here refers to the number of pixels required when the characteristics of the ideal image formed by the optical system on the image plane can be recognized. Usually, the pixel size of a CCD detector is generally around 10 μm, and the $F^{\#}$ (reciprocal of the relative aperture) of a space visible light reconnaissance camera is generally around 10. For such a camera, the resolution corresponding to the ideal cutoff frequency of the optical system is generally a few microns on the image plane [7], so the pixel size of the CCD detector determines the image resolution of the camera in a large extent.

The imaging system using the CCD detector involves many problems related to the fabrication process of the detector, such as signal crosstalk, charge failure, optical system quality, electronic noise, and environmental factors [8, 9]. These factors are numerous and complex. In order to facilitate the study of the problem, the ideal image formed by the optical system on the image plane is defined as the target, so that the problem of the sampling resolution of the CCD detector can be separated for easy research.

1. **Simulation conditions and methods**

The focus here is to evaluate the resolving power of the CCD detector from the recognition of the image features obtained by the CCD, and still use the pattern of the resolution discriminating rate plate commonly used in evaluating the performance of optical remote sensing cameras as the target to evaluate the CCD the resolving power of detector sampling [10]. The advantage of this method is that the target used is consistent with the target used in evaluating the performance of optical remote sensing cameras, and the obtained results are easily included as a part of the performance evaluation of optical remote sensing cameras. Therefore, it has strong practicability and universal significance. Since the resolution discriminant plate is used as the target pattern, the conclusion obtained is equivalent to the MTF value of the square wave signal, which should be multiplied by $\pi/4$ ($\nu > \nu_{\lim}/3$) relative to the sinusoidal signal.

CCD detector imaging is that each pixel of the CCD integrates the energy radiated from the target to each pixel and performs photoelectric conversion. Usually, the difference in the illuminance value obtained on each pixel is generally described by 255 Gy levels, which is the target image formed by the CCD.

In the simulation, each pixel of the CCD is described as a 26×26 dot matrix. Each point represents a piece of light energy. According to this, the light energy obtained by each pixel can be simulated, and the highest light energy is defined as 255. The gray scale can be simulated, which also sets the shape of the pixel of the CCD detector as a square.

The pattern of the resolution discriminating plate is used as the target, which is commonly used in evaluating the performance of optical remote sensing cameras, and its modulation degree is 1. This assumption is reasonable at high contrast [11]. The stripe width of the pattern, the inclination angle and the phase of the stripe and the pixel of the CCD detector as variables, the target pattern is also described with 255 Gy levels to carry out CCD imaging simulation. This simulation is a simulated CCD staring imaging process, that is, a static imaging process.

2. **Simulation results**

A typical result after extensive simulation is shown in Fig. 10.12. The smooth and regular strip on the upper part of each picture is the target picture, and the part with the small square image below is the simulation result of the CCD detector imaging.

There are three small windows below each picture. The first window indicates the inclination angle A of the target strip relative to the column pixels of the CCD detector, the second window indicates the width W of the target strip, and the third window indicates the phase P of the target strip relative to the column pixels of the CCD detector. The size of the CCD pixel in the simulation is 26×26, which only has a relative size to the width of the target strip.

It can be seen from Fig. 10.12a that the simulated picture is more definition and reflects the original target characteristics—black and white strips. However, when the phase parameter P is adjusted, the image becomes blurred as shown in Fig. 10.12b. When $P = 13$, there is no image at all as shown in Fig. 10.12c. This shows that the 26×26 size pixel does not well describe the black and white striping pattern of the same size.

When the width of the strip is increased to $W = 41.6$, Fig. 10.12d is obtained. Although the original features of the target can be seen, it cannot truly reflect the original state of the target. Definition black-gray-black transition state, this state is quantitatively represented by the modulation transfer function MTF, whose MTF = 0.68, this effect is quite considerable, and the width of the strip is 1.6 times that of the pixel at this time.

When $W = 41.6$, if the target strip is given an inclination angle $A = 10°$, the simulation image shown in Fig. 10.12e will be obtained. The broadband Mole stripe effect inclined to the left appears in the picture, which reflects the real characteristics of the target reluctantly.

In Fig. 10.12, the simulation pictures in the two cases of Fig. 10.12f when $W = 36.4$ and Fig. 10.12g when $W = 39$ is also given. The situation is similar, but the value of MTF is smaller, which has appeared Mole stripe effect.

When $W = 46.8$, two pictures are obtained, Fig. 10.12h, i, which clearly reflect the characteristics of the target graphics. At this time MTF = 0.89. In general, the simulation results of $W = 46.8$ reflect the characteristics of the target image well. At this time, the width of the target strip is 1.8 times the size of the CCD pixel.

When $W = 52$, that is, the width of the target strip is twice the size of the CCD pixel, Fig. 10.12j is obtained. At this time, the CCD image is very definition, and its MTF value is about 0.995.

When $W = 57.2$, $A = 45°$ and $A = 60°$, two pictures are obtained respectively, Fig. 10.12k, l , which clearly reflect the characteristics of the target graphics. At this time, MTF = 1, and the width of the target strip is 2.2 times the size of the CCD pixel.

In general, the simulation results of W = 46.8 reflect the characteristics of the target graphics well. At this time, the width of the target strip is 1.8 times the size of the CCD pixel.

Figure 10.12 shows the case where the obtained MTF is small after running various simulations.

Of course, the MTF value mentioned here is the MTF value of the square wave signal, that is, the contrast transfer function (CTF).

3. **Image shift and drift angle of on-orbit imaging**

When using the space optical remote sensing camera to image ground targets on orbits hundreds of thousands of kilometers away, the space optical remote sensing camera works in a large system of high-speed movement such as the carrier platform, the orbit, the earth and the target. The motion will inevitably bring about the relative movement between the CCD detector of the space optical remote sensing camera and the image of the target, this relative movement is called image shift. In order to obtain high-resolution space remote sensing images, it is necessary to study the loss of resolution caused by the image shift. The resolution of the image is affected by the control technology of the carrier platform, the orbital motion of the carrier platform, the rotation of the earth and other motion factors. Therefore, it is necessary to analyze the image shift problem in this dynamic sense [12].

Due to the promotion of military and civilian use, the resolution of space optical remote sensing cameras is getting higher and higher, and its angular resolution has reached the order of μrad. With the improvement of image resolution of space optical remote sensing cameras, the stability requirements of the carrier platform are also increasing correspondingly, when it becomes more and more difficult to achieve this kind of stability, it is naturally proposed that the instability of the carrier platform will have how much of an adverse effect on the image resolution of the space optical remote sensing camera. This is already encountered in the current engineering practice. This is one of the reasons for analyzing the image resolution of the CCD detector in the dynamic sense.

Another aspect of the problem is that only when the degree and law of the influence of the image shift on the image resolution of the CCD detector is recognized, which can be better taken measures to overcome its adverse effect on the image resolution of the CCD detector [13–15].

As mentioned above, the imaging of the CCD detector is a sampling discretization process of the photographed scene, and the influence of image shift is intertwined with the influence of this sampling discretization. When the space optical remote sensing camera is on-orbit imaging, image shift and sampling discretization of CCD detectors always exist and are inseparable. It is necessary to analyze the imaging of CCD detectors by taking them as research objects at the same time, so that the influence of the two can be comprehensively studied.

The form of image shift, with the different orbits and working methods of the space optical remote sensing camera, its image shift forms are also various, which can be expressed as translation, rotation, zoom, distortion, deviation of the integration direction from the image movement direction, etc. many forms. Most of the space detectors are CCD detectors, especially the TDICCD detectors with the function of time delay integration, so only the image shift of the TDICCD detectors is studied here. Because the TDICCD detector uses different pixels to repeatedly expose the same point of the scene during imaging, and accumulate and photoelectrically convert the energy of these exposures, as shown in Fig. 10.13. Its column pixel is the direction of charge integration. During imaging, the charges of the image point a11 are transferred to the a21 pixel, and the charges obtained by re-exposing the same point with the a21 pixel are added, and so on until an1. The same is true for other column pixels. During the imaging process, the transfer rate and direction of the charges must be exactly the same as the moving rate and direction of the corresponding image point of the image, otherwise image shift will occur.

Here mainly analyzes the influence of two forms of image movement, namely translational image movement and deviation of the integration direction and the image movement direction, on the image resolution of the space optical remote sensing camera. Translation image shift refers to the vertical movement of the target image relative to the integration direction of the TDICCD detector. The deviation between the integration direction and the image movement direction refers to the deviation between the column direction of the pixel (the direction of charge transfer) and the moving direction of the target image relative to the detector when the charges of each pixel of the TDICCD detector is imaged during the integral imaging.

This translation and deviation will lead to the decline of MTF, so when the space optical remote sensing system is imaging in orbit, technical measures should be

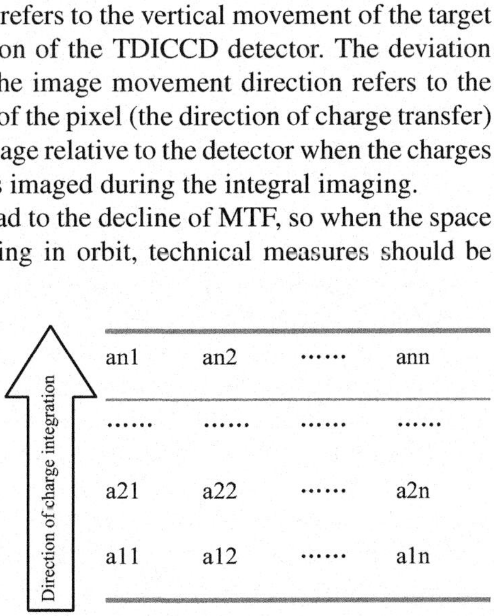

Fig. 10.13 Schematic diagram of TDICCD charge integration

taken to control the deviation of the image translation direction and the pixel column direction during the TDICCD integral imaging.

During on-orbit imaging, the orbital motion of the satellite, the rotational motion of the earth, the attitude error of the satellite, and the installation error of the focal plane are all causes of this deviation. Technically, the deviation angle between the image movement direction and the pixel column direction is called drift angle; the control of drift angle is an important aspect to reduce the image shift of TDICCD on-orbit imaging.

For a detailed discussion of the influence of image shift and sampling discretization on the image quality of TDICCD detectors, please refer to [16].

4. **Results and discussion**

When the MTF value of the other links is high and the target modulation is large, it can be optimistically estimated that the image resolution of the CCD detector is 1.6 times the size of its pixel. At this time, its MTF fell 32%.

In general, the image resolution of the CCD detector is 1.8 times the size of its pixel, which is more moderate, because the larger MTF value at this time has less pressure on the indicators of other parts of the system, and the image better reflects the target image. At this time, its MTF decreased by 11%.

These two conclusions are drawn after a large number of imaging simulations of CCD detectors. They are the cases where the value of MTF decreases the most. Of course, from the perspective of CCD space optical remote sensing camera imaging, the probability of their occurrence should be not large, that is to say, when the image resolution of the CCD detector is 1.8 times the size of its pixel, the maximum value of the MTF drop caused by it is 11%, which should be less than this value in most cases.

If the image resolution of the CCD detector is twice the size of its pixel, the image resolution of the detector will be reduced, which is more conservative.

The image resolution of the CCD detector discussed above is for the purpose of space reconnaissance, and the conclusion is drawn under the condition that the ability to distinguish the characteristics of the target is the criterion. This conclusion can be used for reference in CCD imaging activities such as map mapping, but further analysis is necessary. As can be seen from Fig. 10.12e that the simulated strip width is variable, it is difficult to determine the true width of the target strip.

All in all, because the target used is the pattern of the resolution discriminating plate commonly used in evaluating the performance of optical imaging remote sensing cameras; in the simulation, the interference of other factors is excluded, and only the sampling discretizations are considered as the main research object when the CCD detector is imaging, the discretizations refers to the size of the target described by pixels of a specific size of the CCD detector and the grayscale is digitized; the image resolution of the CCD detector will be defined based on the characteristics of the image identified. Therefore, this conclusion has strong practicability and general significance in evaluating the image resolution performance of imaging optical remote sensing cameras for reconnaissance purposes.

In terms of improving the resolution of space remote sensing imaging, the small particle size of the photosensitive latex of the film is also a powerful factor, especially when it is used for surveying and mapping, it is also a powerful factor to achieve large format, but its technology is complex [17–19], which is cannot be transmitted in real time.

References

1. Jiasheng T. Study on image resolution of CCD space camera based on simulation experiment and computer simulation, 2004.
2. Cangyuan H. Basic theory and application of information optics. Changchun: Changchun Publishing House; 1989. p. p75.
3. Smith SL. Understanding image quality losses due to smear in high-resolution remote sensing imaging systems, Opt Eng. 1999;35(5):821–26.
4. Jingshan J. Space science and applications, Beijing: Science Press.
5. Swets JA. Signal detection and recognition by human observers. New York: Wiley; 1964.
6. Lettington AL, Fairhurst AM, St. Marphy KJ. Visual recognition of blurred shapes. SPIE. 1994;2298:698–707.
7. Fenglin Z, Xuezhu S. Engineering optics. Tianjin: Tianjin University Press; 1990 The second edition.
8. Leachtenauer JC, Malila W, Irvine J, Colburn L, Salvaggio N. General image quality equation: GIQE. Appl Opt. 1997;36:8322–28.
9. Jiasheng T. Model of image sensor performance evaluation and image restoration. Ome Inf. 1999;16(12):13–7.
10. Jiasheng T. Study of the CCD sensor imaging resolution of the imaging reconnaissance. Opt Tech. 2003;29(1):94–95+100.
11. Hua Y, Wenchun J, Yonghong Z, Ying L. Modulation transfer function of CCD camera at Nyquist frequency. Acta Optica Sinica. 2002;22(3):313–6.
12. Fiete RD. Image quality of increased along-san sampling for remote sensing system. Opt Eng. 1999;38(5):815–25.
13. G. C. Holst, Sampling, Aliasing, and Data Fidelity, SPIE, 1996, p145–157.
14. Wang D, Communication Research Center. Comparison of motion-compensated algorithms for frame interpolation. Opt Eng. 2003; 42(2):586–90.
15. Dutton T, Lomheim T, Nelson MD. Survey and comparison of plane MTF measurement techniques. SPIE. 2002;4486:p219-246.
16. Jiasheng T. Study on the imaging quality of TDICCD sensor under image motion and sampling effect. Opto-Electron Eng. 2005;32(8):52–55.
17. Jiasheng T. Film flattening in image remote sensing. Opt Precis Eng. 1998;6(2):72–77.
18. Jiasheng T. Film flattening system in space cameras. Opt Precis Eng. 1998;8(2):150–53.
19. Jiasheng T, Jiaqi W. Numerical modeling of film flattening system in space cameras. Chin J Sci Instrum. 2002;23(2):115–17+20.

The manufacturer's authorised representative in the EU is Springer Nature Customer Service Centre GmbH, Europaplatz 3, 69115 Heidelberg, Germany. If you have any concerns regarding our products, please contact ProductSafety@springernature.com

Printed and bound by CPI Group (UK) Ltd, Croydon, CR0 4YY
01/10/2026
02225900-0018